Technical Draw
with AutoCAD®
A Multidisciplinary Curriculum
for the First Semester

Douglas Smith

Austin Community College

Antonio Ramirez

Austin Community College

PEARSON
Prentice Hall

Upper Saddle River, New Jersey
Columbus, Ohio

Library of Congress Control Number: 2008926482

Editor in Chief: Vernon Anthony
Acquisitions Editor: Jill Jones-Renger
Editorial Assistant: Doug Greive
Production Coordination: Lisa Garboski, bookworks publishing services
Project Manager: Louise Sette
AV Project Manager: Janet Portisch
Operations Specialist: Deidra Schwartz
Art Director: Diane Ernsberger
Cover Designer: Jason Moore
Cover image: istock
Director of Marketing: David Gesell
Senior Marketing Coordinator: Alicia Dysert
Copyeditor: Colleen Brosnan

This book was set in TimesNewRomanPS by Aptara®, Inc. and was printed and bound by Courier Kendallville, Inc. The cover was printed by Coral Graphic Services.

Certain images and materials contained in this publication were reproduced with the permission of Autodesk, Inc., © 2008. All rights reserved. Autodesk and AutoCAD are registered trademarks of Autodesk, Inc., in the U.S.A. and certain other countries.

Disclaimer:
This publication is designed to provide tutorial information about AutoCAD® and/or other Autodesk computer programs. Every effort has been made to make this publication complete and as accurate as possible. The reader is expressly cautioned to use any and all precautions necessary, and to take appropriate steps to avoid hazards, when engaging in the activities described herein.

Neither the author nor the publisher makes any representations or warranties of any kind, with respect to the materials set forth in this publication, express or implied, including without limitation any warranties of fitness for a particular purpose or merchantability. Nor shall the author or the publisher be liable for any special, consequential, or exemplary damages resulting, in whole or in part, directly or indirectly, from the reader's use of, or reliance upon, this material or subsequent revisions of this material.

Pearson Education Ltd., London
Pearson Education Singapore Pte. Ltd.
Pearson Education Canada, Inc.
Pearson Education—Japan

Pearson Education Australia Pty. Limited
Pearson Education North Asia Ltd., Hong Kong
Pearson Educación de Mexico, S.A. de C.V.
Pearson Education Malaysia Pte. Ltd.

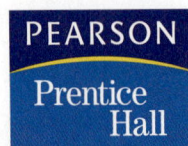

PEARSON
Prentice
Hall

10 9 8 7 6 5 4 3 2 1
ISBN-13: 978-0-13-175122-4
ISBN-10:   0-13-175122-0

THE NEW AUTODESK DESIGN INSTITUTE PRESS SERIES

Pearson/Prentice Hall has formed an alliance with Autodesk® to develop textbooks and other course materials that address the skills, methodology, and learning pedagogy for the industries that are supported by the Autodesk® Design Institute (ADI) software products. The Autodesk Design Institute is a comprehensive software program that assists educators in teaching technological design.

Features of the Autodesk Design Institute Press Series

JOB SKILLS—Coverage of computer-aided job skills, compiled through research of industry associations, job websites, college course descriptions, and the Occupational Information Network database, has been integrated throughout the ADI Press books.

PROFESSIONAL and **INDUSTRY ASSOCIATIONS INVOLVEMENT**—These books are written in consultation with and reviewed by professional associations to ensure they meet the needs of industry employers.

AUTODESK LEARNING LICENSES AVAILABLE—Many students ask how they can get a copy of the AutoCAD® software for their home computer. Through a recent agreement with Autodesk®, Prentice Hall now offers the option of purchasing textbooks with either a 180-day or a 1-year student software license agreement for AutoCAD. This provides adequate time for a student to complete all the activities in the book. The software is functionally identical to the professional license, but is intended for student personal use only. It is not for professional use.

Learning licenses may only be purchased by ordering a special textbook package ISBN. Instructors should contact their local Pearson Professional and Career sales representative. For the name and number of your sales representative, please contact Pearson Faculty Services at 1-800-526-0485.

FEATURES OF *TECHNICAL DRAWING 101* WITH AUTOCAD®

This text presents a curriculum for learning Technical Drawing fundamentals including AutoCAD. It introduces drawing techniques, commands, and procedures that reflect modern and efficient methods of creating drawings using AutoCAD 2008. Features include:

Unit Objectives

- Explain what technical drawings are.
- Explain the terminology used to describe the process of creating technical drawings.
- Explain how technical drawings are produced.
- Explain the training needed to become an engineer, architect, designer, or drafter.
- Describe the process of obtaining employment in the technical drawing field and the qualities that employers seek.
- Describe what career prospects and opportunities, including salary ranges, are available in the field of technical drawing.

Unit Objectives, a bulleted list of learning objectives for each chapter, provide users with a road map of important concepts and practices that will be introduced in the unit.

---

drafter: An individual with specialized training in the creation of technical drawings.

When the design inputs are finished, they a[r]ing the technical drawings for the project. **Drafte**[r] training in the creation of technical drawings. On[e] acquire during their training is the ability to i[n] technical drawings. Drafters usually work close[ly] may include designers, checkers, engineers, arch[itects] technical drawings.

Most drafters use CAD software to prepare

Key Terms are boldfaced and italicized within the running text, briefly defined in the margin, and defined in more detail in the Glossary at the end of the book to help students understand and use the language of the computer-aided drafting world.

TIP To enter an absolute coordinate in releases of AutoCAD prior to Release 2006, or when drawing with the **DYN** setting off in newer releases, you do not need to type the # symbol before entering the X- and Y-coordinates.

To draw the line shown in Figure 4-7, select the **LINE** command icon and at the Specify the first point prompt, type **2,2** and press **<Enter>**. At the Specify the next point prompt, type **#8,7** and press **<Enter>** again. Press **<Esc>** to end the command.

Note:
When defining a relative coordinate that is to the left, or below, the previous point, it is necessary to enter a negative coordinate. This is done by typing a minus sign (−) before the coordinate value. For example, typing -**3,-2** draws a line to a point **3** units to the left on the X-axis and **2** units below the point previously defined.

Relative Coordinates

Relative coordinates are located relative to the last point defined. For example, in Figure 4-8, a line begins at absolute coordinate **1,1** and is drawn to a second point located **6** units along the X-axis and **0** units along the Y-axis relative to the start point (**2,2**). The line continues

Although modern drafters use CAD tools to create drawings, traditional drafting skills such as sketching and blueprint reading are still very important for facilitating communication between drafters and designers.

TIP, **NOTE**, and **JOB SKILLS** boxes highlight additional helpful information for the student.

Because he dropped out of high school, [he] second chance at El Centro College, he would [not] gram at the University of Texas.

SUMMARY

The curriculum of this course is designed to introduce students to the field of technical drawing. This course is a good way to explore whether you possess the interests and aptitudes to pursue a career in which technical drawings are created or interpreted. If you wish to pursue more training,

most comm[on] cialized co[urse] instructor [s] nities and p[rojects]

End-of-Unit material, easily located by shading on page edges, provides:

- Summaries
- Test Questions
- Projects

to help students check their own understanding of important unit concepts.

UNIT EXERCISE

Exercise 3-1: Technical Lettering

Practice lettering the alphabet and numerals in the construction l[ines] shown in Figure 3-46 on the grid sheet located in the back of the b[ook]. Try to match the example lettering style shown.

LETTERING EXERCISE

.250 (1/4) inch Lettering
ABCDEFGHIJKLMNOPQRSTU
1234567890 SECTION A-A FL

Exercises throughout the chapters provide step-by-step walk-through activities for the student, allowing immediate practice and reinforcement of newly learned skills. This is especially helpful in distance education courses.

Figure 3-46
Lettering exercise

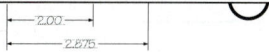

2.00

2.875

UNIT PROJECTS

Project 3-1: Traditional Drafting Project 1 (SI Units)

Directions: On the sheets located at the back of the text,
Front, Top, and Right views of the object in Figure 3-47. This
employs International Units (SI) so use the Metric scale mar
(refer to Figure 3-18) to draw the object **FULL** size (1 = 1)

Projects are organized to allow for application of skills to various fields and are numbered consistently among the chapters for easy back-and-forth reference. The 35 end-of-unit projects require students to use all the commands and skills they have learned cumulatively. Project types include:

- Sketching Multiview Drawings
- Mechanical Drawings—English units
- Mechanical Drawings—Metric units
- ASME Dimensioning Standards
- Architectural Drawings
- Mechanical Assembly Drawings
- Mechanical Detail Drawings
- Creation and use of block libraries
- Architectural Working Drawings
- 3D Modeling

A CD, bound into the textbook, contains student data files.

INSTRUCTOR RESOURCES

The Online Instructor's Resource Manual includes:

- **Check Prints** folder containing AutoCAD dwg files
- **Prototype Drawings** folder containing AutoCAD dwg files
- **3D Models** folder containing AutoCAD dwg files
- **Sketching and Traditional Drafting** folder containing PDF files
- **Unit Lectures** folder containing PowerPoint files
- **Syllabus, Tests, and Quizzes** folder containing docs, PDFs, and PPTs

To access supplementary materials online, instructors need to request an instructor access code. Go to **www.pearsonhighered.com/irc**, where you can register for an instructor access code. Within 48 hours after registering, you will receive a confirming e-mail including an instructor access code. Once you have received your code, go to the site and log on for full instructions on downloading the materials you wish to use.

STUDENT RESOURCES

Companion Website—A Companion Website at **www.prenhall.com/smithramirez** for student access includes an interactive study guide.

Preface

This textbook and the accompanying materials are appropriate for programs whose primary aim is to provide students with the necessary skills for employment as CAD drafters and engineering technicians. This textbook prepares them also for transfer to university engineering programs with more than just a cursory exposure to the creation of technical working drawings. Community college programs and their partners in high school Tech-Prep programs can be confident that their curricula are aligned and equivalent in scope and quality, thus facilitating transfer of credits.

This text focuses on core concepts of Technical Drawing and is tailored for a one semester course (about 80 class hours). It focuses on developing foundation skills that beginning students must master in order to prepare for more advanced courses. We understand that instructors have only about 80 hours of actual class time in a semester to cover this material. The textbook presents the reference materials needed for this course (and does not attempt to supplant the larger reference works currently on the market that may be appropriate for advanced technical drawing courses).

This text is supplemented with all of the instructional support materials needed by faculty members to teach Introduction to Technical Drawing successfully. These supplements provide the instructional materials for an instructor-led course covering the basics of technical drawing including AutoCAD.

Technical Drawing 101 focuses on the creation of 2D working drawings using AutoCAD with a strong emphasis on dimensioning theory and practice. Essentially, this text combines drafting theory *and* 2D AutoCAD fundamentals into one semester. This text provides the foundation for learning more advanced CAD techniques (paper space, Xrefs, etc.) in subsequent courses.

The assignments presented in the text are project oriented and culminate in working drawings. In recognition of their diverse interests, students will create working drawings for both a mechanical assembly and a small architectural project in this course.

An introduction to the fundamentals of creating three-dimensional AutoCAD models is presented in the text. Although students will learn more advanced 3D techniques in later courses, we believe it is important for students in this class to develop a basic understanding of the concept of extruding, subtracting, and unioning geometric shapes.

This text is intended to improve retention of students and to encourage them to explore a career in technical drawing by taking other courses. In testing this text and the supporting materials in our department, we discovered that student and faculty support increased the level of student success and satisfaction. This resulted in more students enrolling in second-level courses such as Intermediate CAD, Residential Drafting, Civil Drafting and Mechanical Drafting.

We also found that students who received better training in the introductory course were more successful in advanced classes. As a result, our department has been able to "raise the bar" for student performance in advanced courses. We feel that this course is an opportunity for our department to make students aware of career opportunities in the technical drawing field, as well as training options available in our department. The increased enrollment in advanced courses has contributed to the overall viability of our department.

TO USERS OF AUTOCAD 2009

This text was written for AutoCAD 2008, but by setting the AutoCAD 2009 workspace to *AutoCAD Classic*, users will find that the screen environment looks almost identical to earlier releases of AutoCAD. The steps involved in setting the AutoCAD 2009 workspace to AutoCAD Classic are covered in Unit 4 of this text. **Appendix E, AutoCAD 2009 Update,** presents many of the other changes, including the new *Ribbon interface*, introduced in AutoCAD 2009.

ACKNOWLEDGMENTS

Tony Ramirez would like to thank his wife Janice for her patience and support during the creation of this book.

Douglas Smith would like to thank his wife Robin and son Carson for their support during the completion of this project. Without their understanding and patience this project would not have been possible.

We would like to thank our students and colleagues whose suggestions and contributions helped shape this text, especially Sam Gideon, Ashleigh Fuller, Elain Dalton, Mischon Olger, Trang Ong, Jacqueline Micci, Matt Wilson, and Quinn Stewart. The authors would also like to thank our acquisitions editor at Pearson Education, Jill Jones-Renger for keeping our spirits up when we felt overwhelmed, and our development editor at bookworks publishing services, Lisa Garboski, for pulling it all together at the end.

We would also like to thank the many reviewers who offered valuable comments and insight: Hollis Driskell, Trinity Valley Community College; Sergio Lujan, Laredo Community College; Margie N. Porter, College of Lake County; Tony Thomas, Amarillo College; Kevin Weston, Rend Lake College; Mel L. Whiteside, Butler Community College; Jack W. Young, Jr., Coastal Bend College; Randy Emert, Clemson University; Susan Johnson, Johnson County Community College; Jonathan D. Combs, Ivy Tech State College; and Jeff Cope, Central Georgia Technical College.

Douglas Smith
Antonio Ramirez

Text Element	Example
Key terms—Boldface and italic on first mention (first letter lowercase, as it appears in the body of the text). Brief definition in margin alongside first mention. Full definition in Glossary at back of book.	Views are created by placing *viewport* objects in the paper space layout.
AutoCAD commands—Bold and uppercase	Start the **LINE** command.
Toolbar names, menu items, and dialog box names—Bold and follow capitalization convention in AutoCAD toolbar or the menu (generally first letter capitalized).	The **Layer Manager** dialog box The **File** menu
Toolbar buttons and dialog box controls/ buttons/input items—Bold and follow the name of the item or the name shown in the AutoCAD tooltip.	Choose the **Line** tool from the **Draw** toolbar. Choose the **Symbols and Arrows** tab in the **Modify Dimension Style** dialog box. Choose the **New Layer** button in the **Layer Properties Manager** dialog box. In the **Lines and Arrows** tab, set the **Arrow size:** to **.125**.
AutoCAD prompts—Dynamic input prompts are italic. Command window prompts use a different font (Courier New). This makes them look like the text in the command window. Prompts follow capitalization convention in AutoCAD prompt (generally first letter capitalized).	AutoCAD prompts you to *Specify first point:* `Specify center point for circle or [3P/2P/Ttr (tan tan radius)]:`
Keyboard input—Bold with special keys in brackets.	Type **3.5 <Enter.↵>**.

Contents

Unit 5 Dimensioning Mechanical Drawings   167

Unit 6   Dimensioning Architectural Drawings   215

Unit 7   Isometric Drawings   223

Technical Drawing

Unit Objectives

- Explain what technical drawings are.
- Explain the terminology used to describe the process of creating technical drawings.
- Explain how technical drawings are produced.
- Explain the training needed to become an engineer, architect, designer, or drafter.
- Describe the process of obtaining employment in the technical drawing field and the qualities that employers seek.
- Describe what career prospects and opportunities, including salary ranges, are available in the field of technical drawing.

INTRODUCTION

Technical drawings are the graphics and documentation (including notes and specifications) used by manufacturers to fabricate electronic and mechanical products and by construction professionals to produce houses, commercial buildings, roads, bridges, and water and waste-water systems. In fact, technical graphics are produced before almost all products are manufactured—from the integrated circuits inside your computer to the buttons on your shirt.

> technical drawing: Term used to describe the process of creating the drawings used in the field of engineering and architecture.

THE ORIGINS OF TECHNICAL DRAWING

Technical drawing is not a new concept; archeological evidence suggests that humans first began creating crude technical drawings several thousand years ago. Through the ages, architects and designers, including Leonardo Da Vinci, created technical drawings. However, a French mathematician, Gaspard Monge, is considered by many to be the founder of modern technical drawing. Monge's thoughts on the subject, *Geometrie Descriptive* (Descriptive Geometry) published around 1799, became the basis for the first university courses. In 1821, the first English-language text on technical drawing, *Treatise on Descriptive Geometry*, was published by Claude Crozet, a professor at the U.S. Military Academy.

Other terms often used to describe the creation of technical drawings are ***drafting***, ***engineering graphics***, ***engineering drawings***, and ***CAD (Computer Aided Design)***.

> drafting: A term often used to describe the creation of technical drawings.
>
> engineering graphics: A term often used to describe the creation of technical drawings.
>
> engineering drawing: A term often used to describe the creation of technical drawings.
>
> CAD (Computer Aided Design): A term often used to describe the creation of technical drawings. Also a term for the software used to create technical drawings. Some popular CAD programs include AutoCAD, SolidWorks, Revit, Pro/ENGINEER, and Inventor.

THE ROLE OF TECHNICAL DRAWING IN THE DESIGN PROCESS

In order to appreciate technical drawing's role in the design process, you must first understand some basics about design process itself.

For most projects, the first phase of a design project is to define clearly the design criteria that the finished design must meet in order to be considered a success. Many designers refer to this phase in the design process as *problem identification*. For example, before designing a house, an architectural designer needs to know what size and style of home the client wants, how many bedrooms and baths, and an idea of the budget for the project. The designer also needs information about the site where the house will be built. Is it hilly or flat? Are there trees, and if so, where are they located? What is the orientation of the site relative to the rising and setting of the sun? These concerns represent just a few of many design parameters that the designer needs to define before beginning the design process.

Once the design problem is clearly defined, the designer begins preparing preliminary designs that can meet the parameters defined during the problem identification phase. During this step, multiple solutions to the design problem may be generated in the form of freehand sketches, formal CAD drawings, or even rendered three-dimensional models. Designers refer to this process of generating many possible solutions to the design problem as the *ideation*, or *brainstorming*, phase of the process.

The preliminary designs are shown to the client to see if the design is in line with the client's expectations. This step allows the designer to clarify the client's needs and expectations. It also is an opportunity for a designer to educate the client about other, possibly better, solutions to the design problem.

After the client decides on a preliminary design that meets the criteria established in the first problem identification phase, the designer begins preparing design inputs that more clearly define the details of the design project. Design inputs may include freehand sketches with dimensional information, detailed notes, or even CAD models. Figure 1-1 shows an example of an architectural designer's sketch of a foundation detail for a house.

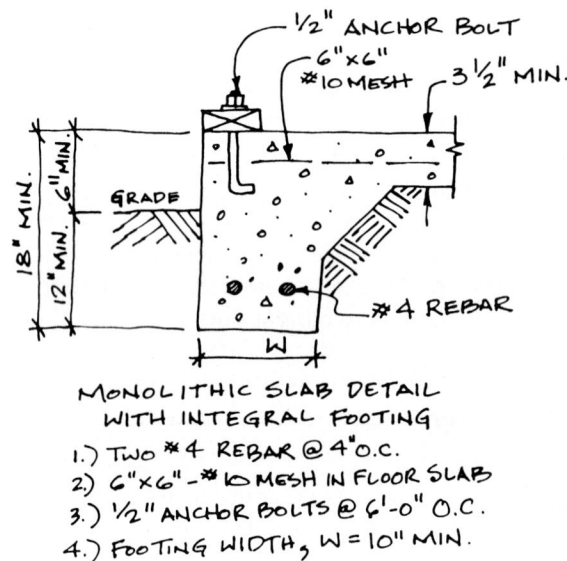

Figure 1-1   Architectural designer's sketch

drafter: An individual with specialized training in the creation of technical drawings.

When the design inputs are finished, they are given to the drafter(s) responsible for preparing the technical drawings for the project. **Drafters** are individuals who have received specialized training in the creation of technical drawings. One of the most important skills that drafters must acquire during their training is the ability to interpret design inputs and transform them into technical drawings. Drafters usually work closely with other members of the design team, which may include designers, checkers, engineers, architects, and other drafters during the creation of technical drawings.

Most drafters use CAD software to prepare the drawings. CAD allows drafters to produce drawings much more quickly than traditional drafting techniques. Popular CAD programs include AutoCAD[®], Autodesk[®], Inventor[TM], SolidWorks[®], and Pro/ENGINEER[®]. CAD software can range in price from several hundred dollars per station to thousands of dollars per station depending on the software. Figure 1-2 shows a CAD drawing prepared from the designer's sketch shown in Figure 1-1.

SECTION A-A
SCALE: 3/8" = 1'-0

MONOLITHIC SLAB W/ INTEGRAL FOOTING

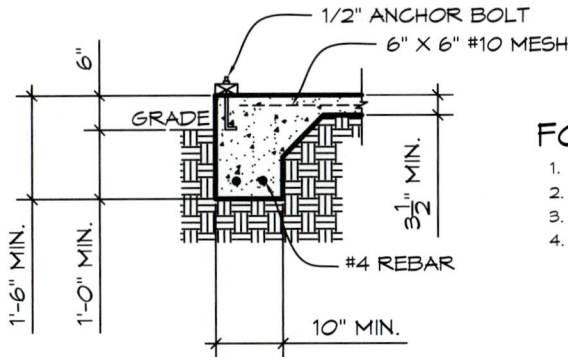

1/2" ANCHOR BOLT
6" X 6" #10 MESH
6"
GRADE
3'-1" MIN. / 2
1'-6" MIN.
1'-0" MIN.
#4 REBAR
10" MIN.

FOUNDATION NOTES:
1. TWO #4 REBAR AT 4' O.C. FOR FOOTINGS
2. 6 X 6-#10 MESH IN FLOOR SLAB
3. 1/2" ANCHOR BOLTS AT 6'-0" O.C.
4. FOOTING WIDTH 10" MINIMUM

Figure 1-2   Detail prepared from architectural designer's sketch in Figure 1-1

Although modern drafters use CAD tools to create drawings, traditional drafting skills such as sketching and blueprint reading are still very important for facilitating communication between drafters and designers.

When the drafter is finished preparing the technical drawings, the designer, or in some cases a *checker*, reviews the drawings carefully for mistakes. If mistakes are found, or if the design has been revised, the drafter will make the necessary corrections or revisions to the drawings. This process is repeated until the construction drawings are considered to be complete. When the entire set of construction drawings is finalized, the drafter and designer(s) put their initials in an area of the drawing called the *title block*.

The finished construction drawings represent the master plan for the project. Everything required to complete the project, from applying for a building permit to securing financing for the project, revolves around the construction drawings. Building contractors use the construction documents to prepare bids for the project, and the winning bidders will use them to construct the building.

Engineering designers follow a similar process when designing products. Most engineering projects begin by defining initial design criteria and progress through the phases of preliminary design, design refinement, preparation of technical drawings, manufacturing, and inspection.

The trend in modern design, whether it is architectural or engineering, is to use CAD tools to create a dynamic, often three-dimensional, database that can be shared by all members of the design team. Increasingly, others in the organization, such as those involved in marketing, finance, or service and repair, will access information from the CAD database to accomplish their jobs.

Catching problems and mistakes during the design and drafting stages of the project can result in huge savings versus correcting mistakes on the job site, or after the project has been built or manufactured. An example is the enormous costs incurred by an automobile manufacturer who has to a recall thousands of cars to correct a design problem versus the cost of catching the problem on the technical drawing before the cars are manufactured.

TRAINING FOR CAREERS IN TECHNICAL DRAWING

Most drafters acquire their training by attending community college or technical school programs that lead to a certificate or associate's degree in Drafting and Design or Computer Aided Design. These programs usually take from one to two years to complete and focus on the skills necessary to work as a drafter in industry, such as drafting techniques, knowledge of drafting standards, and how CAD programs are used to create drawings. Although most employers do not require that drafters be certified, the American Design Drafting Association (ADDA) has established a certification program for drafters. Individuals seeking certification must pass a test, which is administered periodically at ADDA-authorized sites. Some publishers of CAD software also offer certification on their products through authorized training sites.

> **TIP** You can learn more about the American Design Drafting Association (ADDA) by visiting its website at www.adda.org.

Most drafters are full-time employees of architectural and engineering firms. Usually drafters qualify for over-time pay when they work in excess of forty hours. However, some drafters prefer to work as non-employee contractors. Contractors are usually very experienced drafters who often earn higher salaries than direct employees, but have less job stability. Some organizations allow drafters to telecommute and transfer drawing files to the office via the Internet.

designer: Individual who assists engineers or architects with the design process. Designers are often former drafters who have proven their ability to take on more responsibility and decision-making duties.

Designers are often former drafters who have proven their ability to take on more responsibility and decision-making duties. Designers usually earn higher salaries than drafters because they are charged with more responsibility for the design, and even the successful completion, of the project.

To become an engineer or architect, an individual must first earn a bachelor's degree in engineering or architecture from a university program. Bachelor's degree programs generally take four to five years to complete and usually require a mastery of higher level courses in mathematics and physics.

professional engineer (P.E.): An engineer who is licensed by the National Society of Professional Engineers.

American Institute of Architects (AIA): The accrediting body for architects.

National Society of Professional Engineers: The accrediting agency for engineers.

After earning a degree, an engineer may become a *professional engineer* (P.E.), and an architect may become *licensed*, through a process involving both work experience and strenuous professional exams. The accrediting body for architects is the *American Institute of Architects* (AIA). The accrediting agency for engineers is the *National Society of Professional Engineers*.

> **TIP** You can learn more about the American Institute of Architects and the National Society of Professional Engineers by visiting their websites at www.aia.org and www.nspe.org.

Career Paths in Technical Drawing

architectural drafters: Drafters who prepare the drawings used in construction industries.

Architectural drafters work with architects and designers to prepare the drawings used in construction projects. These drawings may include floor plans, elevations, and construction details. Study of construction techniques and materials, as well as building codes, are important to the education of an architectural drafter. Some architectural drafters specialize in residential architecture (houses), whereas others may specialize in commercial architecture (buildings and apartments) or structural drafting (steel buildings or concrete structures). Figures 1-3 and 1-4 show details from architectural drawings.

mechanical drafters: Drafters who prepare detail and assembly drawings of machinery and mechanical devices.

Mechanical drafters work with mechanical engineers and designers to prepare detail and assembly drawings of machinery and mechanical devices. Mechanical drafters are usually trained in basic engineering theory as well as drafting standards and manufacturing techniques. They may be responsible for specifying items on a drawing such as the types of fasteners (nuts, bolts, and screws) needed to assemble a mechanical device or the fit between mating parts. Figure 1-5 shows a detail from a mechanical drawing.

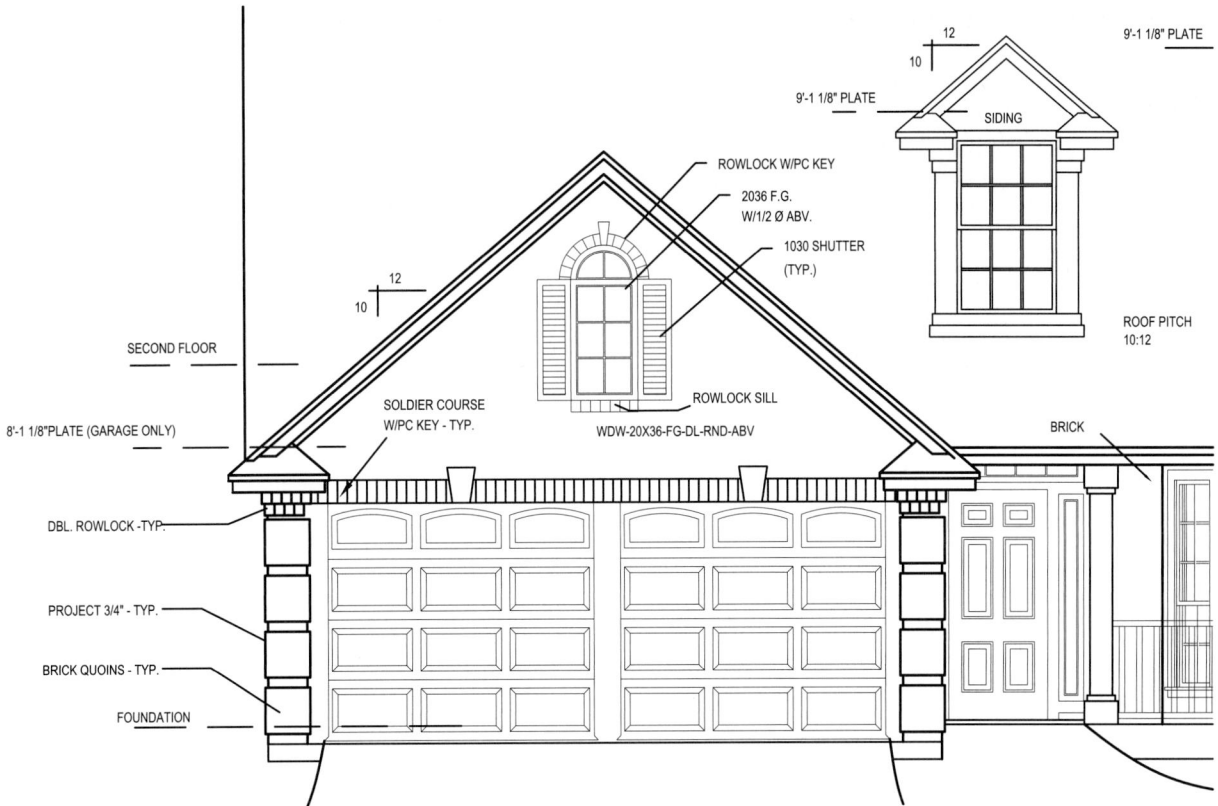

Figure 1-3   Detail from an architectural drawing

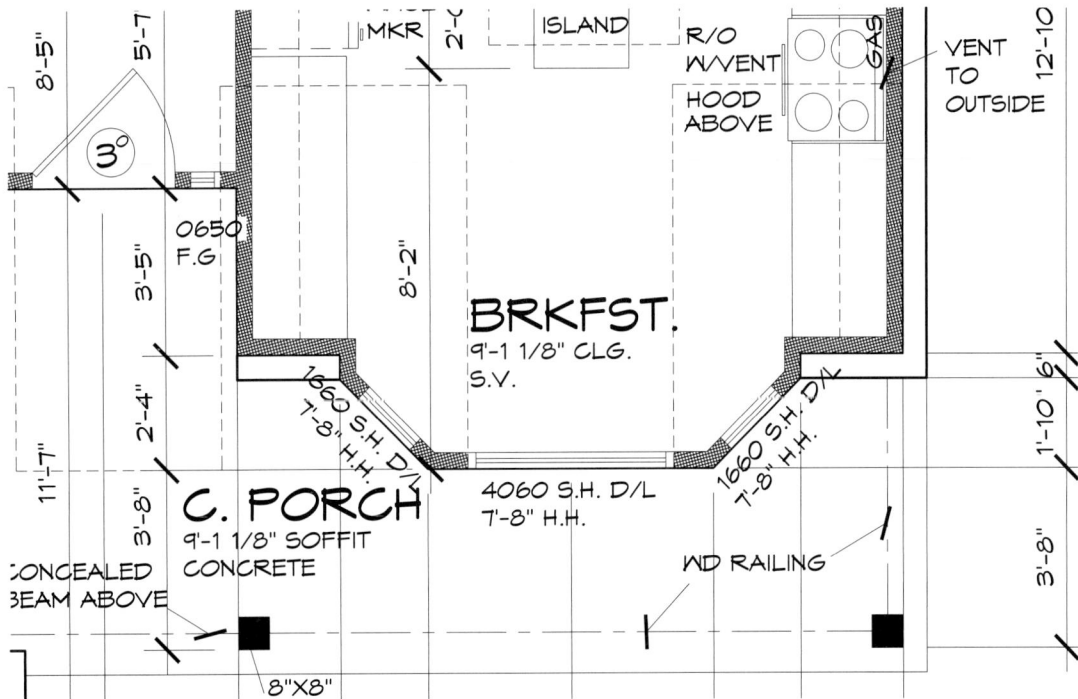

Figure 1-4   Detail from a floor plan

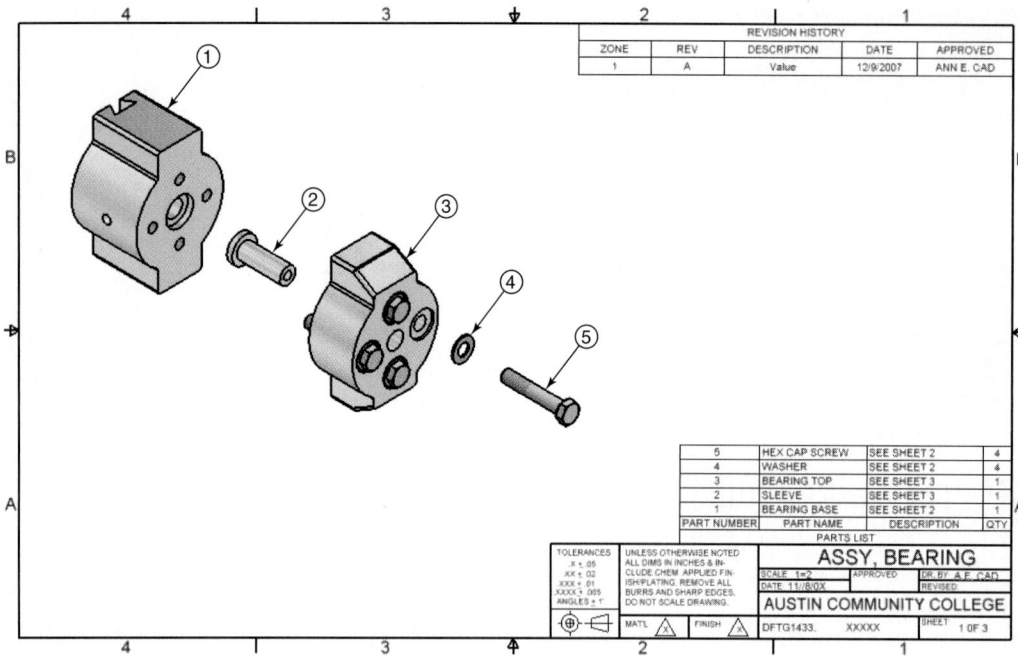

Figure 1-5   Mechanical engineering drawing

Aeronautical or aerospace drafters prepare technical drawings used in the manufacture of spacecraft and aircraft. These drafters often split their duties between mechanical drafting and electrical/electronic drafting and are sometimes referred to as *electro/mechanical drafters*.

Civil drafters and design technicians prepare construction drawings and topographical maps used in civil engineering projects. Civil projects may include roads, bridges, and water and waste-water systems. Civil drafters may also work for *surveying* companies to create site plans and plats for new subdivisions. Figure 1-6 shows an example of a civil drawing.

aeronautical or aerospace drafters: Drafters who prepare engineering drawings detailing plans and specifications used in the manufacture of aircraft and related equipment.

electro/mechanical drafters: Drafters who split their duties between mechanical drafting and electrical/electronic drafting.

civil drafters and design technicians: Drafters who prepare construction drawings and topographical maps used in civil engineering projects.

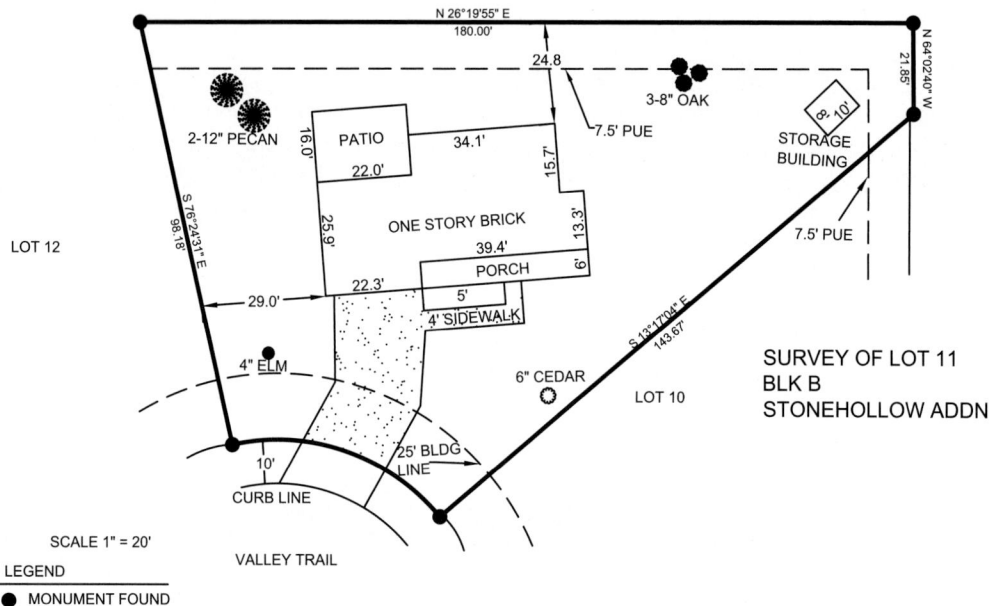

Figure 1-6   Civil engineering drawing

electrical drafters: Drafters who prepare diagrams used in the installation and repair of electrical equipment and building wiring.

Electrical drafters prepare diagrams used in the installation and repair of electrical equipment and the building wiring. Electrical drafters create documentation for systems ranging from low-voltage fire and security systems to high-voltage electrical distribution networks.

Electronics drafters create schematic diagrams, printed circuit board artwork, integrated circuit layouts, and other graphics used in the design and maintenance of electronic (semiconductor) devices. Figure 1-7 shows a detail from an electronic schematic drawing.

Figure 1-8 shows a detail of a printed circuit board (PCB) prepared from the schematic shown in Figure 1-7.

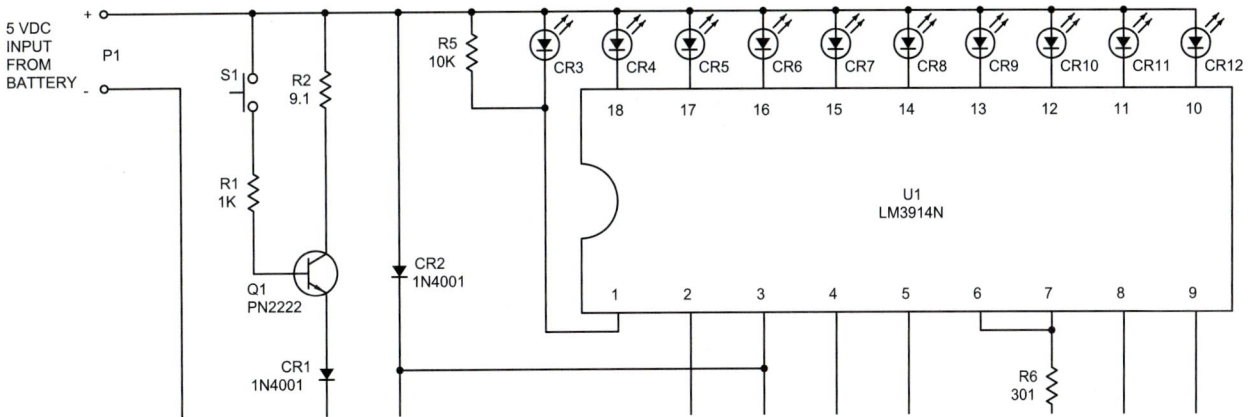

Figure 1-7   Detail from an electronic schematic

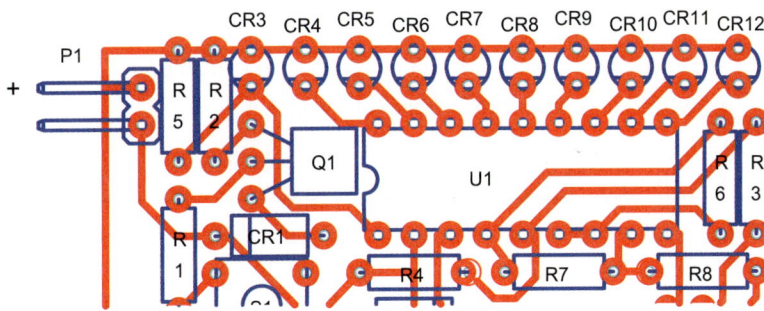

Figure 1-8   Detail of a printed circuit board (PCB) prepared from the schematic shown in Figure 1-7

Pipeline drafters and process piping drafters prepare drawings used in the construction and maintenance of oil refineries, oil production and exploration industries, chemical plants, and process piping systems such as those used in the manufacture of semiconductor devices. Figure 1-9 shows a detail from a process piping drawing.

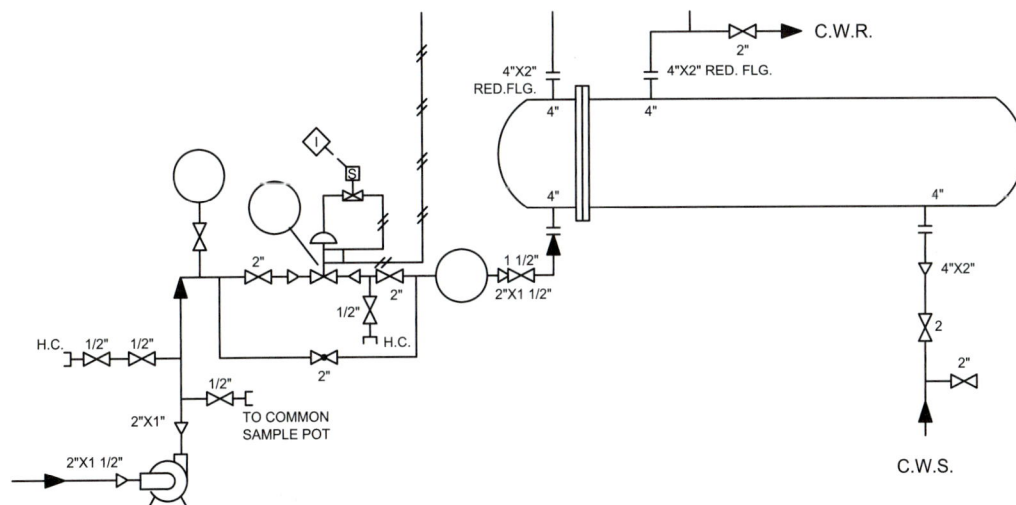

electronics drafters: Drafters who prepare schematic diagrams, printed circuit board artwork, integrated circuit layouts, and other graphics used in the design and maintenance of electronic (semiconductor) devices.

pipeline drafters and process piping drafters: Drafters who prepare drawings used in the construction and maintenance of oil refineries, oil production and exploration industries, chemical plants, and process piping systems such as those used in the manufacture of semiconductor devices.

Figure 1-9   Process piping drawing

Qualities That Employers Want in Drafters

Most employers are very careful in making hiring decisions. Interviews are usually very thorough; in extreme cases, the interview process may take several hours. Candidates may be called back for more than one interview.

When interviewing for a job, the candidate may interview with one person or with the entire design team. In the modern engineering or architectural office, the candidate's attitude regarding work and his or her ability to learn quickly and contribute to the team right away is factored into the hiring decision. Often, tests are administered during the interview to measure proficiency with CAD or the candidate's understanding of necessary concepts, which should have been mastered while in college or during other training.

Typical Interview Questions

- Are you a quick learner (will you contribute to the organization quickly)?
- Are you intelligent, competent, and energetic?
- Will you fit into their team? Do you get along with others?
- Did you make good grades in your major? Will your instructors give you a good recommendation?
- Do you meet deadlines?
- Are you able to work successfully both in groups and individually?
- Did your training prepare you for this job?
- Do you communicate well with others (both verbally and in writing)?
- Do you have good work habits? Are you dependable?
- Will the employer profit from your work?

Job Skills

If you are interested in working in the technical drawing field, develop good work habits while you are in school. Set high standards for your work, strive to create outstanding drawings for your portfolio, and be able to explain how the drawings in your portfolio were created and why. Come to class prepared and on time, meet your deadlines, and try to impress your instructors with your work habits and attitude because often the recommendation of a current, or former, instructor will determine whether you get a job.

Salary Information for Drafters, Architects, and Engineers

Earnings for drafters vary depending on the specialty and geographic area. According to U.S. Department of Labor statistics (June 2006), median annual earnings of architectural and civil drafters were $48,360 ($23.25 per hour). Median annual earnings for architects were $63,328. Median annual earnings for civil engineers were $67,799.

Median annual earnings of mechanical drafters were $43,272 ($21.13 per hour). Median annual earnings for mechanical engineers were $68,316.

Median annual earnings of electrical and electronics drafters were $47,952 ($22.73 per hour). Median annual earnings for electrical engineers were $75,483.

Note:
Salary data taken from the U.S. Department of Labor National Compensation Survey, June 2006.

Job Prospects for Drafters

According to the U.S. Department of Labor, approximately 250,000 people were employed as drafters nationwide in 2004. About half of these jobs were held by architectural and civil drafters, mechanical drafters held about a third of all drafting jobs, and the rest were held by electrical and electronics drafters.

Almost half of all jobs for drafters were in architectural, engineering, and related services firms that design construction projects or perform other engineering work on a contract basis for other industries. More than a quarter of jobs were in manufacturing industries, such as machinery manufacturing, including metalworking and other general machinery; fabricated metal products manufacturing, including architectural and structural metals; computer and electronic products manufacturing, including navigational, measuring, electro-mechanical, and control instruments; and transportation equipment manufacturing, including aerospace products and parts manufacturing, as well as ship- and boat-building. Most of the rest were employed in construction, government, wholesale trade, utilities, and employment services. Only about 6% were self-employed in 2004.

Note:
The median is the middle of a distribution, i.e., half the earnings are above the median and half are below the median.

The Department of Labor's Bureau of Labor Statistics (BLS) predicts that "industrial growth and increasingly complex design problems associated with new products and manufacturing processes will increase the demand for drafting services. Further, drafters are beginning to break out of the traditional drafting role and do work traditionally performed by engineers and architects, thus also increasing demand for drafters."

The BLS goes on to predict that "opportunities should be best for individuals with at least 2 years of postsecondary training in a drafting program that provides strong technical skills, as well as considerable experience with CADD (Computer Aided Design and Drafting) systems. CADD has increased the complexity of drafting applications while enhancing the productivity of drafters. It also has enhanced the nature of drafting by creating more possibilities for design and drafting. As technology continues to advance, employers will look for drafters with a strong background in fundamental drafting principles, a high level of technical sophistication, and the ability to apply their knowledge to a broader range of responsibilities."

TIP   You can find out more about careers in drafting by visiting the U.S. Department of Labor website at www.bls.gov.

THREE TECHNICAL DRAWING SUCCESS STORIES

Jeffrey Briseno first began taking drafting courses at Temple College in Temple, Texas, and later transferred to Austin Community College (ACC) in Austin, Texas. In 1996, before he completed the requirements for his associate's degree at ACC, Jeffrey was hired by Wenzel Engineering in Austin. While at Wenzel, Jeffrey used AutoCAD to prepare site-plan development drawings and on-site wastewater management systems. Jeffrey continued to take classes at night and in 2003 received his Associate of Applied Science degree in Architectural and Engineering Computer Aided Design at ACC. Jeffrey is currently employed by Loomis Austin, a civil engineering firm, where he is involved in subdivision platting, street design, utility design, site-plan development, and land planning. Jeffrey uses AutoCAD® Land Desktop and Autodesk® Civil 3D CAD® software to perform his job functions.

Elain Dalton already had a bachelor's degree from the University of Oregon when she decided to take a drafting course at the University of Houston "just for fun." Using the skills she developed in that class, Elain was able to find a job with Johnson Controls' *Star Wars* defense base in the Marshall Islands (a U.S. territory in the South Pacific) doing facilities operations and master-planning. The job required that Elain split her time between two CAD programs: AutoCAD and Microstation. There was only one small catch, Elain would have to teach herself to use both programs, which she did. When Elain returned to the continental U.S., she took a job with Carrier Corporation and enrolled in the drafting program at San Jacinto Community College in

American Society of Mechanical Engineers (ASME): Publisher of standards for the creation of technical drawing in the United States. *ASME Y14.5M-1994 Dimensioning and Tolerancing* and *ASME Y14.2M Line Conventions and Lettering* are two standards important to drafters.

Pasadena, Texas, earning an Associate of Applied Science degree in 1996. Her work with Carrier Corporation involved mechanical, electrical, and pipe drafting for clients such as Dow Chemical, Vulcan Chemicals, DuPont, and other corporations using commercial heating and cooling systems. She was also responsible for ensuring that the drawings complied with ***American Society of Mechanical Engineers*** standards.

Elain's next position was as a drafter/designer with Lubrizol Chemical Corporation where she used AutoCAD to create mechanical, civil, site mapping, and piping drawings.

At present, Elain operates Dalton Design, a contract drafting services company, has two children, and has taught as an adjunct faculty member in the CAD program at a local community college.

Bryan Lym began his path to a career in architecture in an unusual way—he dropped out of high school. After working at a variety of jobs, he was encouraged by a friend to take the GED test at El Centro College, a community college in Dallas, Texas. Bryan passed his GED and went on to enroll in the architectural drafting program at El Centro. In these drafting classes, Bryan discovered both his aptitudes and interests lay in the field of architecture. Near the end of his second year at El Centro, he was accepted into the School of Architecture at the University of Texas. On his graduation day in 1997, Bryan received two degrees: a Bachelor of Architecture *and* a Bachelor of Science in Architectural Engineering. Following graduation, Bryan held positions with RNL Design, Beck Construction, and Carter and Burgess. While employed at Carter and Burgess, Bryan began taking the licensure exams of the American Institute of Architects, and in 2003, he became a licensed architect. That same year, he began his own firm, Lym Architecture. Today his practice focuses primarily on designs for commercial remodels. His office uses a variety of software including AutoCAD, 3D Studio Max®, Sketch Up, Adobe® Photoshop® and Illustrator®, Word®, Excel®, and Adobe® Acrobat®.

Because he dropped out of high school, Bryan believes that, if he had not been offered the second chance at El Centro College, he would never have been accepted into the architecture program at the University of Texas.

SUMMARY

The curriculum of this course is designed to introduce students to the field of technical drawing. This course is a good way to explore whether you possess the interests and aptitudes to pursue a career in which technical drawings are created or interpreted. If you wish to pursue more training, most community colleges and many universities offer specialized courses in engineering or architectural drawing. Your instructor may also be able to advise you on training opportunities and possible career paths.

UNIT TEST QUESTIONS

Short Answer

1. Name three terms that are used to describe the creation of technical drawings.
2. In which field must a drafter be familiar with floor plans, elevations, and construction details?
3. How can an employer find out about a student's CAD skills, work habits, and dependability?
4. A drafter who divides his or her duties between mechanical drafting and electrical/electronic drafting is known as what kind of drafter?
5. Name the job titles of people who might comprise a design team.

Multiple Choice

1. The acronym AIA stands for what?

   a. Architectural Institute of America
   b. American and International Architects
   c. American Institute of Architects
   d. Association of International Architects

2. According to the U.S. Department of Labor Statistics (2006), which field of drafting has the highest median salary?

   a. Architectural
   b. Mechanical
   c. Electronics
   d. Radiological

3. In which field must a drafter be familiar with roads, bridges, water and waste-water systems, and surveying techniques?

   a. Architectural
   b. Civil
   c. Electronics
   d. Mechanical

4. Which of the following attributes do most employers value?

   a. Ability to fit into to a team
   b. Ability to work independently
   c. Ability to communicate clearly
   d. All of the above

5. Designers have a higher salary than drafters because:

   a. They usually have advanced technical degrees.
   b. They are licensed.
   c. They work longer hours than drafters.
   d. They have a higher degree of responsibility for the success of the project.

Matching

Column A

a. Mechanical

b. Process piping

c. Bachelor's

d. Electronics

e. Associate's

Column B

1. Type of college degree held by most engineers and architects

2. Drafting field concerned with chemical plants and oil refineries

3. Type of college degree held by many drafters

4. Drafting field concerned with screw threads and fasteners

5. Drafting field concerned with design of semiconductor devices

Internet Resources

American Design Drafting Association: *www.adda.org*
American Institute of Architects: *www.aia.org*
American Society for Engineering Education: *www.asee.org*

National Society of Professional Engineers: *www.nspe.org*
U.S. Department of Labor: *www.bls.gov*

UNIT EXERCISES

Exercise 1-1: Locate Bachelor Degree Programs in Architecture or Engineering

Search the Internet for university programs in your geographic area offering a bachelor degree in architecture and/or engineering. Try to locate an on-line degree plan for each major and compare the courses required for each degree.

Exercise 1-2: Locate Associate Degree Programs in CAD

Search the Internet for community college or technical school programs in your geographic area offering an associate's degree in CAD or a related field. Try to locate an on-line degree plan for the program. Compare the courses required for the associate degree to the courses required in the bachelor's degree program.

Multiview Drawing

2

Unit Objectives

- Explain what multiview drawings are and their importance to the field of technical drawing.
- Explain how views are chosen and aligned in a multiview drawing.
- Visualize and interpret the multiviews of an object.
- Describe the line types and line weights used in technical drawings as defined by the *ASME Y14.2M* standard.
- Explain the difference between drawings created with First Angle and Third Angle projection techniques.
- Use a miter line to project information between top and side views.
- Create multiview sketches of objects including the correct placement and depiction of visible, hidden, and center lines.

INTRODUCTION

Multiview drawing is a technique used by drafters and designers to depict a ***three-dimensional (3D) object*** (an object having height, width and depth) as a group of related ***two-dimensional (2D) objects*** (having only width and height, or width and depth, or height and depth). A person trained in interpreting multiview drawings, can visualize an object's three-dimensional (3D) shape by studying the two-dimensional (2D) multiview drawings of the object.

MULTIVIEW DRAWINGS

Figure 2-1 represents a three-dimensional image of a school bus. The 3D image of the bus is very helpful in visualizing its shape, because the viewer can quickly get an idea of the overall height, width, and depth of the bus. However, the 3D view cannot show the viewer all of the sides of the bus, or its true length, width, or height. On the other hand, a multiview drawing (Figure 2-2) can provide the viewer with all of the sides of a bus, represented in its true proportions: width, height, and depth. The six views representing the bus in Figure 2-2—the front, top, right side, left side, back, and bottom—are referred to in technical drawing terminology as the six *regular views*.

multiview drawing: A technique used by drafters and designers to depict a three-dimensional object (an object having height, width, and depth) as a group of related two-dimensional (having only width and height, or width and depth, or height and depth) views.

three-dimensional (3D) object: An object having height, width, and depth.

two-dimensional (2D) object: An object having height and width, width and depth, or height and depth.

regular views: In a multiview drawing, this term refers to an object's front, top, bottom, right, left, and back (or rear) views.

> **TIP**
>
> Although a total of six views is possible using the multiview drawing technique, drafters draw only the views necessary to show clearly all of the features of the object. Dimensional information for the features is added to these views. If no dimensions are placed on a view, the view is probably unnecessary.

Figure 2-1   A three-dimensional image of a school bus

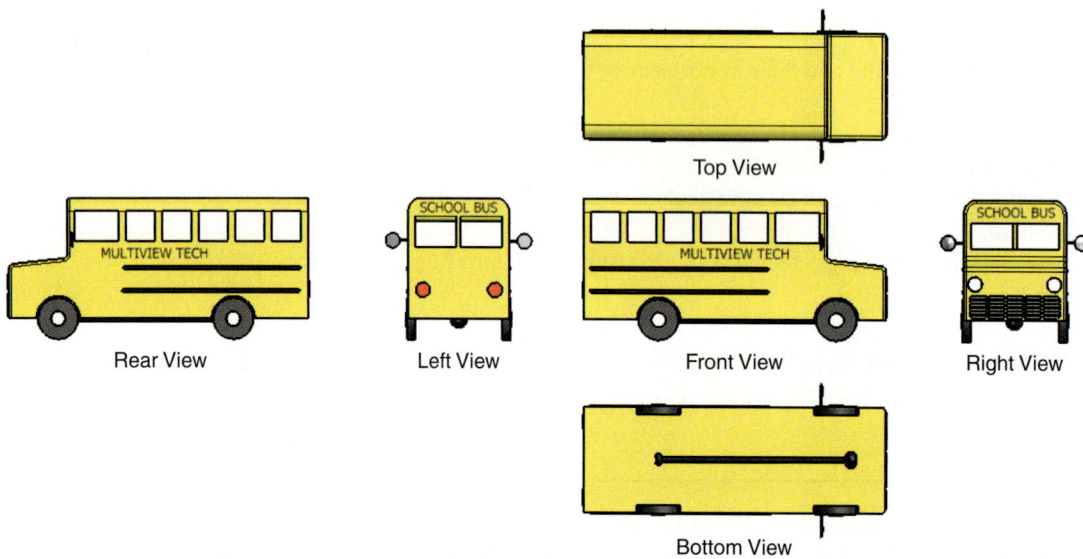

Figure 2-2   The multiviews of the bus depicting the six regular views—front, top, bottom, right, left, and rear

When creating the multiview drawing of the bus, the drafter draws the front, or principal, view first. The top, bottom, right, and left side views are drawn by rotating the bus at 90° intervals relative to the front view. For example, the top view is created by rotating the front view 90° toward the top; the right side view is created by rotating the front view 90° to the right.

TIP   When choosing the front, or principal, view of an object, select the view you would choose if you could show the viewer only one view to describe the object .

VIEW SELECTION AND ALIGNMENT OF MULTIVIEW DRAWINGS

In Figure 2-2, the *side* view of the bus was chosen to be the front, or principal, view because it provides the viewer with the most information about the shape of the bus. The actual *front* of the bus is drawn as the right-side view. The top view is a "bird's-eye" view from directly above the front view. The left-side view shows the back of the bus, and the rear view of the bus is projected from this view. The bottom view is drawn directly below the front view. Features of the bus, such as the headlights, tires, and windows, are aligned in all views.

Because the front view is the principal view of the bus, it is drawn first. The other views are created by *projecting* from the geometry of the front view. For example, after the drafter has measured and drawn the width of the front view, it can be projected to the top and bottom views, so the drafter doesn't need to remeasure the width in those views. Likewise, the height of the front view can be projected to the right- and left-side views. Because drafters can avoid remeasuring features by projecting between views, multiview drawing is an efficient way to create technical drawings.

USING PROJECTION PLANES TO VISUALIZE MULTIVIEWS

If a house were placed inside a glass box as in Figure 2-3, the glass sides of the box would create **projection planes** (also referred to as *viewing planes*). If the three-dimensional geometry of the front, side, and top of the house were projected onto the corresponding two-dimensional projection plane, the resulting 2D image would represent a front, side, or top view as shown in Figures 2-4, 2-5, and 2-6, respectively.

projection planes: An imaginary two-dimensional plane, like a sheet of clear glass, placed parallel to a principal face of the object to be visualized. The object's features (points, lines, planes) are projected perpendicular to the projection plane when visualizing a multiview drawing of the object.

Figure 2-3   House inside glass box

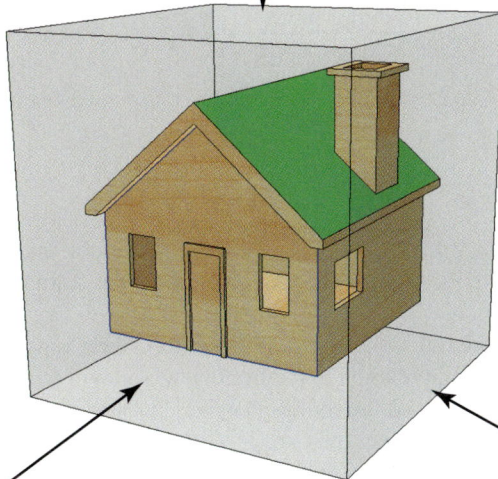

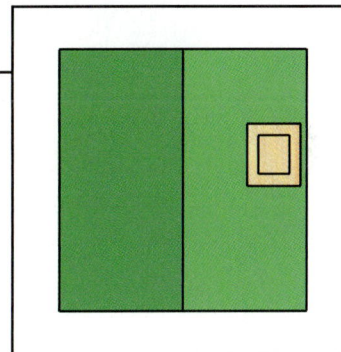

Figure 2-6   The view through the top glass plane shows the top view or roof plan. This view reveals the width and depth of the house

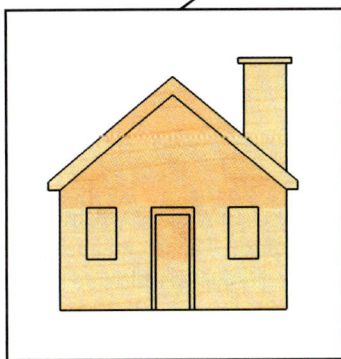

Figure 2-4   The view through the front glass plane shows the front view of the house. This view furnishes the width and height of the house

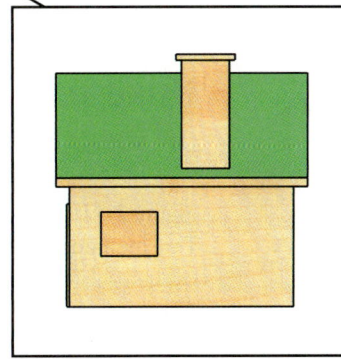

Figure 2-5   The view through the right glass plane shows the right-side view of the house. This view reveals the depth and height of the house

elevation drawings: Drawings that provide information about the exterior details of a building. This information may include roof pitch, exterior materials and finishes, overall heights of features, and window and door styles. All of the dimensions and notations required by workers on the jobsite should be included on this sheet.

In Figure 2-7, the front, top, left-, and right-side views of the house are shown as they would be arranged in a multiview drawing. In the creation of this drawing, the front view was drawn first. Then the top view was drawn directly above the front view. Next, the right and left views were drawn directly to the right and left of the front view respectively. Architects refer to multiview drawings of the exterior of a house as ***elevation drawings***.

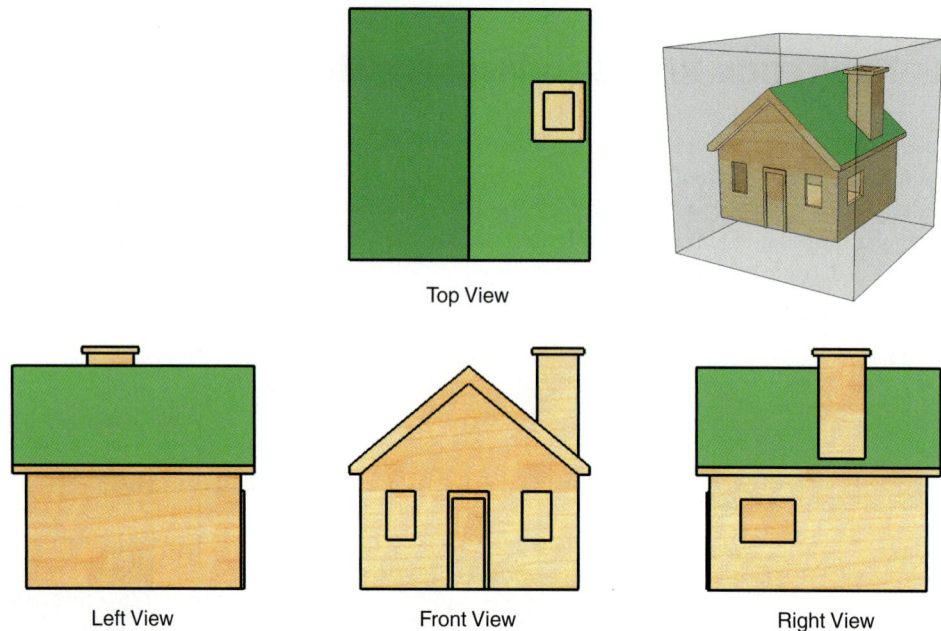

Top View

Left View   Front View   Right View

Figure 2-7   The elevations of a house as they would be arranged in a multiview drawing

Notice how a feature in the front view, such as the peak of the roof, is exactly in line with the top of the roof in both the left and right views. Also note how the features of the chimney are aligned in each of the views.

The planes representing the roof in the right, left, and top views appear as rectangles in the multiviews, but by studying them in relation to the front view, you will see that they actually represent the sloping or ***inclined planes*** of the roof. Since the planes of the roof, as projected to the top and side viewing planes, are sloping, they are not drawn actual, or true, size. In technical drawing, this phenomenon is referred to as ***foreshortening***.

inclined plane: A plane located on an object in a multiview drawing that is sloping and is not perpendicular to the line of sight of the viewer.

foreshortening: A term used to describe the phenomenon that occurs when a feature on an inclined plane is not shown true size, or true shape, in a multiview drawing.

USING THE GLASS BOX TECHNIQUE OF VISUALIZING MULTIVIEWS

Figure 2-7 introduced the concept of placing an object inside a glass box to create viewing planes. This method of visualizing multiviews is known as the *glass box technique*. This technique is often helpful for beginners who are learning the process of visualizing an object's multiviews. The following steps detail the process of using the glass box technique to visualize the multiviews of the object in Figure 2-8.

Step 1. Imagine the object shown in Figure 2-8 is centered inside a glass box and its six regular views are projected out to the glass planes surrounding the object (see Figure 2-9). The plane on which the front view is projected is called the *frontal plane*.

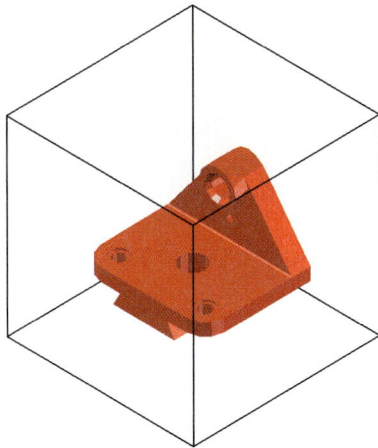

Figure 2-8   Object centered inside glass box

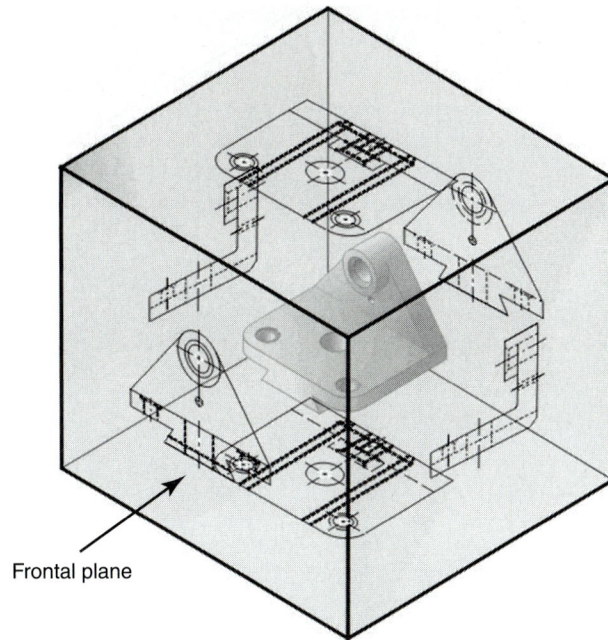

Figure 2-9   Multiviews of object projected onto glass planes

Frontal plane

Step 2. Unfold the glass box as if the four sides of the frontal plane were hinged as shown in Figures 2-10 and 2-11.

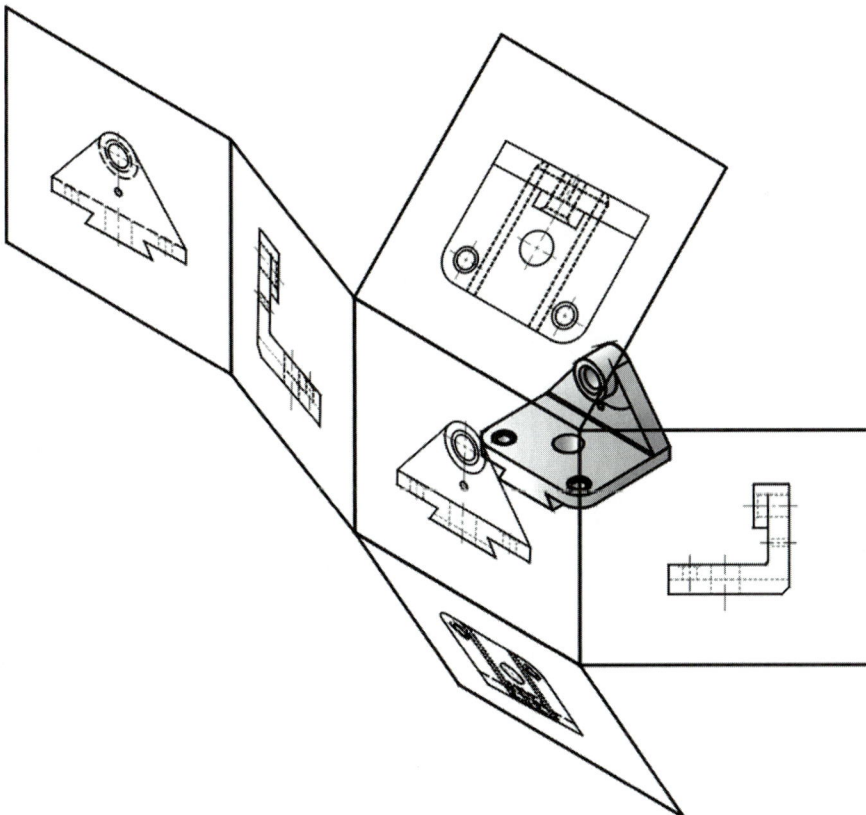

Figure 2-10   Unfolding the glass box

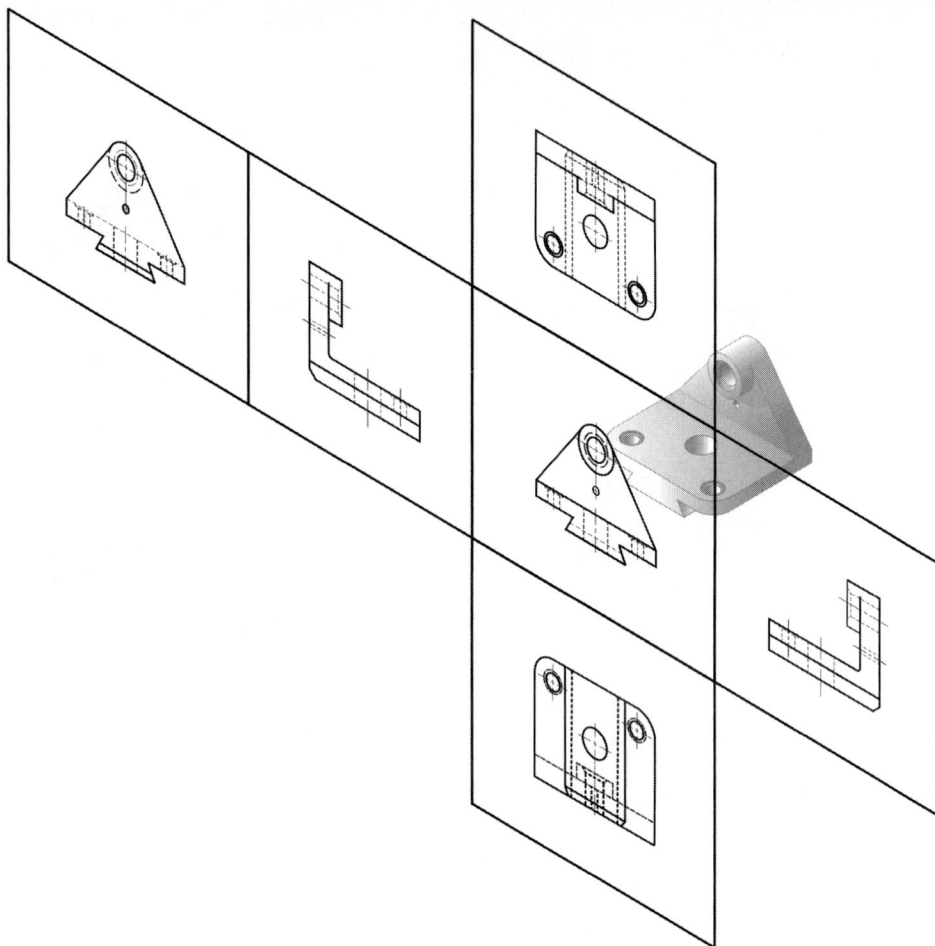

Figure 2-11 Glass box unfolded

Step 3. After the sides of the glass box are unfolded, the six regular views of the object (front, top, bottom, right, left, and rear) are displayed in their "projected" positions as shown in Figure 2-12. Note that the front, right, left and rear views are aligned horizontally and the front, top, and bottom views are in vertical alignment.

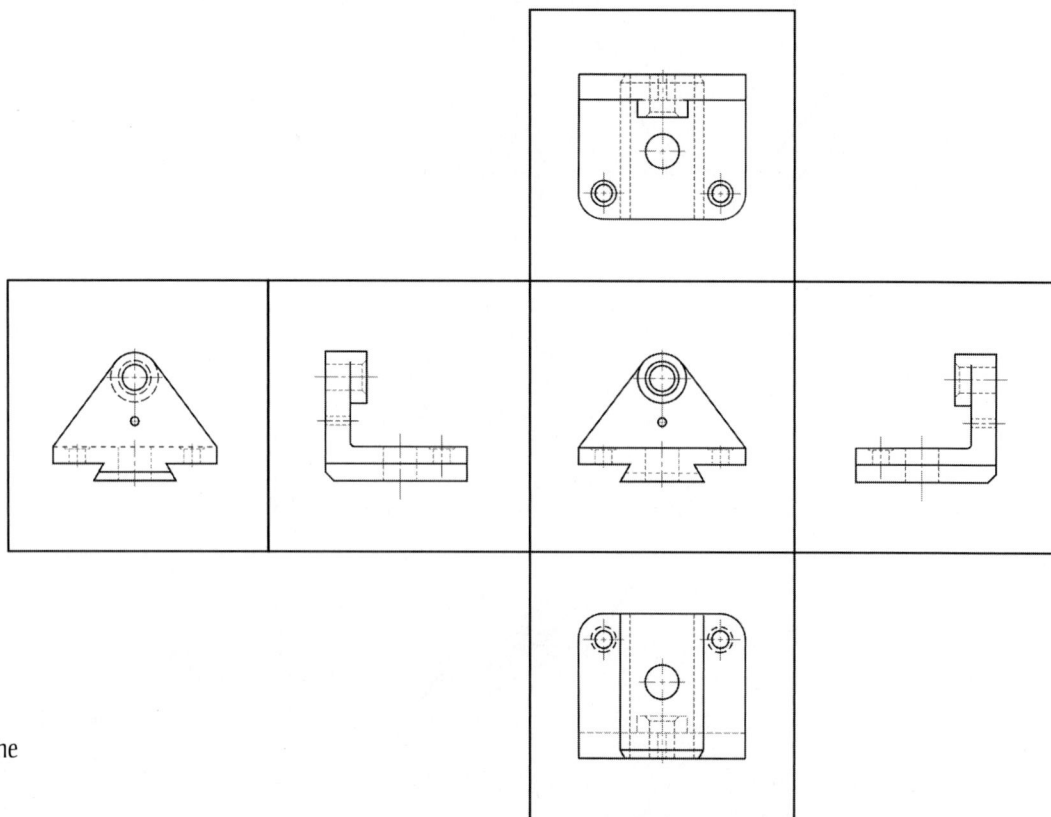

Figure 2-12 The glass box unfolded to display the six regular multiviews of the object

LINE TYPES AND LINE WEIGHTS IN MULTIVIEW DRAWINGS——

The features of an object are shown with differing *line types* and *line weights*. Commonly used line types include *visible lines* which show the visible edges and features of an object, *hidden lines* which represent features that would not be visible, and *center lines* which locate the centers of features such as holes and arcs. The terminology used for the various line types is shown in Figure 2-13.

Line weight refers to the width of the lines in a technical drawing. Standard line types and line weights have been established by the American Society of Mechanical Engineers (ASME). The ASME standard for line conventions and lettering is *ASME Y14.2M-1992*. The line weights for lines specified by this standard for use on technical drawings are shown in Table 2-1.

Note: ASME describes these line thicknesses as the approximate widths.

line types: Include *visible lines* which show the visible edges and features of an object, *hidden lines* which represent features that would not be visible, and *center lines* which locate the centers of features such as holes and arcs. Standard line types have been established by the American Society of Mechanical Engineers (ASME) in *ASME Y14.2M.*

line weight: Refers to the width of the lines in a technical drawing. Standard line weights have been established by the American Society of Mechanical Engineers (ASME) in *ASME Y14.2M.* In this standard, visible lines are drawn 0.6mm wide, while center and hidden lines are drawn .3mm wide.

Table 2-1	ASME Y14.2M Line Thickness Standard
Visible Line = .6mm thick	
Hidden Line = .3mm thick	
Center Line = .3mm thick	
Dimension Line = .3mm thick	
Extension Line = .3mm thick	
Cutting Plane Line = .6mm thick	
Section Line = .3mm thick	

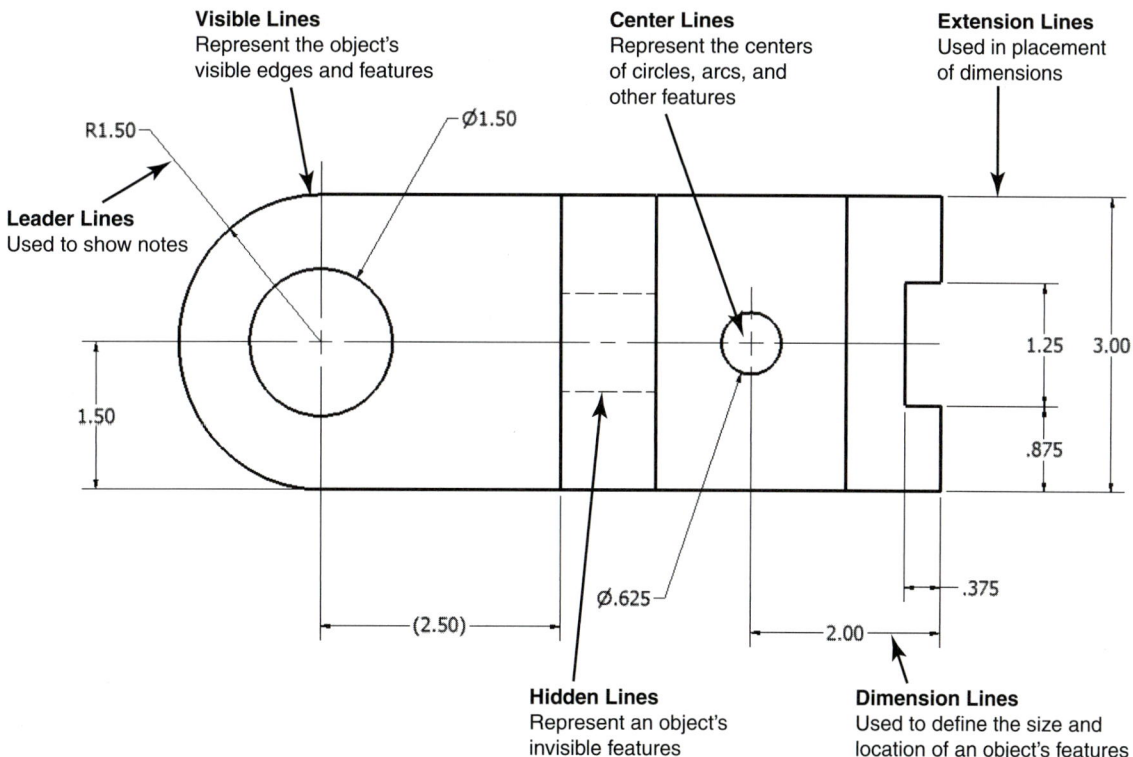

Visible Lines
Represent the object's visible edges and features

Center Lines
Represent the centers of circles, arcs, and other features

Extension Lines
Used in placement of dimensions

Leader Lines
Used to show notes

R1.50   Ø1.50   1.50   (2.50)   Ø.625   2.00   .375   1.25   3.00   .875

Hidden Lines
Represent an object's invisible features

Dimension Lines
Used to define the size and location of an object's features

Figure 2-13   Line type terminology

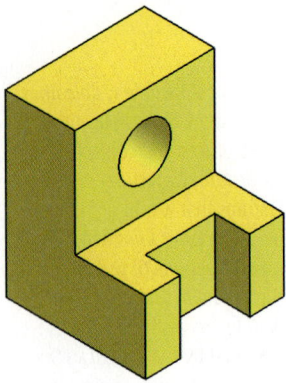

Figure 2-14 Object to be visualized with multiviews

HIDDEN FEATURES AND CENTER LINES IN MULTIVIEW DRAWINGS——

Figure 2-15 shows the six regular views of the object shown in Figure 2-14. Study these examples and note how features that would otherwise be invisible in a view, such as the edges of the hole and slot in the side views, are depicted with hidden lines. Also, note the different ways that the center lines representing the center of the hole are drawn in each view.

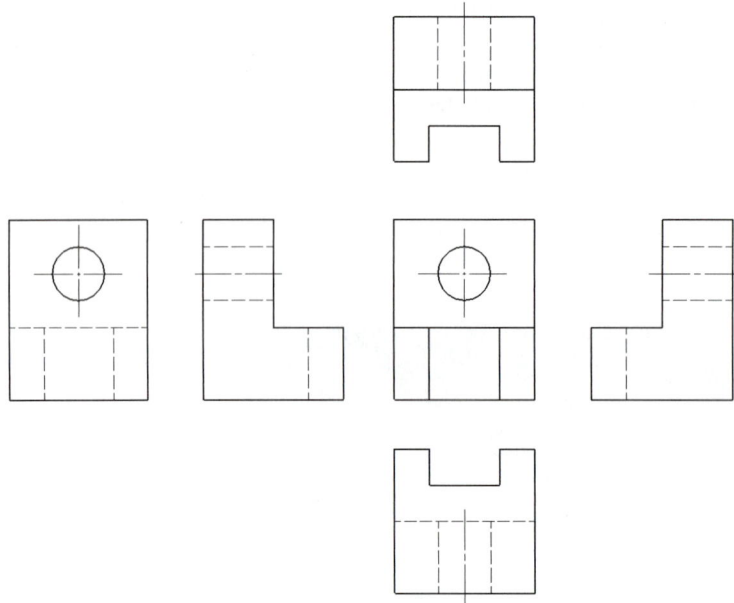

Figure 2-15 The six regular views of the object in Figure 2-14 with visible, hidden, and center lines displayed

USE YOUR IMAGINATION!———————————————————————

As a drafter-in-training, you should develop the ability to use your imagination to visualize the multiviews of an object. Engineering and architecture are fields where a powerful imagination is an important tool for success because most of the objects being designed exist first only in the imagination of the designer(s). The challenge for the design team is to take the design from the imagination stage and turn it into a set of drawings that can be used to make the design a reality.

The following steps document the process of creating a multiview drawing for the object shown in the designer's sketch in Figure 2-16.

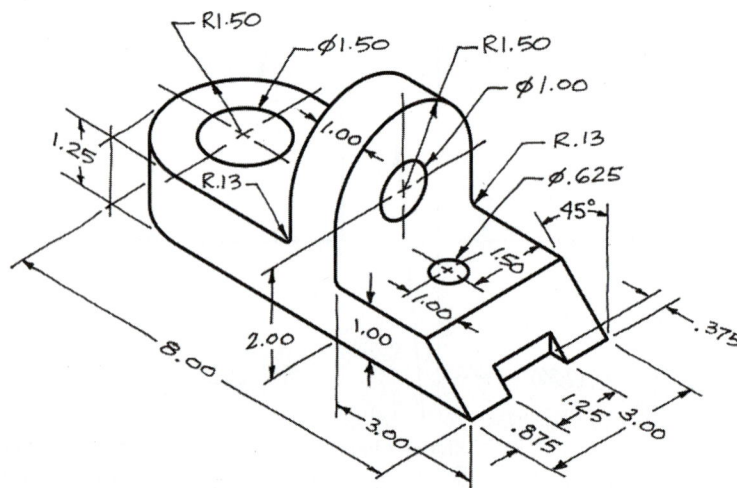

Figure 2-16 The designer's two-dimensional sketch of the object to be visualized

MATERIAL-ALUMINUM 6061

CREATING A MULTIVIEW DRAWING

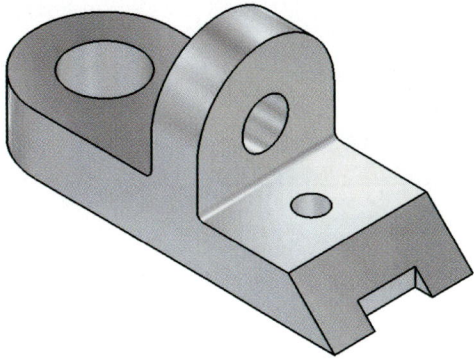

Figure 2-17   The object visualized as a three-dimensional part

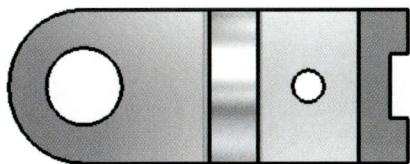

Figure 2-18   The front top of the object

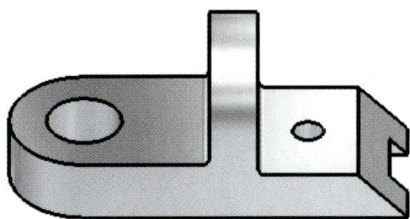

Figure 2-19   Rotating the object toward the top

Step 1. The drafter studies the designer's sketch of the object in Figure 2-16 and imagines how it might look after it has been manufactured (Figure 2-17). Next, the drafter determines the front, or principal, view of the imagined 3D object and how it may be positioned (Figure 2-18).

Then, the drafter rotates the object in his or her mind toward the top (Figure 2-19) until the top, or bird's-eye, view of the principal view is visible as shown in Figure 2-20.

Next, the front view of the object is imagined to be rotated toward the right (Figure 2-21) until the right side view is visualized (Figure 2-22).

This process could be likened to creating a 3D movie of the object in your imagination to facilitate the visualization of the desired views.

Note that the top and right views (Figures 2-20 and 2-22) are drawn at right angles (90°), or perpendicular, to the front view (Figure 2-18).

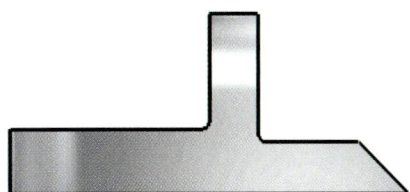

Figure 2-20   The front view of the object

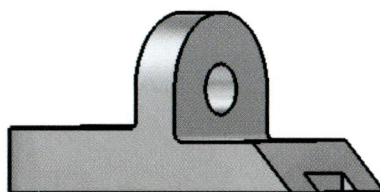

Figure 2-21   Rotating the object toward the right

Figure 2-22   The right view of the object

Step 2. The drafter continues the process begun in Step 1, rotating the object until the six regular views of the object have been visualized (Figure 2-23).

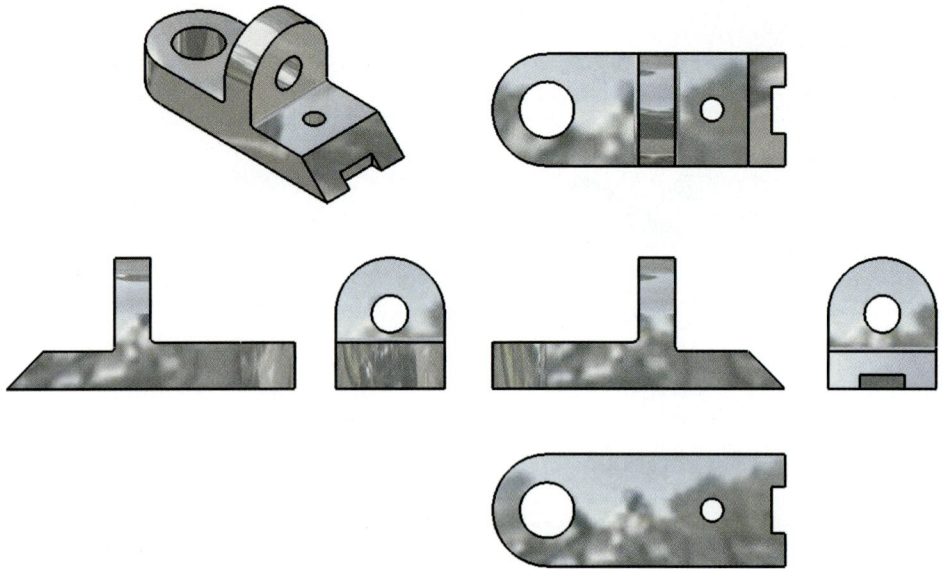

Figure 2-23   The six regular views of the object

Step 3. The drafter visualizes the visible lines of the object (Figure 2-24).

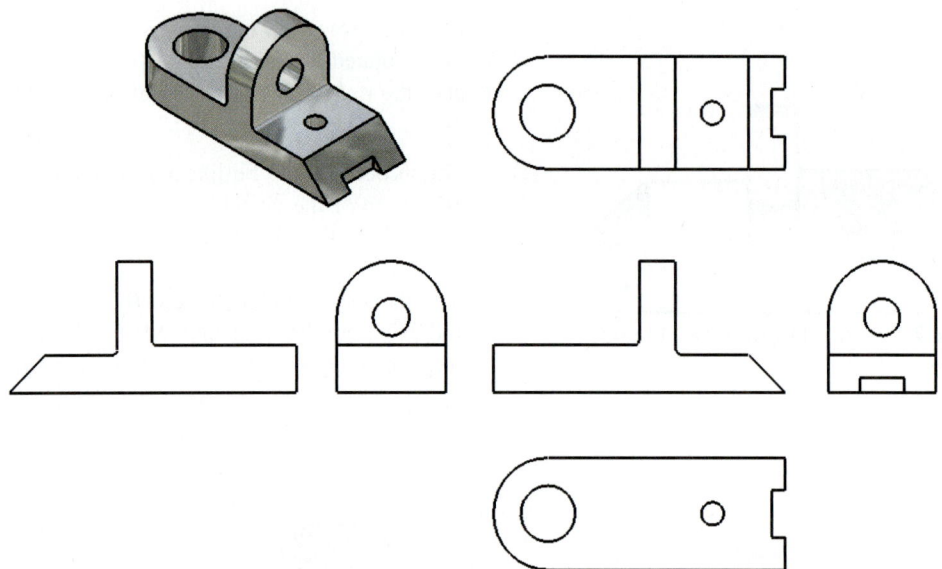

Figure 2-24   The six regular views showing the object's visible lines

Step 4. Next, the drafter visualizes the location of the object's hidden and center lines (Figure 2-25).

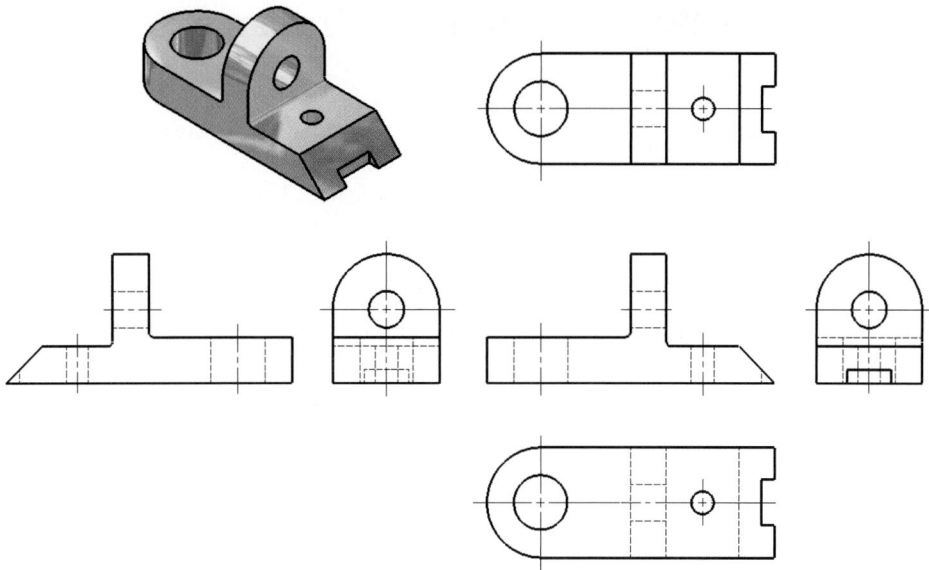

Figure 2-25   The six regular views of the object including hidden and center lines

Step 5. In the last step, the drafter determines which of the six views will be necessary to describe the object and places dimensions on the part (Figure 2-26).

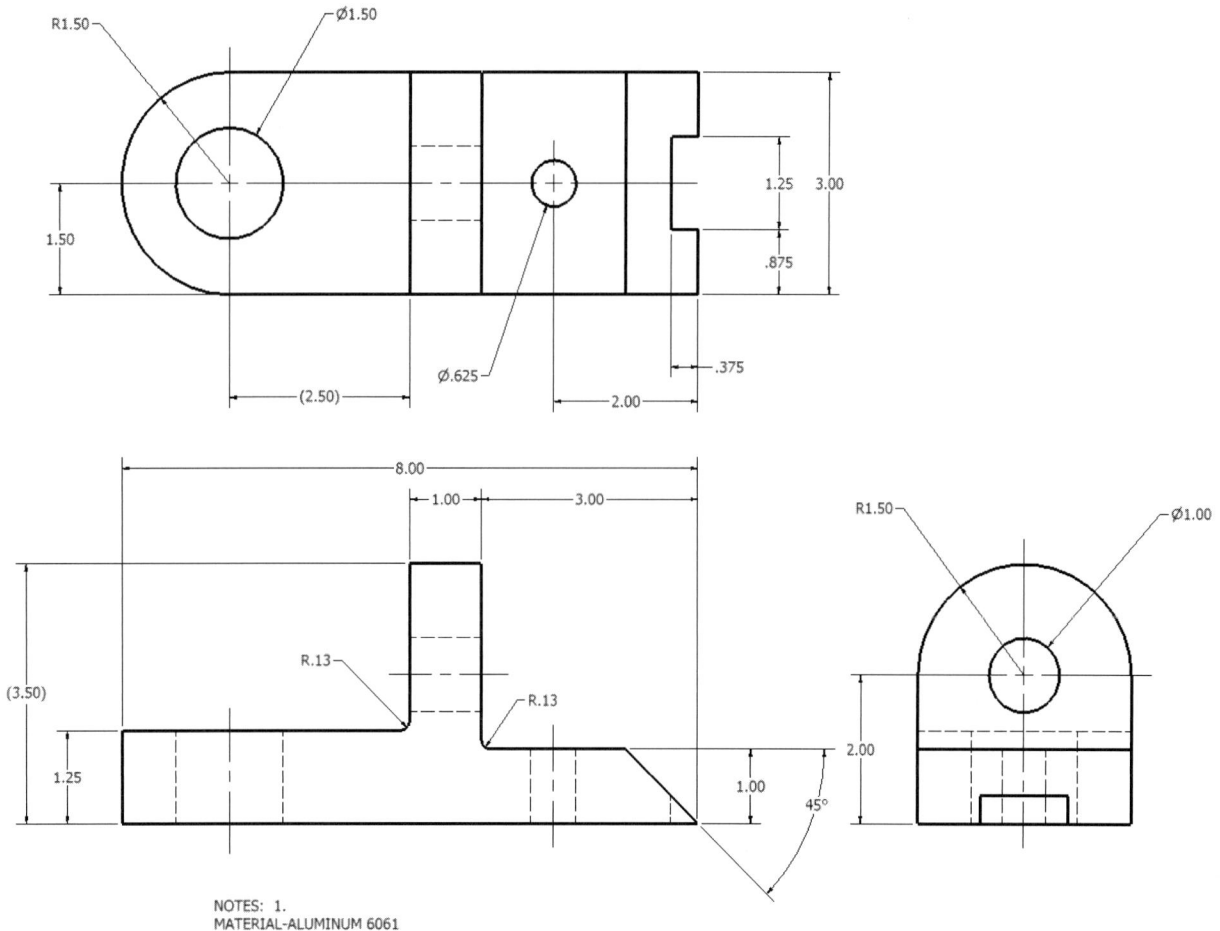

NOTES: 1.
MATERIAL-ALUMINUM 6061

Figure 2-26   The views necessary to describe the object including dimensions

graphic primitives: Geometric shapes such as boxes, cylinders, cones, spheres, wedges, and prisms that can be combined (unioned) or removed from one another (subtracted) to create more complicated shapes.

VISUALIZING THE MULTIVIEWS OF BASIC GEOMETRIC SHAPES

The multiview representations of some basic geometric shapes are shown in Figures 2-27 through 2-40. Shapes such as boxes, cylinders, cones, spheres, wedges, and prisms are often referred to as **graphic primitives** because by combining, or *unioning,* these shapes, or in some cases *subtracting* the geometry of one shape from another one, more complicated shapes can be formed. Graphic primitives could be considered as building blocks used to construct more complex objects. Students who learn to visualize the multiviews of graphic primitives will find it easier to visualize the multiviews of the more complicated shapes formed when they are combined.

Study Figures 2-27 through 2-40 and familiarize yourself with how the front, top, and side views of the graphic primitives and their combinations are drawn, including the placement of hidden and center lines.

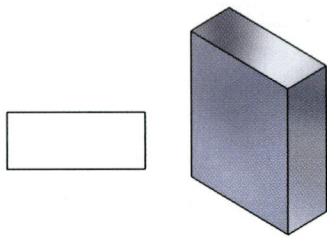

Figure 2-27 Front, top, and right views of a box

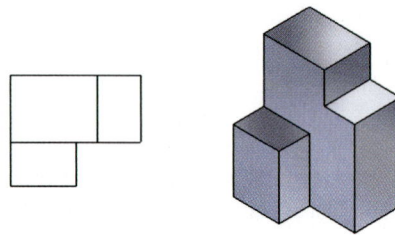

Figure 2-28 Front, top, and right views of a shape formed by unioning and subtracting boxes

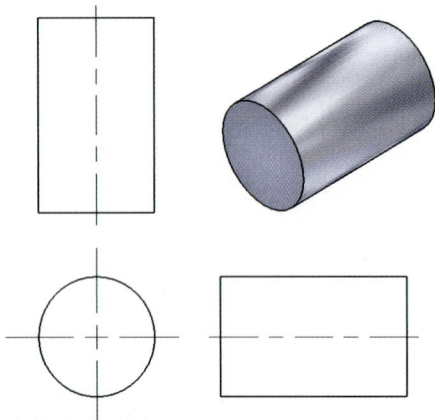

Figure 2-29 Front, top, and right views of a cylinder

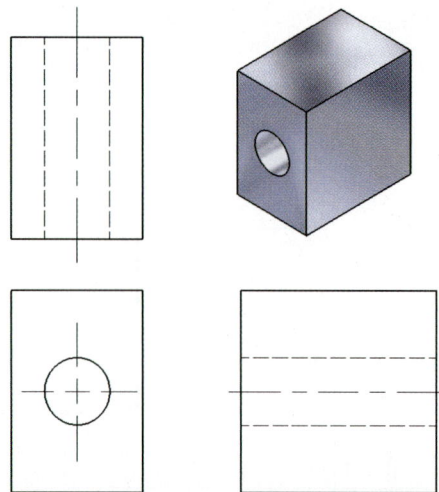

Figure 2-30 Front, top, and right views of a shape formed by subtracting a cylinder from a box

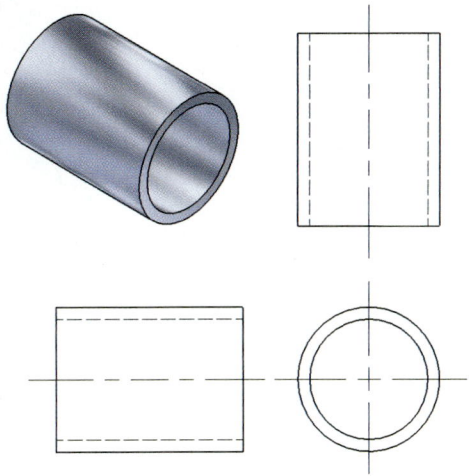

Figure 2-31   Front, top, and left views of a cylinder with a smaller cylinder subtracted from its center

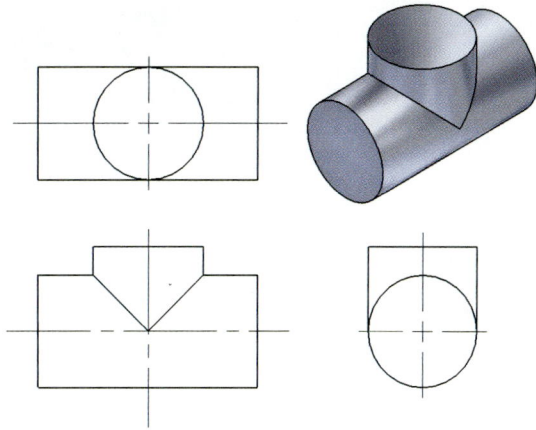

Figure 2-32   Front, top, and right views of a shape formed by the intersection of two cylinders of equal diameter

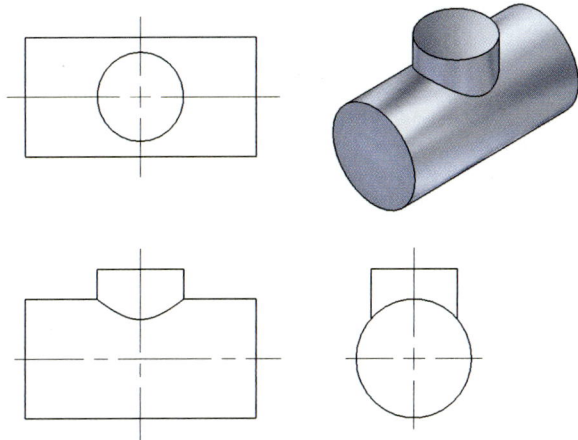

Figure 2-33   The front, top, and right views of the shape resulting from the intersection of two cylinders with unequal diameters

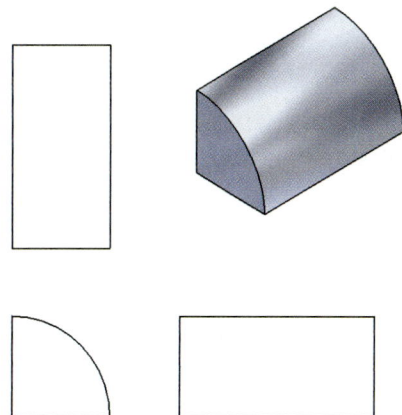

Figure 2-34   Front, top, and right views of a quarter-round shape

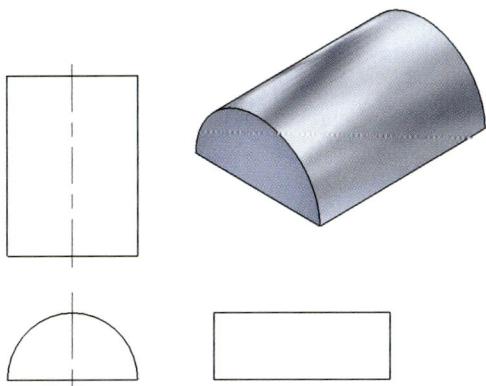

Figure 2-35   Front, top, and right views of a half-round shape

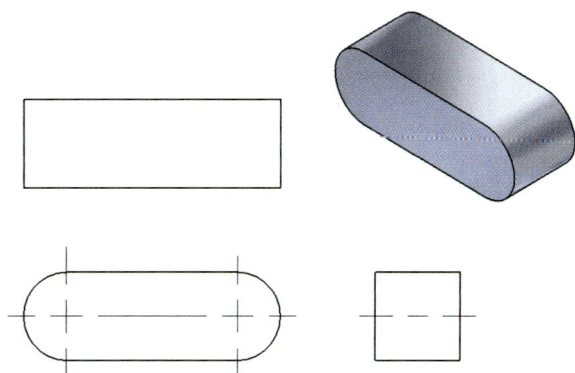

Figure 2-36   Front, top, and right views of a shape resulting from the union of a box and two half-rounds

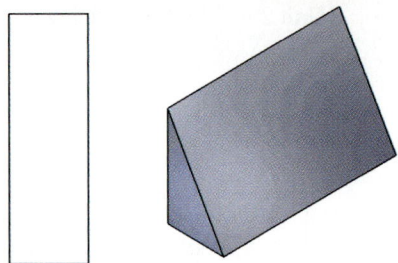

Figure 2-37 Front, top, and right views of a wedge

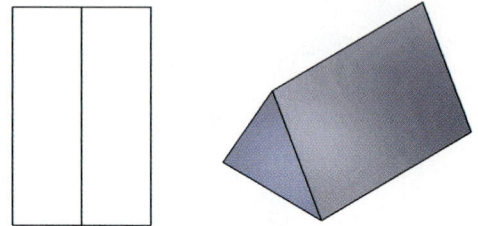

Figure 2-38 Front, top, and right views of a prism

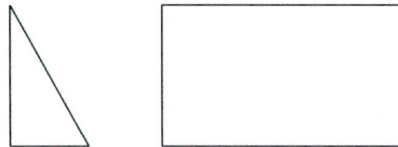

Figure 2-39 Front, top, and left views of a cone

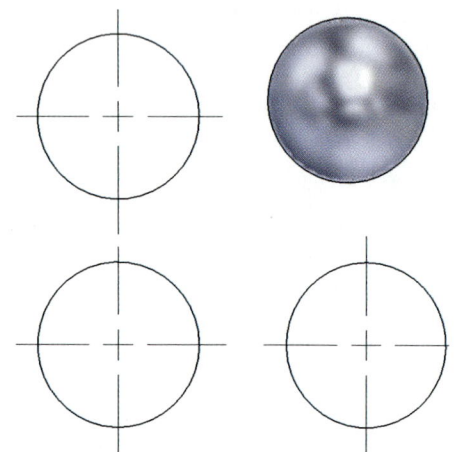

Figure 2-40 Front, top, and right views of a sphere

ORTHOGRAPHIC PROJECTION

Orthographic projection is the technique employed in the creation of multiview drawings to project the size and location of geometric features (points, lines, planes, or other features) from one view to another. Light construction lines are usually drawn between views to facilitate the transfer of this information. Projecting information in this manner is a more efficient way to construct technical drawings than remeasuring the features in each view.

The orthographic projection technique also utilizes a ***miter line*** drawn at 45° which allows geometric information to be projected between the top and side views.

Figure 2-41 shows an example of this technique. Phantom lines illustrate how the size and location of the object's features can be projected from view to view. Note how the 45° miter line is used to facilitate the transfer of information between the top and side views.

UTILIZING ORTHOGRAPHIC PROJECTION TECHNIQUES TO CREATE MULTIVIEW SKETCHES

Step 1. Study the sketch of the part shown in Figure 2-42 and try to imagine it as a three-dimensional object. With this 3D image in mind, visualize the front, top, and right side views.

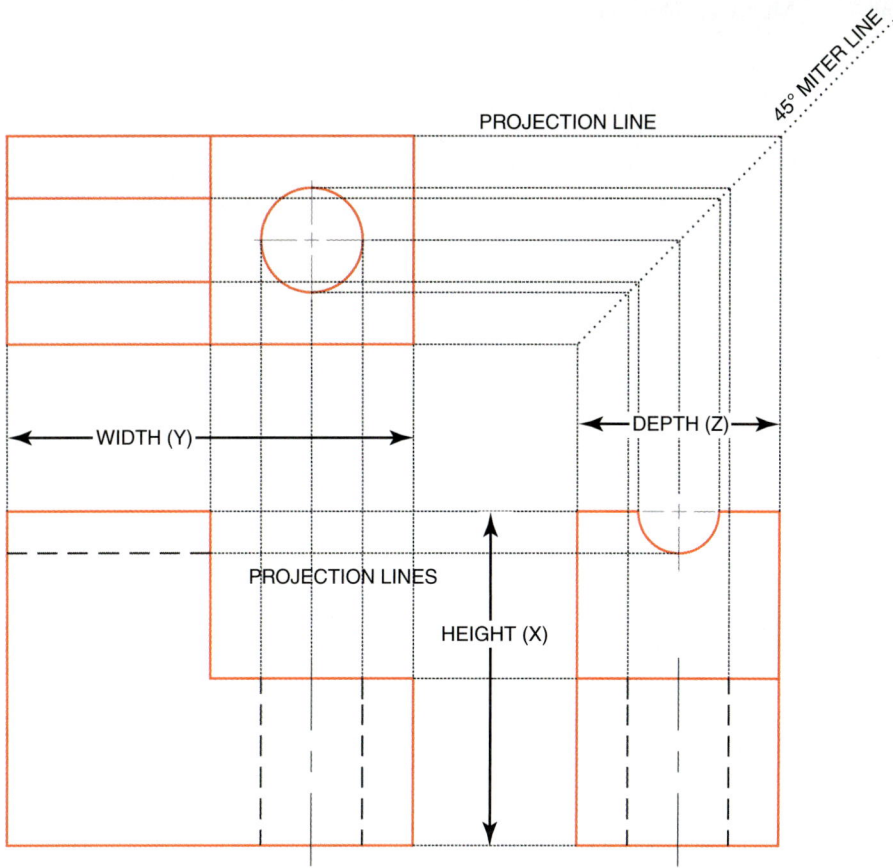

Figure 2-41 Multiview drawing of an object using orthographic projection techniques

PROJECTION LINE

45° MITER LINE

WIDTH (Y)

DEPTH (Z)

PROJECTION LINES

HEIGHT (X)

TOP

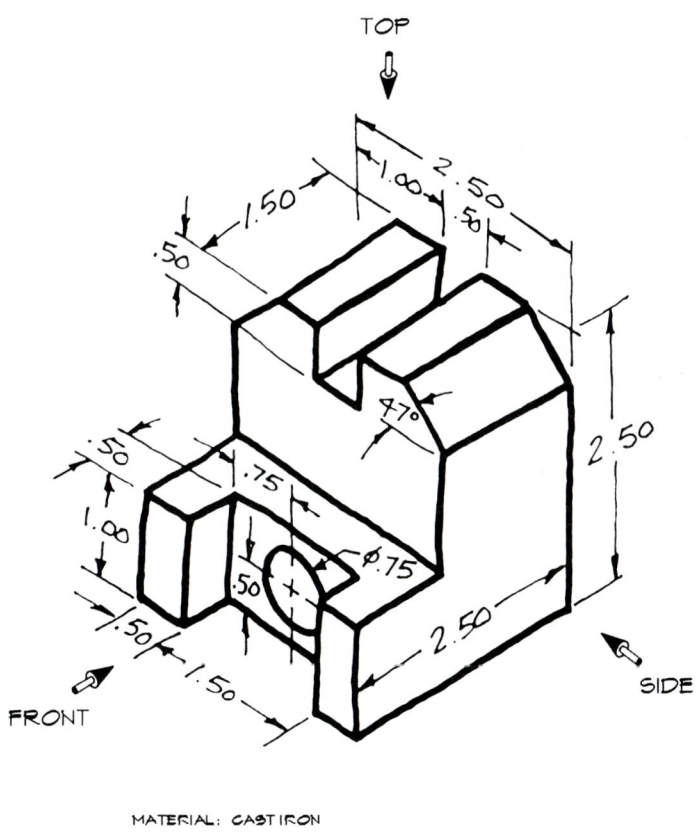

Figure 2-42 Sketch of object to be drawn as a multiview drawing

2.50

1.00

.50

1.50

.50

2.50

47°

.75

ø.75

.50

.50

1.00

.50

1.50

2.50

SIDE

FRONT

MATERIAL: CAST IRON

Note:

The American Society of Mechanical Engineers Standard for creating multiview drawings is ASME Y14.3M-1994.

Step 2. Sketch the front view of the object. Try to sketch the part proportionally to the dimensions specified on the sketch. Extend light construction lines out from the features of the front view to the top and right sides and place a 45° miter line as shown in Figure 2-43.

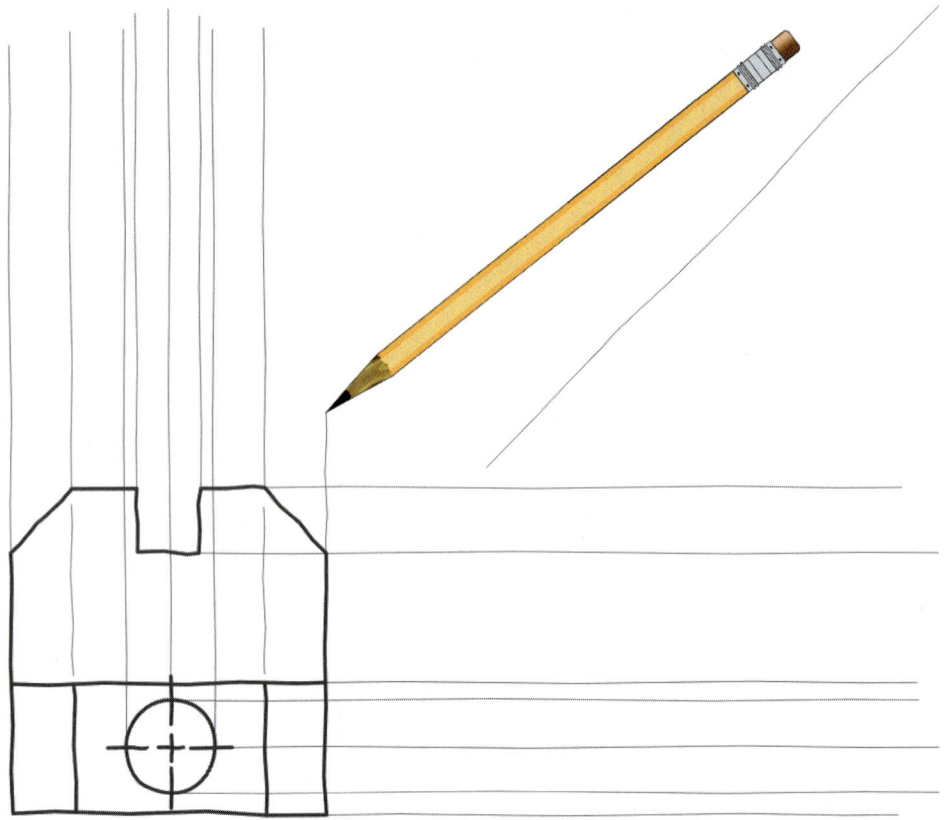

Figure 2-43 Extending construction lines from the features of the front view to the side and top

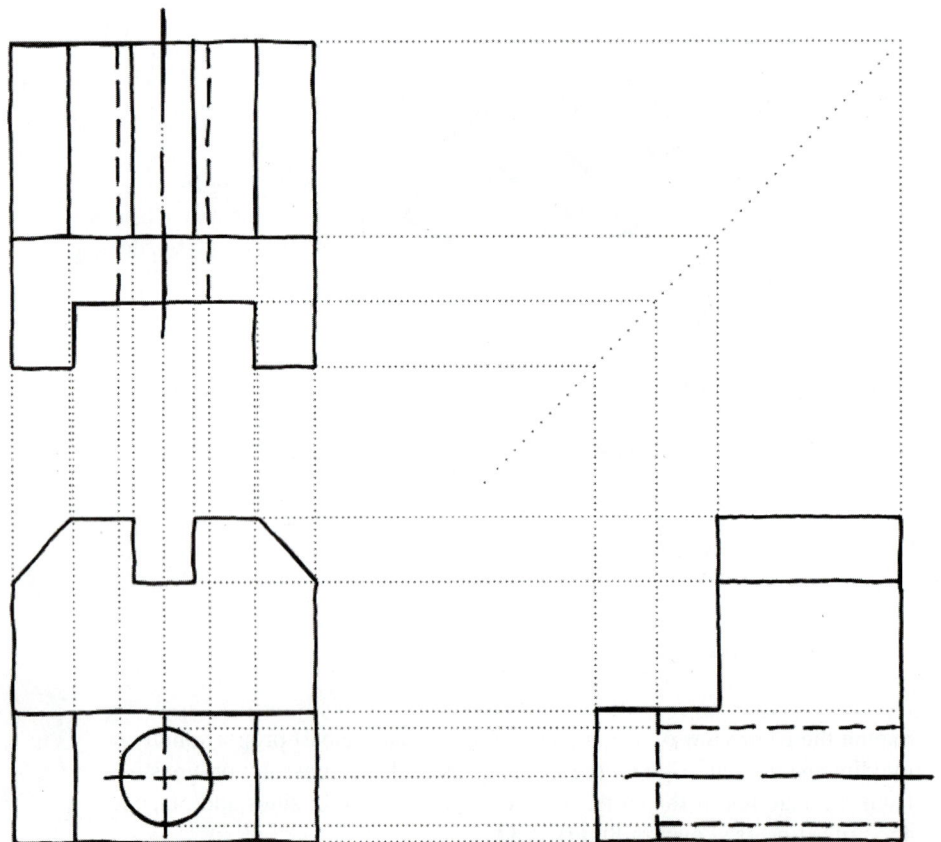

Figure 2-44 Sketching the top and right-side views

Step 3. Sketch the top and right side views of the object as shown in Figure 2-44. Use the construction lines projected from the front view, and construction lines projected through the miter line, to locate the features of each view. Darken the visible, hidden, and center lines as needed. Erase construction lines that appear too dark.

DRAWING OBJECTS TO SCALE

In technical drawings, objects are often drawn *to scale*. This term refers to the relationship between the size of the object in the drawing and the actual size of the object after it is manufactured. Following are four of the scales most commonly used in the creation of mechanical drawings:

- **Full Scale**   This means that the size of the object in the drawing will be the same size as the object after it is manufactured. This is usually only feasible on smaller objects such as machine parts (to draw an average-size house at full scale, you might need a sheet of paper that is 136 feet long by 88 feet wide). When noting on a drawing that the object is drawn full scale, the drafter could write 1 = 1, 1/1, or 1:1.

- **Half Scale**   This means that the size of the object in the drawing is half the size of the object after it is manufactured. The drafter will still place the full-size dimensions on the views of the object so that even though the drawing is half size, the part will be manufactured full size. When noting on a drawing that the object is drawn half scale, the drafter could write 1 = 2, 1/2, .5X, or 1:2.

- **Quarter Scale**   This means that the size of the object in the drawing is one fourth the size of the object after it is manufactured. The drafter will still place the full-size dimensions on the views of the object so that even though the drawing is one-fourth size, the part will be manufactured full size. When noting on a drawing that the object is drawn quarter scale, the drafter could write 1 = 4, 1/4, .25X, or 1:4.

- **Double Scale**   This means that the size of the object in the drawing is twice the size of the object after it is manufactured. The drafter will still place the full-size dimensions on the views of the object so that even though the drawing is twice size, the part will be manufactured full size. This scale is used for smaller objects that would be difficult to dimension if drawn at actual size. When noting on a drawing that the object is drawn double scale, the drafter could write 2 = 1, 2/1, 2:1, or 2X.

DRAWING ARCHITECTURAL PLANS TO SCALE

Following are two of the scales most commonly used in the creation of architectural drawings:

- **Quarter Inch Equals One Foot**   This means that every 1/4″ on the plotted drawing represents a measurement of 1′ on the actual construction project. For example, a wall that is to be built 16′ in length will measure 4″ on the drawing. This allows a drafter to fit a house that is 100′ long and 50′ wide on a sheet of paper measuring only 34″ by 22″. The distance of 100′ will measure only 25″ on the drawing sheet (100 × 1/4″ = 25″), and 50′ will measure 12.5″ on the sheet (50 × 1/4″ – 12.5″). The dimensions on the drawing will be labeled at the actual distance (in feet and inches) required to construct the building full size. When noted on a drawing that the object is drawn to this scale, the drafter would write 1/4″ = 1′ -0″.

- **Eighth Inch Equals One Foot**   This means that every 1/8″ on the plotted drawing will represent a measurement of 1′ on the actual construction project. For example, a wall that is to be built 16′ in length, will measure 2″ on the drawing. This allows a drafter to fit a house that is 200′ long and 100′ wide on a sheet of paper measuring only 34″ by 22″. The distance of 200′ will measure only 25″ on the drawing sheet (200 × 1/8″ = 25″), and 100′ will measure 12.5″ on the sheet (100 × 1/8″ = 12.5″). The dimensions on the drawing will be labeled at the actual distance (in feet and inches) required to construct the building full size. When noted on a drawing that the object is drawn to this scale, the drafter would write 1/8″ = 1′-0″.

DRAWING SHEET SIZES

sheet sizes: Technical drawings are created on standardized sheet sizes. Sheet size varies with the type of drawing and/or the unit of measurement used to create the drawing.

The American Society of Mechanical Engineers and other standards organizations have defined standardized **sheet sizes** for the preparation of technical drawings. These sheet sizes vary depending on the type of drawing and/or the unit of measurement used to create the drawing.

The ASME standard for decimal sheet sizes is *ASME Y14.1-2005*. The sheet sizes defined in this standard begin with an A size sheet which is $11 \times 8.5''$. A B size sheet's dimensions are $17 \times 11''$, which is the equivalent of two A sheets laid side by side. A C size sheet is $22 \times 17''$, which is the equivalent of two B sheets laid side by side. A D size sheet is $34 \times 22''$, which is the equivalent of two C sheets laid side by side. Figure 2-45 illustrates the sheet sizes used in mechanical drawings.

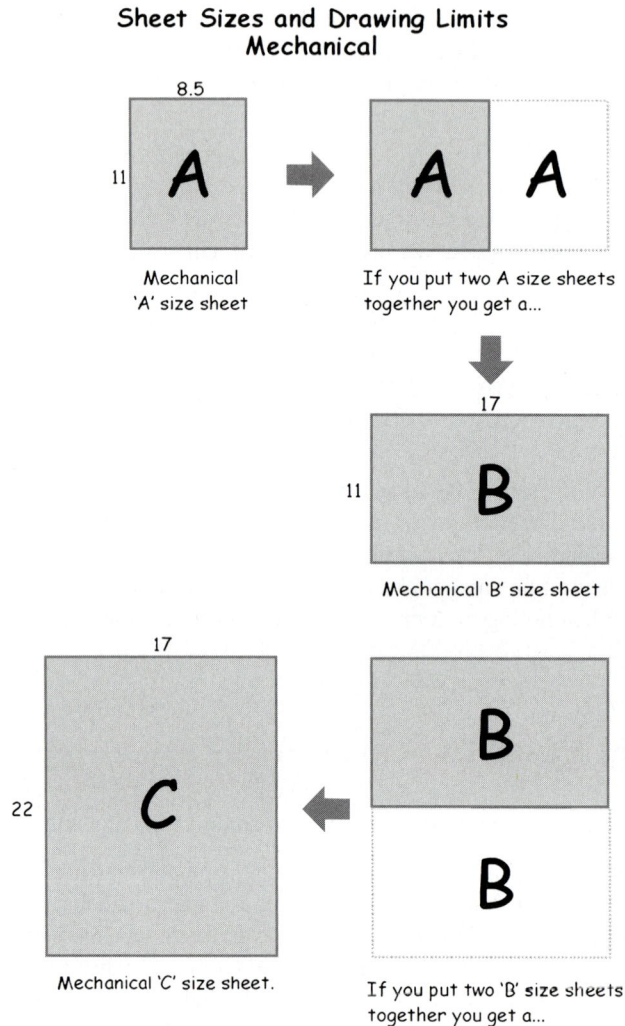

Sheet Sizes and Drawing Limits
Mechanical

Mechanical
'A' size sheet

If you put two A size sheets
together you get a...

Mechanical 'B' size sheet

Mechanical 'C' size sheet.

If you put two 'B' size sheets
together you get a...

Figure 2-45   Sheet sizes for a mechanical drawing

The ASME standard for metric sheet sizes is *ASME Y14.1M-2005*. In this standard, an A4 sheet measures 297×210 millimeters (mm), an A3 sheet measures 420×297 mm, an A2 sheet measures 594×420 mm, an A1 sheet measures 841×594 mm, and an A0 sheet measures 1189×841 mm.

For architectural drawings where inches are used as the unit of measurement, an A sheet measures $12 \times 9''$, a B sheet is $18 \times 12''$, a C sheet is $24 \times 18''$, and a D sheet is $36 \times 24''$.

A high-quality paper known as *vellum*, or tracing paper, is used to plot drawings that are made for the purpose of reproducing by blueprinting. Vellum is a strong, thin paper that allows light to pass through it relatively easily. Vellum can be purchased in rolls 24'' to 36'' in width or in standard sheet sizes. Vellum can also be purchased with preprinted title blocks.

THIRD ANGLE PROJECTION VERSUS FIRST ANGLE PROJECTION

The method of arranging multiviews shown in Figure 2-46, with the top view drawn above the front view and the right side drawn to the right of the front view, is called *Third Angle Projection*. This method is widely used in technical drawings created in the United States. In a Third Angle Projection, the image is projected onto a viewing plane that is located between the object and the viewer.

In many parts of the world, multiviews are arranged using *First Angle Projection* instead of Third Angle Projection. A First Angle Projection is drawn as though the object is between the observer and the projection plane. For this reason, when a drawing is created with First Angle Projection, the right side view appears to the *left* of the front view, and the top view appears *below* the front view, as shown in Figure 2-47.

Note:
The American Society of Mechanical Engineers standard for creating and interpreting multiview drawings using First and Third Angle Projection techniques is ASME Y14.3M-1994.

Third Angle Projection: A technique for arranging multiview drawings with the top view above the front view and the right side drawn to the right of the front view. Commonly used on drawings prepared in North America.

First Angle Projection: A technique for arranging multiview drawings in which the left side view is drawn to the right of the front view and the top view is placed below the front view, and so on. Commonly used on drawings prepared outside North America.

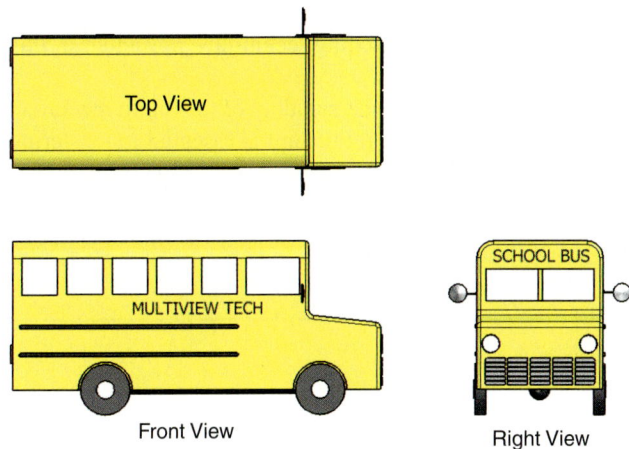

Figure 2-46   Third Angle Projection example

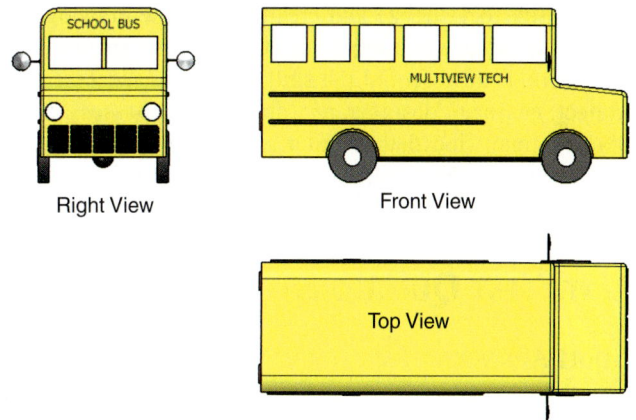

Figure 2-47   Arrangement of views in First Angle Projection

To avoid confusion, it may be necessary to note on the drawing whether First or Third Angle Projection was used to create the drawing. For this reason, symbols have been developed to distinguish between the two types of projection techniques.

Third Angle Projection can be noted on drawings by placing the symbol shown in Figure 2-48 in, or near, the title block.

Note:
The American Society of Mechanical Engineers standard governing the symbols used for First and Third Angle Projection is ASME Y14.1-2005.

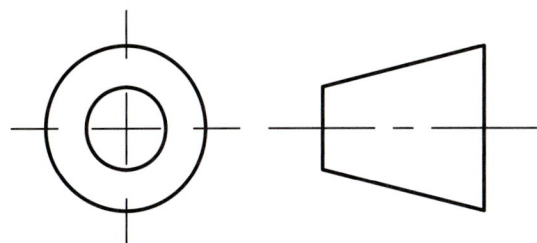

Figure 2-48   Third Angle Projection symbol

The First Angle Projection technique can be noted on drawings with the symbol shown in Figure 2-49. The letters *SI* (Metric International System of Units) indicate that the drawing was prepared using metric units. The unit of measurement commonly used in the creation of mechanical engineering drawings is the millimeter.

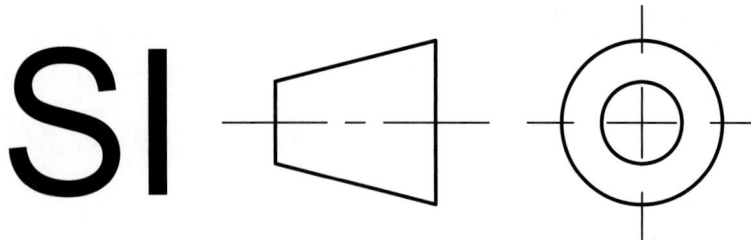

Figure 2-49   First Angle Projection symbol. SI signifies that the drawing was created in metric units

SUMMARY

The ability to visualize and create multiview drawings, as well as the ability to interpret multiview drawings produced by others, is an essential job skill that every successful architect, engineer, designer, and drafter must possess. Some students may find that in order to develop this skill, they will need to put in a significant amount of time practicing the visualization and sketching techniques presented in this unit.

Developing a solid understanding of the multiview drawing techniques presented in this unit is essential to mastering the concepts and drawing assignments you will encounter later in this course.

UNIT TEST QUESTIONS

Short Answer

1. Name the six regular views of an object.
2. How does a drafter determine which view will be the front view?
3. How many views of an object should a drafter draw?
4. What does Quarter Scale mean?
5. Name the standard that controls line thickness in an ASME drawing.

Multiple Choice

1.   What is the sheet size for an architectural D sheet?
   a. 12 × 9
   b. 297 × 210
   c. 36 × 24
   d. 18 × 12

2.   What is the sheet size for a mechanical B size sheet?
   a. 17 × 11
   b. 18 × 12
   c. 841 × 594
   d. 36 × 24

3.   Which ratio indicates that a drawing is full size?
   a. 1 = 1
   b. 1/1
   c. 1:1
   d. All of the above

4.   A C sheet is equal in size to two of these sheets laid side by side:
   a. A
   b. B
   c. D
   d. None of the above

5.   What does 1/4″ = 1′-0″ represent on an architectural drawing of a house?
   a. The scale of the drawing is 1 = 4.
   b. One foot on the construction site equals 1/4 inch on the drawing.
   c. The house should be built one-quarter scale.
   d. A distance of 100 feet will measure 12.5 inches on the drawing.

Matching

Column A

a. 2 = 1

b. .3mm

c. .3mm

d. 1 = 2

e. .6mm

Column B

1.   Ratio indicating a drawing is half scale

2.   ASME line thickness for a visible line

3.   Ratio indicating a drawing is double scale

4.   ASME line thickness for a center line

5.   ASME line thickness for a hidden line

Internet Resources

American Society of Mechanical Engineers: *www.asme.org*

UNIT EXERCISES

The following sketching exercises are designed to help you develop multiview sketching and visualization skills.

Directions

1. On the grid sheets located at the back of the text, sketch the front, top, and side views of the objects in Multiview Sketching Exercises 2-1 through 2-6. The black arrows on each sketching exercise identify the view of the object to sketch as the front view.

2. Begin each sketch by counting the number of grids that define the features of the front view and transfer these distances to the grid sheet. Start the front view in the darkened corner located in the lower left corner of each numbered grid box (see the example shown on the Exercise 2-1 Grid Sheet). Begin the top and right views in the darkened corners above, or to the right, of the front view

3. Take advantage of the miter line drawn in each grid box to transfer information between the top and right views whenever possible.

Note:
Remove the grid sheets by carefully tearing along the perforation. If you need extra grid sheets, print the Multiview Grid Sheet.pdf file located on the Student Resource CD in the Sketching and Traditional Drafting Sheets folder.

If you have trouble with a sketching problem, you may find that referring to Figures 2-27 through 2-40 is helpful. Also, do not hesitate to ask your instructor for assistance. This activity may seem difficult at first, but keep working at it, because through practice it is possible for you to develop this important drafting skill.

Exercise 2-1

On the grid sheet located at the back of the text, sketch the front, top, and right-side views of the objects shown on the next page. Begin views in the dark corners shown in the grid.

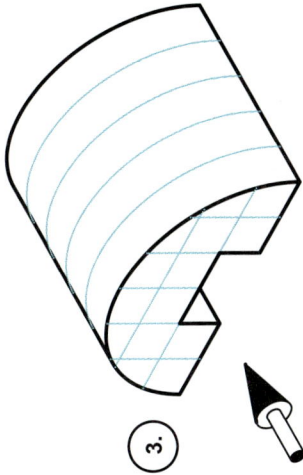

3.

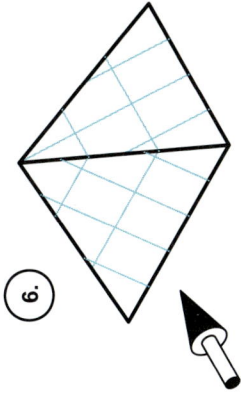

6.

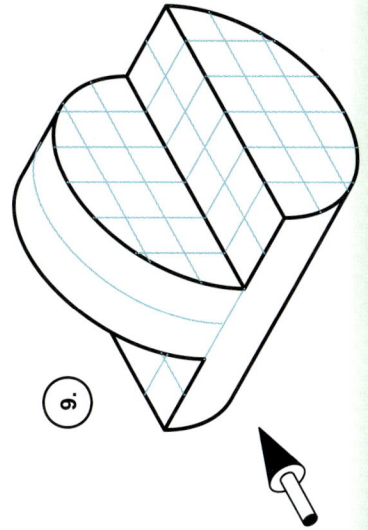

9.

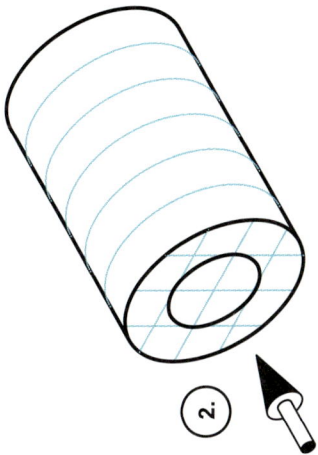

2.

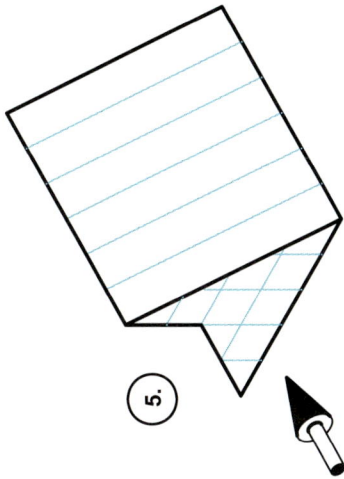

5.

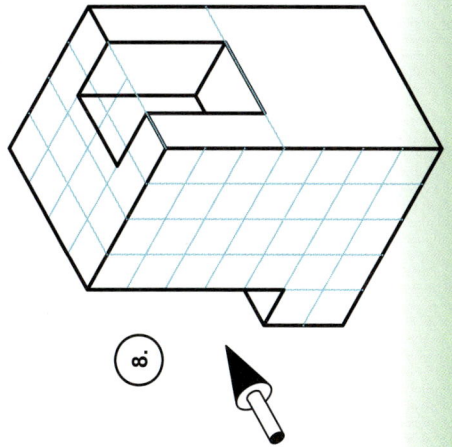

8.

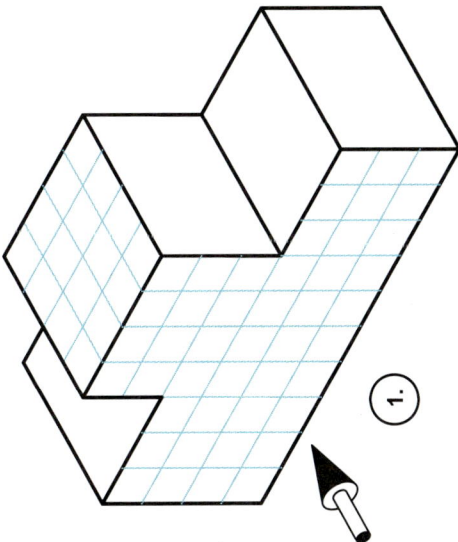

1.

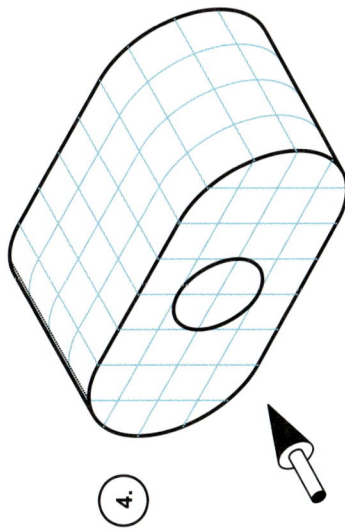

4.

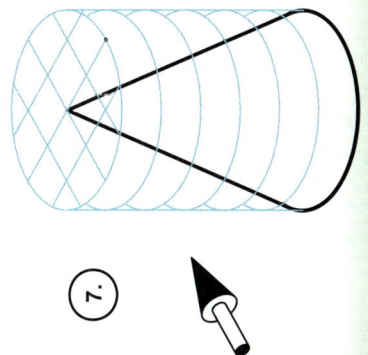

7.

Exercise 2-2

On the grid sheet located at the back of the text, sketch the front, top, and right-side views of the objects shown on the next page. Begin views in the corners shown in the grid.

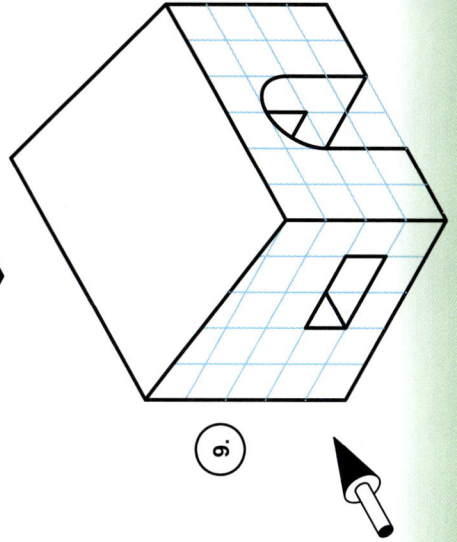

(3)

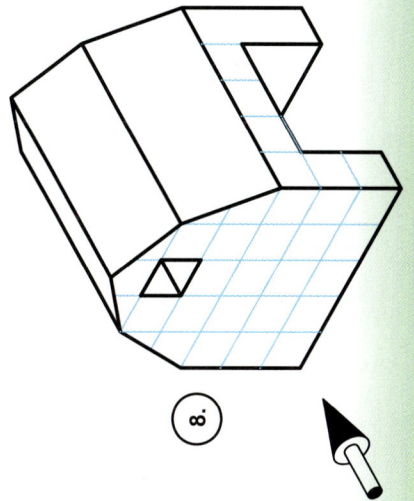

(6)

(9)

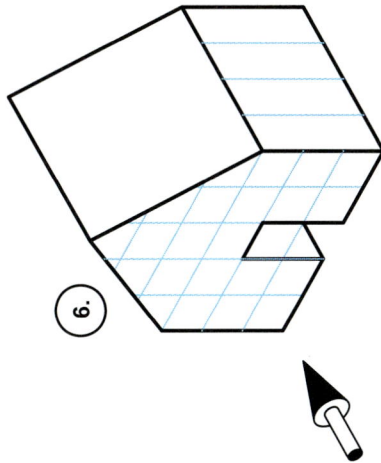

(2)

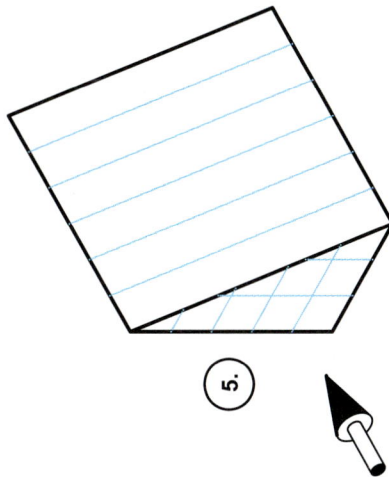

(5)

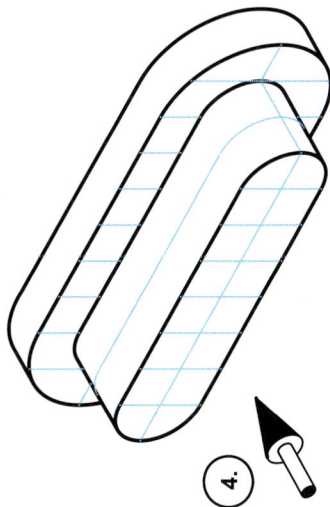

(8)

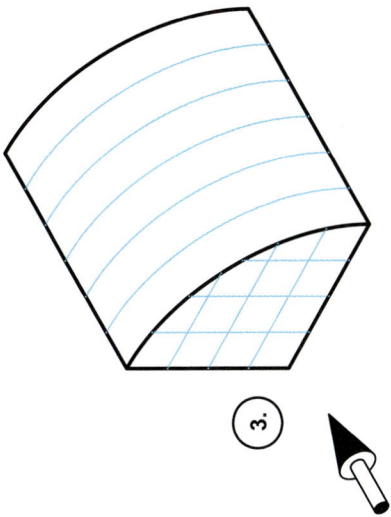

(1)

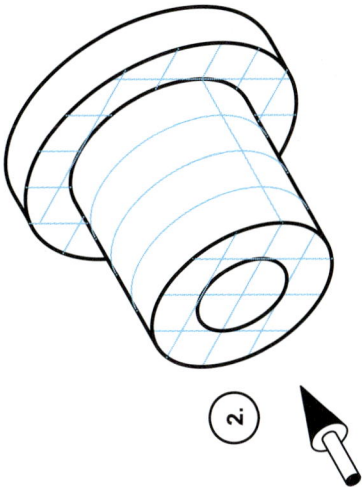

(4)

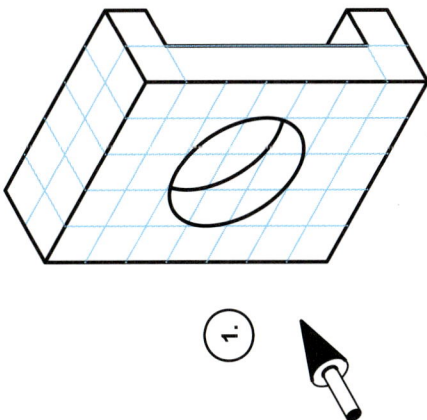

(7)

Exercise 2-3

On the grid sheet located at the back of the text, sketch the front, top, and right-side views of the objects shown on the next page. Begin views in the corners shown in the grid.

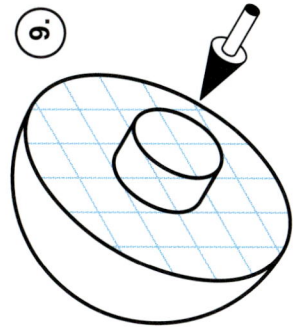

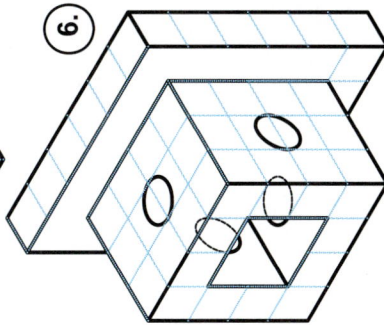

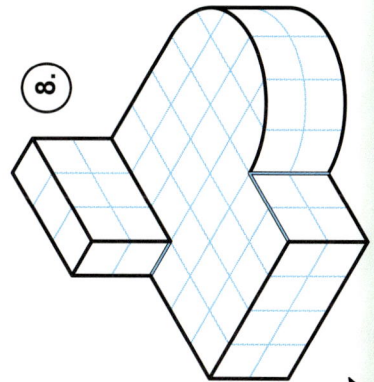

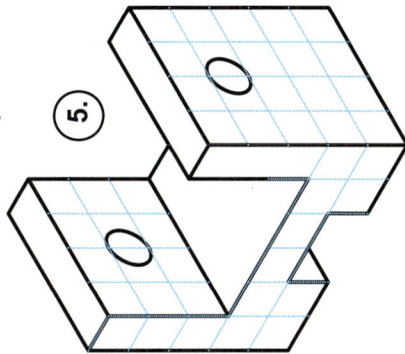

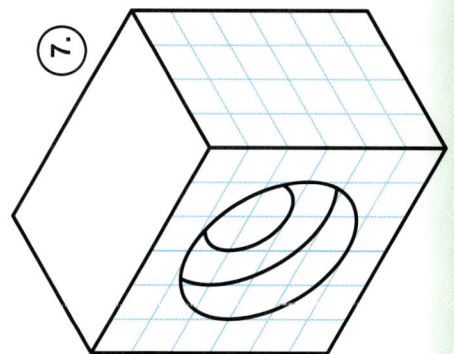

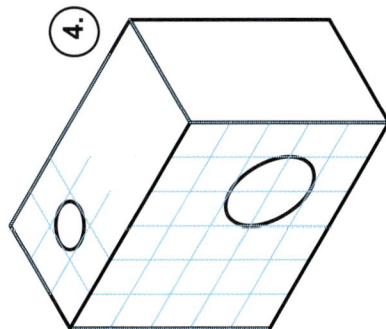

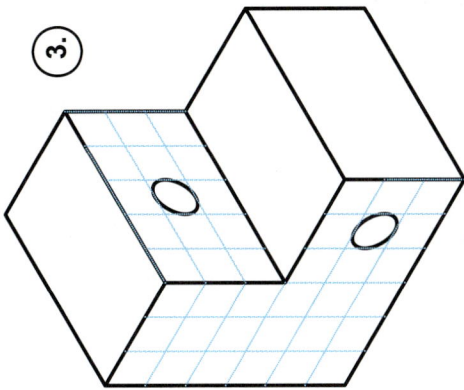

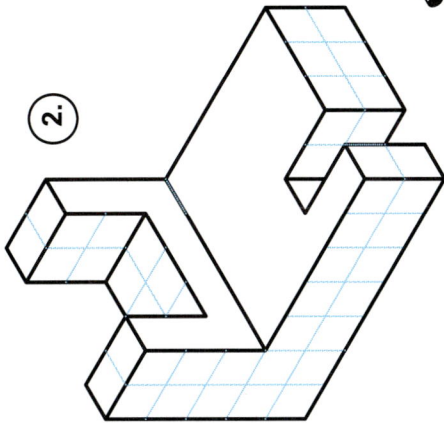

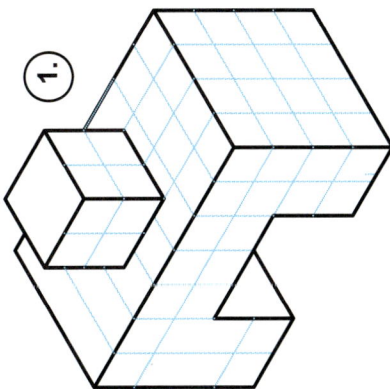

Exercise 2-4

On the grid sheet located at the back of the text, sketch the front, top, and right-side views of the objects shown on the next page. Begin views in the corners shown in the grid.

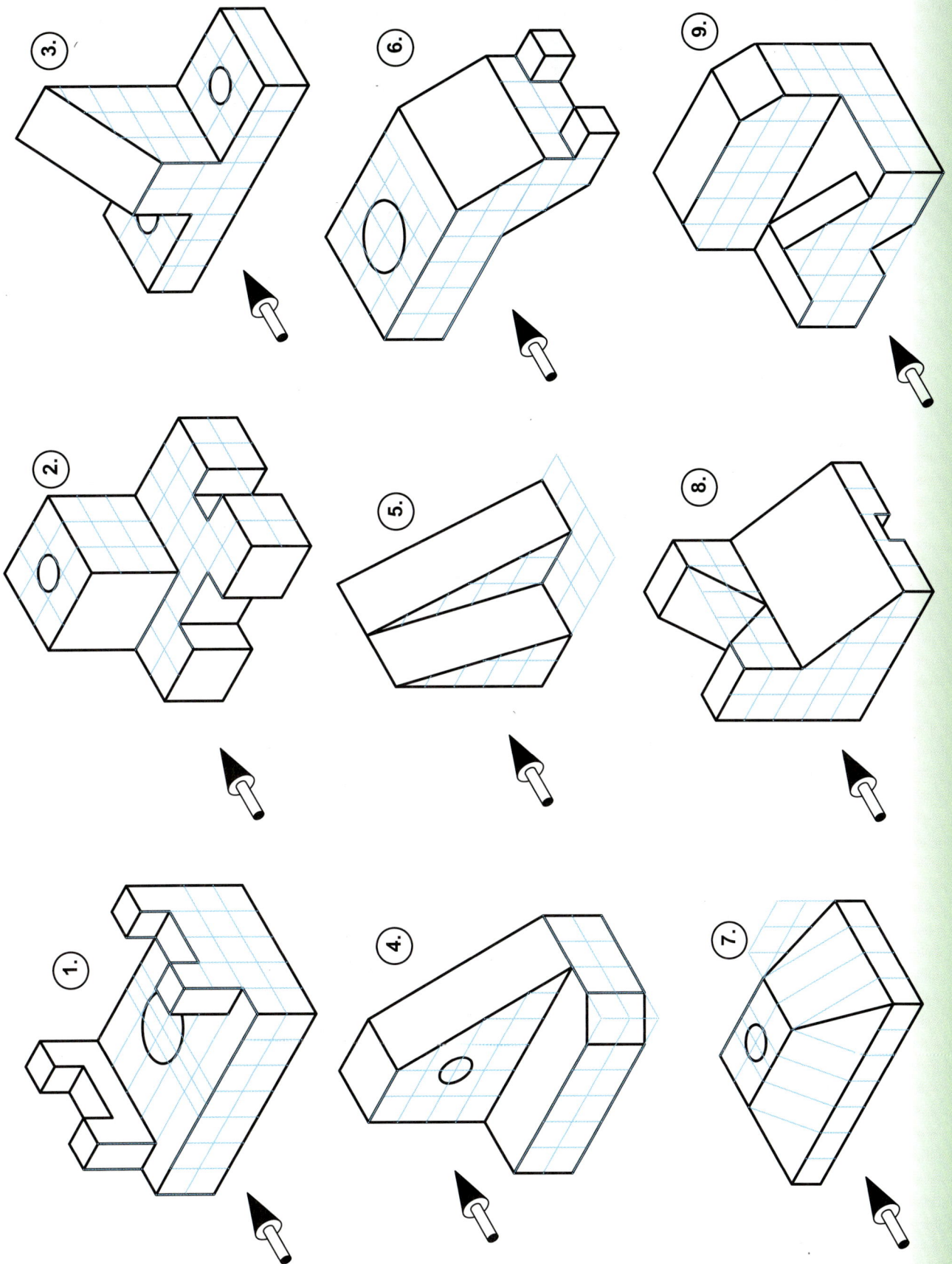

3.

6.

9.

2.

5.

8.

1.

4.

7.

Exercise 2-5

On the grid sheet located at the back of the text, sketch the front, top, and right-side views of the objects shown on the next page. Begin views in the corners shown in the grid.

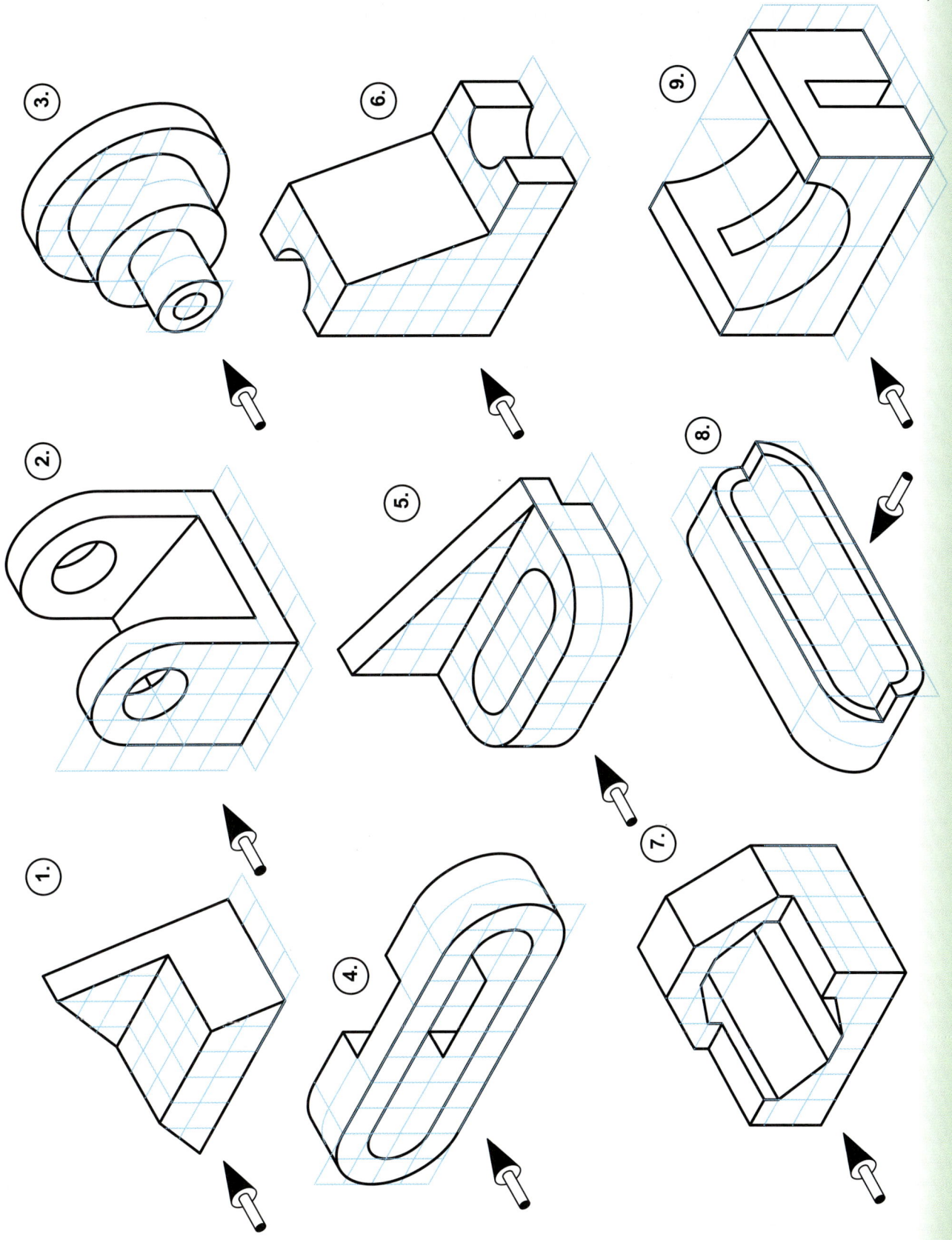

Exercise 2-6

On the grid sheet located at the back of the text, sketch the front, top, and right-side views of the objects shown on the next page. Begin views in the corners shown in the grid.

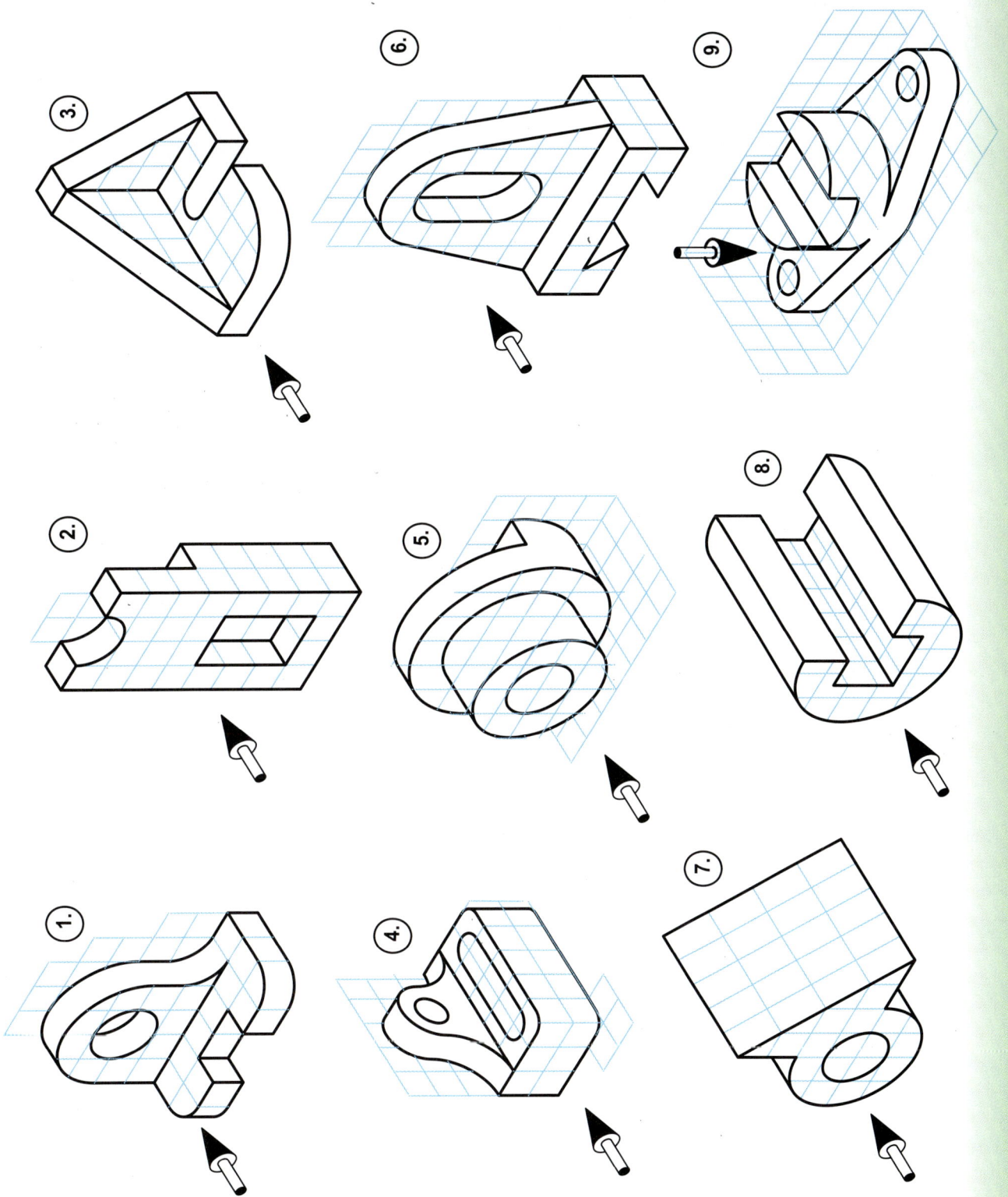

45

REFERENCES

ASME Y14.2M-1992 Line Conventions and Lettering
ASME Y14.1-2005 Decimal Inch Drawing Sheet Size and Format
ASME Y14.3M-1994 Multiview and Sectional View Drawings
ASME Y14.1M-2005 Metric Drawing Sheet Size and Format

Traditional Drafting Tools and Techniques

3

Unit Objectives

- Describe the tools and techniques used in traditional drafting.
- Use technical pencils, straightedges, triangles, scales, protractors, and templates to construct the geometry of technical drawings.
- Read a conversion table to convert between decimal, fractional, and metric units.
- Use traditional drafting tools to create multiview drawings of objects including the correct placement and depiction of visible, hidden, and center lines.
- Hand-letter notes and dimensions on technical drawings that are clear and legible.

INTRODUCTION

Before Computer Aided Design revolutionized the way technical drawings are produced, drafters and designers sat at drawing tables and used ***traditional drafting tools*** like technical pens, straight-edges, triangles, scales, protractors, and templates to draw on sheets of vellum or mylar (a thin sheet of plastic with a matte surface).

In today's engineering or architectural office, it would be rare for a drafter to create a drawing in the traditional way, but many of the techniques developed by traditional drafters, such as orthographic projection, are still used to create 2D CAD drawings. The traditional tools discussed in this unit can be purchased through drafting supply outlets.

traditional drafting tools: The tools that were used to create technical drawings before CAD techniques became the standard. These tools include parallel straight-edges, drafting machines, drafting boards, drafting triangles, protractors, circle and ellipse templates, and technical pens and pencils.

TRADITIONAL DRAFTING TOOLS AND TECHNIQUES

Shown in Figures 3-1 through 3-3 are examples of traditional drafting equipment. A drafting machine, or a parallel straightedge, attached to the top of a drawing table allows a drafter to draw horizontal lines that are parallel to each other. Another tool that can be used to draw parallel horizontal lines is a T-square.

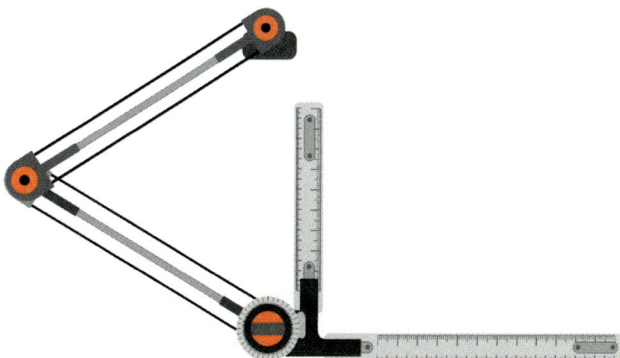

Figure 3-1 Drafting machines can be easily adjusted for drawing variable angles. Drafting machines were once common in engineering offices

Figure 3-2   Drawing table with drafting machine

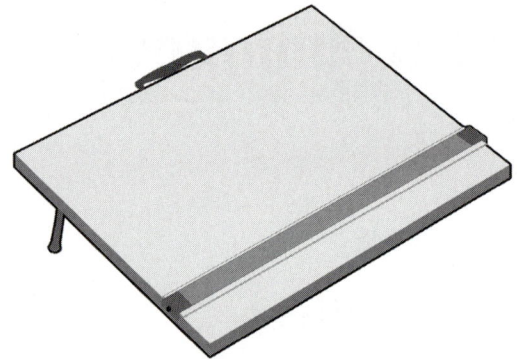

Figure 3-3   Parallel straightedge attached to the top of a portable drawing table

TECHNICAL PENCILS AND PENS

Professional grade technical pens and pencils are often used by design professionals to create technical drawings and sketches (see Figure 3-4).

Leads are inserted after removing both the top cap and the eraser

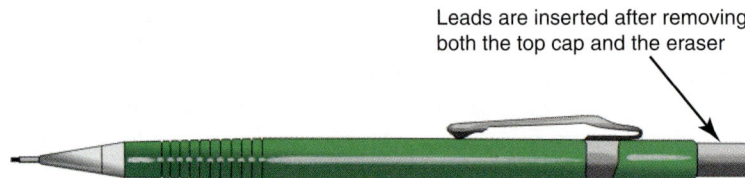

Figure 3-4   Technical pencil

lead hardness grade: The scale that defines the hardness of graphite pencil leads. Soft leads range between 2B and 7B. Medium leads include 3H, 2H, H, F, HB, and B. Hard leads range between 4H and 9H.

Technical pens and pencils come in differing line widths (.3mm, .5mm, .7mm, or .9mm). Leads for technical pencils are available in a variety of hardness grades, defined by the ***lead hardness grade***, depending on the type of work to be performed. Table 3-1 shows the lead (graphite) hardness grades and the appropriate application for the leads in each hardness range.

Table 3-1   Lead Hardness Grades
Hard leads (4H–9H) are useful for laying out construction lines on a technical drawing.
Medium leads (3H, 2H, H, F, HB, and B) are often used for general drafting and sketching (2H is a popular medium lead).
Soft leads (2B–7B) are often used for shading and rendering drawings.

BEGINNING A TRADITIONAL DRAFTING PROJECT

Figure 3-5 illustrates the proper technique for attaching a sheet of vellum to the top of a drawing table. Align the bottom edge of the sheet to the top edge of the drafting machine arm, parallel straightedge, or T-square, and tape all four corners to the table top.

Horizontal lines are drawn along the top edge of the straight-edge as shown in Figure 3-6. Right-handed drafters would hold the straightedge in place with their left hand when drawing a horizontal line.

Figure 3-5 Aligning the bottom edge of a sheet of vellum to the top edge of a parallel straightedge

Figure 3-6 Drawing a horizontal line

DRAFTING TRIANGLES

Triangles, such as the ones shown in Figures 3-7 and 3-8, provide drafters with angles commonly used in technical drawings: 30°, 45°, 60°, and 90°. Triangles are usually made of acrylic plastic and are available in a variety of sizes.

TIP Some triangles are designed for drawing with graphite and others for drawing with pen and ink. Inking triangles have a beveled, or stepped, edge to prevent the ink from running under the edge of the triangle and smearing.

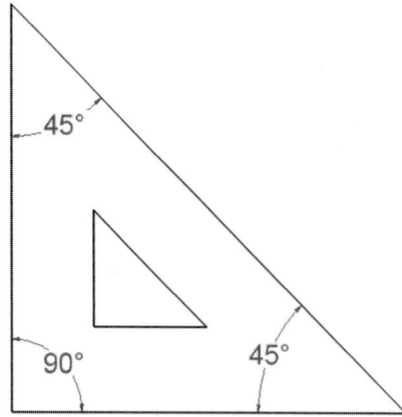

Figure 3-7   45° triangle

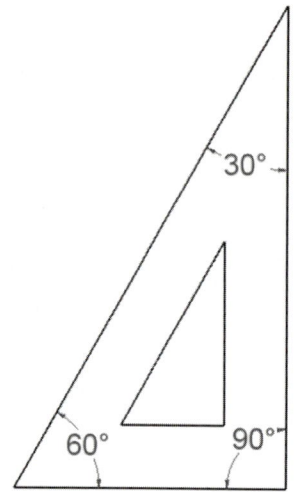

Figure 3-8   30°/60° triangle

DRAWING LINES WITH TRIANGLES AND PARALLEL STRAIGHTEDGES

To draw horizontal and vertical lines that are mutually perpendicular (90°), drafters use a triangle in conjunction with a parallel straightedge.

To draw vertical lines, the triangle should be placed on top of the straightedge as in Figure 3-9. Hold both the straightedge and triangle with your left hand while drawing the vertical line. In this way you are assured that a vertical line will be drawn at a 90° angle relative to lines drawn along the top of the parallel straightedge. You can use either the 30°/60° triangle (as shown) or the 45° triangle.

Beginning students often erroneously believe that holding a triangle as vertical as their eyes can position it (known as "eye-balling") and drawing a line will give them a perpendicular vertical line, when in fact, the only way to draw an accurate vertical line is to make sure the triangle rests on the straightedge as shown in Figure 3-9. However, there are occasions when the triangle is not positioned on the straightedge, for example, when connecting two points or the ends of two lines, when the angle of the resulting line is not equal to 30°, 45°, 60°, or 90°.

Figure 3-10 illustrates how you can draw a line connecting the end points of vertical and horizontal lines. The desired angle does not match one of the triangle's normal angles (30°,

Figure 3-9   Drawing a vertical line

Figure 3-10   Drawing a line between the endpoints of two lines.

45°, 60°, or 90°). In this case, the triangle would be floated and aligned with the ends of the vertical and horizontal lines. The line is drawn between the points along the edge of the triangle.

Placing the 30°/60° triangle on the top of the straightedge as in Figure 3-11 allows you to draw lines 30° from horizontal. These lines can be drawn sloping either to the right (as shown) or to the left by flipping the triangle over.

Figure 3-12 shows lines being drawn at a 60° angle from horizontal. By flipping the triangle over, lines can be drawn sloping to the left or right as needed.

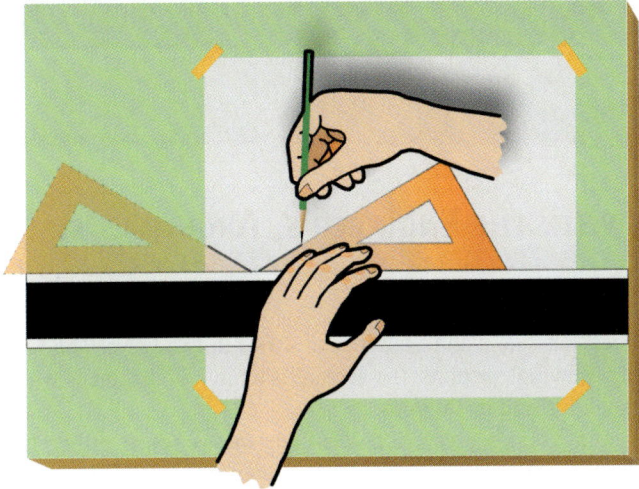

Figure 3-11   Drawing a 30° line

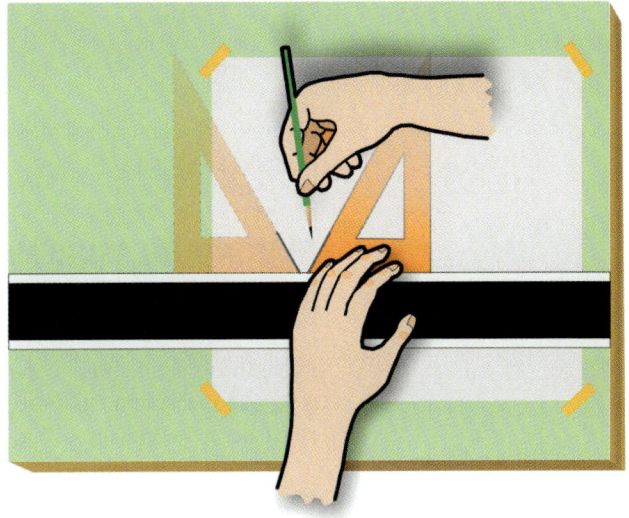

Figure 3-12   Drawing a 60° line

The 45° triangle can be used to draw 45° lines sloping to either the left or right as in Figure 3-13.

Using two triangles in combination allows you to draw lines at 15° increments. In Figure 3-14, the 45° and 30°/60° triangles are combined to draw lines at 75° and 165°.

Figure 3-13   Drawing a 45° line

Figure 3-14   Combining triangles to produce 75° and 165° lines

In Figure 3-15, the 45° and 30°/60° triangles are combined to draw lines at 105° and 15°.

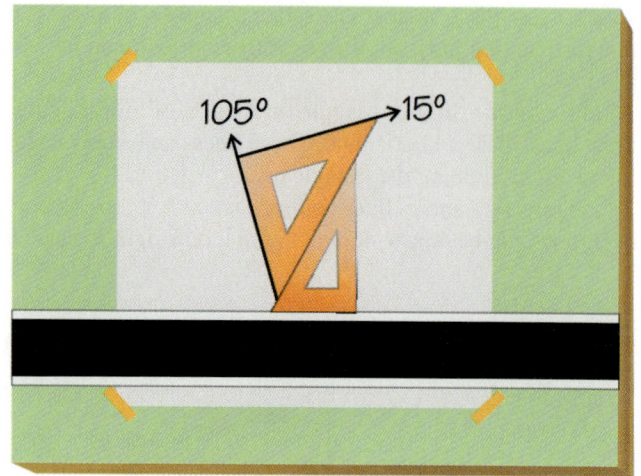

Figure 3-15 Combining triangles to produce 15° and 105° lines

Making Measurements with the Engineer's, Architect's, and Metric Scales

In engineering and architectural offices, designers and drafters use scales in two ways. The first is to take measurements from existing drawings or plots; the second is to lay out distances when constructing a drawing using traditional drafting techniques.

Depending on the type of drawing being created, a designer may choose either the **Engineer's, Architect's, or Metric** scales to measure distances (a *Combination* scale is also available that has a mix of the most commonly used scales).

Reading the Engineer's Scale

The Engineer's scale is used by both mechanical and civil engineers. The marks on the scale may be interpreted differently depending on the discipline. In Figure 3-16, the Engineer's 10 scale is used to measure decimal inches. In Figure 3-17, the Engineer's 10 scale is used to measure distances in feet.

Engineer's, Architect's, and Metric scales: Precision measurement instruments used to make measurements during the creation, or interpretation, of technical drawings.

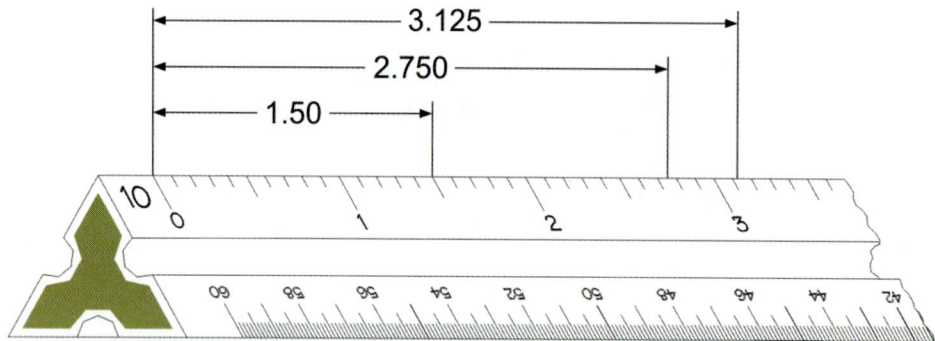

Figure 3-16 The Engineer's 10 scale showing decimal inches

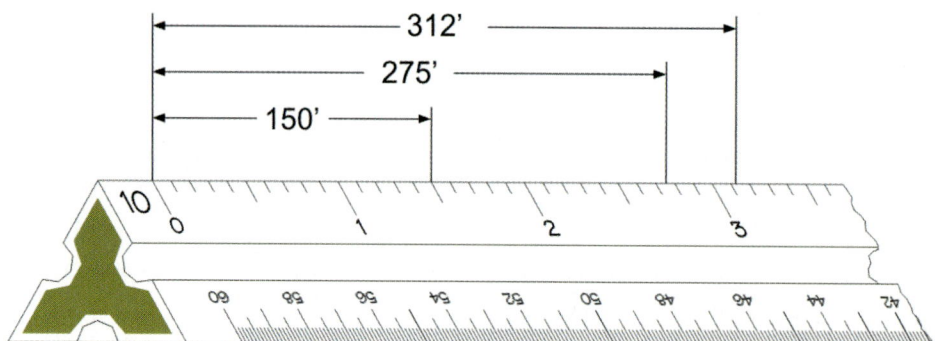

Figure 3-17 The Engineer's 10 scale showing feet. This scale would be interpreted as 1″ = 100′

Reading the Metric Scale

Mechanical engineers can use the Metric scale shown in Figure 3-18 to measure distances in millimeters. In Figure 3-18, each small mark on the scale equals one millimeter.

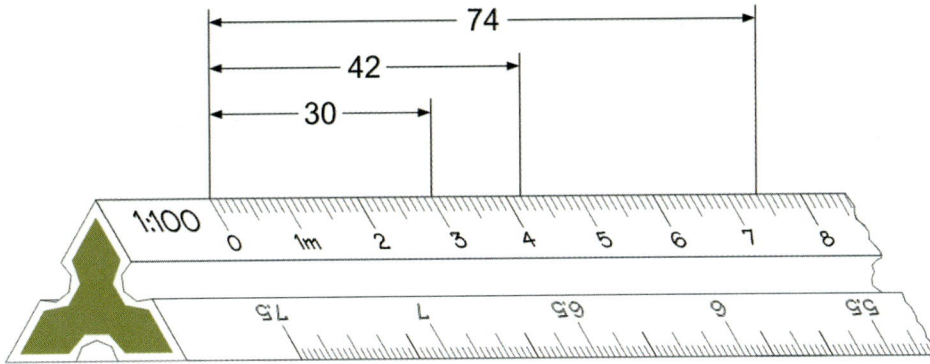

Figure 3-18   The Metric scale used to measure millimeters

Reading the Architect's Scale

The Architect's scale is used to measure feet and inches on a floor plan. The Architect's 1/4 scale shown in Figure 3-19 would be interpreted as 1/4″ = 1′-0″. This means that 1/4 inch measured on the drawing would equal one foot at the construction site.

Figures 3-20, 3-21, and 3-22 show examples of other commonly used architectural scales.

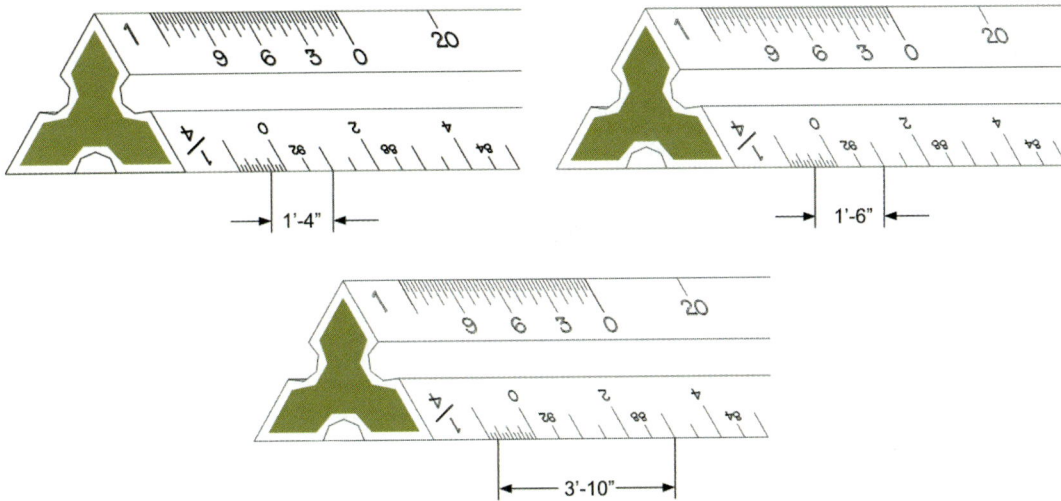

Figure 3-19   The Architect's 1/4 scale illustrating several different measurements of feet and inches

The secret to using these scales is to align the numbers with the zero. Once you have done this it becomes relatively easy to determine the required ticks for measuring.

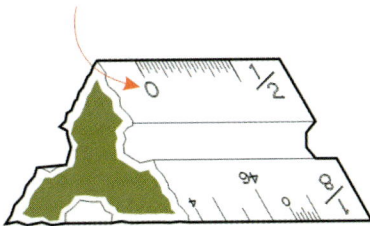

Each mark on the 1/8th scale is worth two inches (2,4,6,8, etc.)

Figure 3-20   Interpreting inch marks on the Architect's 1/2 and 1/8 scales

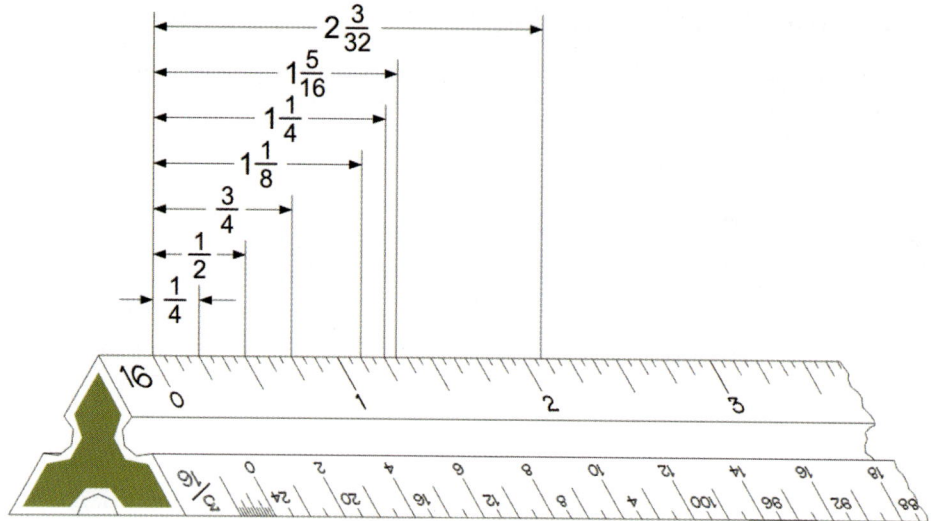

Figure 3-21 The Architect's 16 scale can be used to measure fractional inches

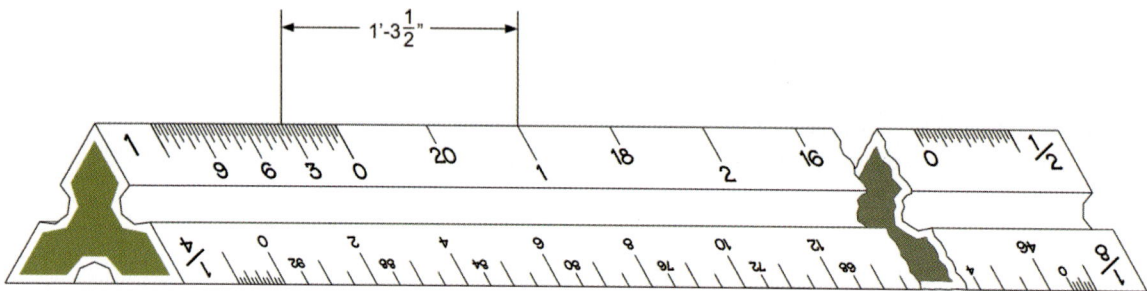

The scale along this edge - if read from the right - would be 1/8"=1'-0"

Figure 3-22 The Architect's 1 scale can be used to measure feet and inches on a floor plan. The scale below would be interpreted as 1" = 1'-0"

> **TIP** To avoid having to search for a desired scale on a triangular scale, use a binder clip to mark your scale. For example, if using the 10 scale on an Engineer's scale, attach the binder clip to the 50 scale. When you pick up your scale, the binder clip will orient you to the 10 scale.

CONVERTING UNITS OF MEASUREMENT

Drafters must sometimes convert from one unit of measurement to another. Some commonly used conversion factors are as follows:

- Fractional inches can be converted to decimal inches by dividing the numerator (the top number) by the denominator (the bottom number).
- Decimal inches can be converted to millimeters by multiplying them by 25.4.
- Millimeters can be converted to decimal inches by dividing them by 25.4.

Table 3-2 is useful for quickly finding the equivalent value between the various units listed.

READING THE PROTRACTOR

A protractor is a tool used to lay out angles on a drawing. Full circle protractors are divided into 360° while half-moon protractors (see Figure 3-23) are divided into 180°. Both are divided in 10° increments. Note how the protractor in Figure 3-23 is divided from 0° to 180° in both the clockwise and counterclockwise directions.

Table 3-2   Fractional Inches to Decimal Inches to Millimeters Conversion Table for Values Up to One Inch

Fractional Inch	Decimal Inch	Metric (mm)	Fractional Inch	Decimal Inch	Metric (mm)
1/64	.015625	0.3969	33/64	.515625	13.0969
1/32	.03125	0.7938	17/32	.53125	13.4938
3/64	.046875	1.1906	35/64	.546875	13.8906
1/16	.0625	1.5875	9/16	.5625	14.2875
5/64	.078125	1.9844	37/64	.578125	14.6844
3/32	.09375	2.3813	19/32	.59375	15.0813
7/64	.109375	2.7781	39/64	.609375	15.4781
1/8	.1250	3.1750	5/8	.6250	15.8750
9/64	.140625	3.5719	41/64	.640625	16.2719
5/32	.15625	3.9688	21/32	.65625	16.6688
11/64	.171875	4.3656	43/64	.671875	17.0656
3/16	.1875	4.7625	11/16	.6875	17.4625
13/64	.203125	5.1594	45/64	.703125	17.8594
7/32	.21875	5.5563	23/32	.71875	18.2563
15/64	.234375	5.9531	47/64	.734375	18.6531
1/4	.250	6.3500	3/4	.750	19.0500
17/64	.265625	6.7469	49/64	.765625	19.4469
9/32	.28125	7.1438	25/32	.78125	19.8438
19/64	.296875	7.5406	51/64	.796875	20.2406
5/16	.3125	7.9375	13/16	.8125	20.6375
21/64	.328125	8.3344	53/64	.828125	21.0344
11/32	.34375	8.7313	27/32	.84375	21.4313
23/64	.359375	9.1281	55/64	.859375	21.8281
3/8	.3750	9.5250	7/8	.8750	22.2250
25/64	.390625	9.9219	57/64	.890625	22.6219
13/32	.40625	10.3188	29/32	.90625	23.0188
27/64	.421875	10.7156	59/64	.921875	23.4156
7/16	.4375	11.1125	15/16	.9375	23.8125
29/64	.453125	11.5094	61/64	.953125	24.2094
15/32	.46875	11.9063	31/32	.96875	24.6063
31/64	.484375	12.3031	63/64	.984375	25.0031
1/2	.500	12.700	1	1.000	25.400

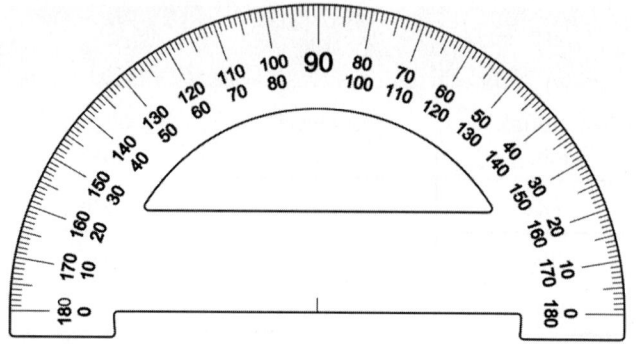

Figure 3-23   A half-moon, or 180° protractor

Figures 3-24 through 3-27 illustrate some of the ways the protractor can be used to measure angles.

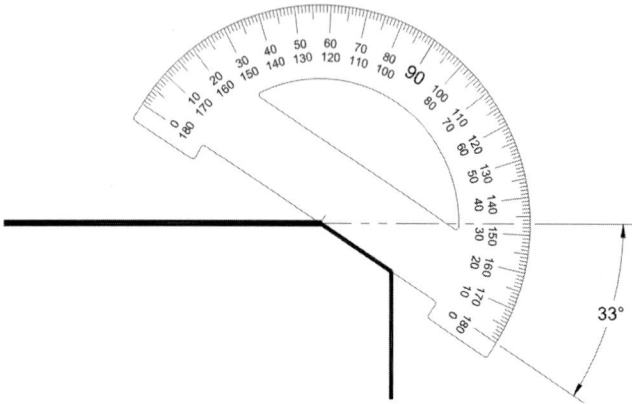

Figure 3-24   A 33° angle using the inner dial

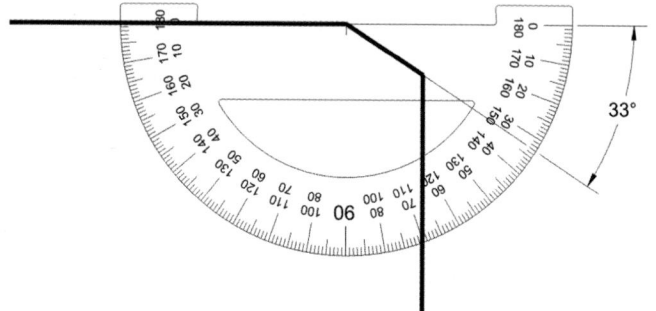

Figure 3-25   A 33° angle using the outer dial

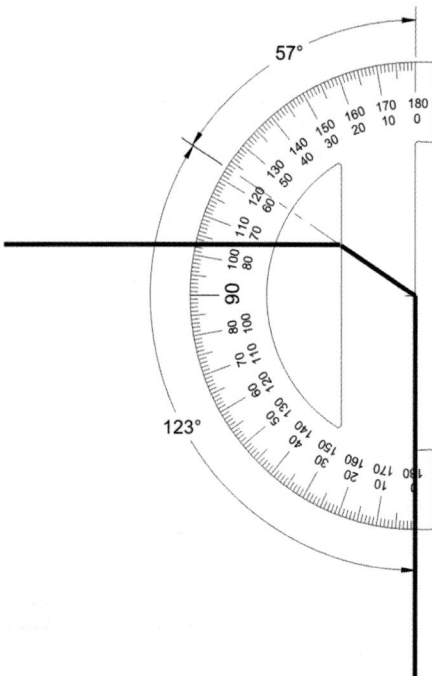

Figure 3-26   A 57° angle measured from vertical

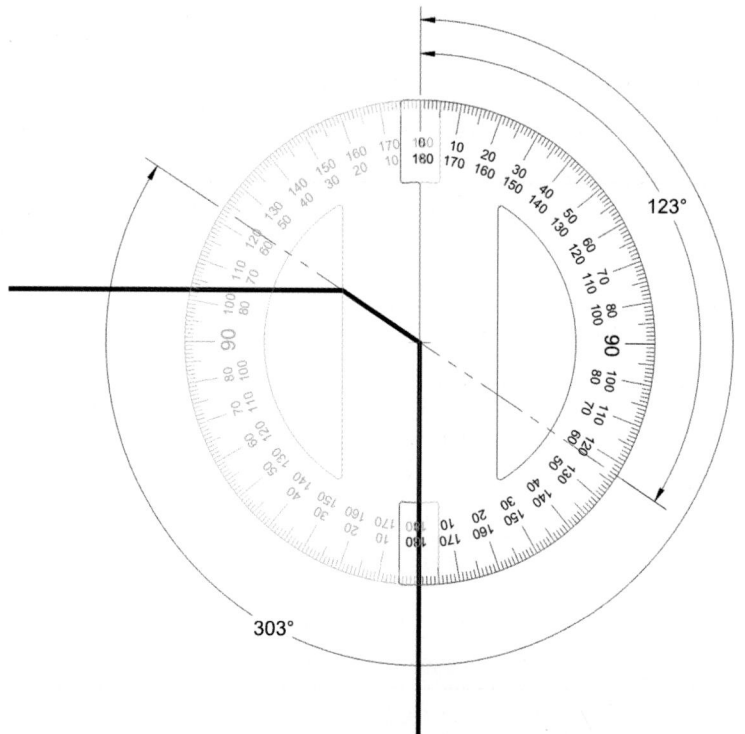

Figure 3-27   Measuring an angle greater than 180° using a half-moon protractor

CIRCLE TEMPLATE

Circle templates come in a wide range of units (decimal inches, fractional inches, and millimeters) and diameters (see Figure 3-28). Figure 3-29 illustrates the steps in drawing a circle with the circle template.

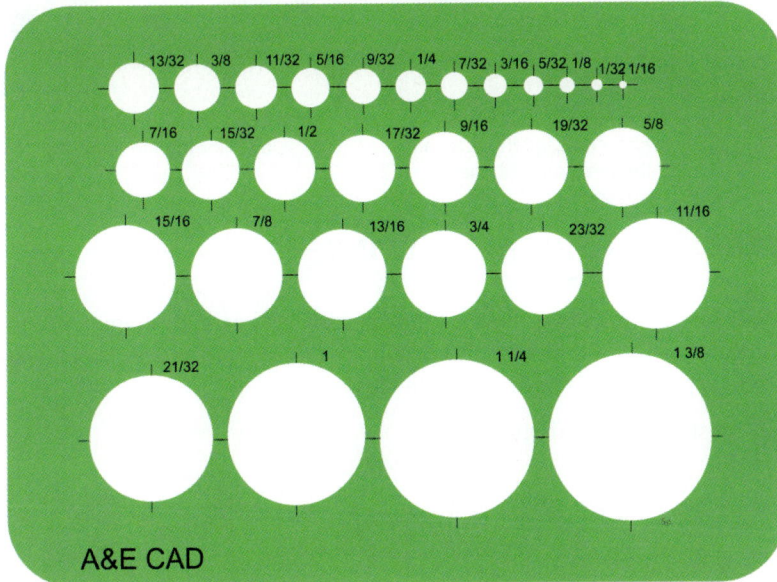

Figure 3-28   A circle template

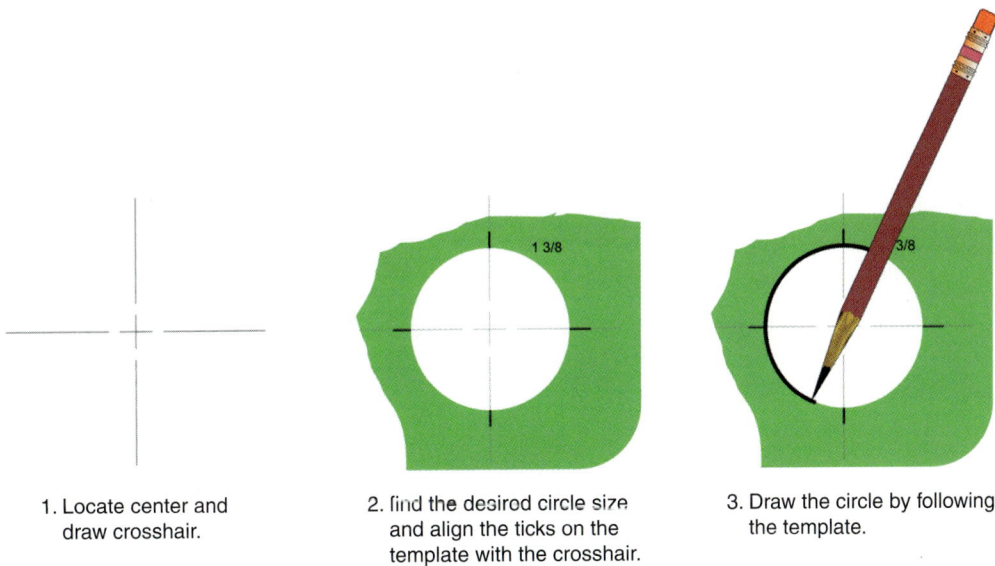

1. Locate center and draw crosshair.

2. find the desired circle size and align the ticks on the template with the crosshair.

3. Draw the circle by following the template.

Figure 3-29   Using the circle template

ISOMETRIC ELLIPSE TEMPLATE

On isometric drawings, circles are represented as ellipses. Isometric ellipse templates allow drafters to place ellipses on isometric drawings quickly. Figure 3-30 illustrates the steps in drawing an ellipse with this template.

A. Disregard the maximim diameter ticks unless using them to align the objects horizontally.

B. Disregard the minimum diameter ticks unless using them to align the objects vertically.

1 ½

1 ½

1. Locate center and draw crosshair at 30° angles.

2. Find the desired ellipse size and align the ticks on the template with the crosshair.

3. Draw the isometric ellipse by following the template.

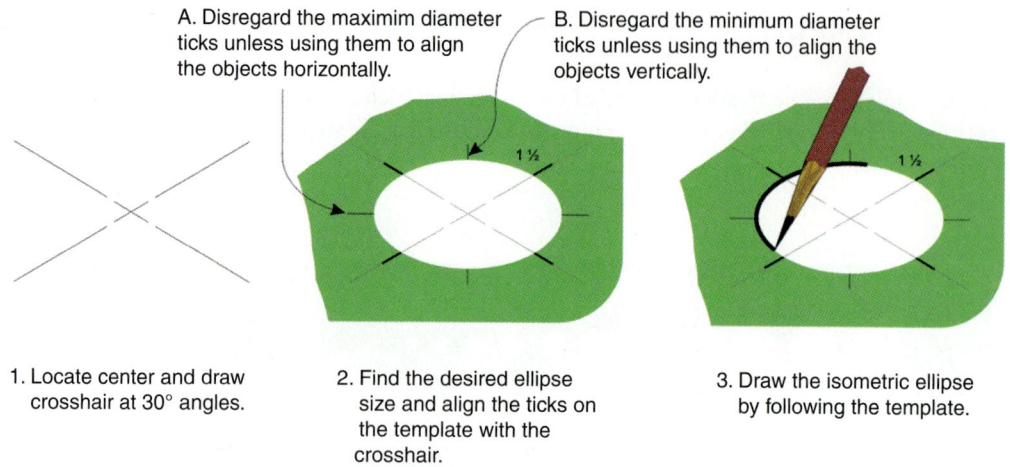

Figure 3-30   Using the isometric ellipse template

STEPS IN CONSTRUCTING A SIMPLE DRAWING

Steps 1 through 8 illustrate how a drafter may use the triangle, scale, and parallel straightedge to construct a simple technical drawing. At the end of this unit are drafting projects in which you will have an opportunity to apply the techniques shown in these steps.

Step 1. Draw a light 30° construction line (Figure 3-31).

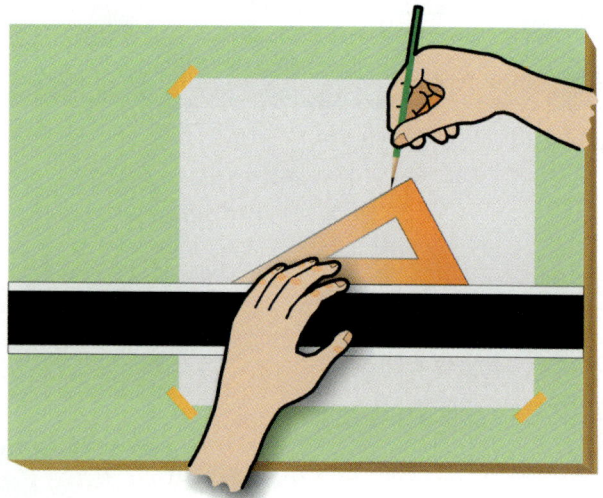

Figure 3-31   Drawing a 30° construction line

Step 2. Align the scale along the construction line and place light "tick marks" to denote the desired measurement (Figure 3-32).

Figure 3-32   Making a measurement along the 30° construction line

Step 3. Align the edge of the 30°/60° triangle with the construction line and draw a dark line along the top edge of the triangle between the tick marks located in Step 2 (Figure 3-33).

Step 4. The darkened visible line is drawn the desired distance (Figure 3-34).

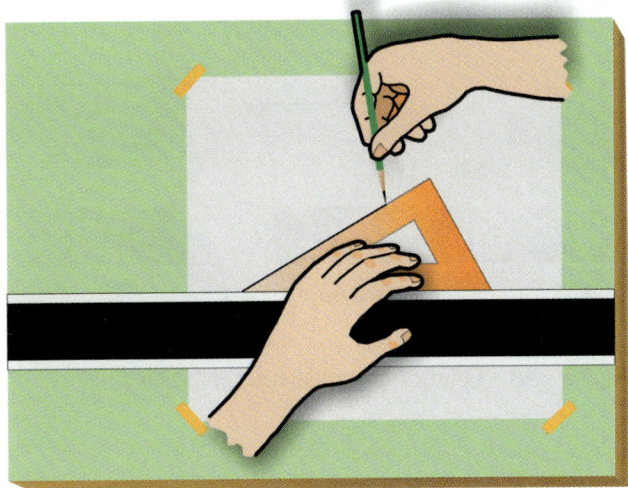

Figure 3-33   Darkening the line

Figure 3-34   The line drawn to the desired length

Step 5. Slide the horizontal bar until it is aligned with the lower tick mark and lightly draw a horizontal construction line (Figure 3-35).

Step 6. Measure along the horizontal construction line and mark off the required distance with a small tick mark (Figure 3-36).

Figure 3-35   Drawing a horizontal construction line

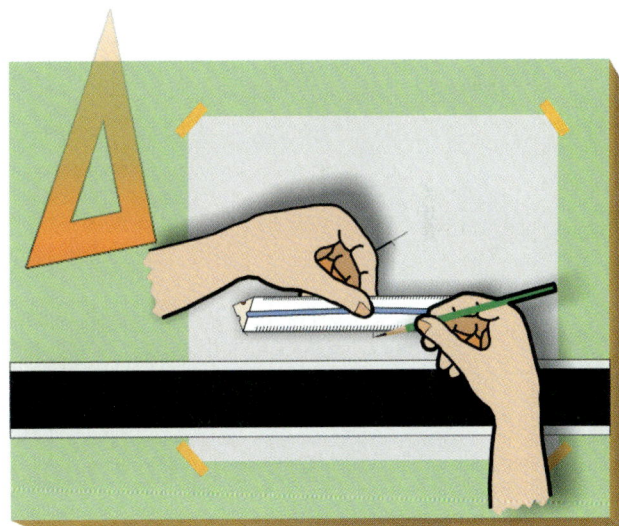

Figure 3-36   Making a measurement along the horizontal construction line

Step 7. Draw a dark visible line between the tick marks (Figure 3-37).

Step 8. Because the angle between the ends of the lines does not match an angle on either the 45° or 30°/60° triangle, move the triangle until it is aligned with the endpoints of each line and draw a dark line connecting them (Figure 3-38). The completed drawing is shown in Figure 3-39.

Figure 3-37   Darkening the horizontal line

Figure 3-38   Connecting the endpoints of the lines

Figure 3-39   The finished drawing

STEPS IN CONSTRUCTING A MULTIVIEW DRAWING

The steps involved in constructing a multiview drawing with traditional tools are illustrated in Figure 3-40 through Figure 3-44.

Step 1. Study the width, depth, and height of the object to be drawn and use these dimensions to determine the location of the views. Views are usually equally spaced both horizontally and vertically as shown in Figure 3-40. A formula for determining the horizontal spacing is to take the width of the sheet, subtract the width of the front view and the depth of the side view, and divide the remainder by 3. To find the vertical spacing, take the height of the sheet, subtract the height of the front view and the depth of the top view, and divide the remainder by 3.

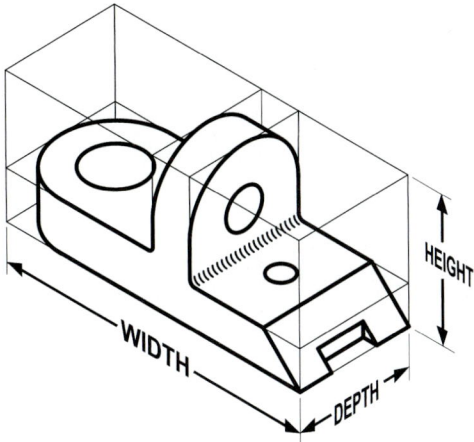

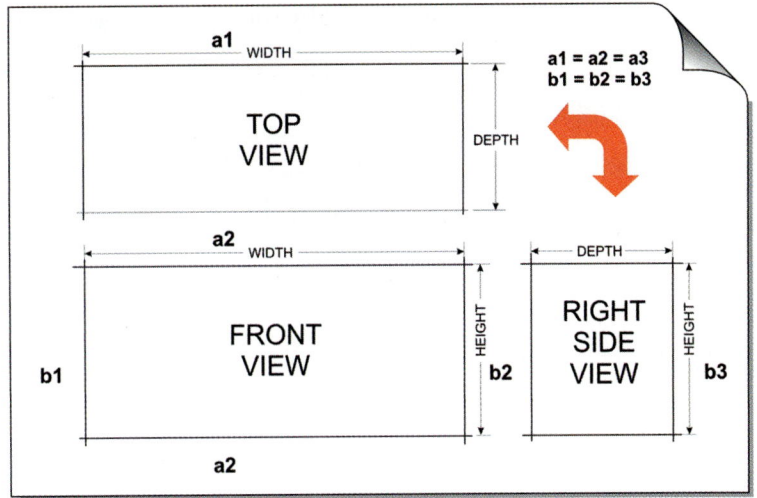

Figure 3-40   Determining view spacing

Step 2. The width, height, and depth of the views are laid out with light construction lines (Figure 3-41).

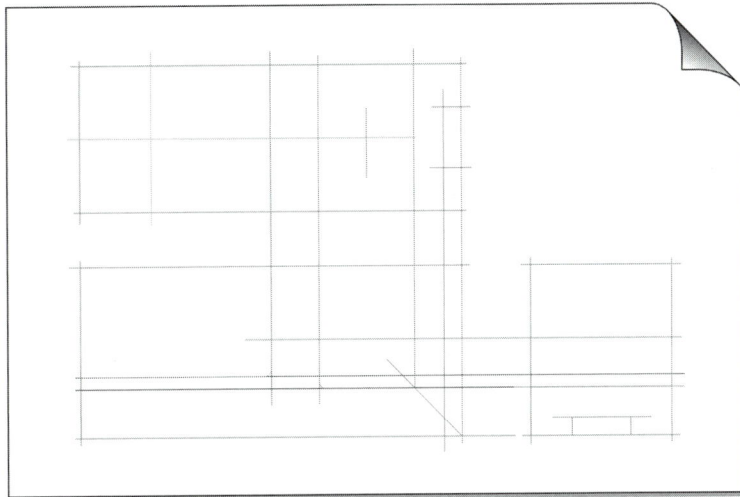

Figure 3-41   Rough layout of views

Step 3. A miter line is added so that orthographic projection techniques can be used to project between the top and side views. Circles and arcs are drawn in lightly with a circle template or compass. Information is projected between views with light construction lines (Figure 3-42).

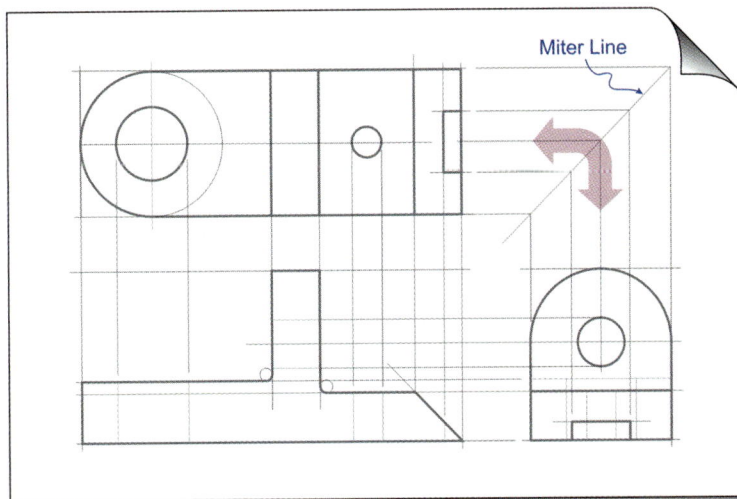

Figure 3-42   Adding miter line and circles

Step 4. Hidden and center lines are added (Figure 3-43).

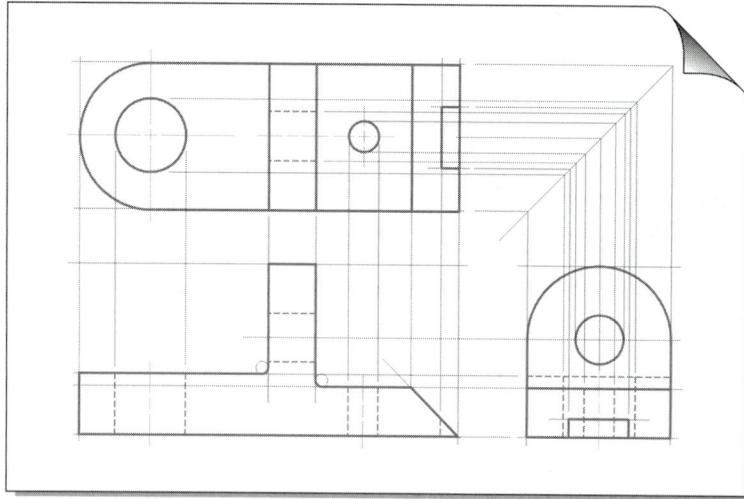

Figure 3-43   Adding hidden and center lines

Step 5. Construction lines are erased, and the visible, hidden, and center lines are darkened to complete the drawing (Figure 3-44).

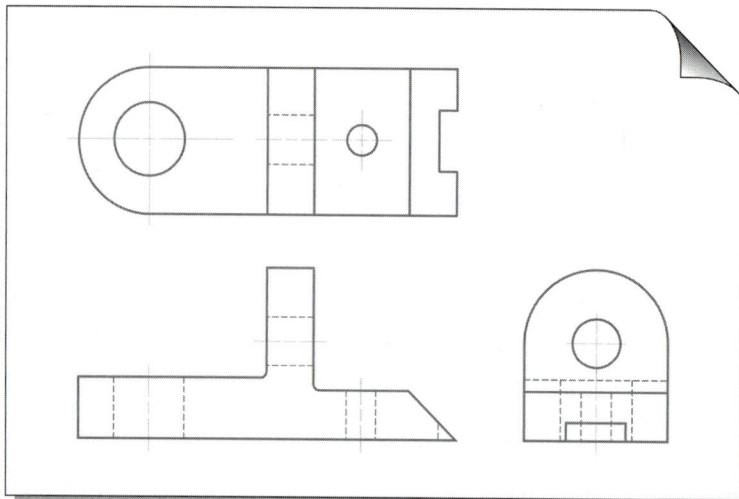

Figure 3-44   Finished drawing

It is best to draw all the views using construction lines. If you make a mistake, it is easier to erase a construction line than a darkened visible, hidden, or center line. After you have drawn the views with construction lines, darken the Top view first, then the Front view, and the Side view last. Darkening your lines in this order allows you to work away from your completed views and minimizes smudging.

TECHNICAL LETTERING

technical lettering: Freehand lettering that is added to a technical drawing or sketch. Technical lettering should be legible and consistent with regard to style.

Every technical drawing text is required by law to expound on the importance of good *technical lettering* skills. Well, not really, but technical drawing teachers, as sworn defenders of an ancient art form, feel an obligation to convey the importance of this skill to their students. So here goes.

During the design process, drafters and designers often make sketches which contain hand-written notes and dimensions. It is very important that any hand-lettered text on a drawing be neat, uniform, and legible. In fact, lettering that does not have these qualities would be considered unprofessional in most engineering or architectural offices. This is because poor lettering on a design sketch could result in dimensions or notes being incorrectly depicted on a technical drawing which could lead to a costly mistake on a construction site or manufacturing floor. It might also cost a designer his or her job.

Hopefully, that last sentence caught your attention so you realize that developing good lettering skills is very important to your success as a drafter/designer. Good lettering is a skill that is only mastered through practice. Over time, most drafters and designers develop their own unique lettering style. Some drafters, especially in the architectural field, develop a lettering style that could be considered an art form. Although you may never develop a style that qualifies as "art," with practice you can develop a style that is neat, legible, and uniform.

Developing a Technical Lettering Style

Study the examples of the letters and numerals shown in Figure 3-45. Notice how the characters are uniform with regard to height, width and style. The angle of vertical strokes should be consistent, and each character should be clear and legible.

In the guidelines provided in Figure 3-45, practice lettering the alphabet and numerals. Try to match the lettering style shown in the example. Lettering should be dark, so press down hard with the pencil when making the strokes that form a letter or numeral.

Figure 3-45   *Example of a good technical lettering style*

SUMMARY

Although traditional drafting tools are rarely used in modern engineering and architectural offices, many of the same techniques developed by traditional drafters are still used in the creation of drawings with CAD. For example, the location of points and planes is still projected between multiviews; however, instead of using a drafting triangle, an AutoCAD drafter uses a drafting setting called *Ortho* to draw perfectly straight horizontal and vertical lines. An understanding of how angles are measured with a protractor facilitates drawing angles with AutoCAD.

Drafters and designers often use scales to take measurements from plotted drawings (which is not always a good idea, by the way), and an understanding of how to interpret an Engineer's, Architect's, or Metric scale is useful in understanding how scaling applies to AutoCAD drawings.

The creation of freehand sketches with legible lettering remains an important skill in today's design office. Every professional drafter/designer, whether traditional or CAD, should develop a freehand lettering style that is neat, uniform, and legible to facilitate communication between designers, or between designers and their clients.

UNIT TEST QUESTIONS

Short Answer

1. Name three tools used by traditional drafters to draw parallel horizontal lines.

2. Drafting triangles come in what angles?

3. How is a triangle used with a parallel straightedge to draw a vertical line that is perpendicular to a horizontal line?

4. What is the range of lead hardness for general drafting and sketching?

5. Name three qualities that technical lettering should possess.

6. What unit of measurement is represented by each small increment on the 1:100 Metric scale as interpreted by a mechanical engineer?

7. Which scale is used by mechanical engineers to measure drawings in decimal inches?

8. Into how many degrees is a half-moon protractor divided?

9. Name three lead widths in which technical pencils can be purchased.

10. What is the multiplier used to convert decimal inches to millimeters?

Internet Resources

Pentel: *www.pentel.com*

Staedtler: *www.staedtler.com*

UNIT EXERCISE

Exercise 3-1: Technical Lettering

Practice lettering the alphabet and numerals in the construction lines shown in Figure 3-46. Print the Lettering Plate.pdf file located on the Student Resource CD in the *Sketching and Traditional Drafting Sheets* folder.

LETTERING EXERCISE

.250 (1/4) Inch Lettering
A B C D E F G H I J K L M N O P Q R S T U V W X Y Z
1 2 3 4 5 6 7 8 9 0   SECTION A-A   FLOOR PLAN

.156 (5/32) Inch Lettering
A B C D E F G H I J K L M N O P Q R S T U V W X Y Z
1 2 3 4 5 6 7 8 9 0   SCHEMATIC OF FM TUNER

.125 (1/8) Inch Lettering
A B C D E F G H I J K L M N O P Q R S T U V W X Y Z
1 2 3 4 5 6 7 8 9 0   SCHEMATIC   WALL SECTION   INTERIOR ELEVATIONS

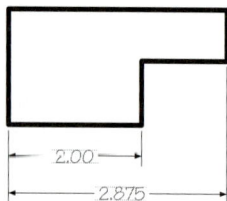

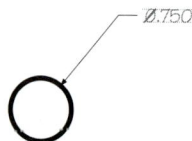

Figure 3-46
Lettering exercise

UNIT PROJECTS

Project 3-1: Traditional Drafting Project 1 (SI Units)

Directions: On the sheet located at the back of the text, draw the Front, Top, and Right views of the object in Figure 3-47. This drawing employs International Units (SI) so use the Metric scale marked 1:100 (refer to Figure 3-18) to draw the object **FULL** size (1 = 1).

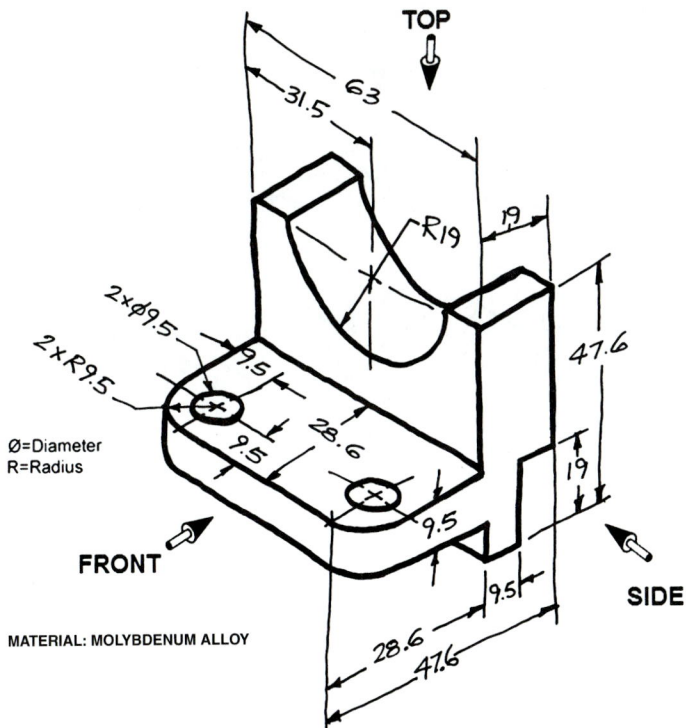

Figure 3-47   Traditional Drafting Project 1

Add dimensions to the views as instructed by your teacher. Letter your name, the material of the part, and the scale in the guidelines provided (refer to Figure 3-48).

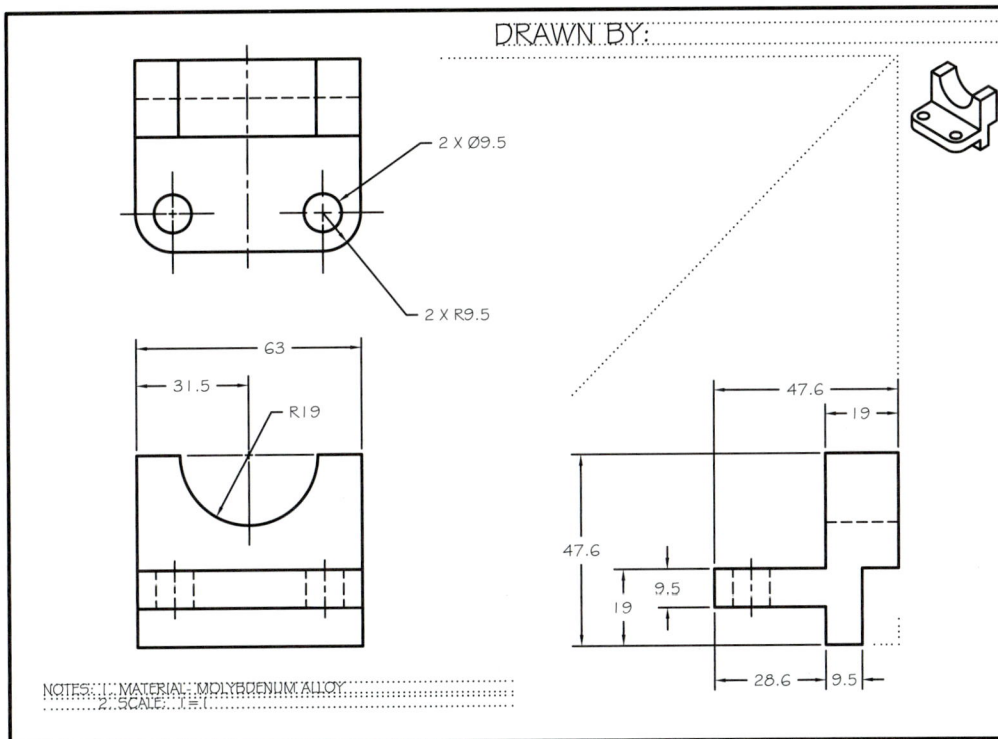

Figure 3-48   Finished project

Project 3-2: Traditional Drafting Project 2

Directions: On the sheet located at the back of the text, draw the Front, Top and Right view of the object in Figure 3-49. Using the Engineer's 10 scale (refer to Figure 3-16), draw the object **3/4** size (**Note:** To convert the dimensions in Figure 3-49 to 3/4 size, multiply them by .75).

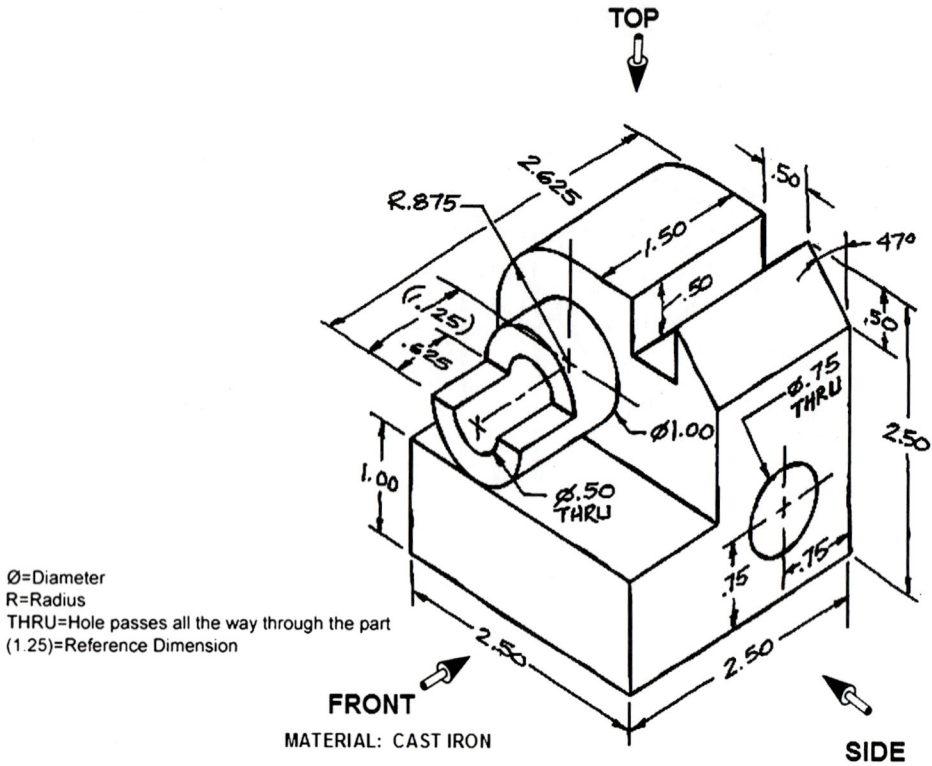

Ø=Diameter
R=Radius
THRU=Hole passes all the way through the part
(1.25)=Reference Dimension

MATERIAL: CAST IRON

Figure 3-49   Traditional Drafting Project 2

Add dimensions to the views as instructed by your teacher. Letter your name, the material of the part, and the scale in the guidelines provided (refer to Figure 3-50).

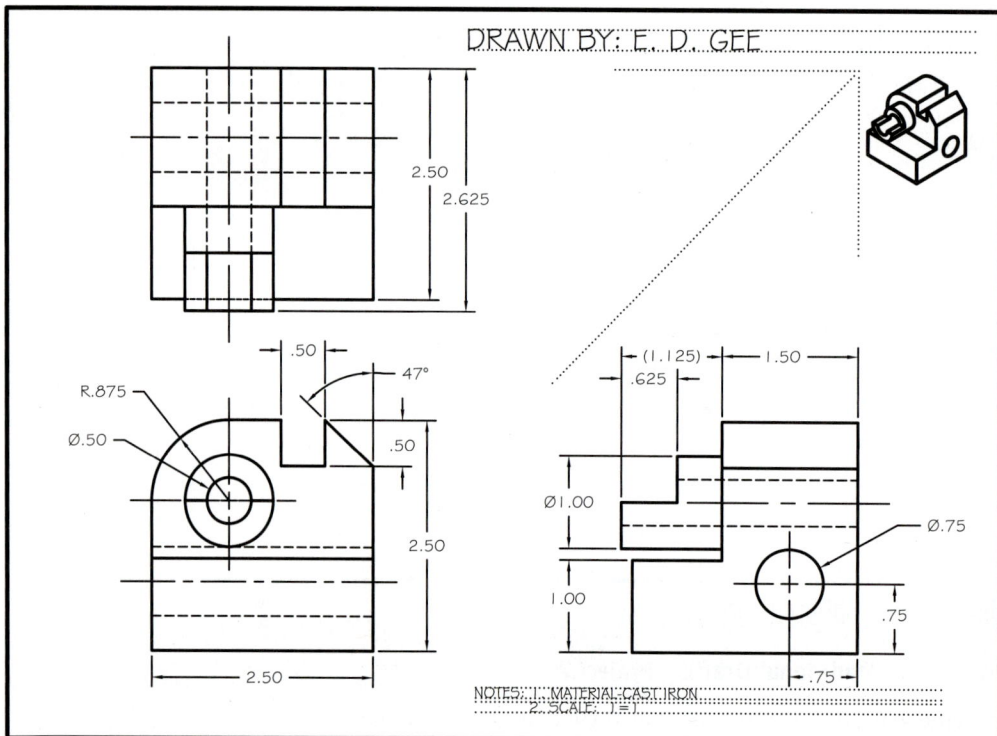

Figure 3-50   Finished project

Optional Traditional Drafting Projects

The drafting projects on the following pages are designed to give you additional practice applying the tools and techniques of traditional drafting.

Student Files

Directions: Follow your instructor's directions to print copies of the Traditional Drafting Sheet.pdf file located on the Student Resource CD in the Sketching and Traditional Drafting Sheets folder. Then draw the Front, Top, and Right views of the objects in Figures 3-51 through 3-53. Use the appropriate scale for the units provided in the designer's sketch.

Add dimensions to the views as instructed by your teacher. Letter your name, the material of the part, and the scale in the guidelines provided.

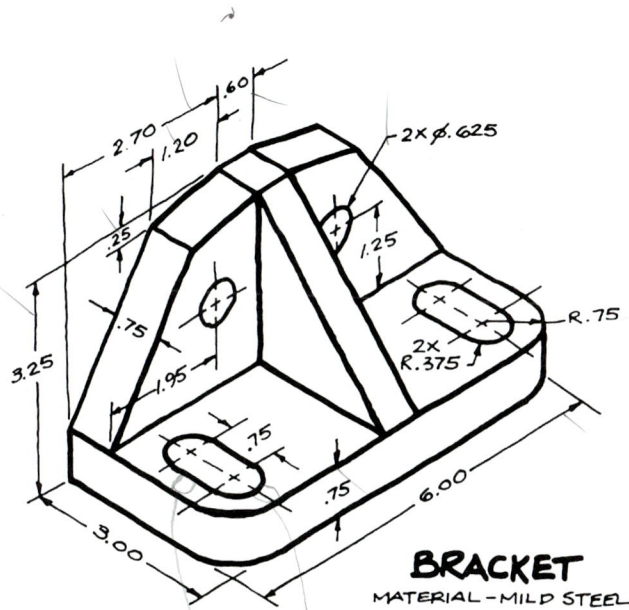

BRACKET
MATERIAL – MILD STEEL

Figure 3-51   Traditional Drafting Project 3

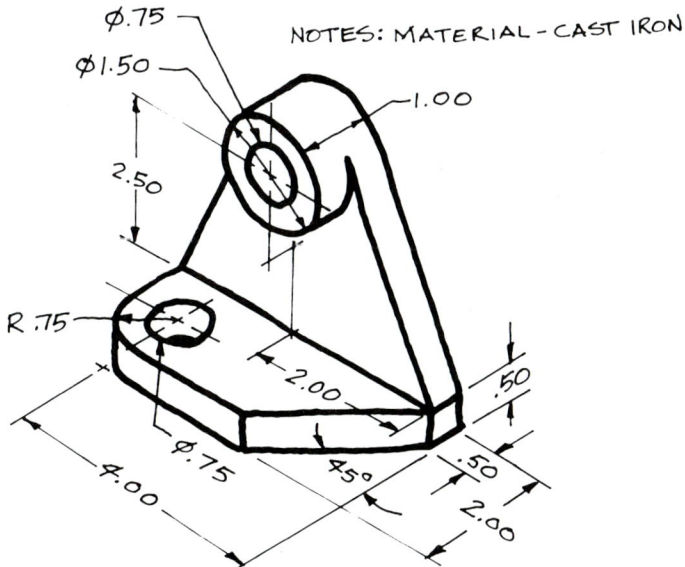

NOTES: MATERIAL – CAST IRON

Figure 3-52   Traditional Drafting Project 4

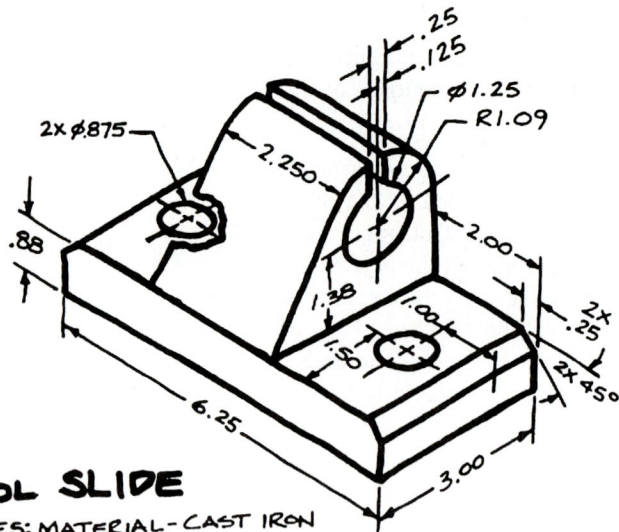

Figure 3-53   Traditional Drafting Project 5

Computer Aided Design Basics

Unit Objectives

- Describe the AutoCAD screen layout
- Perform an AutoCAD drawing setup, including setting units, limits, layers, linetypes, and lineweights.
- Explain the coordinate systems used to create AutoCAD drawings (absolute, relative, and polar coordinates).
- Create and edit AutoCAD drawings using the commands found on the Draw, Modify, and Inquiry toolbars.
- Employ Object Snaps to facilitate construction of AutoCAD drawings.
- Use the Properties tool to inquire about, or change the properties of, an entity.
- Add text to a drawing and edit the text including text style.
- Create a floor plan for a small cottage.
- Create multiview drawings of machine parts.
- Plot AutoCAD drawings.

INTRODUCTION

In most engineering and architectural offices, drafters and designers produce technical drawings using Computer Aided-Design (CAD) systems. A CAD system consists of a personal computer (PC) or workstation coupled with a CAD software program. One of the most widely used CAD software programs is called *AutoCAD*. AutoCAD was one of the first CAD programs that could run on a PC. Autodesk, the company that publishes Auto-CAD software, reports that there are over six million AutoCAD users worldwide. The price for a single station of AutoCAD for a professional user is about $3000, but a student version is available for much less through student software outlets.

There are many other CAD programs on the market as well. Some CAD programs are designed to perform work in a specialized area. In mechanical design, Inventor, ProE, and Solidworks are three of the principal CAD programs; in electronic design, Cadence and Mentor are widely used. In the civil and architectural fields, Land Desktop, Civil 3D, Microstation, and Revit are popular CAD programs.

BEGINNING AN AUTOCAD DRAWING

Use the mouse's left-click button to double-click on the **AutoCAD** icon located on the desktop of your computer. This will launch the AutoCAD program.

A new AutoCAD drawing will open similar to the one shown in Figure 4-1. Moving the mouse causes the cursor to move in the graphics window of the drawing.

Study the AutoCAD Screen Layout shown in Figure 4-2 and acquaint yourself with the terminology used to describe its features. Your instructor will call your attention to these various

Note:

This text assumes that the AutoCAD **Workspaces** mode is set to AutoCAD Classic, and that the **DYN** (Dynamic Input) button on the Status Bar is turned **On**. Refer to Figure 4-1 to locate these elements. Your instructor will guide you in the steps required to put these settings into effect.

For users of AutoCAD 2009, follow these steps to set the workspace to AutoCAD Classic.

- Clink on the **Menu Browser Button** (the red A in the upper left corner of the AutoCAD screen) to display the pulldown menu.
- From the pulldown menu, select **Tools**, then **Workspaces**, and click on **AutoCAD Classic**.
- Select the **Dynamic Input** icon on the **Status Bar** to turn on **Dynamic Input**.

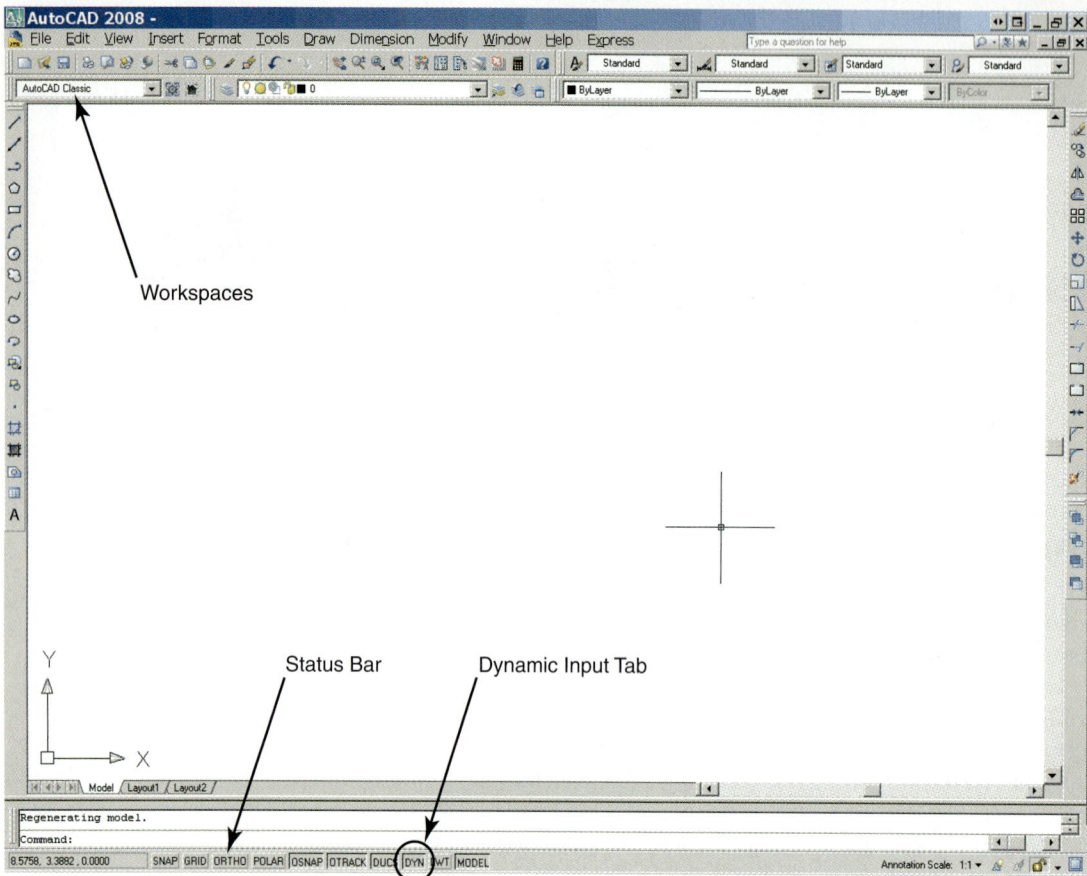

Figure 4-1 The AutoCAD 2008 Environment

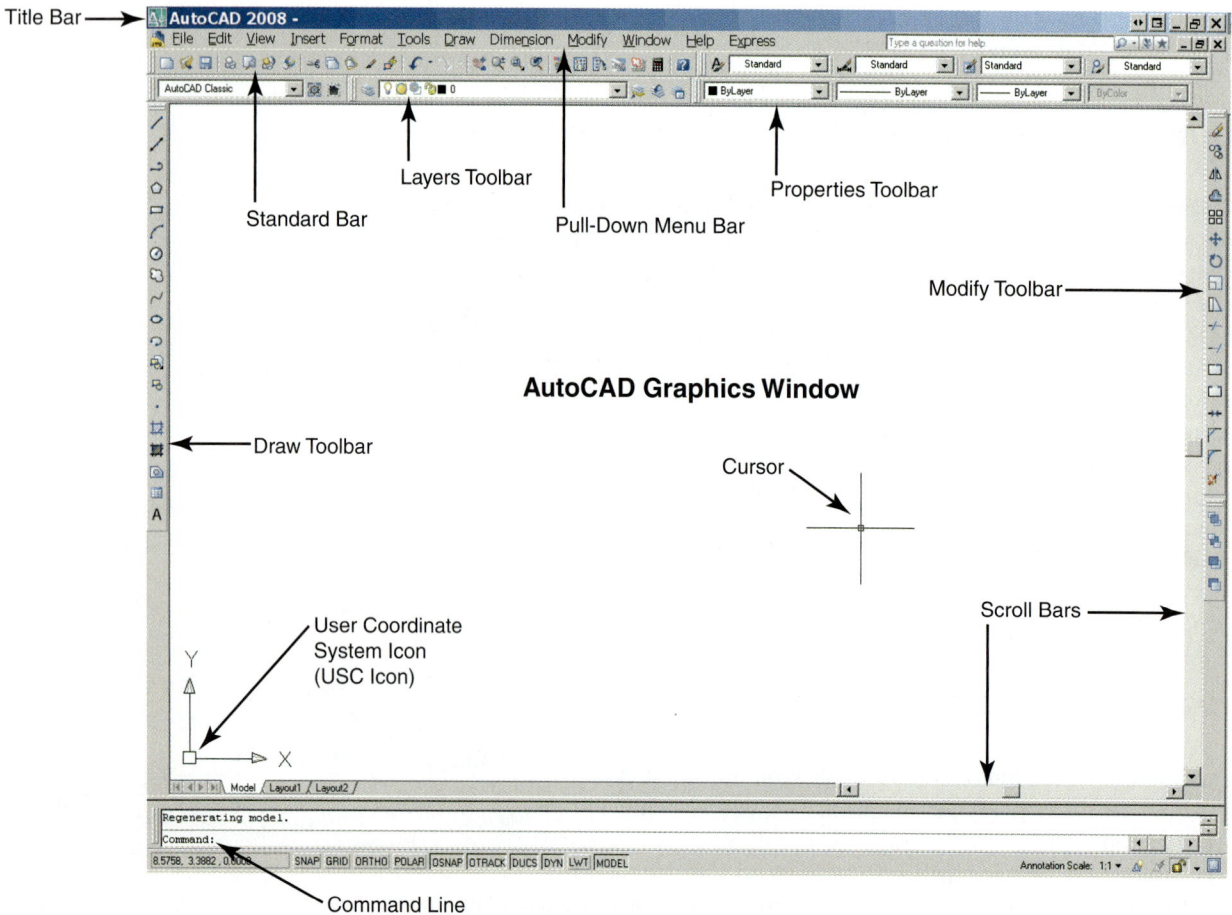

Figure 4-2 AutoCAD 2008 Screen Layout

locations as you proceed with your CAD training. This text refers to these menus and toolbars as well. Find the **Command Line** noted in Figure 4-2; it is very important for beginners to refer frequently to the command line because it offers important prompts and cues necessary to complete an AutoCAD command successfully.

To access the pulldown menu shown in Figure 4-2 in AutoCAD 2009, click on the **Menu Browser** button (the red A in the upper left corner of the screen), or type **Menubar** on the command line and change the setting to **1**. For more information on AutoCAD 2009, see Appendix E.

DRAWING YOUR FIRST LINE WITH AUTOCAD

Step 1. Use the left-click button of your mouse to select the **LINE** icon from the **Draw** toolbar or type **Line** at the command line and press **<Enter>**.

Step 2. Move the cursor into the graphics area and pick a point with the left-click button, then move the mouse to a new point and pick again. Congratulations, you've drawn your first line (see Figure 4-3)! By continuing to pick points, you can add to the line. When you are finished, press the **<Esc>** (escape) key to end the **LINE** command. Drawing lines to random points in this way is easy; drawing lines to exact points is a little more complicated. For this, you'll need to understand Cartesian Coordinates.

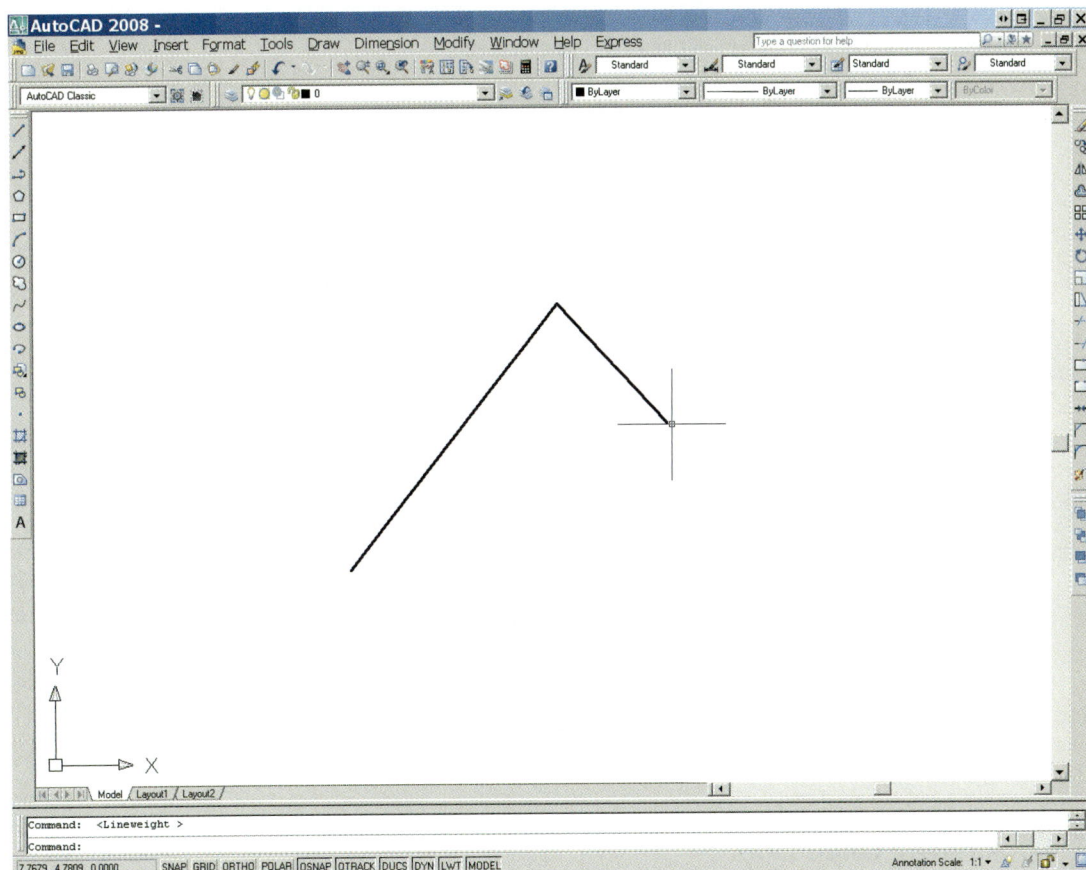

Figure 4-3   Lines drawn in AutoCAD's graphics window

LOCATING POINTS ON THE CARTESIAN COORDINATE SYSTEM

AutoCAD employs the *Cartesian Coordinate System* to define the exact location of points in the graphics window. In the Cartesian Coordinate system, a **0,0** (zero, zero) point is established as the origin point. The first zero represents the start point of measurements along the

Cartesian Coordinate System: A system of locating points along *X*- and *Y*-axes relative to a starting point representing "Zero X" and "Zero Y." Named for its originator, René Descartes.

X-axis (horizontal), and the second zero represents the start point of measurements along the *Y*-axis (vertical). All other points are located along the *X*- and *Y*-axes using **0,0** as the starting point. In Figure 4-4, the **0,0** point is located in the lower left corner. The other coordinate points labeled on the grid refer to each point's location measured along the *X*- and *Y*-axes relative to **0,0**.

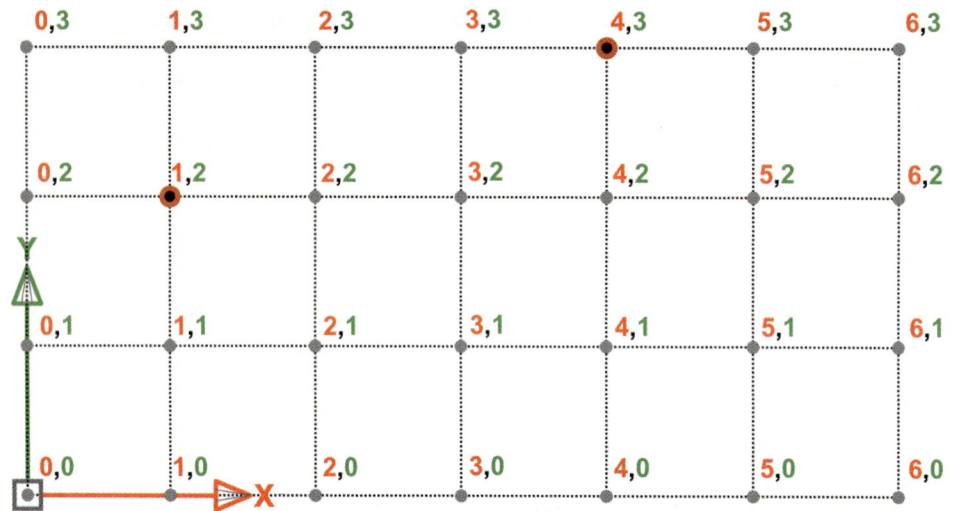

Figure 4-4   Points on the Cartesian Coordinate System

Locate the coordinate point labeled **1,2** in Figure 4-4. This coordinate is located on the grid by starting at the **0,0** origin in the lower left corner of the grid and measuring over 1 unit along the *X*-axis and up 2 units along the *Y*-axis. The *X* and *Y* values are separated with a comma. CAD drafters refer to this point as **1,2**.

Next, locate the coordinate labeled **4,3** in Figure 4-4. This coordinate is found by starting at the **0,0** origin in the lower left corner of the grid and measuring over 4 units along the *X*-axis and up 3 units along the *Y*-axis. CAD drafters refer to this point as **4,3**. Lines drawn in two dimensions have a start and an endpoint. Both points are defined by their respective *X*- and *Y*-coordinates.

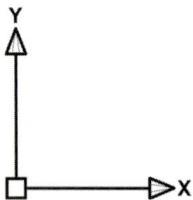

Figure 4-5   UCS icon

User Coordinate System (UCS): The point in an Auto-CAD drawing where the *X*-, *Y*-, and *Z*-axes intersect (0,0,0). This point is noted in the graphics window with the UCS icon.

absolute coordinates: Points that are located along the *X*-, *Y*-, and *Z*-axes that are relative to a point defined as 0,0,0. In an AutoCAD drawing, 0,0,0 is usually located in the lower left corner of the graphics window.

relative coordinates: Points that are located along the *X*-, *Y*-, and *Z*-axes that are relative to the last point defined.

polar coordinates: Coordinates defined by a length and an angle that are relative to the last point defined.

The User Coordinate System (UCS) Icon

AutoCAD represents the **0,0** point in the graphics window by placing the icon shown in Figure 4-5 in the lower left corner of the graphics window. This icon is called the ***User Coordinate System (UCS)*** icon. The visibility of this icon can be controlled by typing **UCSICON** at the command line, pressing **<Enter>**, and selecting **On** or **Off** from the settings listed. This icon orients the CAD operator to AutoCAD's **0,0** point.

AutoCAD uses several types of coordinate systems to specify the location of points, although each system has its basis in Cartesian coordinates. AutoCAD terminology refers to these other coordinate systems as ***absolute coordinates***, ***relative coordinates***, and ***polar coordinates***. AutoCAD drafters must be familiar with each system.

Absolute Coordinates

In AutoCAD terminology, points that are relative to point **0,0** (usually located in the lower left corner of the AutoCAD screen) are referred to as *absolute coordinates*. In Figure 4-6, a red line begins at absolute coordinate **2,2** and is drawn to coordinate **7,3**, then it is drawn to **10,6** and ends at coordinate **4.5,7**. Because each of these points is located relative to **0,0** as measured along the *X*- and *Y*-axes, all of these coordinates would be considered absolute coordinates.

Drawing a Line with Absolute Coordinates.   To draw the line shown in Figure 4-6 using absolute coordinates (assuming the **DYN** tab has been selected on the Status Bar), select the **LINE**

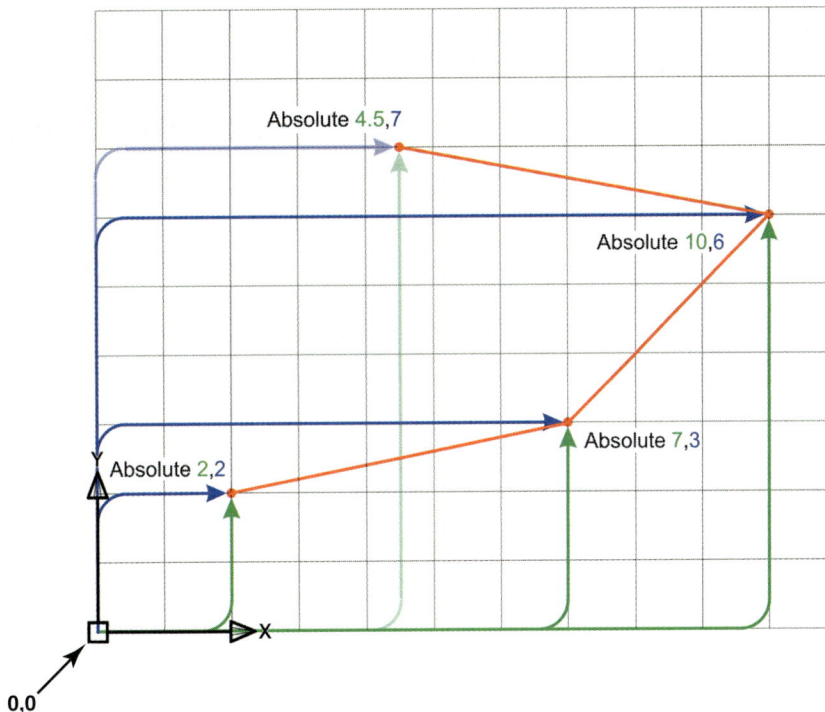

Figure 4-6 Absolute coordinates

command icon from the **Draw** toolbar and at the Specify the first point prompt, type **2,2** and press **<Enter>**. At the Specify the next point prompt, type **#7,3** and press **<Enter>**. To continue the line from **7,3** to absolute coordinate **10,6**, type **#10,6** and press **<Enter>**. To finish the line, type **#4.5,7** and press **<Enter>**. Pressing the **<Esc>** key will discontinue the **LINE** (or any other) command.

TIP Typing the # sign directs AutoCAD to locate points using absolute coordinates when the **DYN** setting is on.

TIP To enter an absolute coordinate in releases of AutoCAD prior to Release 2006, or when drawing with the **DYN** setting off in newer releases, you do not need to type the # symbol before entering the X- and Y-coordinates.

To draw the line shown in Figure 4-7, select the **LINE** command icon and at the Specify the first point prompt, type **2,2** and press **<Enter>**. At the Specify the next point prompt, type **#8,7** and press **<Enter>** again. Press **<Esc>** to end the command.

Note: When defining a relative coordinate that is to the left, or below, the previous point, it is necessary to enter a negative coordinate. This is done by typing a minus sign (−) before the coordinate value. For example, typing **-3,-2** draws a line to a point **3** units to the left on the X-axis and **2** units below the point previously defined.

Relative Coordinates

Relative coordinates are located relative to the last point defined. For example, in Figure 4-8, a line begins at absolute coordinate **1,1** and is drawn to a second point located **6** units along the X-axis and **0** units along the Y-axis relative to the start point **(2,2)**. The line continues

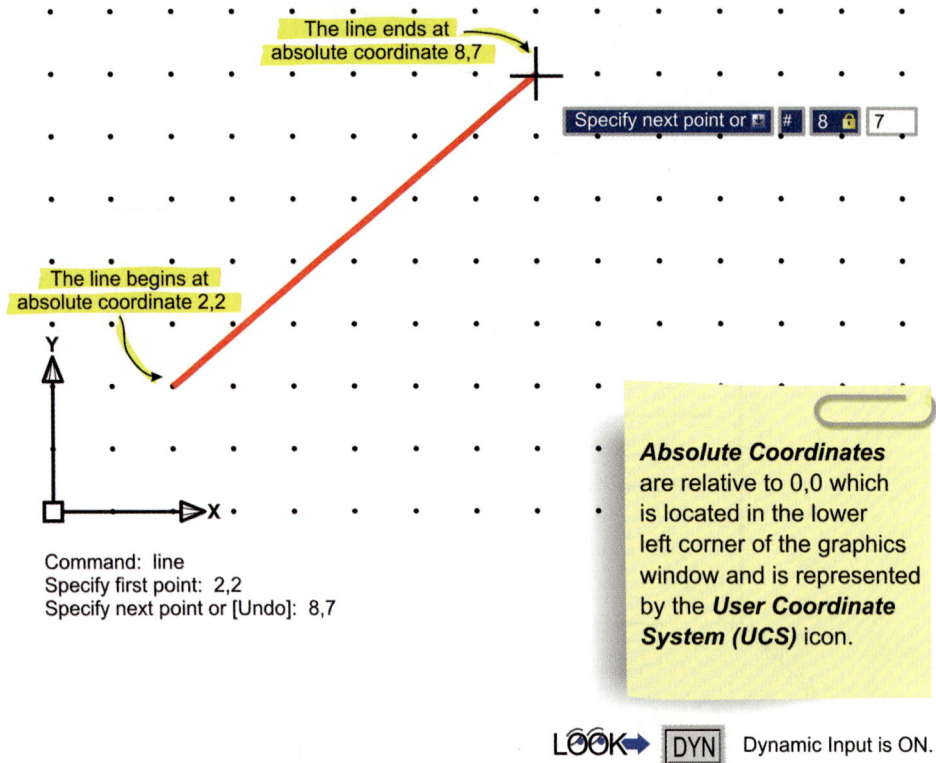

The line ends at absolute coordinate 8,7

Specify next point or # 8 🔒 7

The line begins at absolute coordinate 2,2

Command: line
Specify first point: 2,2
Specify next point or [Undo]: 8,7

Absolute Coordinates are relative to 0,0 which is located in the lower left corner of the graphics window and is represented by the **User Coordinate System (UCS)** icon.

LOOK➡ DYN Dynamic Input is ON.

Figure 4-7   Line drawn by entering absolute coordinates

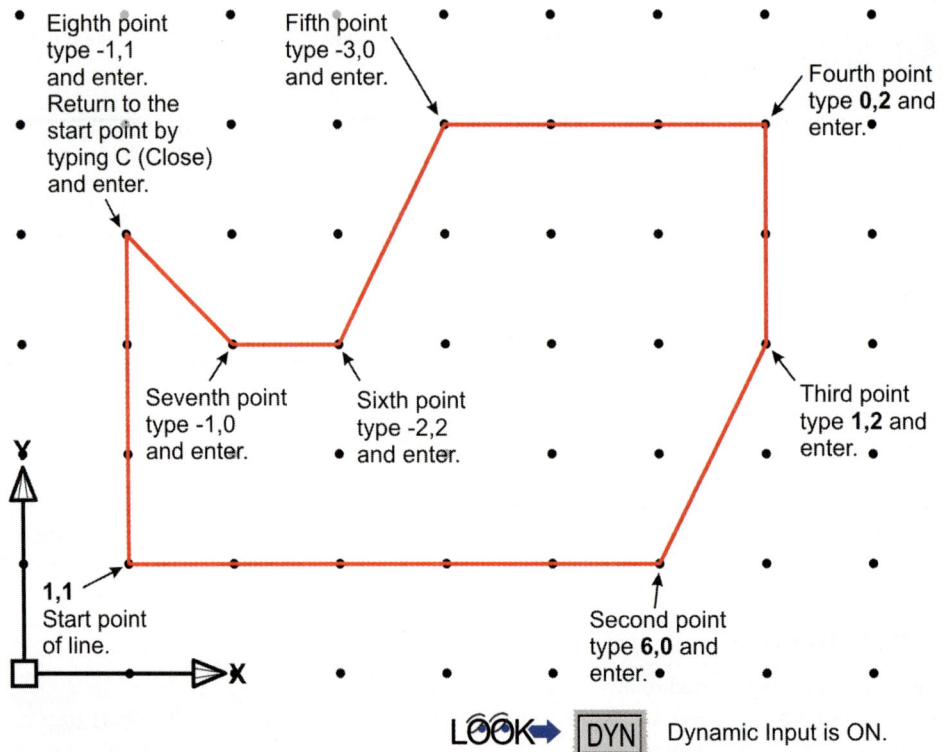

Eighth point type -1,1 and enter. Return to the start point by typing C (Close) and enter.

Fifth point type -3,0 and enter.

Fourth point type **0,2** and enter.

Seventh point type -1,0 and enter.

Sixth point type -2,2 and enter.

Third point type **1,2** and enter.

1,1 Start point of line.

Second point type **6,0** and enter.

LOOK➡ DYN Dynamic Input is ON.

Figure 4-8   Line defined with relative coordinates

to a third point located **1** unit along the *X*-axis and **2** units along the *Y*-axis relative to the second point. The line continues to a fourth point located **0** units along the *X*-axis and **2** units along the *Y*-axis relative to the third point. The line continues in this fashion until it returns to the start point. With the exception of the start point of the first line, each point is located *relative* to the previously defined point.

> **TIP**
>
> To enter a relative coordinate in releases of AutoCAD prior to Release 2006, or when drawing with the **DYN** setting off in newer releases, you must first type the @ symbol before entering the *X* and *Y* distance. For example: typing **@6,5** draws a line to a point located **6** units on the *X*-axis and **5** units on the *Y*-axis relative to the last point entered.

Drawing a Line with Relative Coordinates.   Select the **LINE** command icon and type **2,2** for the first point and press **<Enter>**. This will begin the line at absolute coordinate **2,2**. At the Specify the next point prompt, type **6,5** and press **<Enter>** again. The line will begin at absolute coordinate **2,2** and be drawn to a point located **6** units along the *X*-axis and **5** units along the *Y*-axis relative to **2,2** (see Figure 4-9).

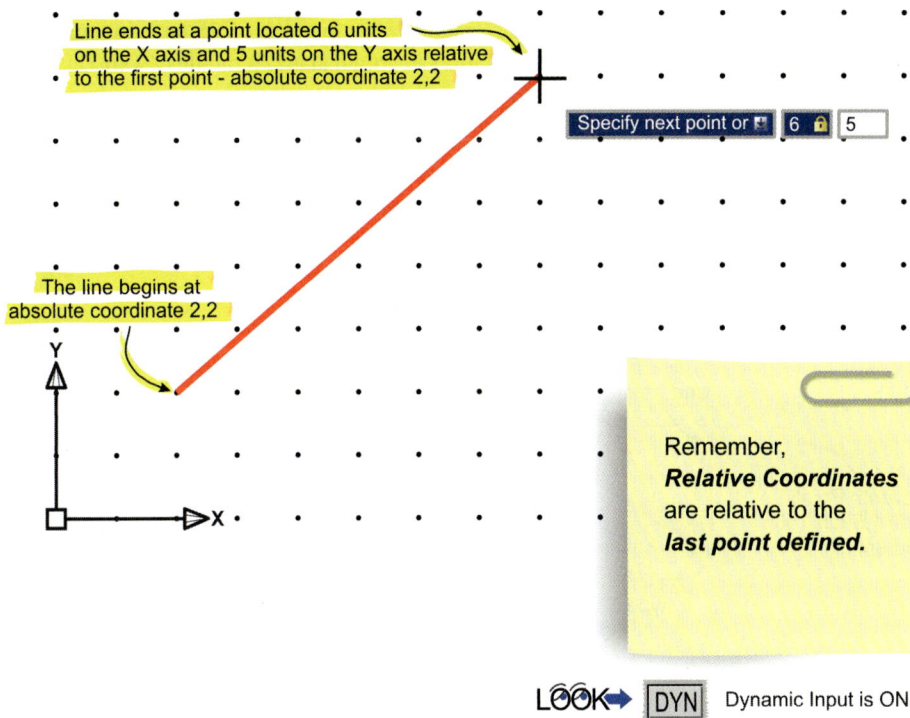

Line ends at a point located 6 units on the X axis and 5 units on the Y axis relative to the first point - absolute coordinate 2,2

Specify next point or ▣ 6 🔒 5

The line begins at absolute coordinate 2,2

Remember,
Relative Coordinates
are relative to the
last point defined.

LOOK➡ DYN   Dynamic Input is ON.

Figure 4-9   A line drawn with relative coordinates

Polar Coordinates

In order to understand polar coordinates, you first have to understand two things about how AutoCAD measures angles:

1. *East* (as on a compass) is considered zero degrees.

2. Angles are measured *counterclockwise* (see Figure 4-10).

Polar coordinates are defined with a length *and* an angle and are located relative to the last point you entered. When specifying a polar coordinate, it is necessary to type the length of the line, press the **<Tab>** key, and enter the desired angle. For example, entering **10 Tab 30** would

To lay out a 45 degree angle in AutoCAD, you would begin at East (0 degrees) and turn counterclockwise 45 degrees.

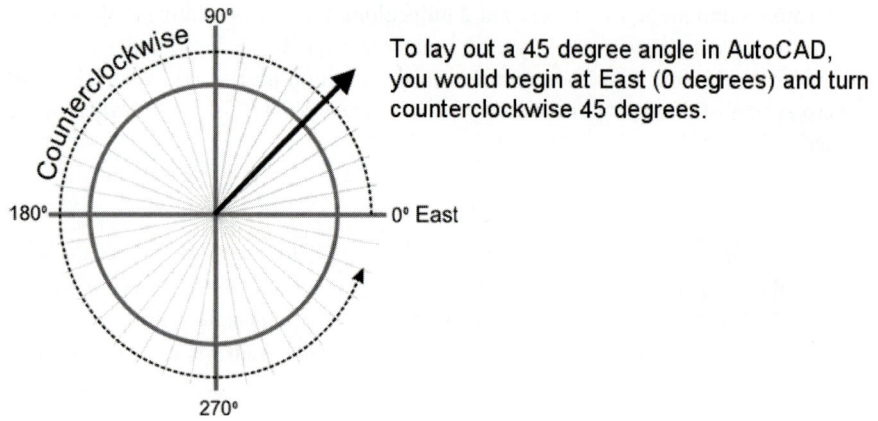

Figure 4-10   Angles in AutoCAD

draw a line **10** units long at a **30°** angle relative to the previous point defined (remember that AutoCAD measures angles counterclockwise and East is **0°**). Pressing **<Tab>** switches Auto-CAD's coordinate entry mode from linear to angular. In Figure 4-11, the first line was begun at absolute coordinate **1,1** and drawn to a second point that was **6** units in length along a **0°** angle (**6 Tab 0**). The second line begins at the last point and is drawn to a point **2.25** units in length at a **60°** angle (**2.25 Tab 60**). The third line is drawn **2** units in length at a **120°** angle (**2 Tab 120**). The line continues in this fashion until it ends at the seventh point.

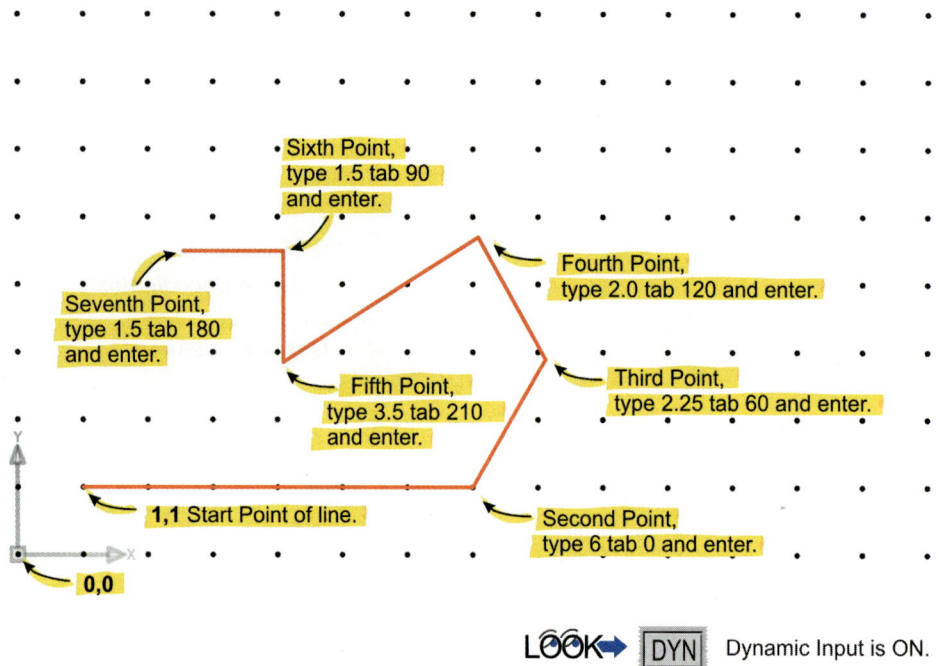

Figure 4-11   Lines drawn with polar coordinates

TIP   To enter a polar coordinate in releases of AutoCAD prior to Release 2006, or when drawing with the **DYN** setting off in newer releases, you must first type the @ symbol to enter a polar coordinate. Instead of pressing **<Tab>**, type the < symbol; for example, **@4<90**.

Drawing a Line with Polar Coordinates.   Select the **LINE** command icon and type **2,2** for the location of the first point and press **<Enter>**. At the Specify the next point prompt, type **6 Tab 45** and press **<Enter>** again. This will result in a line beginning at absolute coordinate **2,2** that is drawn **6** units in length at a **45°** angle (see Figure 4-12).

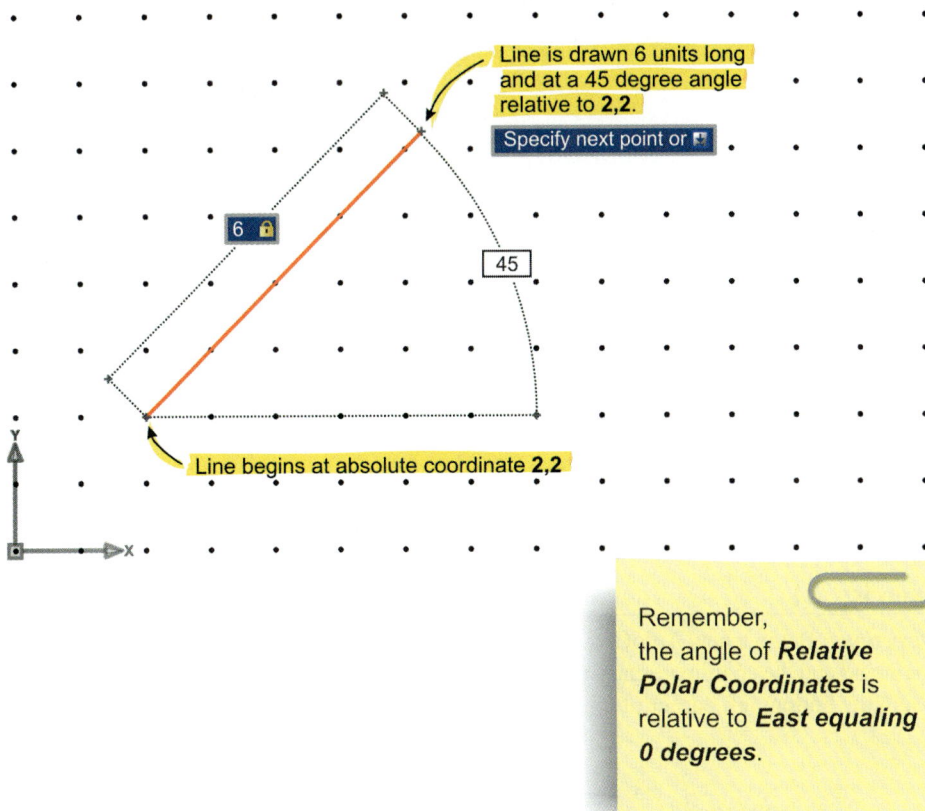

Line is drawn 6 units long and at a 45 degree angle relative to **2,2**.

Specify next point or

6

45

Line begins at absolute coordinate **2,2**

Remember, the angle of ***Relative Polar Coordinates*** is relative to ***East equaling 0 degrees***.

LOOK   DYN   Dynamic Input is ON.

Figure 4-12   Line drawn with polar coordinates

DIRECT ENTRY METHOD OF DRAWING LINES

Another method of drawing lines is through *direct entry*. This is the quickest and easiest way to draw horizontal and vertical lines. To use this method, turn **On** the **Ortho** button located on the Status Bar. Next, begin the **LINE** command and type in an absolute coordinate as the start point for the line. Then move the mouse in the desired X or Y (positive or negative) direction and type in the length of the line and press **<Enter>**. In Figure 4-13(a), the drafter turned **Ortho On** and began a line at **1,1**, then moved the cursor to the right (or positive X), typed in **2.5**, and pressed

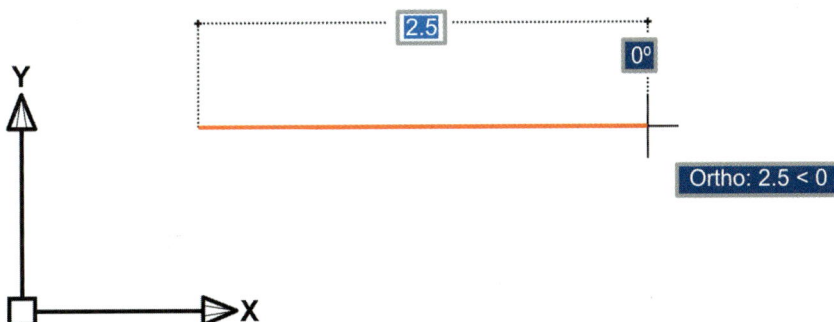

2.5

0°

Y

Ortho: 2.5 < 0

X

Figure 4-13(a)   Direct entry method of drawing a horizontal line

<Enter>. The resulting horizontal line is drawn **2.5** units to the right of the start point. In Figure 4-13(b), the drafter continues the line by moving the mouse in the positive *Y* direction (or up), typing **2**, and pressing <**Enter**>. The resulting line will be 2 units long and perfectly vertical. This method can also be used to draw lines of defined lengths at preset angles by turning **On** the **Polar** button located on the Status Bar.

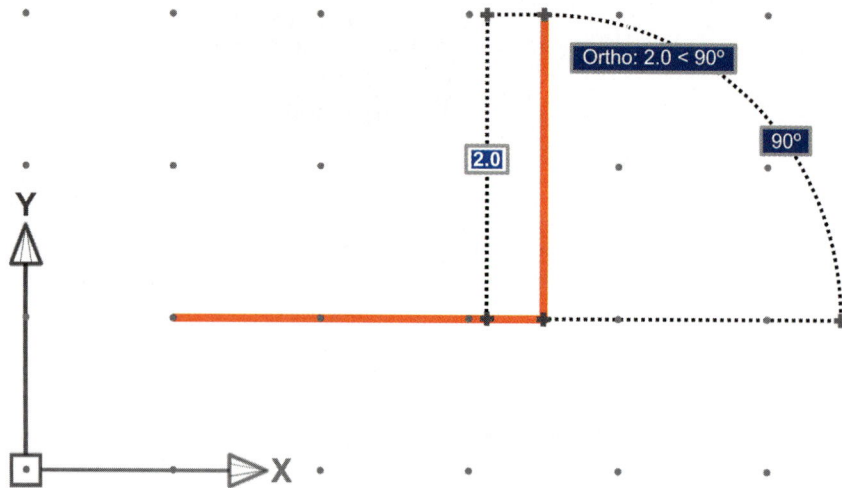

Figure 4-13(b)   Direct entry method of drawing a vertical line

Exercise 4-1

Directions: Begin a new AutoCAD drawing (default to the **acad.dwt** template). Using the **LINE** command and the coordinate entry methods described earlier, draw the object shown in Figure 4-14. Begin the bottom left corner of the object at absolute coordinate **2,2** and draw the first line in the

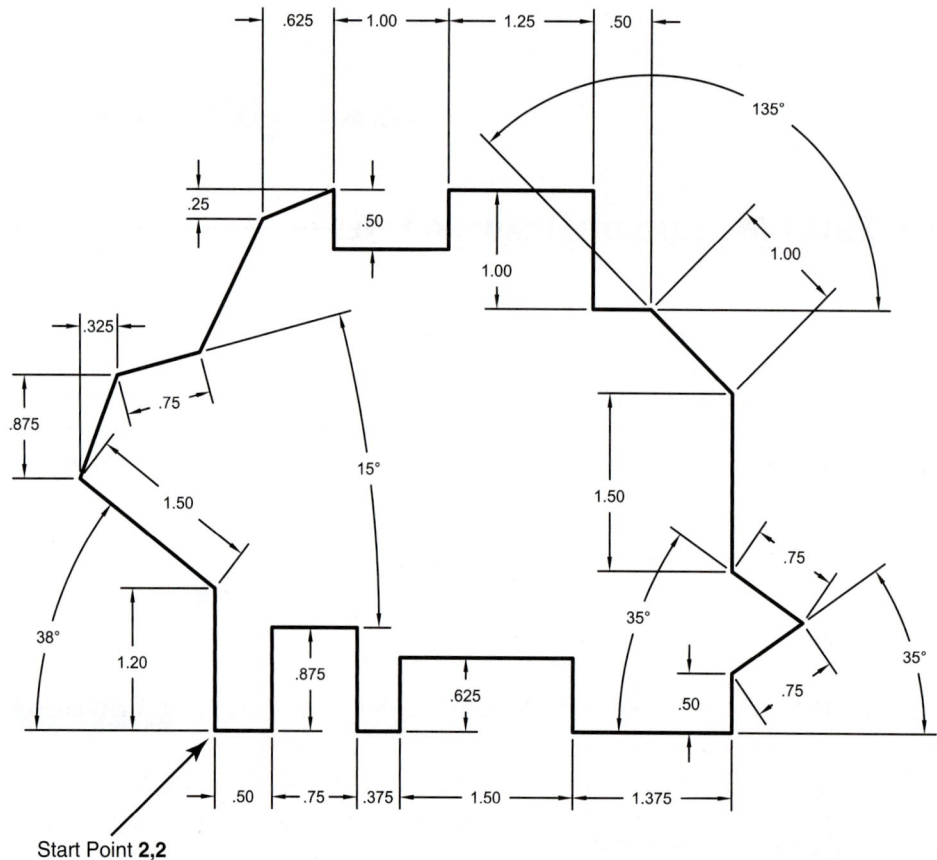

Figure 4-14   Exercise 4-1   Start Point **2,2**

positive *X* direction. Continue drawing lines until you are unable to continue due to lack of coordinate information (21 contiguous lines total). At this point, begin a new line from absolute coordinate **2,2** and draw in the positive *Y* direction. Continue drawing lines in this manner until you are unable to continue due to lack of coordinate information (4 contiguous lines total). To complete the drawing, draw a line connecting the endpoints of the two sets of lines. If you need assistance with this exercise, ask your instructor for help.

SETTING THE ENVIRONMENT FOR AUTOCAD DRAWINGS

Drawing Units

Before beginning an AutoCAD drawing, a drafter must first determine the appropriate *drawing units*, or units of measurement, for the type of drawing being created. For example, for architectural drawings, architectural units (feet and fractional inches) would be appropriate. For civil engineering drawings, engineering units (feet and decimal inches) would be appropriate; for mechanical engineering drawings, decimal units would be chosen.

drawing units: The units of measurement to be used in the creation of an AutoCAD drawing. For example, in an architectural drawing, one unit may equal one foot, whereas in a mechanical drawing, one unit might equal one inch or one millimeter. The drawing units should be set before beginning a drawing or defining the drawing limits.

Setting Drawing Units

Step 1. Open the **Drawing Units** dialog box by choosing the **Format** pull-down menu and selecting **Units** (Figure 4-15).

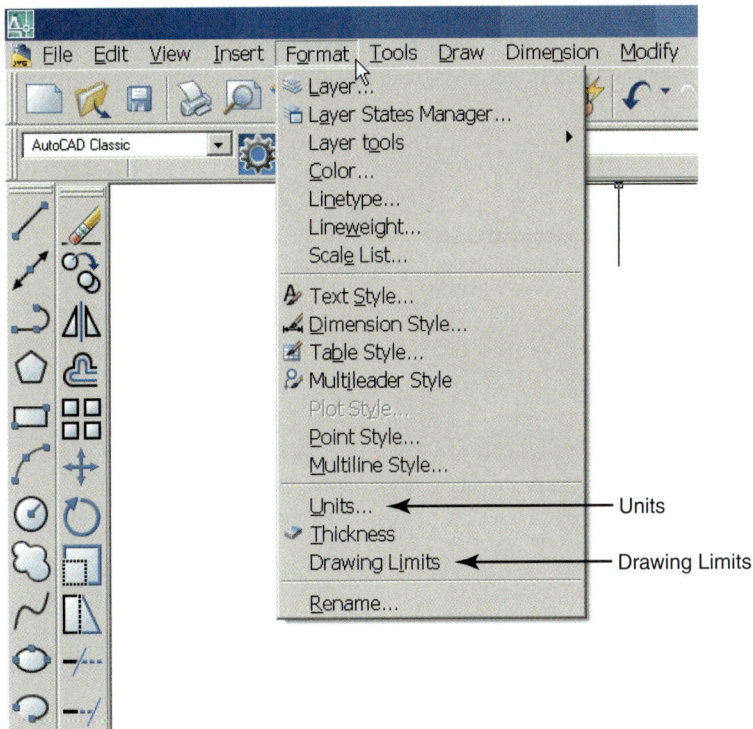

Figure 4-15   Selecting drawing units and limits from the format menu

Step 2. Select the type of units (decimal, engineering, architectural, fractional, or scientific) in **Length Type:** (Figure 4-16).

Step 3. Select the level of precision (the number of decimal places or fractional precision) for entering units in **Precision:** below **Length Type:**.

In Figure 4-16, the drawing units are set to **Decimal,** which means that coordinates will be entered, and displayed, in decimal units. **Precision:** for entering and displaying data is set to four decimal places.

Decimal Limits-
Unit of measurement is usually decimal inches but can be used for metric units (millimeters) as well.

Precision

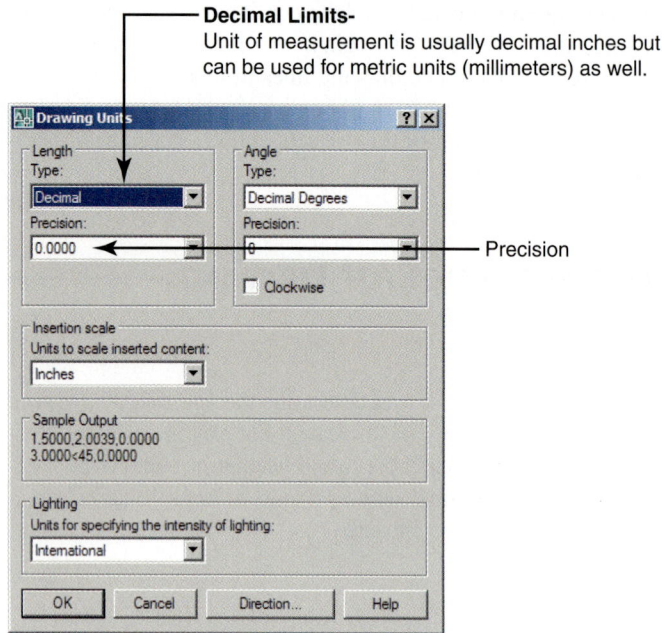

Figure 4-16   Drawing Units dialog box—Decimal units

In Figure 4-17, the drawing units are set to **Architectural**, which means that coordinates will be entered, and displayed, in feet and fractional inches. **Precision:** for entering and displaying data is set to 1/16″.

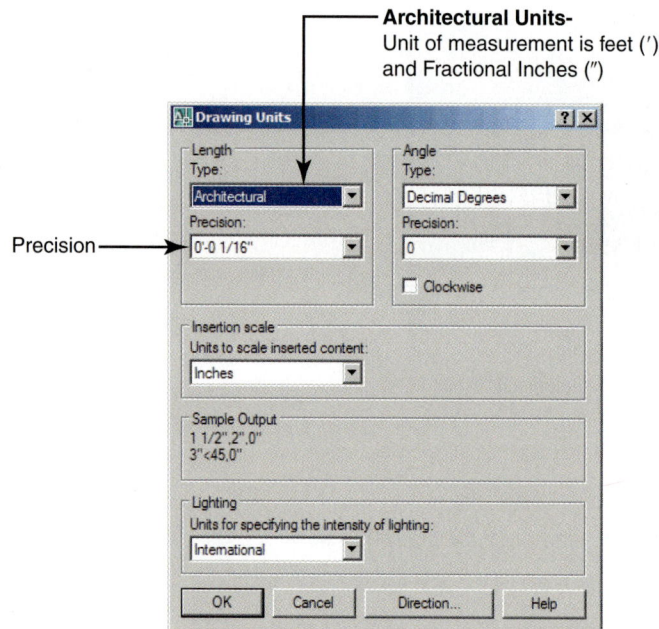

Architectural Units-
Unit of measurement is feet (′) and Fractional Inches (″)

Precision

Figure 4-17   Drawing Units dialog box—Architectural units

In Figure 4-18, the drawing units are set to **Engineering**, which means that coordinates will be entered, and displayed, in feet and decimal inches. **Precision:** for entering and displaying data is set to four decimal places.

Engineering Units-
Unit of measurement is feet (′)
and Decimal Inches (″)

Precision

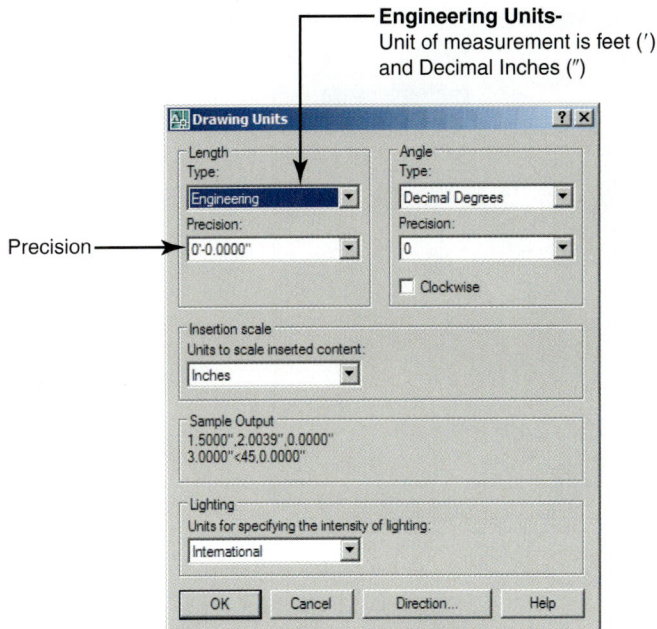

Figure 4-18   Drawing Units dialog box—Engineering units

Setting Angle Type.   After selecting the length type and precision, drafters select the angle type for the drawing. Several options are available: **Decimal Degrees**, **Degrees/Minutes/Seconds**, **Grads**, **Radians**, and **Surveyor's Units**.

Step 1. In **Angle Type:**, select from the options (decimal, deg/min/sec, grads, radians, or Surveyor's units). See Figure 4-19.

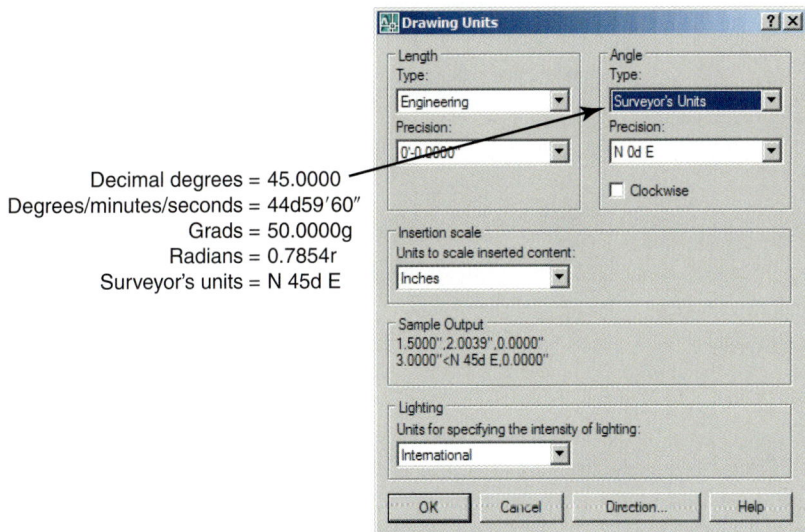

Decimal degrees = 45.0000
Degrees/minutes/seconds = 44d59′60″
Grads = 50.0000g
Radians = 0.7854r
Surveyor's units = N 45d E

Figure 4-19   Drawing Units dialog box—Setting angle type

Step 2. In **Precision:**, select the measurement of angles (below **Angle Type:**). See Figure 4-20.

Setting the Direction of Angle Measurement.   Selecting the **Direction...** button in the **Drawing Units** dialog box opens the **Direction Control** dialog box (see Figure 4-21). The **Base Angle** setting affects the starting point for measuring angles, polar coordinates, and polar tracking. **East** (the default direction), is usually assigned as the base angle; however, a base angle other than East can be set as the direction for 0° by selecting either **North**, **West**, **South**, or **Other**.

Precision for angular measurement.

The default setting for angle measurement is counterclockwise. If you would like to measure angles clockwise select this box.

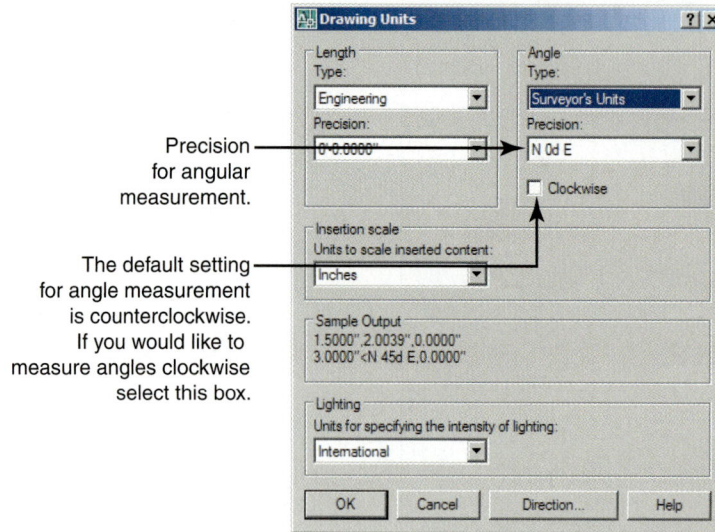

Figure 4-20 Drawing Units dialog box—Setting precision and direction for angular measurement

Figure 4-21 Setting direction for measuring angles in AutoCAD

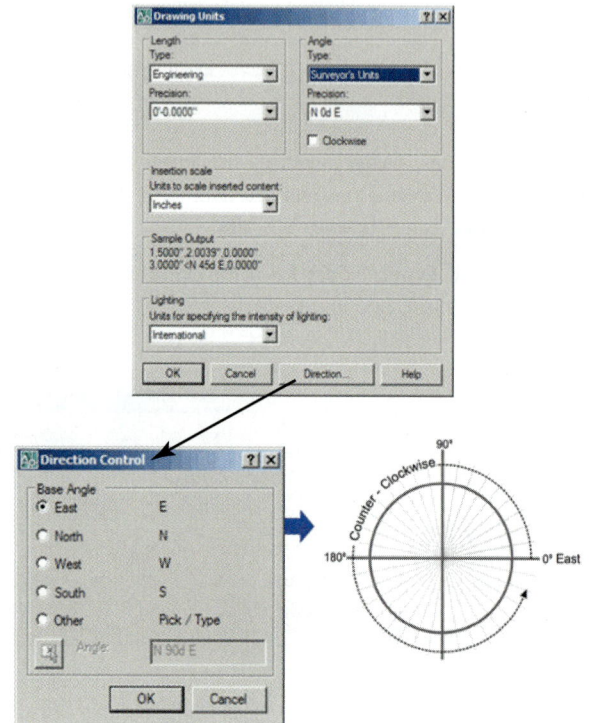

drawing limits: The limits of an AutoCAD drawing define its drawing area; this is comparable to selecting the sheet size for the drawing. When setting limits, you are prompted to specify the lower left and upper right corners of the drawing area. In most cases, the lower left corner will default to **0,0**, and the upper right corners will be defined by typing in the coordinates of the corresponding sheet size. For example, if using decimal units, an A size sheet limits would be **0,0** and **12,9**; a B size sheet limits would be **0,0** and **17,11**; a C size sheet limits would be **0,0** and **22,17**; and a D sheet limits would be **0,0** and **34,22**.

Drawing Limits

Setting the *drawing limits* defines the drawing area; this is comparable to selecting the sheet size for the drawing. Limits should be set *after* the units of the drawing have been set because the value for the limits will be displayed in the current units. When setting limits, you are prompted to specify the lower left and upper right corners of the drawing area. In most cases, the lower left corner is defaulted to **0,0**, and the upper right corner is defined by typing in the coordinates of the corresponding sheet size. For example, if using decimal units, an A size sheet limits are **0,0** and **12,9**; a B size sheet limits are **0,0** and **17,11**; a C size sheet limits are **0,0** and **22,17**; and a D sheet limits are **0,0** and **34,22**.

Setting Drawing Limits

Step 1. Open the **Drawing Limits** dialog box by choosing the **Format** pull-down menu and selecting **Drawing Limits** as shown in Figure 4-15.

```
Command: Limits
Reset Model space limits:
Specify lower left corner or [ON/OFF] <0.0000,0.0000>:

Specify upper right corner <12.0000,9.0000>:
```

Figure 4-22 Command line displaying default limits when decimal units are in effect

Step 2. When prompted to Specify lower left corner (the default limits of **0,0** will be displayed as shown in Figure 4-22), press <**Enter**> to accept **0,0** as the lower left limit.

Step 3. When prompted to Specify upper right corner (the default limits of **12,9** will be displayed as shown in Figure 4-22), type in the coordinates for a different sheet size, for example, **24,18** and press <**Enter**> (see Figure 4-23).

Note:

Although it is possible to define a coordinate other than 0,0 as the lower left limit, it is seldom done.

The drawing area will update to the new limits, but a **VIEW** command named **ZOOM-ALL** must be performed in order to see the new limits displayed in the graphics window. To perform a **ZOOM-ALL**, type **Z**, press <**Enter**>, and then type **A** and press <**Enter**>.

```
Command: Limits
Reset Model space limits:
Specify lower left corner or [ON/OFF] <0.0000,0.0000>:
Specify upper right corner <12.0000,9.0000>: 24,18

Command:
```

Figure 4-23 Command line displaying limits of 0,0 and 24,18

The limits of a drawing are dependent on the units of measurement assigned to the drawing. Therefore, the limits assigned are based on the sheet sizes that are appropriate for drawings created with the assigned units. Tables 4-1, 4-2, and 4-3 show the **Limits** settings for various sheet sizes relative to the units of measurement for the drawing (decimal, architectural, or engineering).

Table 4-1   Limits Settings for Decimal and Metric Units
Limits Based on ASME Y14.1 Decimal Sheet Sizes
A Sheet Limits = 0,0 and 11,8.5
B Sheet Limits = 0,0 and 17,11
C Sheet Limits = 0,0 and 22,17
D Sheet Limits = 0,0 and 34,22
E Sheet Limits = 0,0 and 44,34
Limits Based on ASME Y14.1M Metric Sheet Sizes
A4 Sheet Limits = 0,0 and 297,210
A3 Sheet Limits = 0,0 and 420,297
A2 Sheet Limits = 0,0 and 594,420
A1 Sheet Limits = 0,0 and 841,594
A0 Sheet Limits = 0,0 and 1189,841

Table 4-2	Limits Settings for Architectural Units	
Limits for Scale of: 1/4″ = 1′-0″		
A Sheet Limits = 0′,0′ and 48′,36′		
B Sheet Limits = 0′,0′ and 72′,48′		
C Sheet Limits = 0′,0′ and 96′,72′		
D Sheet Limits = 0′,0′ and 144′,96′		
Limits for Scale of 1/8″ = 1′-0′		
A Sheet Limits = 0′,0′ and 96′,72′		
B Sheet Limits = 0′,0′ and 144′,96′		
C Sheet Limits = 0′,0′ and 192′,144′		
D Sheet Limits = 0′,0′ and 288′,192′		

Table 4-3	Limits Settings for Engineering Units
Limits for Scale of: 1″ = 100′	
A Sheet Limits = 0′,0′ and 1200′,900′	
B Sheet Limits = 0′,0′ and 1800′,1200′	
C Sheet Limits = 0′,0′ and 2400′,1800′	
D Sheet Limits = 0′,0′ and 3600′,2400′	

layers: In AutoCAD drawings, lines and other entities are drawn on layers. Think of layers as sheets of clear glass layered one on top of the other. A layer can have its own properties such as color, linetype, or lineweight.

Figure 4-24   Layer Properties Manager icon

Layers

In AutoCAD drawings, lines and other entities are drawn on *layers*. Think of layers as sheets of clear glass layered one on top of the other. A layer can have its own color, linetype, or lineweight assigned to it.

When you begin an AutoCAD drawing from scratch, it contains only one layer, **Layer 0** (zero). If more layers are needed, they must be created. The steps involved in creating new layers follow.

Step 1. Click on the **Layer Properties Manager** icon located on the **Layers** toolbar (see Figure 4-24).

Step 2. When the **Layer Properties Manager** dialog box shown in Figure 4-25 opens, click the **New** button.

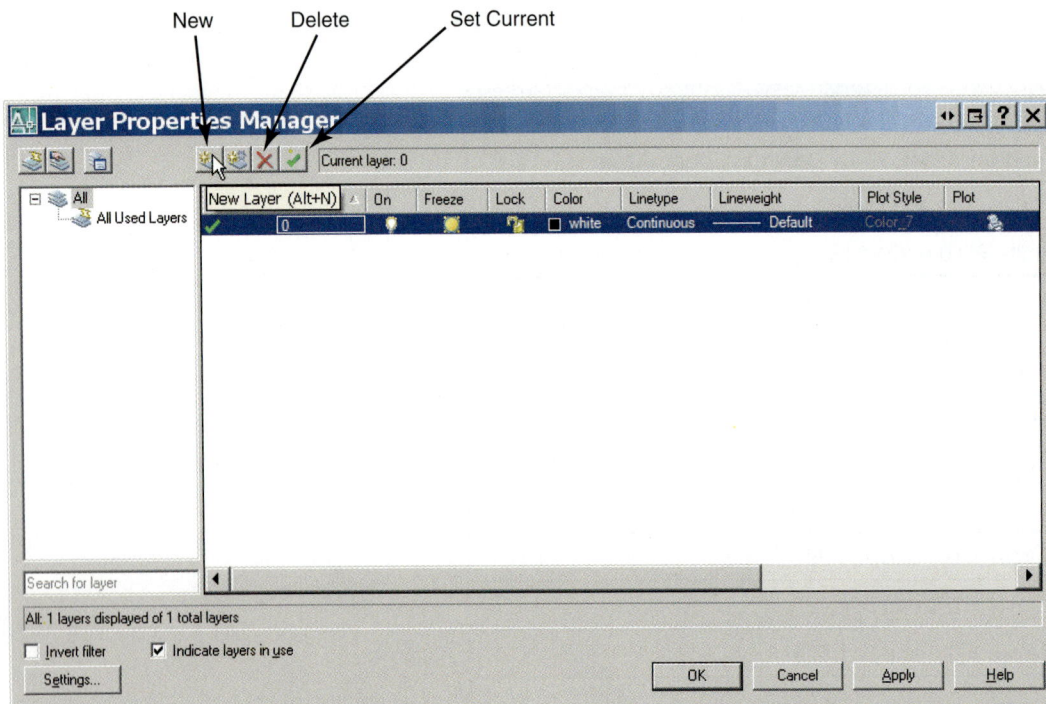

Figure 4-25   Layer Properties Manager dialog box

Step 3. Select the new layer and replace its default name, **Layer 1**, with the new layer name.

Step 4. Repeat Step 3 to create other layers. When all the new layers have been created, click **OK**.

Figure 4-26 shows the layers created for a mechanical drawing.

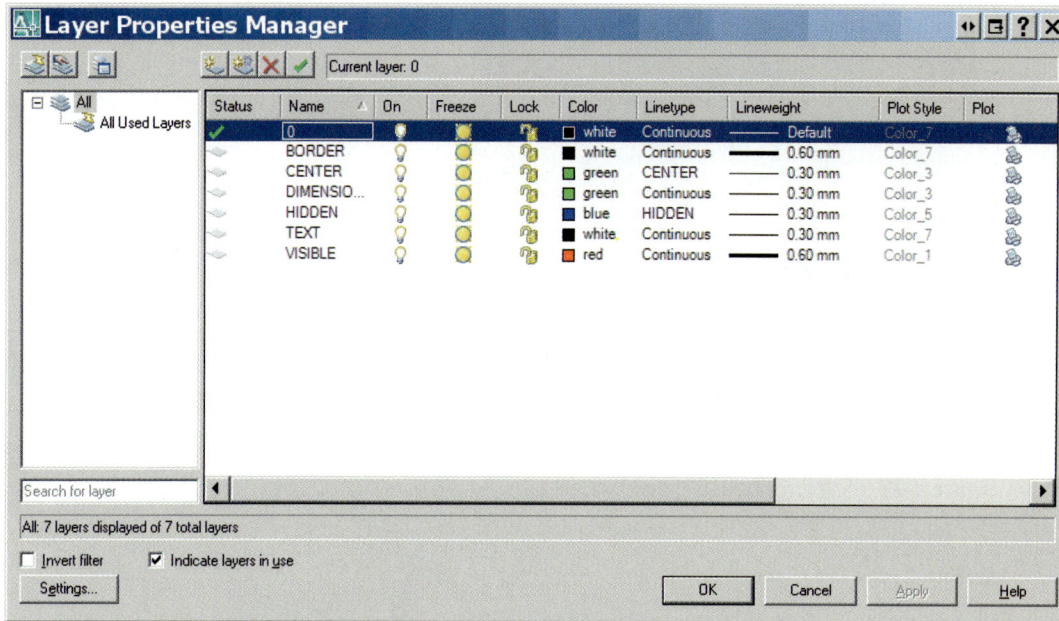

Figure 4-26   Layer Properties Manager Layer examples

Setting Layer Color

Step 1. Click on the **Layer Properties Manager** icon shown in Figure 4-27.

Step 2. Select the layer to which you want to assign a new color and choose on the color assigned to the layer beneath the **Color** column in the dialog box as shown in Figure 4-28.

Figure 4-27   Layer Properties Manager icon

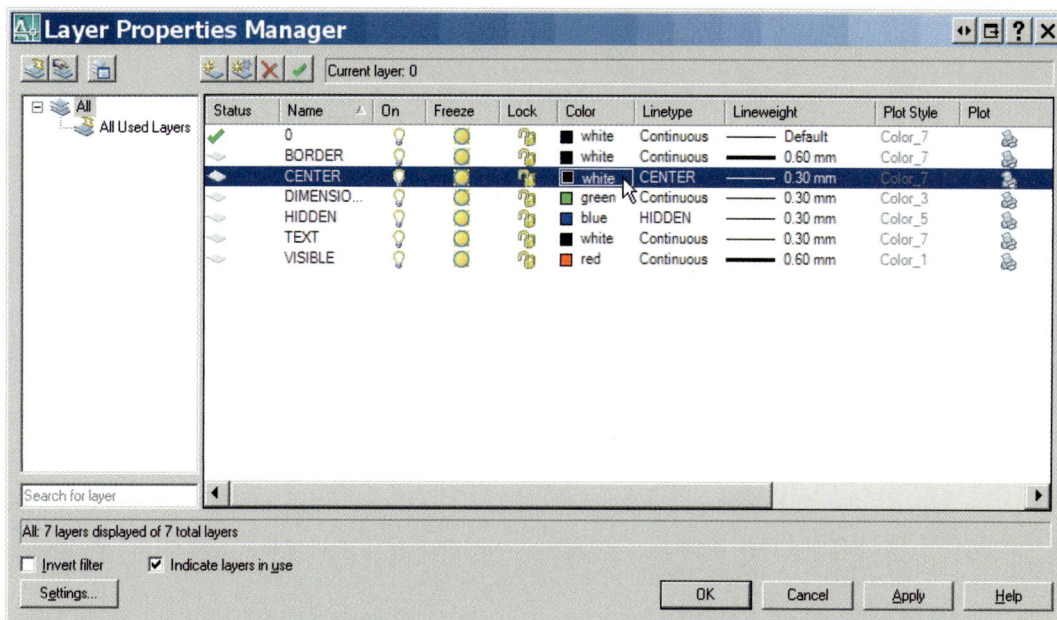

Figure 4-28   Layer Properties Manager dialog box

Step 3. When the **Select Color** dialog box shown in Figure 4-29 opens, select the desired tile from the color palette and click **OK**.

Figure 4-29 Select Color dialog box

Setting Layer Linetype

Step 1. Click on the **Layer Properties Manager** icon shown in Figure 4-30.

Figure 4-30 Layer Properties Manager icon

Step 2. Select the layer to which you want to assign a new linetype and click on the linetype name shown beneath the **Linetype** column. See Figure 4-31.

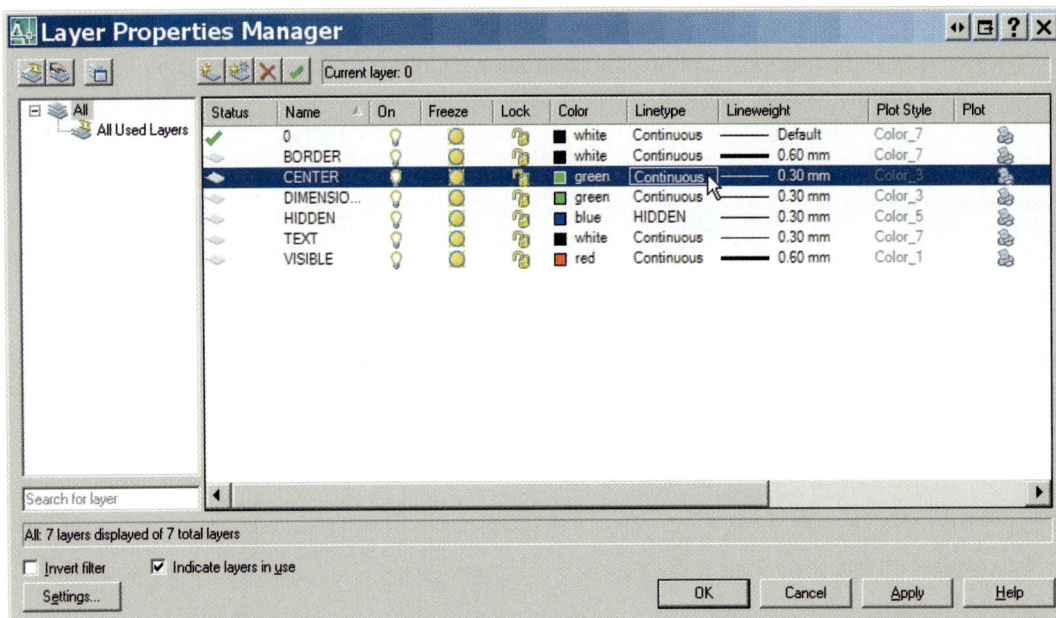

Figure 4-31 Layer Properties Manager dialog box

Step 3. The **Select Linetype** dialog box shown in Figure 4-32 will open. If you do not see the desired linetype listed, click on the **Load...** button.

Step 4. The **Load or Reload Linetypes** dialog box shown in Figure 4-33 will open. Scroll through the linetypes. Select the linetype you wish to load and click **OK**.

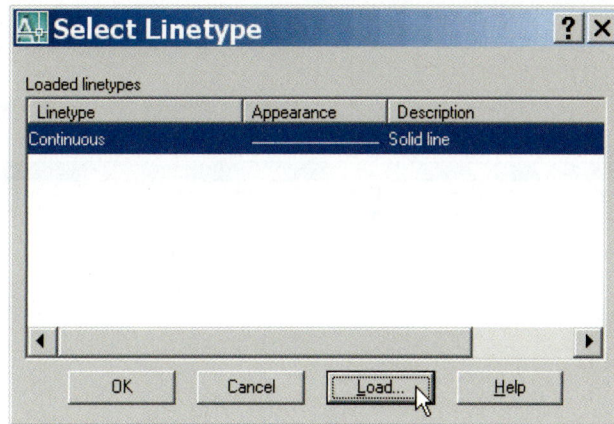

Figure 4-32   Select Linetype dialog box

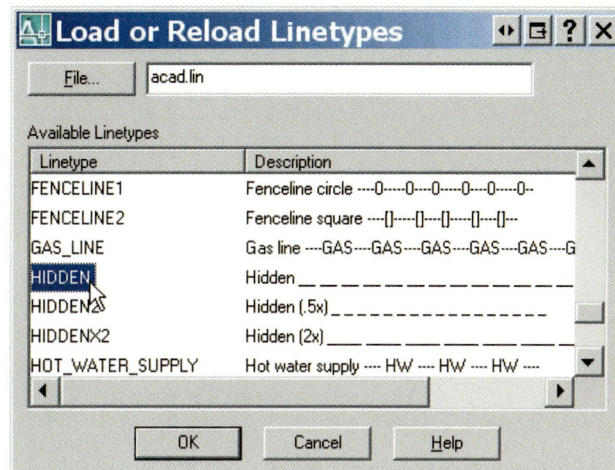

Figure 4-33   Load or Reload Linetypes dialog box

Step 5. Select the newly loaded linetype from the **Select Linetype** dialog box and click **OK**. The new linetype will be assigned to the layer selected in Step 2. See Figure 4-34.

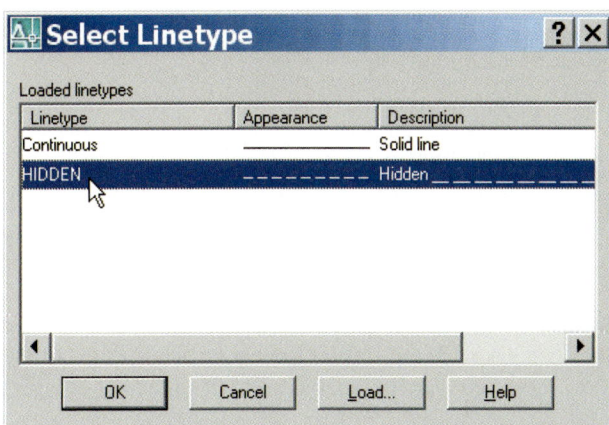

Figure 4-34   Select Linetype dialog box

Setting Layer Lineweight

Step 1. Click on the **Layer Properties Manager** icon. See Figure 4-35.

Step 2. When the **Layer Properties Manager** dialog box opens, select the layer to which you want to assign a new lineweight and select the lineweight setting shown beneath the **Lineweight** tab. See Figure 4-36.

Figure 4-35   Layer Properties Manager icon

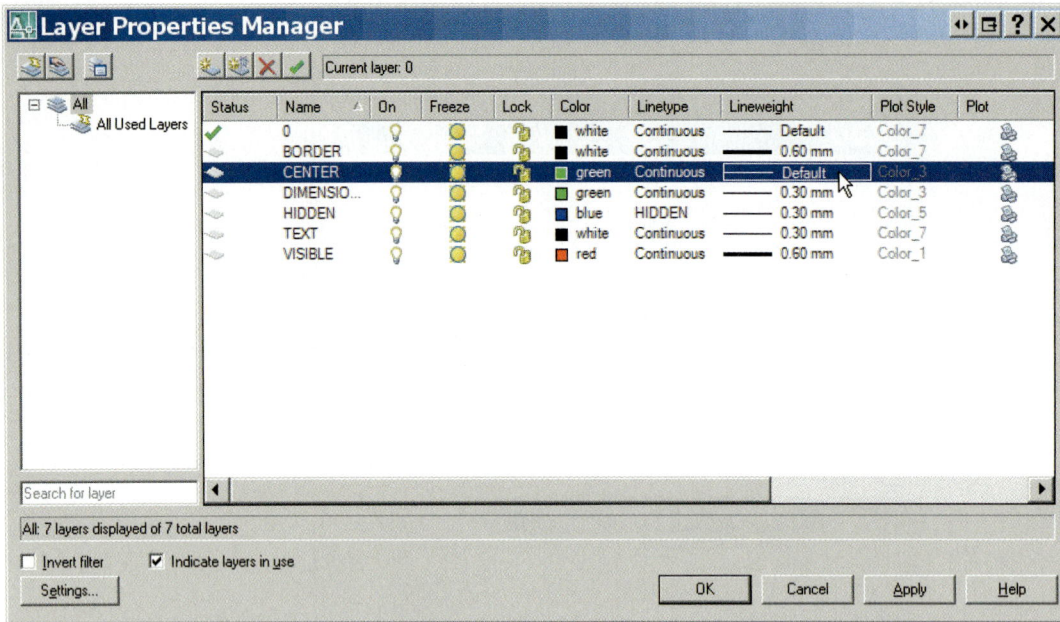

Figure 4-36   Layer Properties Manager dialog box

Step 3. When the **Lineweight** dialog box opens, scroll through and select the desired line thickness in which you want the layer to be printed and click **OK**. See Figure 4-37.

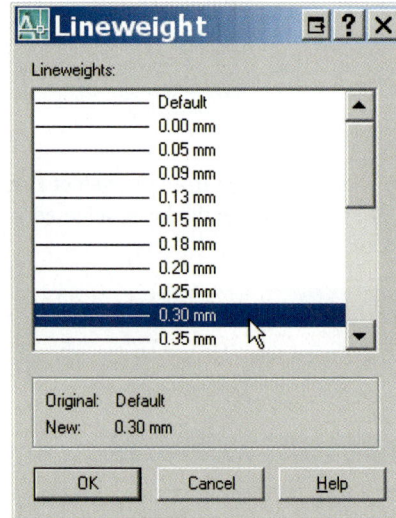

Figure 4-37   Lineweight dialog box

Setting the Current Layer.   In an AutoCAD drawing, you can draw only on the current layer. To make a different layer current, select the down arrow in the **Layers** toolbar located on the **Object Properties** toolbar. Then left-click on the layer you want to make current from the list of layers shown. See Figure 4-38.

> **TIP**
> An entity drawn on one layer can be moved to a different layer simply by picking the entity in the graphics window, and selecting the down arrow in the **Layers** toolbar located on the **Object Properties** toolbar, and picking the layer to which you want the entity to be moved.

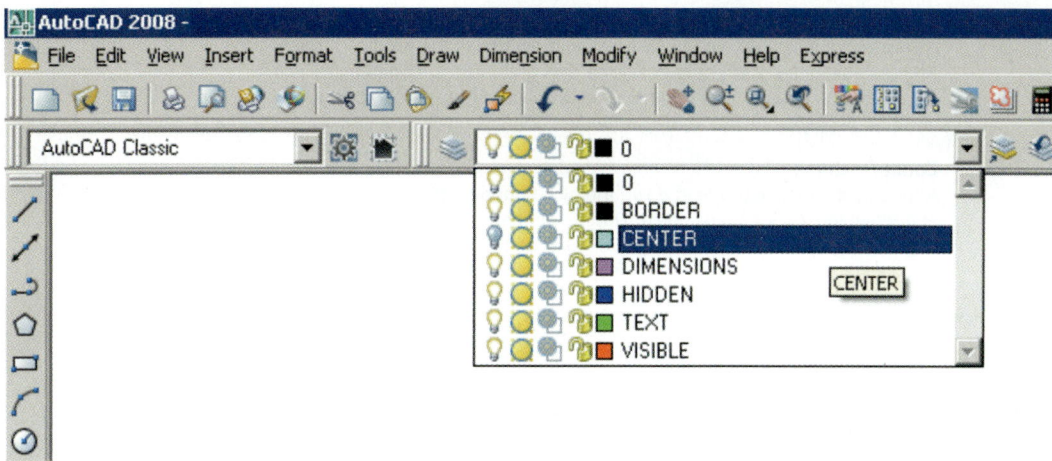

Figure 4-38   Layers toolbar

Controlling Layer Visibility.   Visibility of a drawing's layers can be controlled in two ways: either turning the layers *off* or *freezing* them. This is particularly useful if you need an unobstructed view of an area of the drawing, or if you are working in detail on a particular layer or set of layers. Construction lines are often drawn on layers that are later turned off, or frozen, because entities on these layers are not plotted.

Turning Layers Off

Select the down arrow in the **Layers** toolbar and turn a layer **Off** by selecting the yellow light bulb next to the layer name. Layers that are off will display the dark bulb symbol (see Figure 4-38).

Freezing Layers.   Select the down arrow in the **Layers** toolbar and freeze a layer by clicking on the **Sun** symbol next to its name. When the layer is frozen, the **Sun** symbol will be replaced with a snowflake. Freezing, and *thawing* (unfreezing), layers takes a little more time than turning layers on and off because this operation causes the drawing to be regenerated.

> **TIP**   The current layer can be turned off, but it cannot be frozen.

ZOOM AND PAN COMMANDS

The **ZOOM** command allows users to view a drawing up close or far away. Zooming does not actually change the scale of entities in the drawing (this is accomplished with the **SCALE** command), just their magnification in the graphics window. The **PAN** command allows users to reposition the view of the drawing in the graphics window. Panning does not change the location of entities in the drawing (this is accomplished with the **MOVE** command), just the viewer's point of view. The **PAN** and **ZOOM** commands are located on AutoCAD's **Standard** toolbar. For a detailed explanation of these important viewing tools, see Figure 4-39.

> **TIP**   AutoCAD's **ZOOM** icons are also located on the **Zoom** toolbar.

Pan
Allows users to reposition the view of the drawing in the graphics area.
After selecting this icon, hold down the pick button on the pointing device.
This locks the cursor to its current location. By moving the cursor, the
drawing display is moved in the same direction.

Zoom Realtime
After selection this icon, hold down the pick button on the pointing device,
and move the device up or down. The drawing magnification will become
larger as the mouse moves up, and smaller as the mouse moves down.

Zoom Previous
Displays the previous view(s) up to the 10 previous views.

Zoom Window
Displays the area specified by two diagonal corners of a rectangular window

Zoom Dynamic
Displays a portion of the drawing inside a view box which may be resized or moved
dynamically

Zoom Scale
Displays the view at a specified scale factor. Enter a scale factor in (nX or nXP).
Enter a value followed by x to specify the scale relative to the current view.

Zoom Center
Displays a window defined by a center point and a magnification value or height.

Zoom Object
Displays a view with the largest possible magnification that includes all of the
objects selected.

Zoom In
Magnifies the current view 2X each time it is selected.

Zoom Out
Magnifies the current view .5X each time it is selected.

Zoom All
Displays either the user-defined grid limits or the drawing extents, whichever view
is larger.

Zoom Extents
Displays a view with the largest magnigfication that includes all of the objects in the
drawing, including objects on layers that are turned off but not objects on frozen layers.

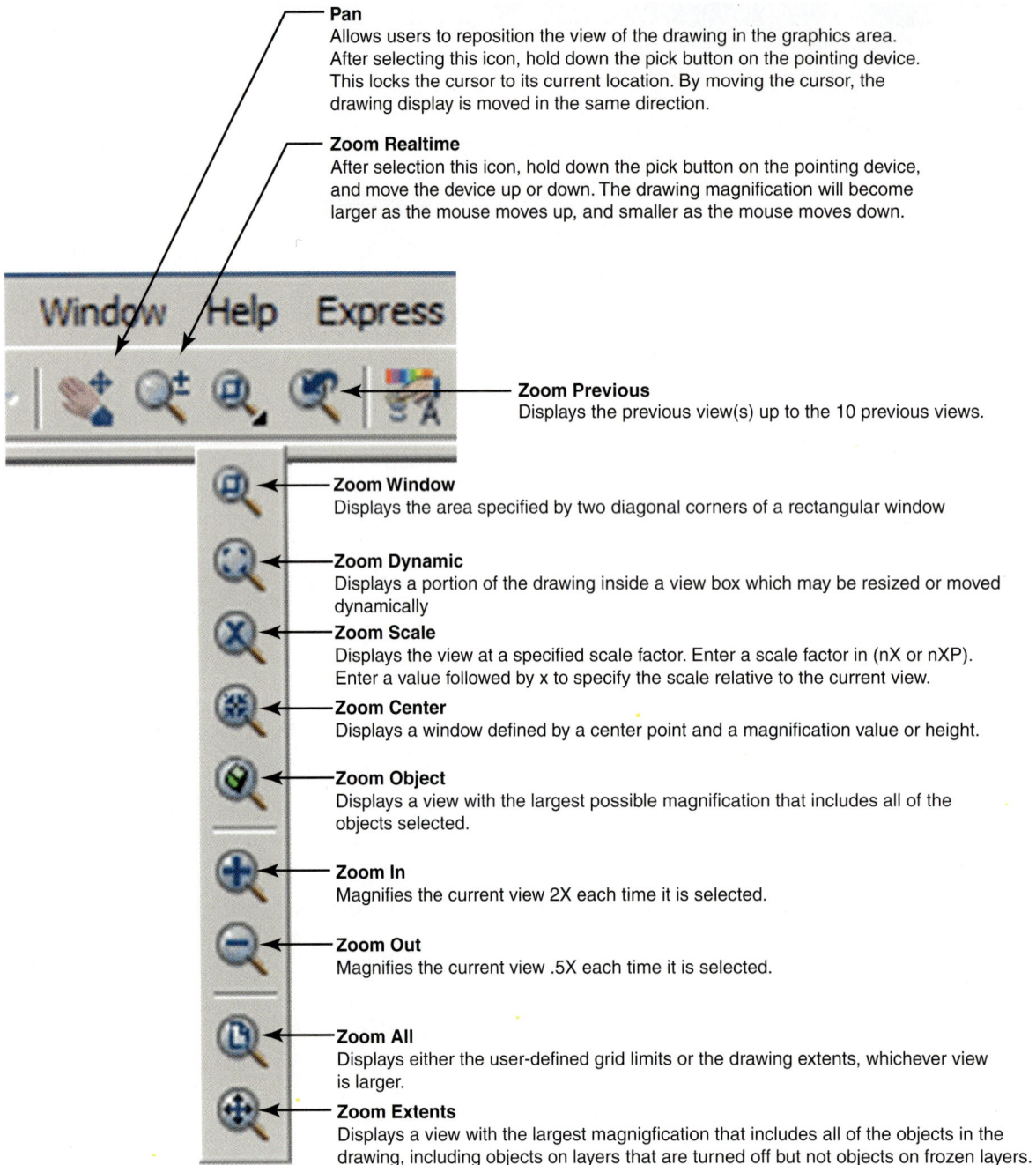

Figure 4-39 Zoom and Pan tools

> Many AutoCAD commands have a *command alias*. A command alias is a shortcut that can be
> typed to begin a command. For example, a quick way to perform a **ZOOM WINDOW** is to
> type ZOOM's alias, **Z** (upper or lower case), and press <**Enter**>. Then type **W** (for Window),
> press <**Enter**>, and pick two points to define the area to be zoomed into. Likewise, a quick
> way to perform a **ZOOM ALL** is to type **Z**, press <**Enter**>, type **A**, and press <**Enter**>.

AUTOCAD TOOLBARS

AutoCAD commands can be invoked by choosing the appropriate icon from the Menu Bar, the
Standard Bar or from command toolbars such as **Draw**, **Modify**, or **Dimension**. The quickest
method of locating a toolbar is to move the cursor onto any existing toolbar or icon on the Auto-

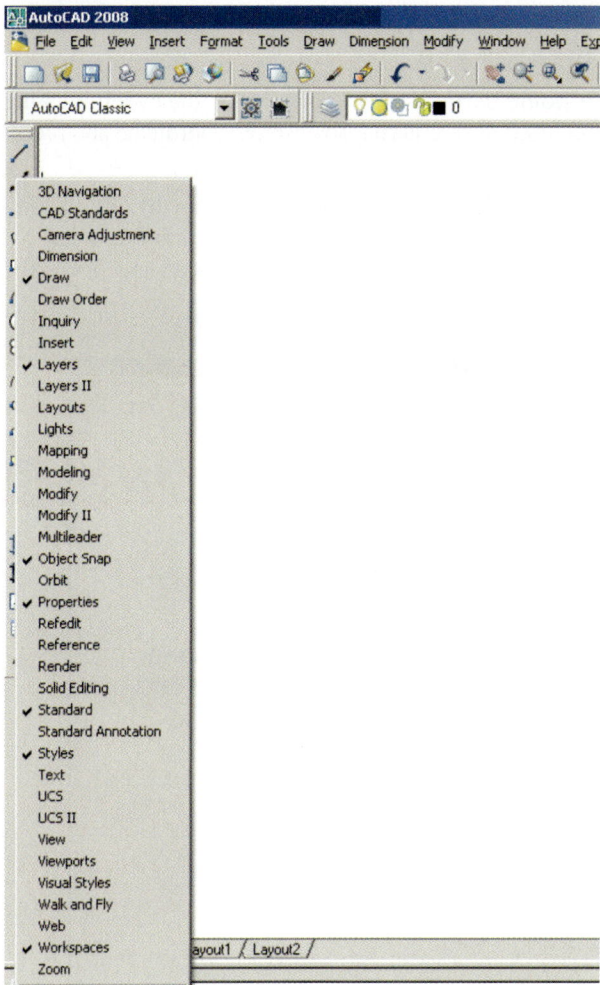

Figure 4-40   AutoCAD toolbars

CAD screen and right-clicking. A list of available toolbars will appear as shown in Figure 4-40. From this list, select the name of the toolbar that you wish to open by left-clicking on it. After the toolbar opens, you can drag it to a different location on the screen or dock the toolbar along the edges of the graphics window.

Draw Toolbar

The *Draw commands* on the AutoCAD **Draw** toolbar are shown in Figure 4-41. Some of the commands located on this toolbar, such as **LINE**, **CIRCLE**, **ARC**, and **MULTILINE TEXT**, are used more frequently than others; however, all the commands on this toolbar are useful, and you should familiarize yourself with each of them. Video tutorials for the commands on the **Draw** toolbar are located on the Student Resource CD in the AutoCAD Tutorial Videos folder.

> **TIP**
> Press <**Esc**> to cancel an AutoCAD command. Pressing <**Enter**> will return you to the last AutoCAD command used.

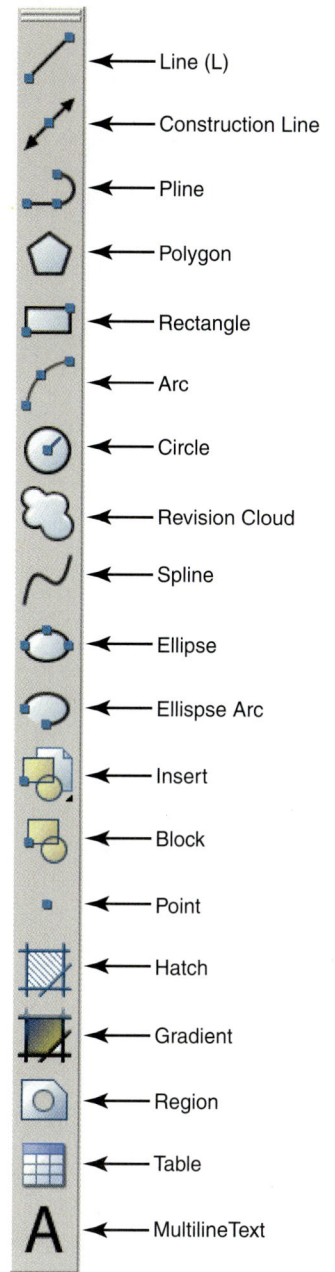

LINE Command.   The icon for the **LINE** command is shown in Figure 4-42(a). This command is used to draw lines in the graphics window. Lines can be drawn by using absolute coordinates (see Figure 4-7), relative coordinates (see Figures 4-8 and 4-9), or polar coordinates (see Figures 4-11 and 4-12), or by direct entry (see Figures 4-13[a] and 4-13[b]).

Figure 4-41   AutoCAD Draw toolbar

Draw commands: The commands used to place geometry in an AutoCAD drawing. These commands are located on the **Draw** toolbar and include **LINE**, **CIRCLE**, **ARC**, and **MULTILINE TEXT**.

Figure 4-42(a)   Line icon

LINE Command Tutorial

Step 1. Select the **LINE** icon from the **Draw** toolbar.

Step 2. When prompted to Specify start point, define the start point of the line by left-clicking to select a point in the graphics window, or by entering an absolute coordinate and pressing **<Enter>**.

Step 3. When prompted to Specify next point, define the next point of the line by left-clicking to select a point in the graphics window, or by entering an absolute, relative, or polar coordinate and pressing **<Enter>**. You can continue to define lines in this manner or end the command by pressing **<Esc>** or **<Enter>**. See Figure 4-42(b).

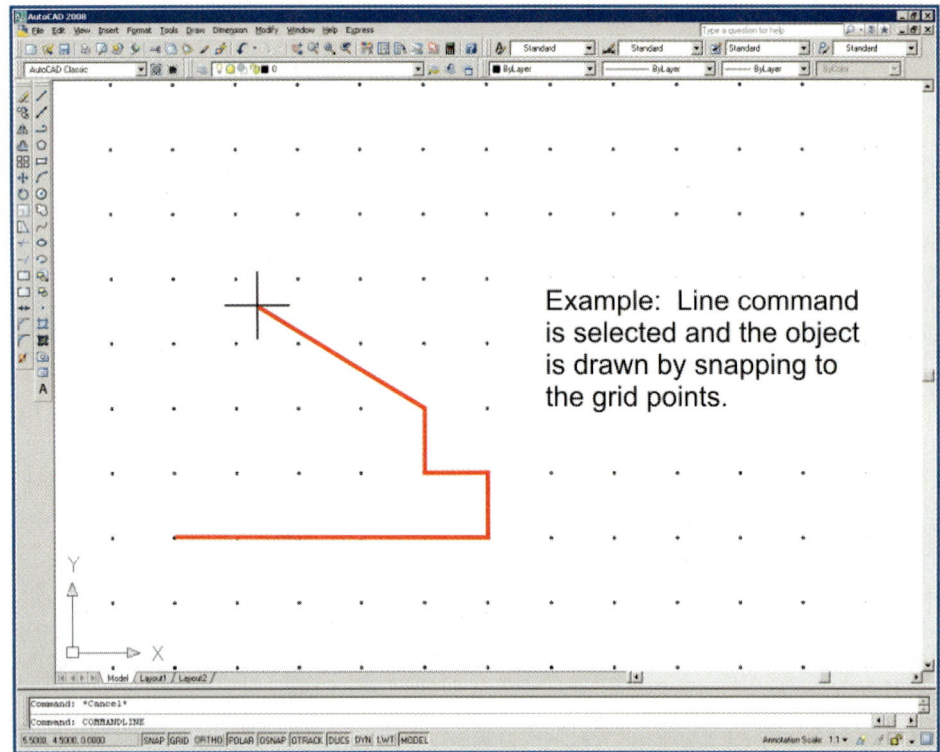

Example: Line command is selected and the object is drawn by snapping to the grid points.

Figure 4-42(b)   LINE command

> **TIP** You can also find the **LINE** command in the **Draw** pull-down menu. The command alias is **L**.

CONSTRUCTION LINE Command.   The icon for the **CONSTRUCTION LINE** command is shown in Figure 4-43(a). This command creates lines that extend to infinity which can be placed on the drawing to facilitate the construction of other objects.

Figure 4-43(a)   Construction Line icon

Construction Line

CONSTRUCTION LINE Command Tutorial

Step 1. Select the **CONSTRUCTION LINE** icon from the **Draw** toolbar.

Step 2. When prompted to Specify a point, type **H** and press **<Enter>** to place a horizontal construction line, type **V** and press **<Enter>** to place a vertical construction line, or type **A**

and press **<Enter>** and an angle value at the Enter angle of xline prompt to place a construction line at an angle. Press **<Enter>** after entering the value for an angle.

Step 3. At the Specify through point, select a point on the screen through which the construction line is to be drawn. You can continue to pick points for placement of other construction lines or end the command by pressing **<Esc>** or **<Enter>**. See Figure 4-43(b).

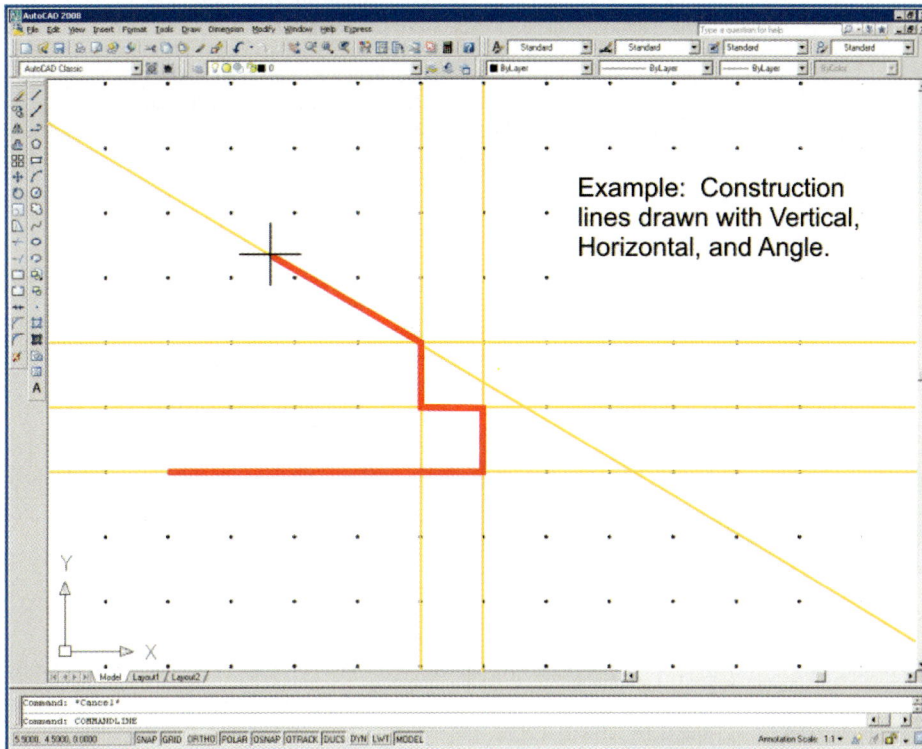

Figure 4-43(b)   CONSTRUCTION LINE command

> **TIP**
>
> You can also find the **CONSTRUCTION LINE** command in the **Draw** pull-down menu. The command alias is **XL**.

The PLINE Command.   The icon for the **POLYLINE** command is shown in Figure 4-44(a). This command creates continuous lines that may vary in width and shape.

POLYLINE Command Tutorial

Step 1. Select the **POLYLINE** icon from the **Draw** toolbar.

Step 2. When prompted to Specify start point, type absolute coordinate **1,8** and press **<Enter>**.

Step 3. When prompted to Specify next point or [Arc/Halfwidth/Undo/Width], move your mouse to the right (with **Polar Tracking** turned on), and type **4.5** and press **<Enter>**. See Figure 4-44(b).

Step 4. When prompted to Specify next point or Arc/Halfwidth/Undo/Width], type **A** for **Arc** and press **<Enter>**.

Step 5. When prompted to Specify endpoint of arc or [Angle/Center/Close/Direction/Halfwidth/Line/Radius/Undo/Width], type in **W** for **Width** and press **<Enter>**.

Step 6. When prompted to Specify starting width (0.0000), type **.06** and press **<Enter>**. When prompted to Specify ending width <0.0600>, press **<Enter>** to accept the default lineweight. Notice that the lineweight has changed.

Polyline

Figure 4-44(a)   Polyline icon

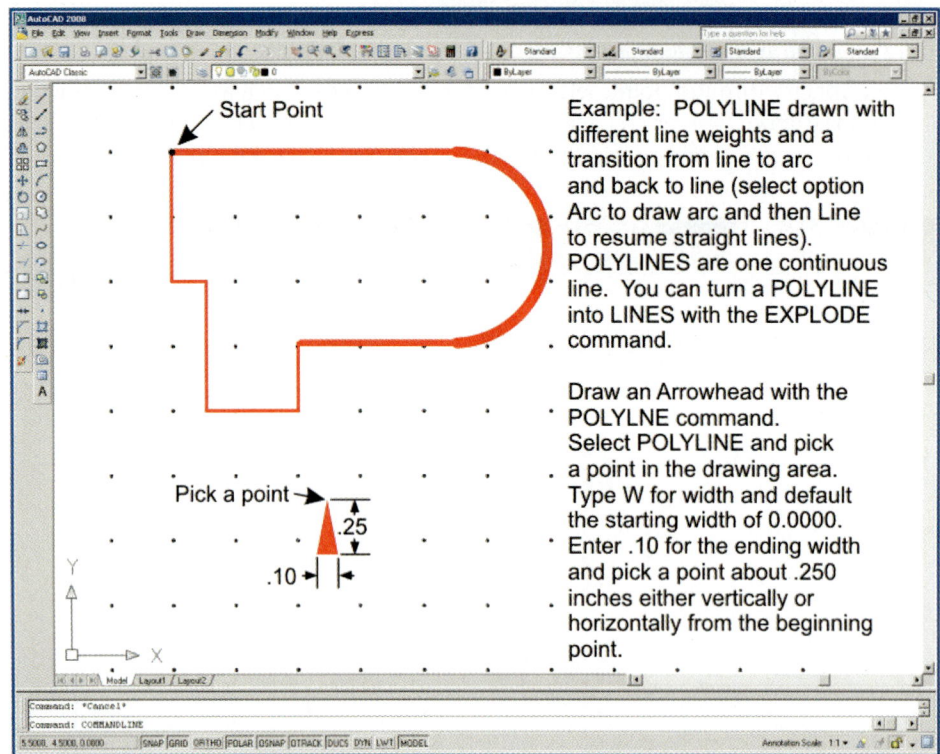

Figure 4-44(b)   POLYLINE command

Step 7. When prompted to Specify endpoint of arc or [Angle/Center/Close/Direction/ Halfwidth/Line/Radius/Undo/Width], ensure that **Polar Tracking** is tracking at **270 degrees** and type **2** for the distance and **<Enter>**.

Step 8. When again prompted to Specify endpoint of arc or [Angle/Center/Close/Direction/Halfwidth/Line/Radius/Undo/Width], type in **L** for **Line** and press **<Enter>**.

Step 9. When prompted to Specify next point or [Arc/Halfwidth/Undo/Width], type **W** for **Width** and press **<Enter>**.

Step 10. When prompted to Specify starting width (0.0600) type **0**. When prompted with Specify ending width (0.0000) and press **<Enter>**.

Step 11. Continue drawing polylines at either a 0 width or changing to different widths. Press **<Enter>** to end the command.

TIP You can also find the **POLYLINE** command in the **Draw** pull-down menu. The command alias is **PL**.

Figure 4-45(a)   Polygon icon

POLYGON Command.   The icon for the **POLYGON** command is shown in Figure 4-45(a). This command is used to create multisided shapes whose sides are of equal length.

POLYGON Command Tutorial

Step 1. Select the **POLYGON** icon from the **Draw** toolbar.

Step 2. When prompted to Enter number of sides, enter a value and press **<Enter>**.

Step 3. At the Specify center of polygon prompt, select a point on the screen by left-clicking or typing a coordinate. See Figure 4-45(b).

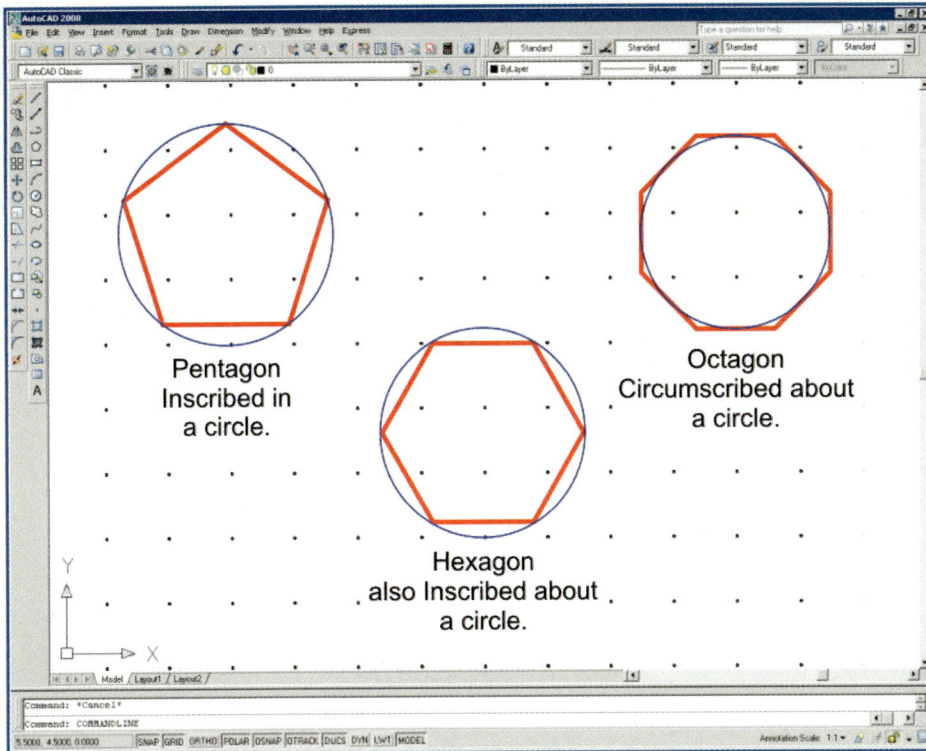

Figure 4-45(b)   POLYGON command

Step 4. When prompted to Enter an option [Inscribed in circle/Circumscribed about circle], type either an **I** for inscribed or a **C** for circumscribed, and press **<Enter>**.

Step 5. When prompted to Specify radius of circle, enter a value for the radius of the circle in which the polygon will be inscribed inside or circumscribed about, and press **<Enter>**.

> **TIP**
>
> You can also find the **POLYGON** command in the **Draw** pull-down menu. The command alias is **POL**.

RECTANGLE Command.   The icon for the **RECTANGLE** command is shown in Figure 4-46(a). This command is used to create continuous line rectangles.

RECTANGLE Command Tutorial

Step 1. Select the **RECTANGLE** icon from the **Draw** toolbar.

Step 2. When prompted to Specify first corner point or [Chamfer/Elevation/Fillet/Thickness/ Width], type **1,2** and press **<Enter>**. See Figure 4-46(b).

Step 3. When prompted to Specify other corner point or [Area/Dimensions/Rotation], type **6,2** (with **Dynamic** on), or **@6,2** and press **<Enter>**.

Rectangle

Figure 4-46(a)   Rectangle icon

> **TIP**
>
> You can also find the **RECTANGLE** command in the **Draw** pull-down menu. The command alias is **REC**.

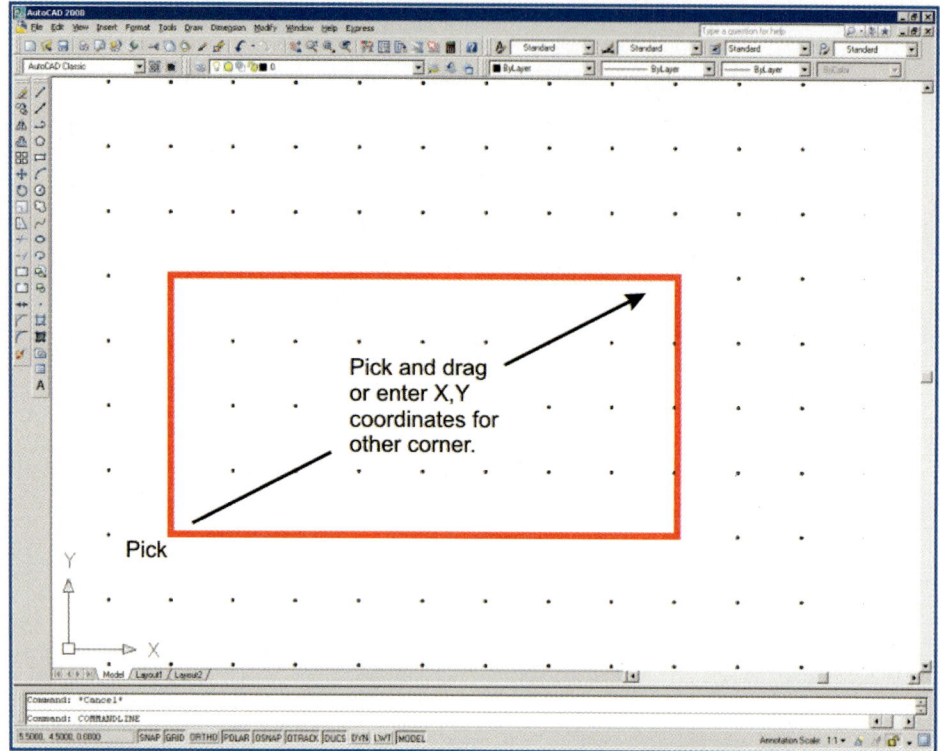

Figure 4-46(b) REC-TANGLE command

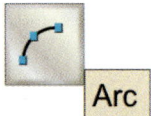

Figure 4-47(a) Arc icon

ARC Command. The icon for the **ARC** command is shown in Figure 4-47(a). This command can create arcs using 11 different methods. Note: Arcs are drawn counterclockwise.

ARC Command Tutorial (Start, End, Radius Method)

Step 1. Select the **Draw** pull-down menu, highlight **Arc**, and pick **Start**, **End**, **Radius**.

Step 2. When prompted to Specify start point of arc or [Center], type **4,5** and press **<Enter>**.

Step 3. When prompted to Specify end point of arc, type **8,5** and press **<Enter>**.

Step 4. When prompted to Specify radius of arc, type **3** and press **<Enter>**. See Figure 4-47(b).

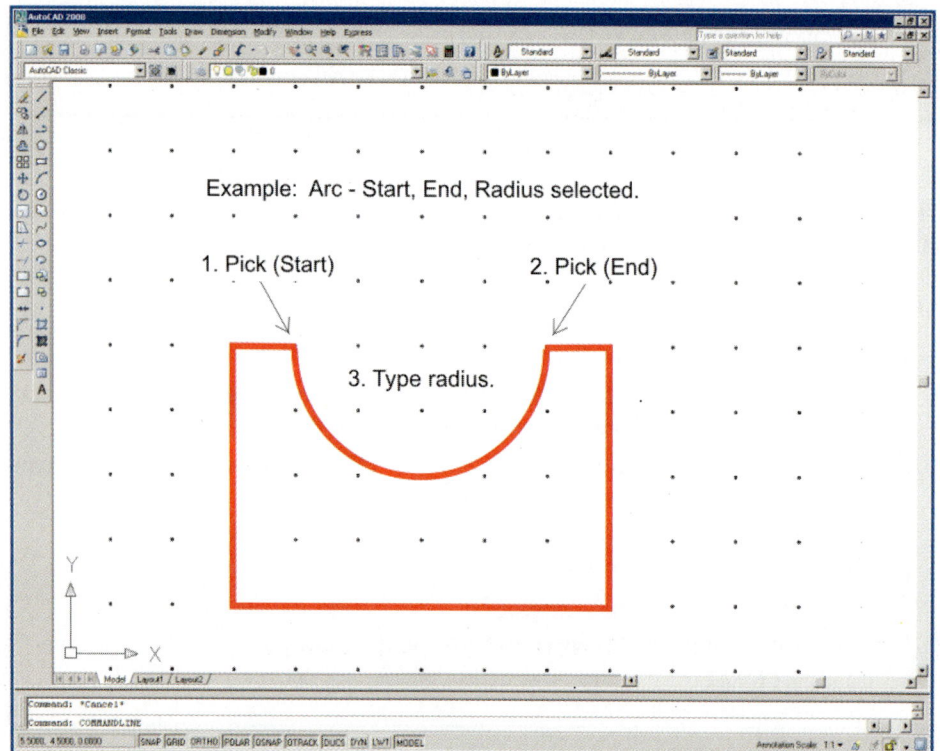

Figure 4-47(b) ARC command

Step 5. Try drawing another arc by typing in **8,5** for the **Start** and **4,5** for the **End** with a radius of **3**.

> You can also find the **ARC** command in the **Draw** pull-down menu. The command alias is **A**.
>
> **TIP**

CIRCLE Command.   The icon for the **CIRCLE** command is shown in Figure 4-48(a). This command is used to draw a circle after prompting you to select (or enter) the center point and the radius or diameter.

Figure 4-48(a)   Circle icon

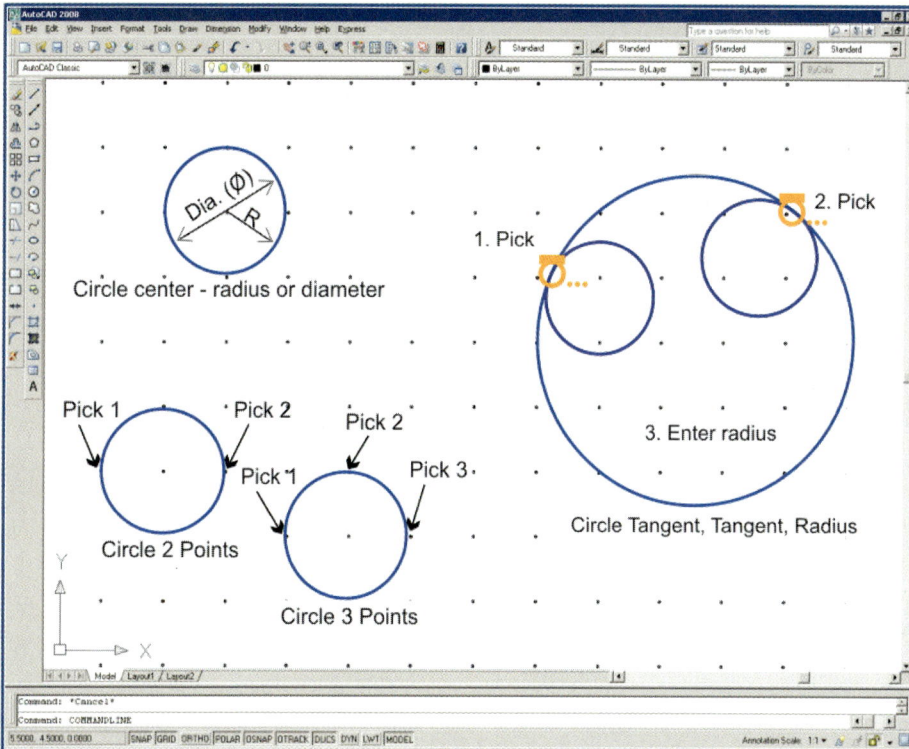

CIRCLE Command Tutorial

Step 1. Select the **CIRCLE** icon from the **Draw** toolbar.

Step 2. When prompted to Specify center point of circle or [3P/2P/Ttr (tan tan radius)], type **2,7** and press **<Enter>**.

Step 3. When prompted to Specify radius of circle or [Diameter], type **1** and press **<Enter>**. See Figure 4-48(b).

Figure 4-48(b)   CIRCLE command

Step 4. Try it again except type **6,7** for the center and press **<Enter>**. When prompted to Specify radius of circle or [Diameter], type **D** for **Diameter**.

Step 5. When prompted to Specify diameter of circle <2.0000>, type **1** and press **<Enter>**.

> You can also find the **CIRCLE** command in the **Draw** pull-down menu. The command alias is **C**.
>
> **TIP**

Figure 4-49(a) Revcloud
icon

REVISION CLOUD Command. The icon for the **REVISION CLOUD** command is shown in Figure 4-49(a). This command creates a cloud shape made of polyline arcs. Revision clouds are placed on drawings to draw attention to an area of the drawing.

REVISION CLOUD Command Tutorial

Step 1. Select the **REVISION CLOUD** icon from the **Draw** toolbar.

Step 2. When prompted to Specify start point or [Arc length/Object/Style], pick a point on the screen and draw the **REVISION CLOUD** either clockwise or counterclockwise. Selecting a point close to the beginning point will close the cloud. See Figure 4-49(b).

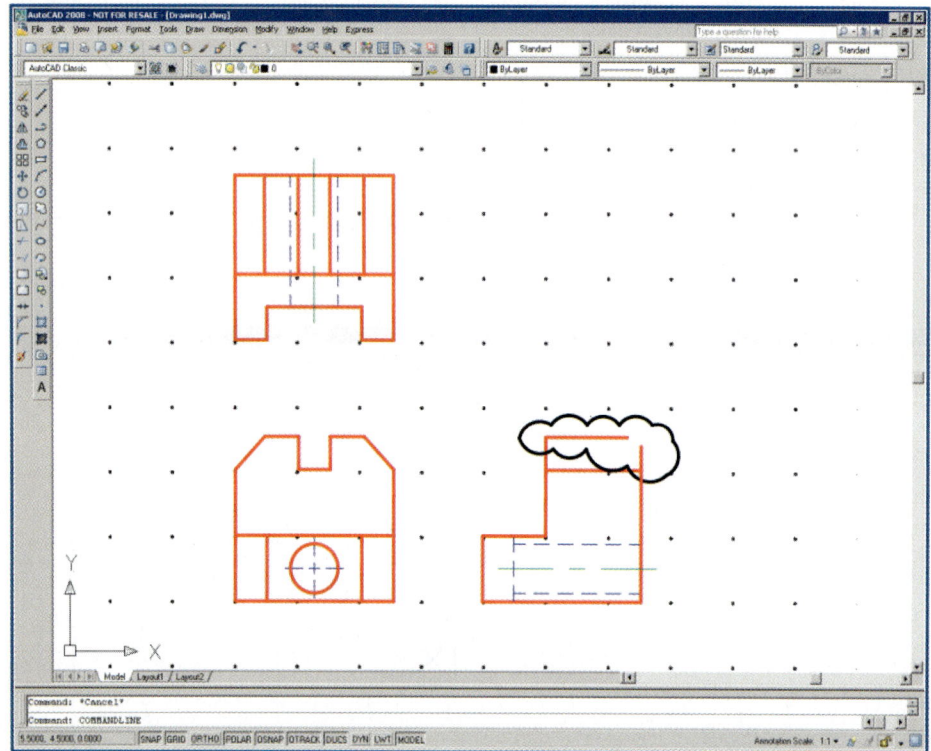

Figure 4-49(b) REVCLOUD command

> You can also find the **REVISION CLOUD** command in the **Draw** pull-down menu.
> **TIP**

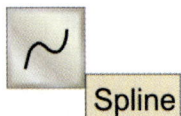

Figure 4-50(a) Spline
icon

SPLINE Command. The icon for the **SPLINE** command is shown in Figure 4-50(a). This command creates a nonuniform spline curve.

SPLINE Command Tutorial

Step 1. Select the **SPLINE** icon from the **Draw** toolbar.

Step 2. When prompted to Specify first point or [Object], type **2,2** and press **<Enter>**.

Step 3. When prompted to Specify next point, type **4,5** and press **<Enter>**.

Step 4. When prompted to Specify next point, type **6,2** and press **<Enter>**.

Step 5. When prompted to Specify next point, type **8,5** and press **<Enter>**.

Step 6. When prompted to Specify next point, type **10,2** and press **<Enter>**.

Step 7. When prompted to Specify next point, press <**Enter**>.

Step 8. When prompted to Specify start tangent, press <**Enter**>.

Step 9. When prompted to Specify end tangent, press <**Enter**> to finish the **SPLINE** curve. See Figure 4-50(b). Note: The user must press <**Enter**> three times to complete the **SPLINE** command.

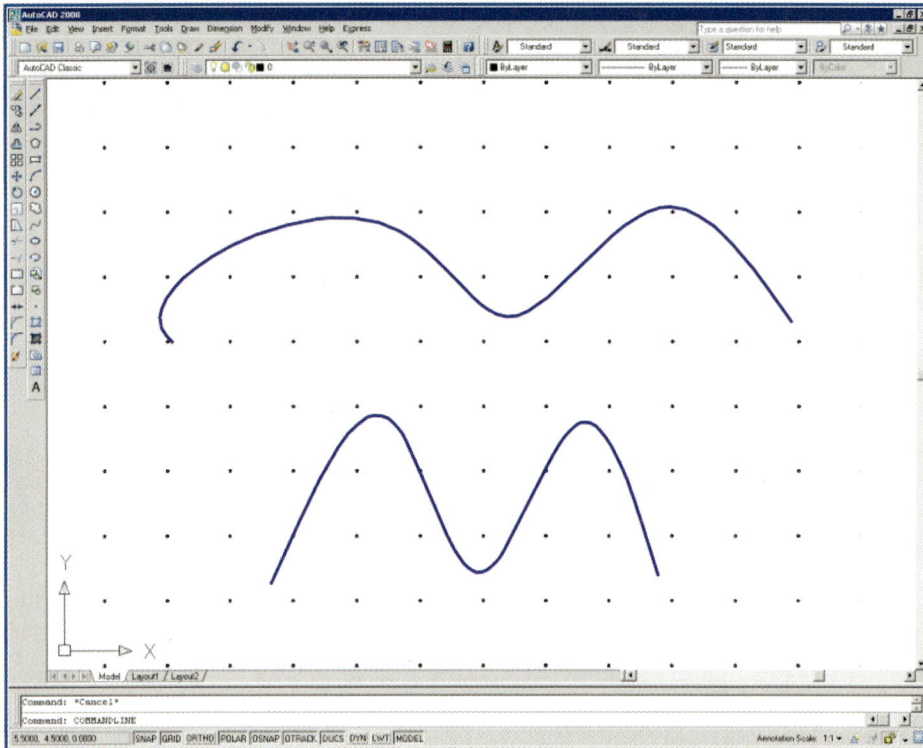

Figure 4-50(b)   SPLINE command

> You can also find the **SPLINE** command in the **Draw** pull-down menu. The command alias is **SPL**.

ELLIPSE Command.   The icon for the **ELLIPSE** command is shown in Figure 4-51(a). This command is used to create an ellipse. An ellipse is defined by two axes: a long axis that defines its length (called the major axis) and a shorter axis that defines its width (called the minor axis).

Figure 4-51(a)   Ellipse icon

ELLIPSE Command Tutorial

Step 1. Select the **ELLIPSE** icon from the **Draw** toolbar.

Step 2. At the Specify axis endpoint of ellipse prompt, pick the start point of the major axis of the ellipse. See Figure 4-51(b).

Step 3. At the Specify other endpoint of axis prompt, pick the location of the endpoint of the major axis, or define the endpoint with a coordinate, and press <**Enter**>.

Step 4. At the Specify distance to other axis prompt, drag the mouse in the direction desired for the minor axis and enter a distance for half the length of the minor axis and press <**Enter**>.

> You can also find the **ELLIPSE** command in the **Draw** pull-down menu. The command alias is **EL**.

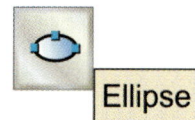

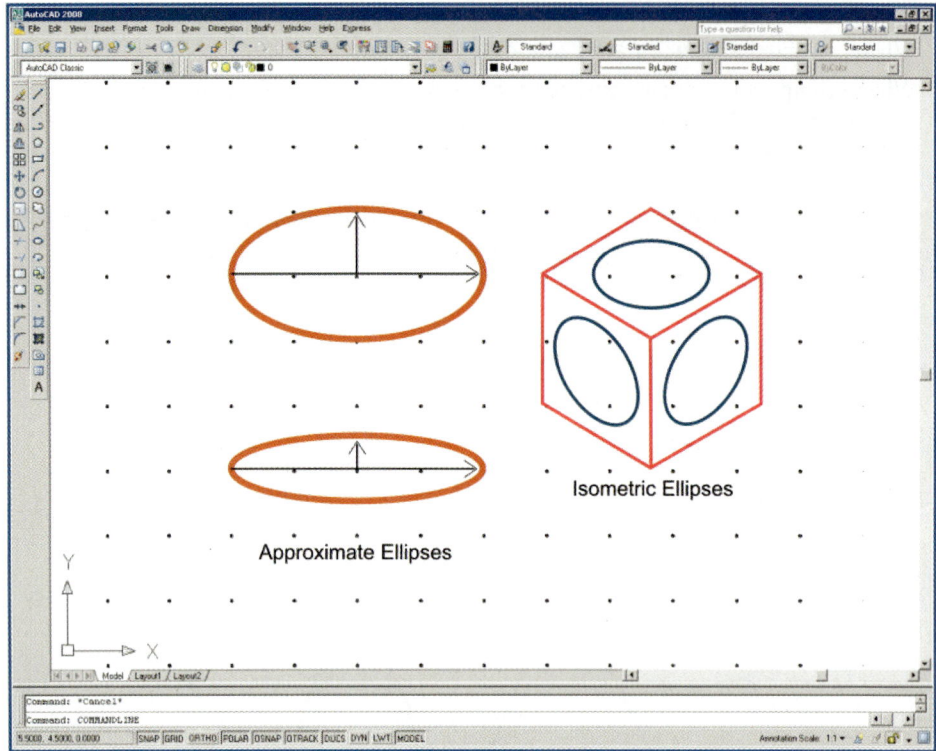

Figure 4-51(b)   ELLIPSE
command

Figure 4-52(a)   Ellipse
Arc icon

ELLIPSE ARC Command.   The icon for the **ELLIPSE ARC** command is shown in Figure 4-52(a). This command is used to draw an elliptical arc. An elliptical arc is defined by the length of its major axis and the endpoints of the break in the ellipse.

ELLIPSE ARC Command Tutorial

Step 1.   Select the **ELLIPSE ARC** icon from the **Draw** toolbar.

Step 2.   At the Specify axis endpoint of elliptical arc prompt, pick the start point of the major axis of the ellipse. See Figure 4-52(b).

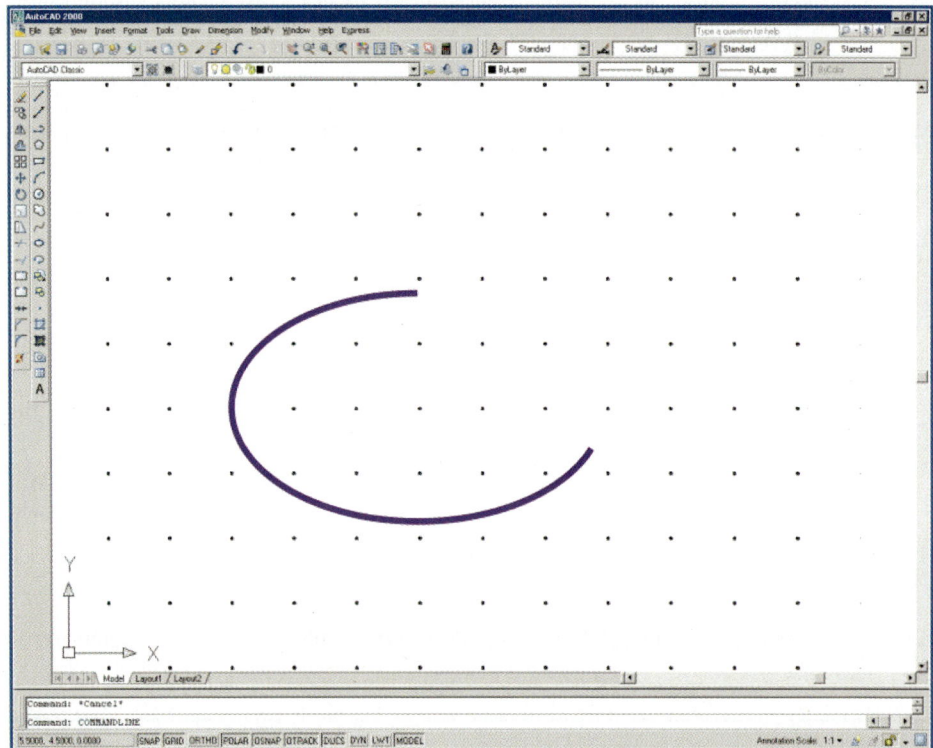

Figure 4-52(b)   ELLIPSE
ARC command

Step 3. At the Specify other endpoint of axis prompt, pick the location of the endpoint of the major axis, or define the endpoint with a coordinate, and press **<Enter>**.

Step 4. At the Specify distance to other axis prompt, drag the mouse in the direction desired for the minor axis and enter a distance equal to half the length of the minor axis and press **<Enter>**.

Step 5. At the Specify start angle prompt, enter an angle relative to the point defined in Step 4 where the elliptical arc should begin.

Step 6. At the Specify end angle prompt, enter an angle relative to the point defined in Step 5 where the elliptical arc should end.

> **TIP** Consulting the angle noted in the tooltip window next to the cursor is helpful when defining the start and end angles for an elliptical arc.

INSERT BLOCK Command.   The icon for the **INSERT BLOCK** command is shown in Figure 4-53. This command is used to place a named block, or a drawing, into the current drawing.

INSERT BLOCK Tutorial.   The steps involved in using the **INSERT BLOCK** command are presented in Unit 10 of this text.

Insert Block

Figure 4-53   Insert Block icon

> **TIP** You can also find the **INSERT BLOCK** command in the **Draw** pull-down menu. The command alias is **I**.

BLOCK Command.   The icon for the **BLOCK** command is shown in Figure 4-54. A block is a named image defined within a drawing file that can be inserted into the drawing whenever it is needed. An example of this would be a symbol for a door that is used multiple times in the creation of a floorplan.

BLOCK Command Tutorial.   The steps involved in using the **BLOCK** command are presented in Unit 10 of this text.

Make Block

Figure 4-54   Block icon

> **TIP** You can also find the **BLOCK** command in the **Draw** pull-down menu. The command alias is **B**.

POINT Command.   The icon for the **POINT** command is shown in Figure 4-55(a). This command places a single point on a drawing. The point style is defined by selecting the **Format** pull-down menu and selecting **Point Style**. When the **Point Style** dialog box opens, select the tile containing the desired style and click **OK**.

Point

Figure 4-55(a)   Point icon

POINT Command Tutorial

Step 1. Select the **POINT** icon from the **Draw** toolbar.

Step 2. When prompted to Specify a point, type **4.5,4.5** and press **<Enter>**. A point (a dot) will be placed at the coordinate point defined.

Step 3. To change the point style, select the **Format** pull-down menu and select **Point Style**. See Figure 4-55(b).

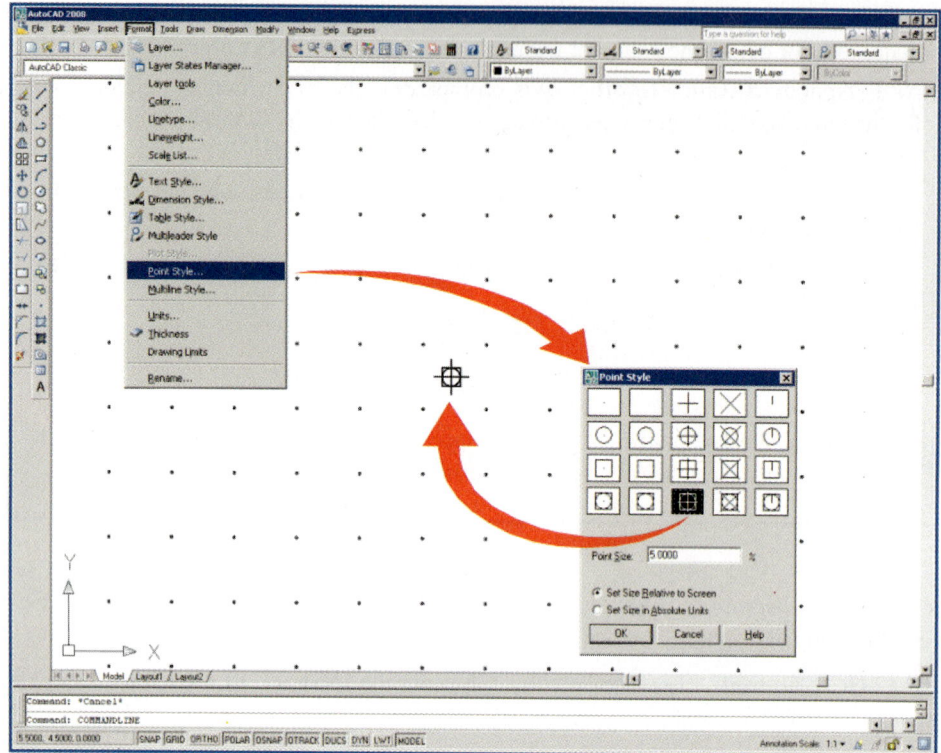

Figure 4-55(b) POINT
command

Step 4. When the **Point Style** dialog box opens, pick the tile whose symbol looks like a box with a center mark inside of it and click **OK**. The original point will change to the new point style definition.

Step 5. Select **Format**, **Point Style** and practice placing another point with a different style.

> **TIP** You can also find the **POINT** command in the **Draw** pull-down menu. The command alias is **PO**.

HATCH Command. The icon for the **HATCH** command is shown in Figure 4-56. This command fills an enclosed area, or selected objects, with a hatch pattern, a solid fill, or a gradient fill.

Figure 4-56 Hatch icon

HATCH Command Tutorial. The steps involved in using the **HATCH** command are presented in Unit 8 of this text.

> **TIP** You can also find the **HATCH** command in the **Draw** pull-down menu. The command alias is **H**.

GRADIENT Command. The icon for the **GRADIENT** command is shown in Figure 4-57(a). This command specifies a fill that allows a smooth transition from a darker color to a lighter one.

GRADIENT Command Tutorial

Step 1. Draw a **RECTANGLE**.

Figure 4-57(a) Gradient
icon

Step 2. Select the **GRADIENT** icon from the **Draw** toolbar.

Step 3. Choose any of the **GRADIENT** styles shown in the thumbnails in the dialog box. See Figure 4-57(b).

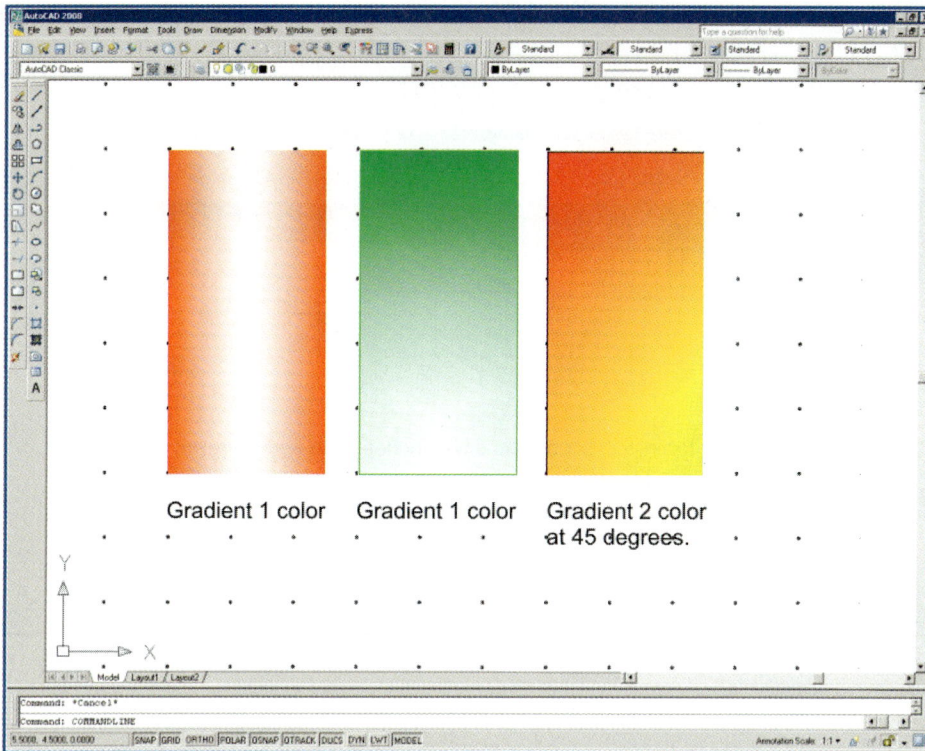

Gradient 1 color   Gradient 1 color   Gradient 2 color at 45 degrees.

Figure 4-57(b)   GRADIENT command

Step 4. Select the **Add: Pick points** button in the dialog box.

Step 5. When prompted to Pick internal point or [Select objects/remove Boundaries], pick a point inside the **RECTANGLE** and press **<Enter>**.

Step 6. Click **OK**.

> **TIP**
> You can also find the **GRADIENT** command in the **Draw** pull-down menu. The command alias is **GD**.

REGION Command.   The icon for the **REGION** command is shown in Figure 4-58. Regions are two-dimensional closed shapes or loops. Regions are often created with closed lines or polylines.

REGION Command Tutorial.   The steps involved in using the **REGION** command are presented in Unit 13 of this text.

Figure 4-58   Region icon

> **TIP**
> You can also find the **REGION** command in the **Draw** pull-down menu. The command alias is **REG**.

TABLE Command.   The icon for the **TABLE** command is shown in Figure 4-59(a). A table is an object that displays data in rows and columns.

Figure 4-59(a)   Table icon

Figure 4-60 Multiline Text icon

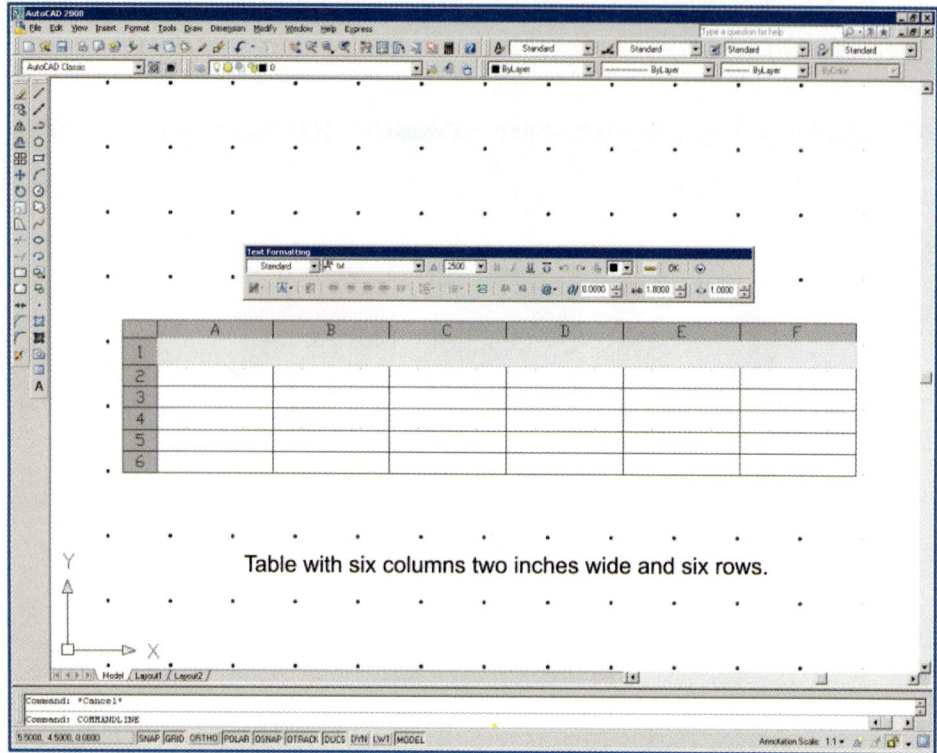

Figure 4-59(b) TABLE command

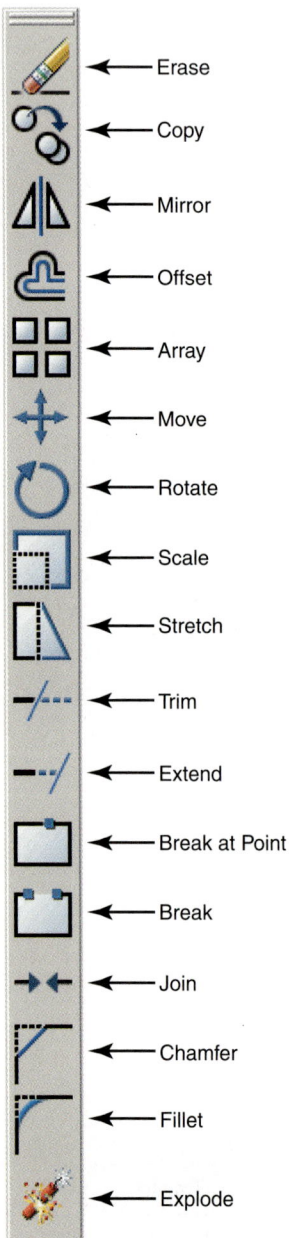

Figure 4-61 Modify toolbar

TABLE Command Tutorial

Step 1. Select the **TABLE** icon from the **Draw** toolbar.

Step 2. When the **Insert Table** dialog box opens, enter the number of columns and rows to be created, as well as their respective widths and height, in the **Column & Row Settings** area. Click **OK** when these values have been entered.

Step 3. At the Specify insertion point prompt, select a point in the graphics window to insert the table. When the **Text Formatting** dialog box opens, enter the content for the top cell. Click **OK** when the content has been entered. Content for other cells can be added or edited by double-clicking inside a cell with the left-click button the mouse and using the **Text Formatting** options. Click **OK** when the content has been entered. See Figure 4-59(b).

> **TIP** You can also find the **TABLE** command in the **Draw** pull-down menu. The command alias is **TB**.

MTEXT Command. The icon for the **MTEXT** command is shown in Figure 4-60. This command is used to place one or more paragraphs of multiline text into the field of a drawing. Saved text from other formats can also be inserted into a drawing using this command.

MTEXT Command Tutorial. The procedures involved in placing text with the **MTEXT** command are described later in this chapter.

> **TIP** You can also find the **MTEXT** command in the **Draw** pull-down menu. The command alias is **T**.

MODIFY Toolbar

The *modify commands* on the **Modify** toolbar are shown in Figure 4-61. Although you may find yourself using a few of the commands on this toolbar, such as **MOVE**, **COPY**, **TRIM**, and **OFFSET**, much more frequently than some of the others, *all* of the commands on this toolbar are useful, and you should familiarize yourself with each of them. Video tutorials for the commands on the **Modify** toolbar are located on the Student Resource CD in the AutoCAD Tutorial Videos folder.

There's a saying among CAD drafters, "Never draw anything twice." What they mean by this is that drafters should use commands like **COPY**, **MOVE**, and **ROTATE** to create technical drawings more quickly and efficiently. The **MODIFY** toolbar has many tools that speed up the drafting process.

ERASE Command.   The icon for the **ERASE** command is shown in Figure 4-62(a). This command is used to remove objects from a drawing.

ERASE Command Tutorial

Step 1. Select the **ERASE** icon from the **Modify** toolbar.

Step 2. When prompted to Select objects, you can select them by left clicking and selecting in the following ways: **Window** (picking left to right), **Crossing Window** (picking right to left), or by typing **Crossing Polygon** (type **CP**), and **Fence** (type **F**). When you are finished selecting the objects to erase, press <**Enter**> to complete the command. See Figure 4-62(b).

> **TIP** The **OOPS** command can be used to restore erased objects (just type **OOPS** on the command line and press <**Enter**>).

> **TIP** Objects can also be removed from a drawing by selecting them and pressing the **Delete** key. You can also find the **ERASE** command in the **Modify** pull-down menu. The command alias is **E**.

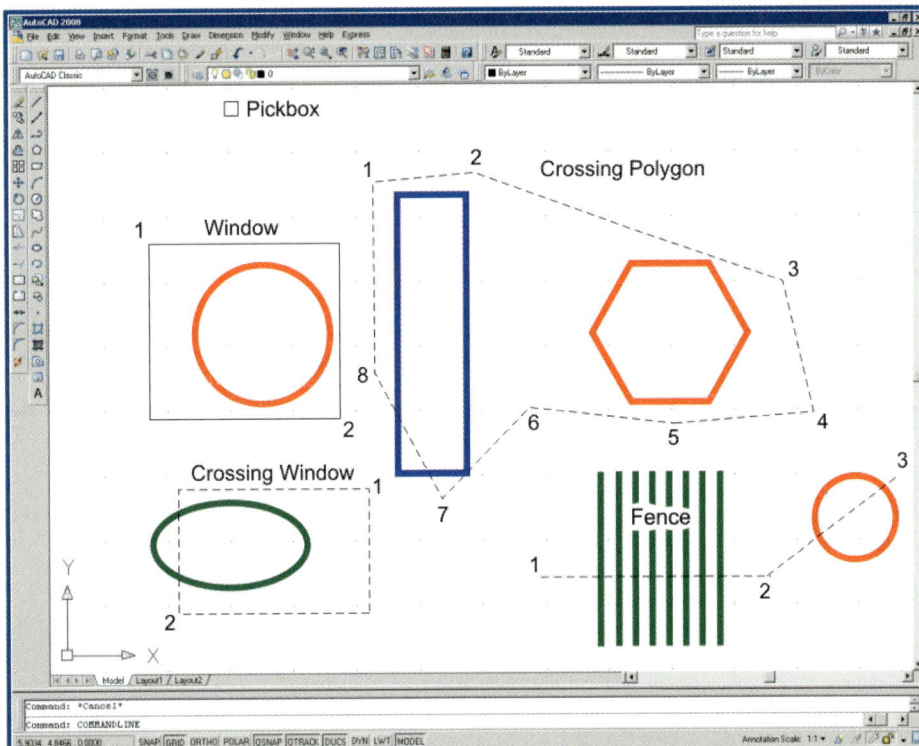

Modify commands: The commands used to modify the geometry of an AutoCAD drawing. These commands are located on the **Modify** toolbar and include **ERASE, MOVE, COPY, OFFSET, ROTATE**, and **SCALE**.

Figure 4-62(a)   Erase icon

Figure 4-62(b)   ERASE command

Figure 4-63(a) Copy icon

COPY Command. The icon for the **COPY** command is shown in Figure 4-63(a). This command is used to create a copy of a selected object or objects.

COPY Command Tutorial

Step 1. Select the **COPY** icon from the **Modify** toolbar.

Step 2. When prompted to Select objects, pick the objects you would like to copy by selecting them either individually, or with a window, and press **<Enter>**. See Figure 4-63(b).

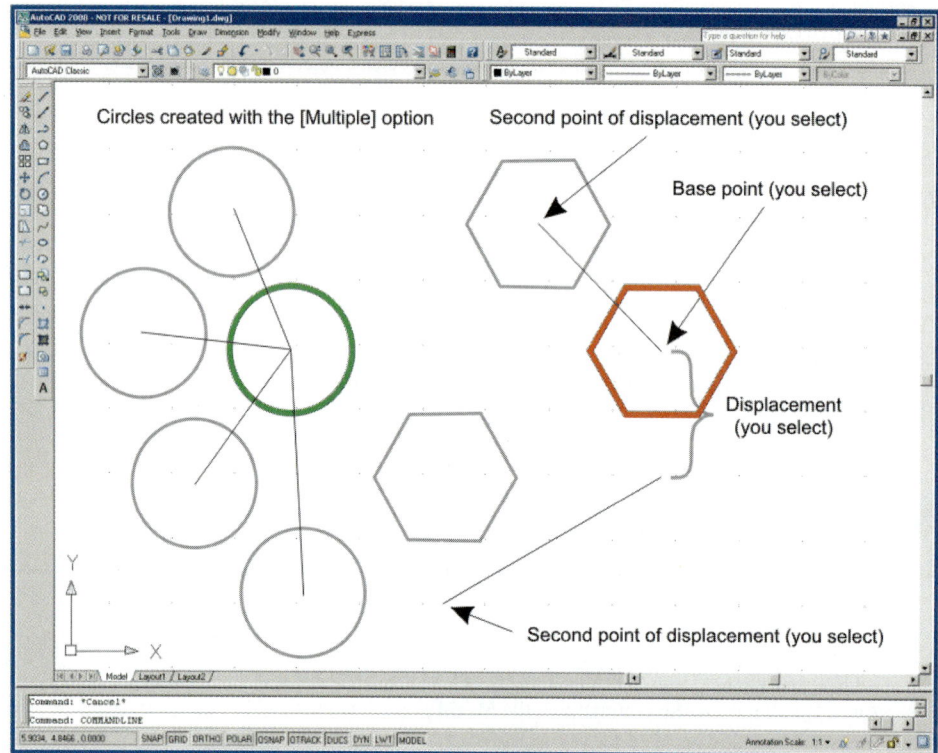

Figure 4-63(b) COPY command

Step 3. When prompted to Select the Base point, pick a point on the object. The base point can be thought of as a handle on the object to be copied.

Step 4. When prompted to define the Displacement, pick a point at a specific distance from the original object or by entering an absolute, relative, or polar coordinate. You can continue to place copies by picking more points. Press **<Esc>** or **<Enter>** to end the command.

> **TIP** You can also find the **COPY** command in the **Modify** pull-down menu. The command alias is **CO**.

MIRROR Command. The icon for the **MIRROR** command is shown in Figure 4-64(a). This command creates a mirror image of an object around an axis called a *mirror line*.

MIRROR Command Tutorial

Step 1. Select the **MIRROR** icon from the **Modify** toolbar.

Figure 4-64(a) Mirror icon

Step 2. At the select objects prompt, pick the objects you would like to mirror by selecting them either individually, or with a window, and press **<Enter>**. See Figure 4-64(b).

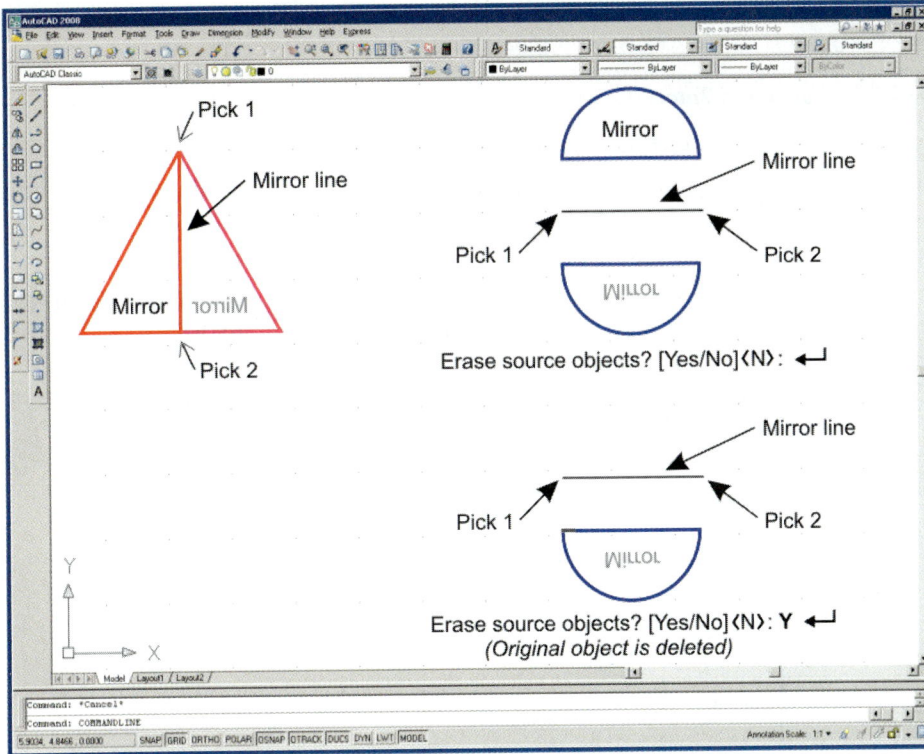

Figure 4-64(b) MIRROR command

Step 3. When prompted to Specify first point of mirror line, define the first point of the mirror axis by selecting a point in the graphics window where you want the mirror axis to begin.

Step 4. When you are prompted to Select the second point of the mirror line, select a second point in the graphics window defining the other end of the mirror axis.

> **TIP**
> Set **Ortho** to **On** when defining a vertical, or horizontal, mirror line.

Step 5. When prompted to Erase source objects? [Y/N], press **<Enter>** to retain the source object (the prompt's default is **No**), or type a **Y** and press **<Enter>** to erase the object being mirrored.

> **TIP**
> To prevent text from being mirrored, type **MIRRTEXT** on the command line, press **<Enter>**, and set the value to **0** and press **<Enter>**.

> **TIP**
> You can also find the **MIRROR** command in the **Modify** pull-down menu. The command alias is **MI**. **MIRROR** is also a **GRIPS** option.

Figure 4-65(a) Offset icon

OFFSET Command. The icon for the **OFFSET** command is shown in Figure 4-65(a). The offset command creates a new object whose shape parallels the shape of the selected object. Offsetting a circle or an arc creates a larger or smaller circle or arc (depending on which side you specify for the offset) that is concentric to the original circle or arc.

OFFSET Command Tutorial

Step 1. Select the **OFFSET** icon from the **Modify** toolbar.

Step 2. When prompted to Select the offset distance, type in the value of desired offset distance and press **<Enter>**.

Step 3. When prompted to Select object to offset, select the object by left-clicking.

Step 4. At the Specify point on side to offset prompt, pick a point on the side of the object where you want the new object to be created. See Figure 4-65(b).

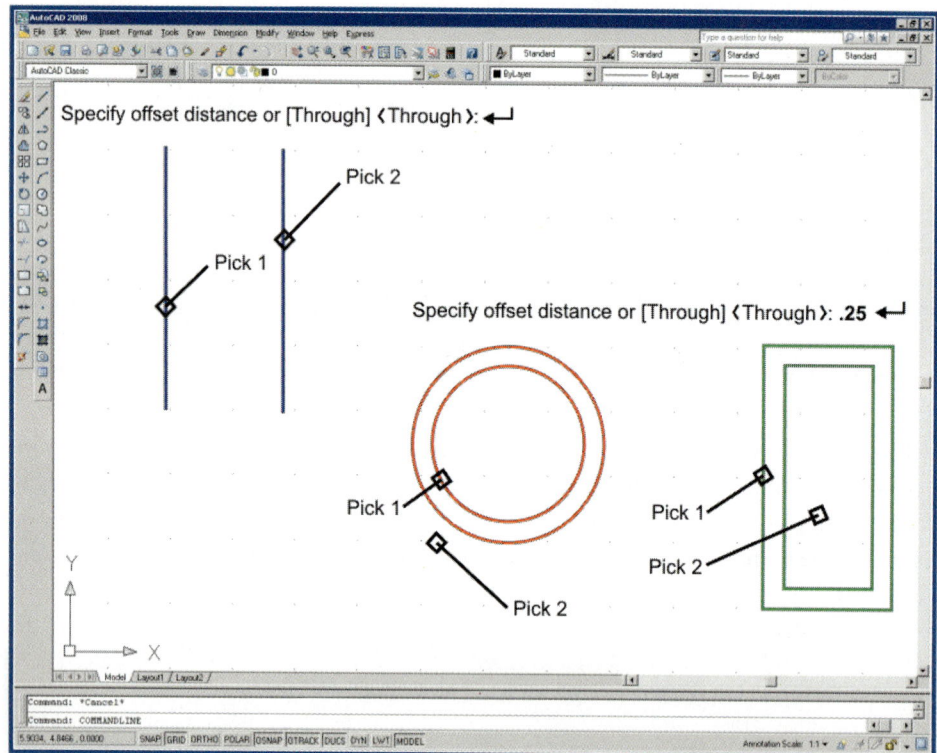

Figure 4-65(b)   OFFSET command

> **TIP** You can also find the **OFFSET** command in the **Modify** pull-down menu. The command alias is **O**.

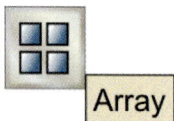

ARRAY Command.   The icon for the **ARRAY** command is shown in Figure 4-66(a). This command creates multiple copies of objects in either a rectangular or a circular pattern. A circular array is referred to as a *Polar Array*.

Figure 4-66(a)   Array icon

ARRAY Command Tutorial (Rectangular Option)

Step 1. Select the **ARRAY** icon from the **Modify** toolbar.

Step 2. From the **Array** dialog box, select **Rectangular Array**.

Step 3. Select the **Select Objects** button and pick the objects to be arrayed. Press **<Enter>** when the objects to be arrayed have been selected.

Step 4. Enter the number of **Rows** (which run horizontally) and **Columns** (which run vertically).

Step 5. Enter the **Offset distance** between the **rows** and **columns** (**Note:** negative values can be entered) and the **Angle of array** (the default angle is **0**).

Step 6. Click **OK**. See Figure 4-66(b).

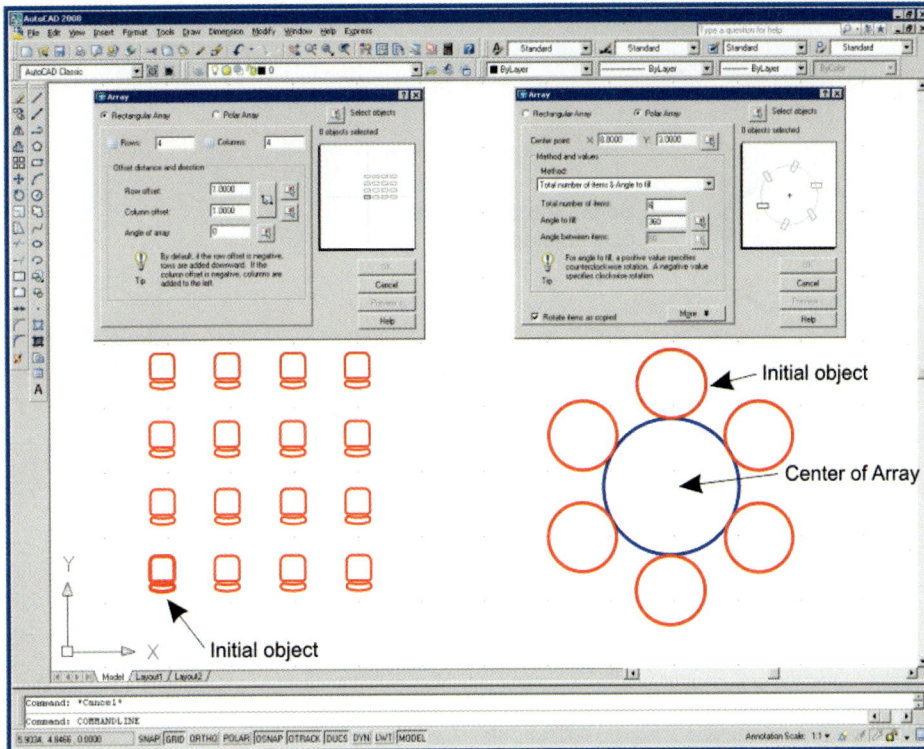

Figure 4-66(b) ARRAY command

ARRAY Command Tutorial (Polar Option)

Step 1. Select the **ARRAY** icon from the **Modify** toolbar.

Step 2. From the **Array** dialog box, select **Polar Array**.

Step 3. Select the **Select Objects** button and pick the objects to be arrayed. Press **<Enter>** when the objects to be arrayed have been selected.

Step 4. Define the **Center Point** for the array by entering *X* and *Y* coordinates, or by left-clicking a point in the graphics window.

Step 5. Enter the **Total number of items** to be arrayed and the **Angle to fill** (to array in a full circle, enter **360**). Note: Items are arrayed in a counterclockwise direction.

Step 6. Check the box next to **Rotate objects as copied** if this is desired.

Step 7. Click **OK**. See Figure 4-66(b).

> **TIP** You can also find the **ARRAY** command in the **Modify** pull-down menu. The command alias is **AR**.

MOVE Command.   The icon for the **MOVE** command is shown in Figure 4-67(a). This command is used to move existing objects to a new location.

Figure 4-67(a) Move icon

MOVE Command Tutorial

Step 1. Select the **MOVE** icon from the **Modify** toolbar.

Step 2. When prompted to Select objects, pick the objects you would like to move by selecting them either individually or with a window, and press **<Enter>**.

Step 3. When prompted to Select the Base point, pick a point on the object. The base point is like a handle on the object to be moved.

Step 4. When prompted to define the Displacement, pick a point at a specific distance from the original object or enter an absolute, relative, or polar coordinate. See Figure 4-67(b).

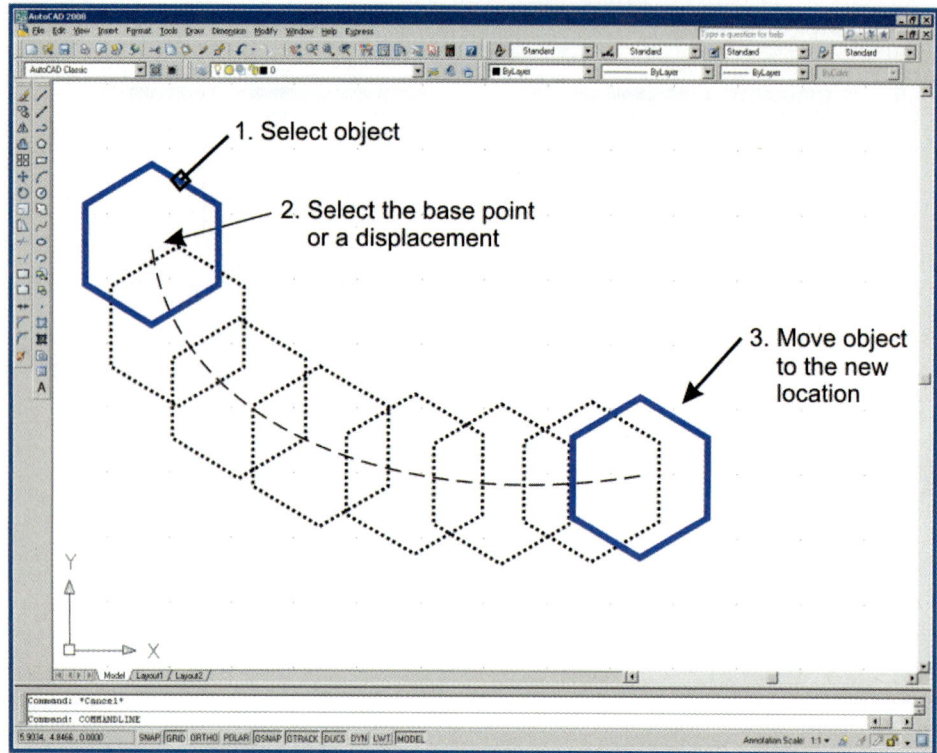

Figure 4-67(b)   MOVE command

> You can also find the **MOVE** command in the **Modify** pull-down menu. The command alias is **M**. **MOVE** is also a **GRIPS** option.
>
> TIP

Rotate

Figure 4-68(a)   Rotate icon

ROTATE Command.   The icon for the **ROTATE** command is shown in Figure 4-68(a). This command is used to rotate objects about a base point. The rotation angle is based on the initial position of the object to be rotated and the selected base point. With **Ortho** set to **On**, rotations are limited to 90° angles only.

ROTATE Command Tutorial

Step 1.   Select the **ROTATE** icon from the **Modify** toolbar.

Step 2.   When prompted to Select objects, pick the objects you would like to rotate by selecting them either individually or with a window, and press **<Enter>**.

Step 3.   When prompted to Select the Base point, select a point on the object. The base point is the pivot point around which the rotation will occur.

Step 4.   When prompted to Specify the rotation angle, enter a value and press **<Enter>**. See Figure 4-68(b).

Note:
Objects are rotated counter-clockwise.

> You can also find the **ROTATE** command in the **Modify** pull-down menu. The command alias is **RO**. **ROTATE** is also a **GRIPS** option.
>
> TIP

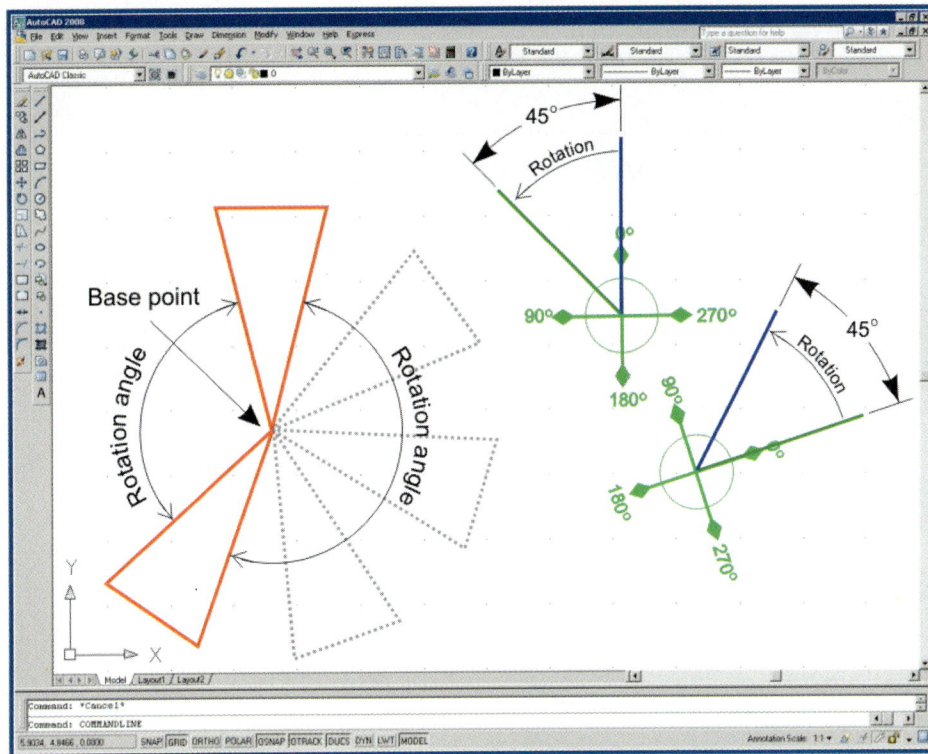

Figure 4-68(b) ROTATE command

SCALE Command. The icon for the **SCALE** command is shown in Figure 4-69(a). This command is used to change the size of objects.

You can scale up or down by either dragging or entering a scale factor. Scale factors are as follows: .5 = half size, 2 = twice the size, .75 = 3/4 size, etc.

Figure 4-69(a) Scale icon

SCALE Command Tutorial

Step 1. Select the **SCALE** icon from the **Modify** toolbar.

Step 2. When prompted to Select objects, pick the objects you would like to scale by selecting them either individually or with a window, and press **<Enter>**.

Step 3. When prompted to Select the Base point, pick a point on the object. The objects that are scaled will expand or contract relative to this point.

Step 4. When prompted to Specify scale factor, enter a value (for example: .5 = half size, 2 = twice the size, .75 = 3/4 size, etc.) and press **<Enter>**. See Figure 4-69(b).

> **TIP** You can also find the **SCALE** command in the **Modify** pull-down menu. The command alias is **SC**. **SCALE** is also a **GRIPS** option.

STRETCH Command. The icon for the **STRETCH** command is shown in Figure 4-70(a). This command can be used to lengthen or shorten objects. It can also distort objects. To stretch an object, you can either drag the base point to a new location or enter coordinates.

STRETCH Command Tutorial

Step 1. Select the **STRETCH** icon from the **Modify** toolbar.

Step 2. When prompted to Select objects, pick the objects you would like to stretch with a *crossing window* (define the window by picking from right to left) and press **<Enter>**.

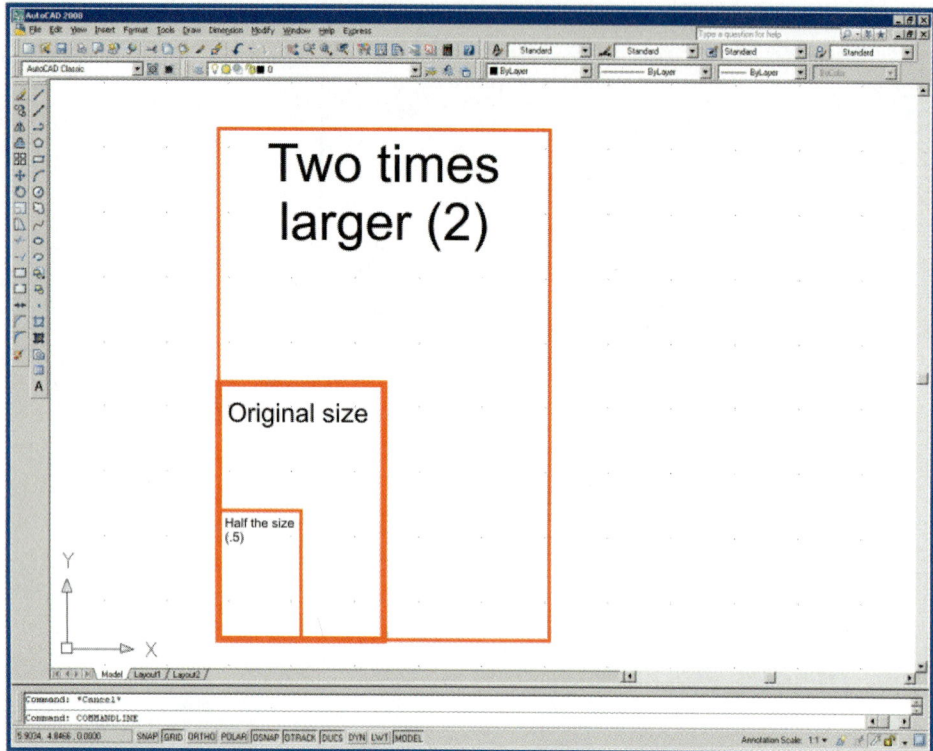

Figure 4-69(b) SCALE command

Figure 4-70(a) Stretch icon

Step 3. When prompted to Specify base point, select a point located on the object to be stretched.

Step 4. When prompted to Specify second point, define the point you wish the object to stretch to by picking a point with the mouse, or entering an absolute, relative, or polar coordinate. See Figure 4-70b.

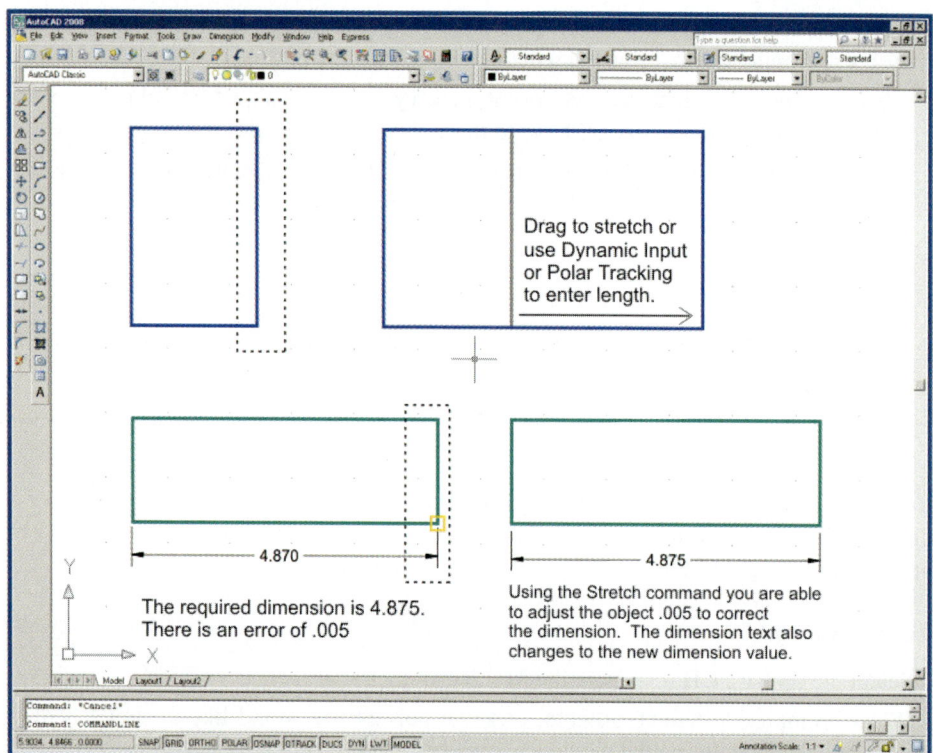

Figure 4-70(b) STRETCH command

TIP You can also find the **STRETCH** command in the **Modify** pull-down menu. The command alias is **S. STRETCH** is also a **GRIPS** option.

TRIM Command.   The icon for the **TRIM** command is shown in Figure 4-71(a). This command is used to trim objects to an intersection or cutting edge defined by another object. Two trim methods are available: *quick trim* and *regular trim*.

TRIM Command Tutorial (Quick Trim Option)

Step 1. Select the **TRIM** icon from the **Modify** toolbar.

Step 2. When prompted to Select cutting edges, press <**Enter**>. This will make every entity a cutting edge.

Note: Objects that intersect a cutting edge, but do not extend beyond it, cannot be trimmed. They should be erased instead.

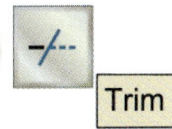

Figure 4-71(a)   Trim icon

TIP Cutting edges can be selected individually with the mouse instead if you prefer. This is known as a *regular* trim.

Step 3. When prompted to Select object to trim, pick the object to be trimmed on the side of the cutting edge you want to trim to.

Step 4. Press <**Enter**> to end the command. See Figure 4-71(b).

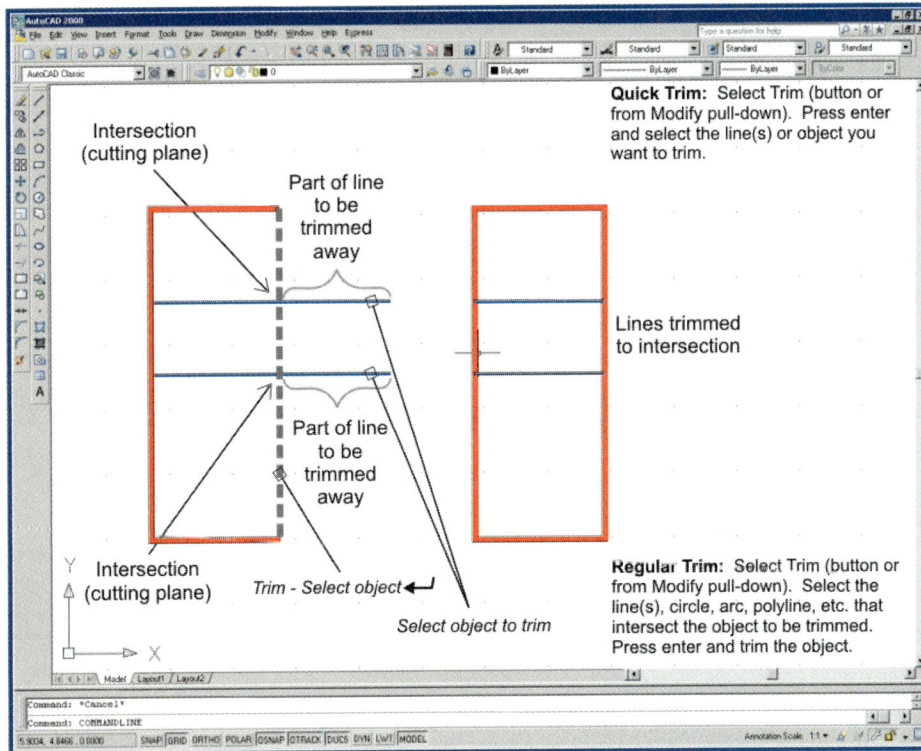

Figure 4-71(b)   TRIM command

TIP You can also find the **TRIM** command in the **Modify** pull-down menu. The command alias is **TR**.

Figure 4-72(a) Extend icon

EXTEND Command. The icon for the **EXTEND** command is shown in Figure 4-72(a). This command is used to extend a line or object to meet another object or line (called a *boundary edge*). Objects that can be extended are arcs, elliptical arcs, lines, polylines (2D and 3D), and rays.

EXTEND Command Tutorial

Step 1. Select the **EXTEND** icon from the **Modify** toolbar.

Step 2. When prompted to Select boundary edge, press **<Enter>**. This will make all entities a boundary edge.

Step 3. When prompted to Select object to extend, select the object to be extended near the end that you would like to extend.

Step 4. Press **<Esc>** or **<Enter>** to end the command. See Figure 4-72(b).

Note:
Objects that do not intersect a boundary edge cannot be extended.

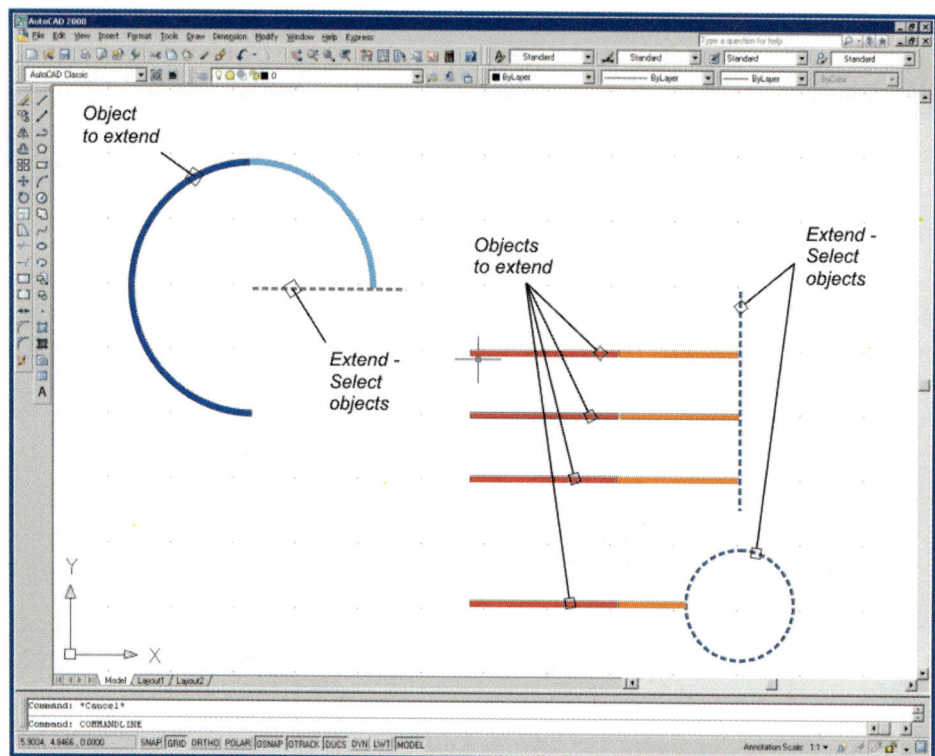

Figure 4-72(b) EXTEND command

You can also find the **EXTEND** command in the **Modify** pull-down menu. The command alias is **EX**.

TIP

BREAK AT POINT Command. The icon for the **BREAK AT POINT** command is shown in Figure 4-73. This command is used to break a line into two separate, but collinear, lines at the selected point.

Figure 4-73 Break at Point icon

BREAK AT POINT Command Tutorial

Step 1. Select the **BREAK AT POINT** icon from the **Modify** toolbar.

Step 2. When prompted to Select object to break, select the line to be broken.

Step 3. When prompted to Select first break point, select the point on the line where you would like the break to occur.

BREAK Command.   The icon for the **BREAK** command is shown in Figure 4-74(a). This command is used to break an object between two points.

Figure 4-74(a)   Break icon

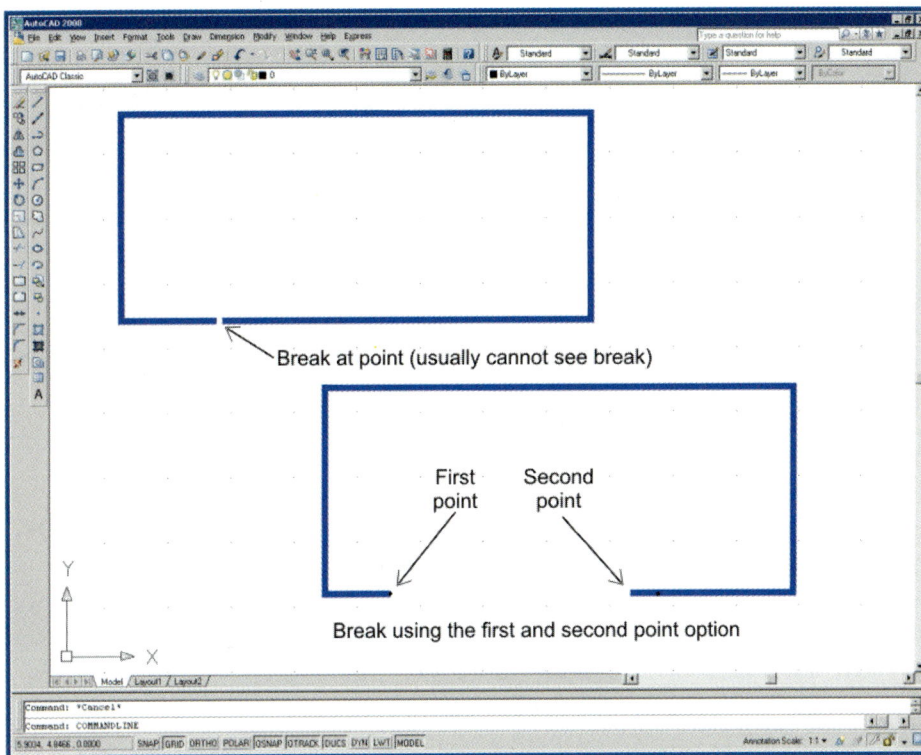

BREAK Command Tutorial

Step 1. Select the **BREAK** icon from the **Modify** toolbar.

Step 2. When prompted to Select object to break, select the object to be broken.

Step 3. Type **F** for **First Point** and press **<Enter>**.

Step 4. When prompted to Select first break point, select the object to be broken at the start point of the desired break.

Step 5. When prompted to Select second break point, select the endpoint of the desired break. See Figure 4-74(b).

Break at point (usually cannot see break)

First point   Second point

Break using the first and second point option

Figure 4-74(b)   BREAK command

TIP   You can also find the **BREAK** command in the **Modify** pull-down menu. The command alias is **BR**.

JOIN Command.   The icon for the **JOIN** command is shown in Figure 4-75(a). This command is used to join objects (collinear lines, for example, can have a gap between them) into one object.

Figure 4-75(a)   Join icon

JOIN Command Tutorial

Step 1. Select the **JOIN** icon from the **Modify** toolbar.

Step 2. When prompted to Select source object, select the first line of a set of collinear lines.

Step 3. When prompted to Select lines to join to source, select the collinear line(s) that you would like to join to the source line. Press **<Enter>** to complete the command. See Figure 4-75(b).

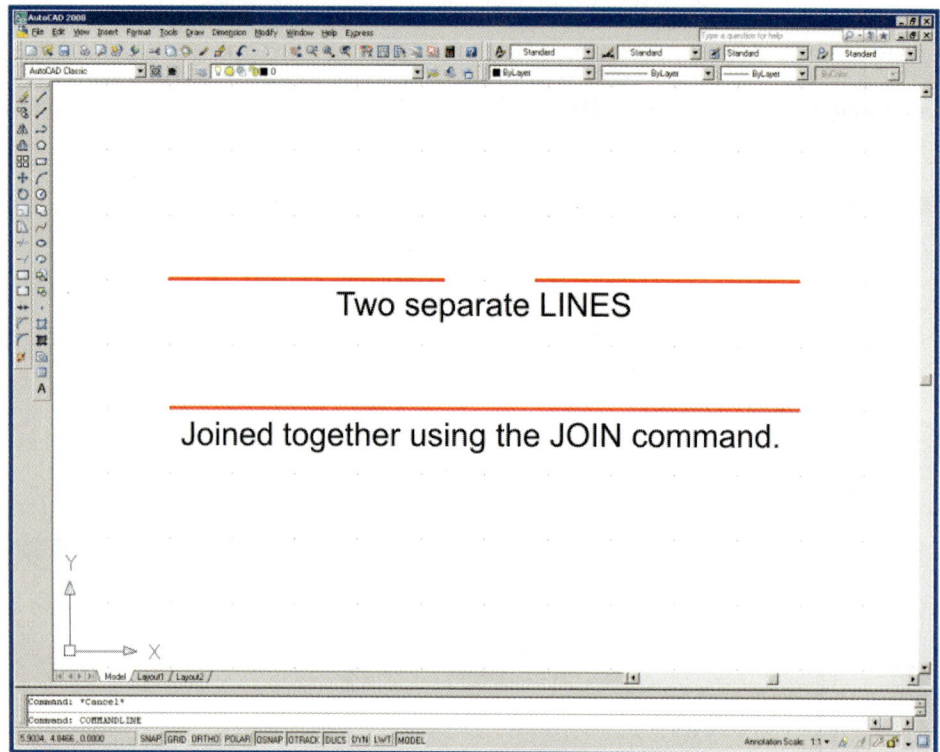

Figure 4-75(b) JOIN command

> You can also find the **JOIN** command in the **Modify** pull-down menu. The command alias is **J**.
>
> **TIP**

Chamfer

Figure 4-76(a) Chamfer icon

CHAMFER Command. The icon for the **CHAMFER** command is shown in Figure 4-76(a). This command is used to bevel the corners of objects and lines. The **Distance** option allows you to enter a chamfer distance for both sides of the beveled corner. The **Angle** option allows you to enter a distance and an angle for the beveled corner. The **Polyline** option will bevel the corners of the polyline.

CHAMFER Command Tutorial

Step 1. Select the **CHAMFER** icon from the **Modify** toolbar.

Step 2. Type **D** (for distance) and press **<Enter>**.

> To enter a distance and an angle for a chamfer, type **A** (for angle) instead of **D** and enter the desired chamfer distance and angle in response to the prompts.
>
> **TIP**

Step 3. When prompted to Specify first chamfer distance, enter the distance to bevel the first edge and press **<Enter>**.

Step 4. When prompted to Specify second chamfer distance, enter the distance to bevel the second edge and press **<Enter>**.

Step 5. When prompted to Select first line, pick the first line to be beveled near the end to be beveled.

Step 6. When prompted to Specify second line, pick the second line to be beveled. See Figure 4-76(b).

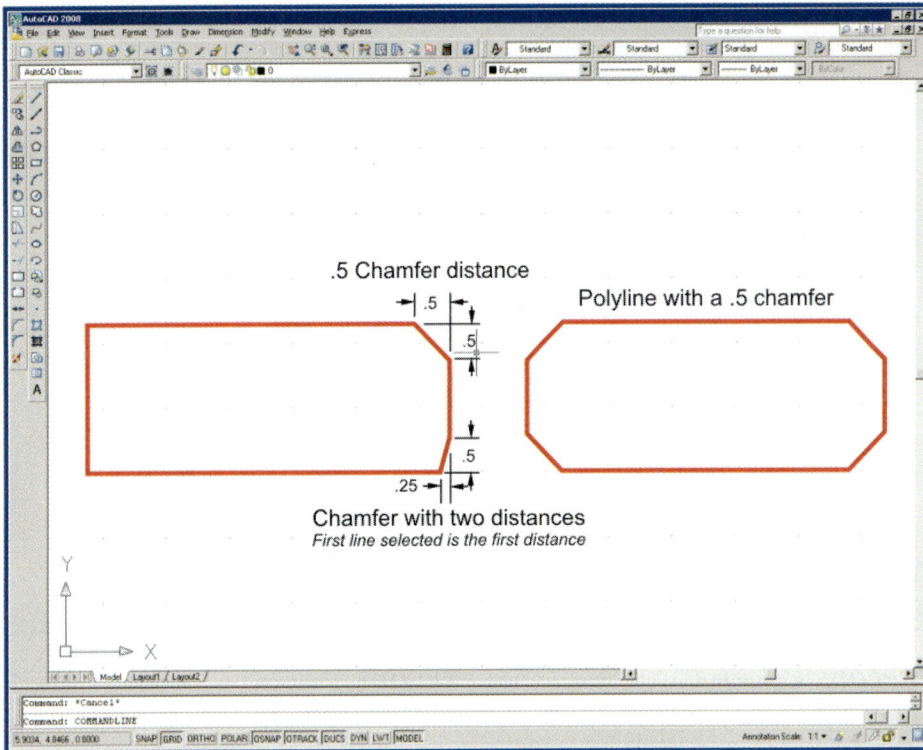

.5 Chamfer distance

Polyline with a .5 chamfer

Chamfer with two distances
First line selected is the first distance

Figure 4-76(b)   CHAMFER command

You can also find the **CHAMFER** command in the **Modify** pull-down menu. The command alias is **CHA**.

FILLET Command.   The icon for the **FILLET** command is shown in Figure 4-77(a). This command is used to create fillets (rounded inside corners) and rounds (rounded outside corners) on objects. The **Polyline** option creates fillets and rounds on the corners of a polyline.

FILLET Command Tutorial

Step 1. Select the **FILLET** icon from the **Modify** toolbar.

Step 2. Type **R** (for radius) and press **<Enter>**.

Step 3. When prompted to Specify fillet radius, enter the radius of the fillet and press **<Enter>**.

Step 4. When prompted to Select first object, pick the first line to be filleted near the end to be filleted.

Step 5. When prompted to Select second object, pick the second line to be filleted. See Figure 4-77(b).

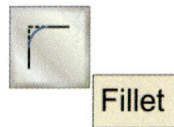

Fillet

Figure 4-77(a)   Fillet icon

You can also find the **FILLET** command in the **Modify** pull-down menu. The command alias is **F**.

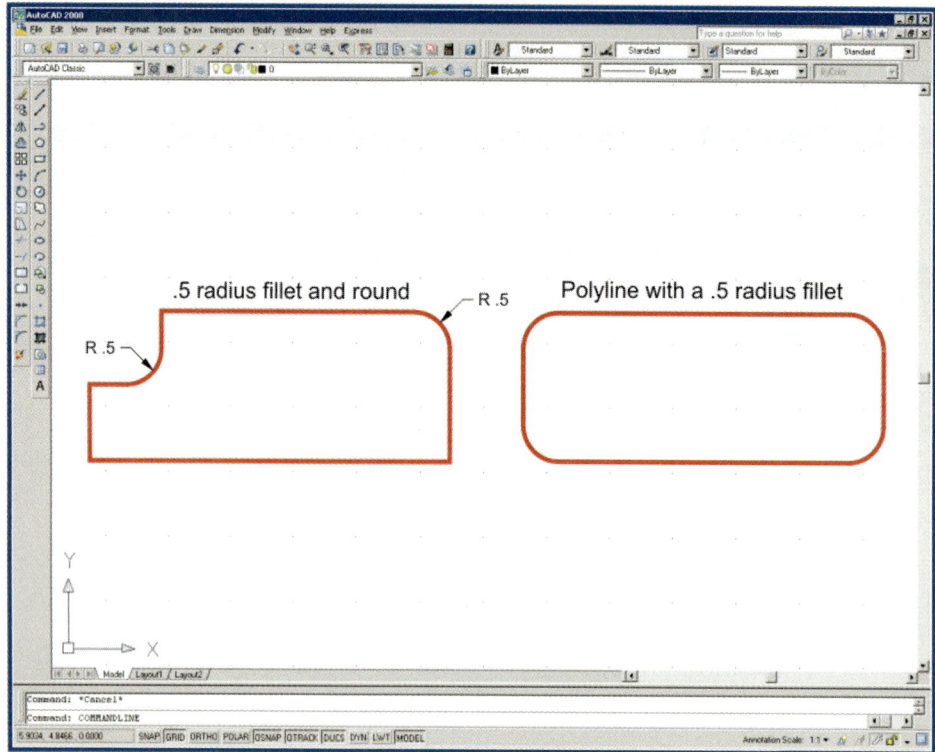

Figure 4-77(b) FILLET command

Explode

Figure 4-78(a) Explode icon

EXPLODE Command.   The icon for the **EXPLODE** command is shown in Figure 4-78(a). This command breaks compound objects, such as rectangles, polygons, polylines, and blocks, into separate entities—usually for the purpose of editing them.

EXPLODE Command Tutorial

Step 1. Select the **EXPLODE** icon from the **Modify** toolbar.

Step 2. When prompted to Select objects, select the object(s) to be exploded.

Step 3. Press <**Enter**> to complete the command. See Figure 4-78(b).

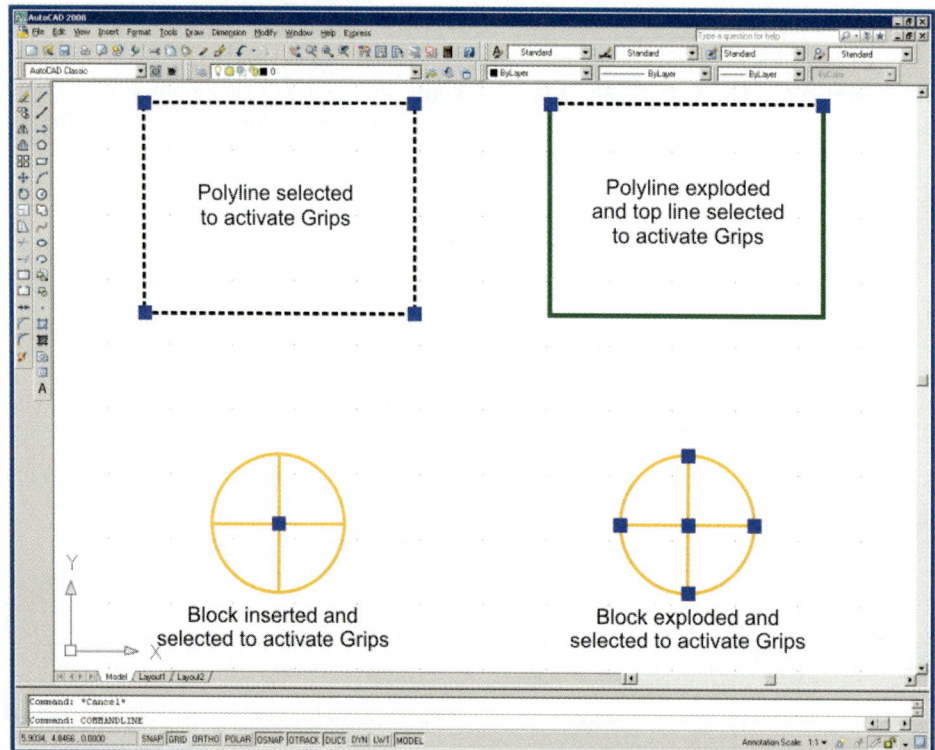

Figure 4-78(b) EXPLODE command

> **TIP**
>
> You can also find the **EXPLODE** command in the **Modify** pull-down menu. The command alias is **X**.

Text Toolbar

The **Text** toolbar gives drafters the options to create and edit text on drawings. Open this toolbar by moving the cursor onto any existing toolbar or icon on the AutoCAD screen, right-clicking, and selecting the **Text** toolbar from the list of toolbars by left-clicking. The icons on this toolbar are identified in Figure 4-79.

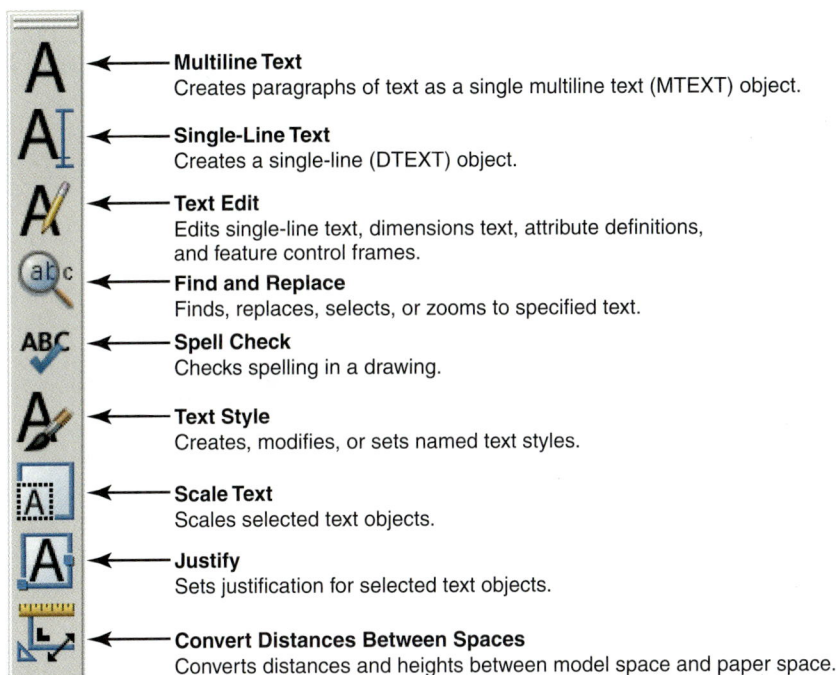

Multiline Text
Creates paragraphs of text as a single multiline text (MTEXT) object.

Single-Line Text
Creates a single-line (DTEXT) object.

Text Edit
Edits single-line text, dimensions text, attribute definitions, and feature control frames.

Find and Replace
Finds, replaces, selects, or zooms to specified text.

Spell Check
Checks spelling in a drawing.

Text Style
Creates, modifies, or sets named text styles.

Scale Text
Scales selected text objects.

Justify
Sets justification for selected text objects.

Convert Distances Between Spaces
Converts distances and heights between model space and paper space.

Figure 4-79   Text toolbar

Placing Text on a Drawing.   Two ways to place text in the field of a drawing are **MTEXT** (multi-line text) and **DTEXT** (dynamic or single-line text).

MTEXT Command.   Select the **MTEXT** icon shown in Figure 4-79 from the **Text** toolbar, or type **MTEXT** and press <**Enter**>, to start the command. You will then be prompted to Specify the first and Opposite corner of a text box by picking two points in the graphics window. After defining the corners of the text box, the **Text Formatting** dialog box will open. Enter the desired text into the text window and click **OK** (see Figure 4-80). By highlighting the text, the *text style* and height can also be changed in this dialog box.

> text style: The characteristics of text used in a drawing such as font name, height, width factor, and oblique angle. These values are determined by the values set in the **Text Style** dialog box.

2 X R.25

Figure 4-80   Text Formatting dialog box

DTEXT Command. Select the **DTEXT** icon as shown in Figure 4-79 from the **Text** toolbar, or type **DTEXT** and press **<Enter>**, to start the command. You are prompted to select the text's **start point**, **height**, and **rotation angle**. When these text specifications have been defined, a cursor will appear. At the cursor, you can enter the desired text. To complete the command, you must press **<Enter>** twice. See Figure 4-81.

```
Command: dtext
Current text style:  "Standard"  Text height:  0.2000  Annotative:  No
Specify start point of text or [Justify/Style]:
Specify height <0.2000>:
Specify rotation angle of text <0>:
```

Figure 4-81 DTEXT command line prompts

Editing Text

Step 1. Pick on the **Text Edit** icon seen in Figure 4-79, or type **ED** and press **<Enter>**.

Step 2. Select the text or dimension to edit by picking on it. If the annotation selected was originally created with the **MTEXT** or **Dimensioning** tools, the **Multiline Text Editor** dialog box will appear (see Figure 4-82). Make the changes to the text in the field of the **Multiline Text Editor** dialog box and click **OK**.

> **Note:**
> The font and height of the text can also be changed while in the **Text Editor** mode by first highlighting the text and selecting a new text style, font or height.

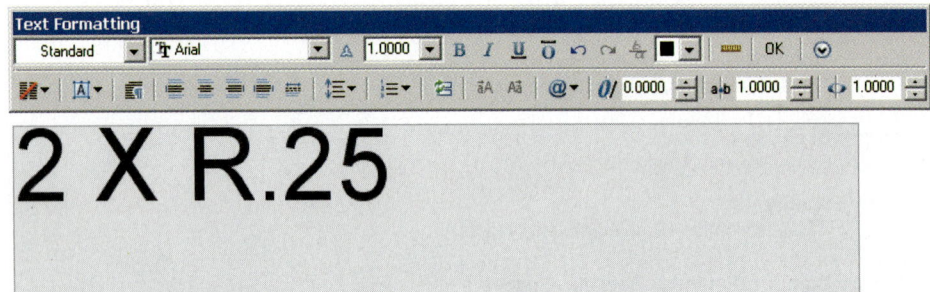

Figure 4-82 Multiline text editor dialog box

Figure 4-83 Single-line text edit box

If the annotation to be edited was originally placed with the **DTEXT** command, the **Text Edit** box will appear as shown in Figure 4-83. Place the cursor in this box to begin editing.

Controlling Text Style. The characteristics of text used in a drawing, such as font name, height, width factor, and oblique angle, are determined by the values set in the **Text Style** dialog box. Drafters can either default to the settings of the **Standard** style, edit the **Standard** style, or create a new text style.

Changing Text Style Settings

Step 1. Select the **Text Style** icon shown in Figure 4-79 or pick **Text Style** from the **Format** pull-down menu. The **Text Style** dialog box will open as shown in Figure 4-84.

Step 2. Select the **Standard** style name or select the **New...** button to create a new style name.

Step 3. To change the font, select the **Font Name** you want to assign from the list of font styles. For mechanical drawings, a Gothic font, such as Arial, is appropriate. For architectural drawings, an architectural font, such as Stylus BT or City Blueprint, may be more appropriate. The value for the height, width factor, and oblique angle can also be changed (see Figure 4-85). After the desired edits have been made to the text style, select **Apply**. The new text style can also be assigned to dimension text by changing the **Text style** setting in the **Text** tab of the **Dimension Style Manager** (see Figure 5-54).

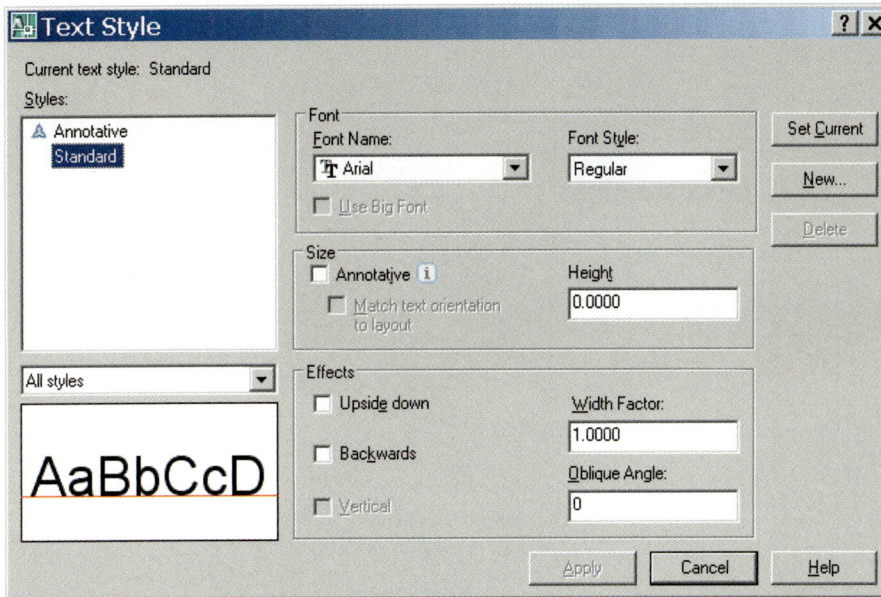

Figure 4-84   Text Style dialog box

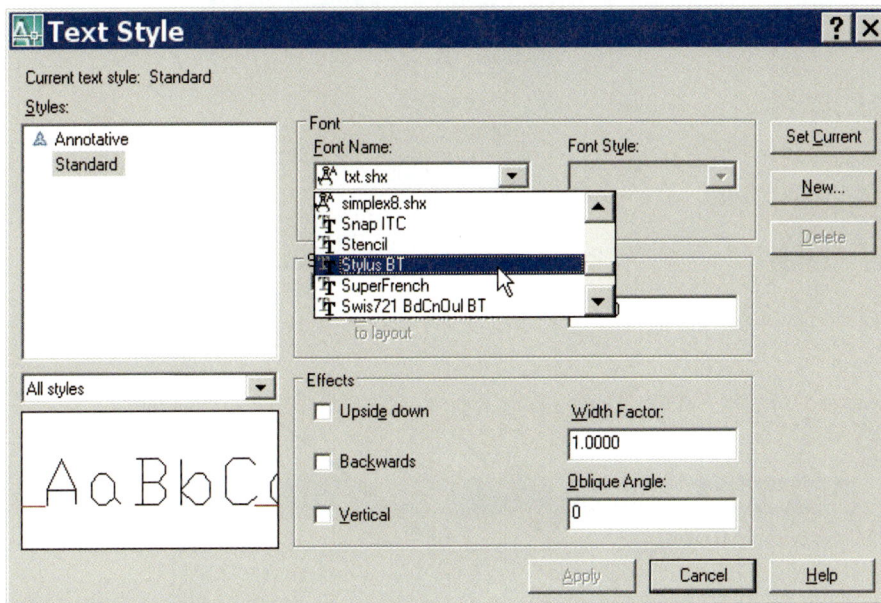

Figure 4-85   Selecting a new Font Name from the list in the Text Style dialog box

> **TIP**
> When creating a new text style, it is best to leave the text height set to **0.0000**; otherwise, all text will default to the height defined in the Text Style dialog box.

DRAFTING SETTINGS DIALOG BOX

Drafting Settings are used to define AutoCAD's **Grid**, **Snap**, **Polar Tracking**, **Running Object Snaps**, and **Dynamic Input** settings.

Open the **Drafting Settings** dialog box by selecting **Drafting Settings** from the **Tools** pull-down menu (see Figure 4-86), or by right-clicking on the **Snap**, **Grid**, **Ortho**, **Polar**, **Osnap**, or **Otrack** tab located on the Status Bar and selecting **Settings**.

The **Drafting Settings** dialog box has four tabs to choose from: **Snap and Grid**, **Polar Tracking**, **Object Snap**, and **Dynamic Input** (see Figure 4-87).

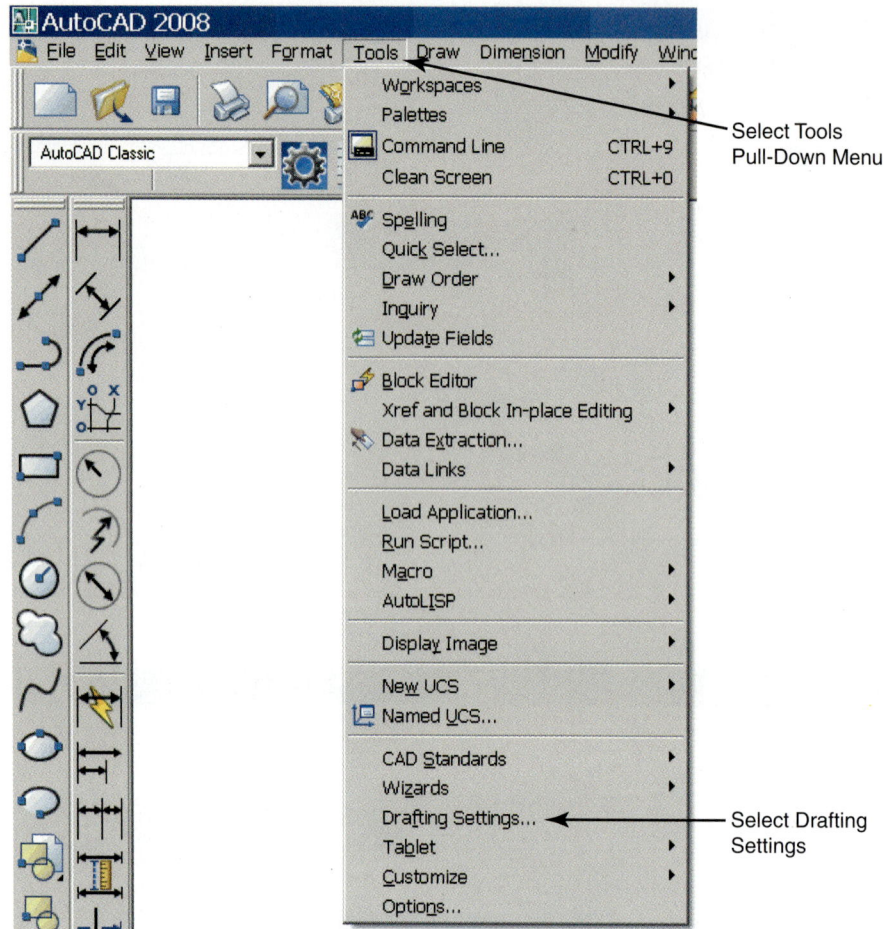

Figure 4-86 Opening the Drafting Settings dialog box

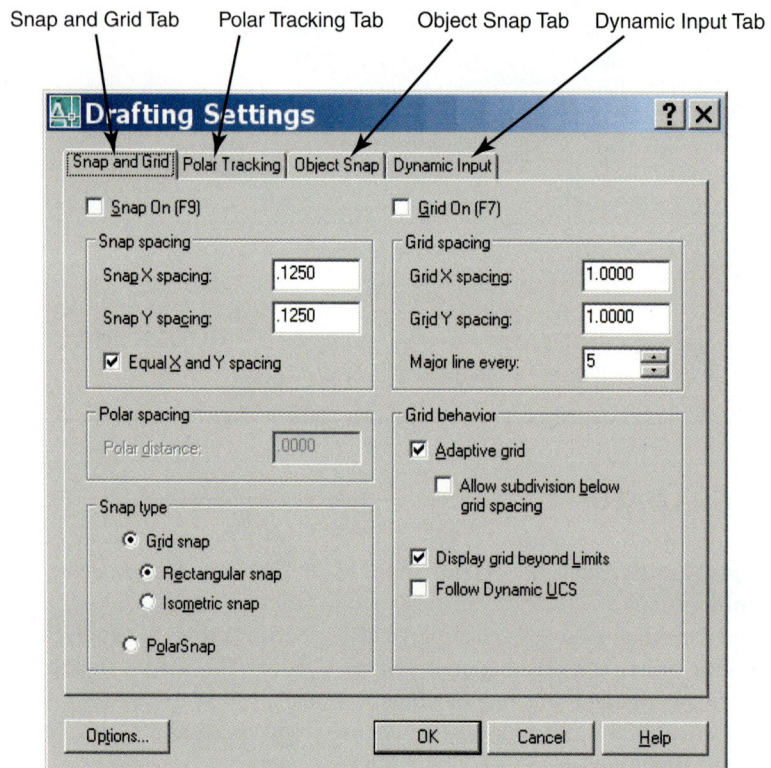

Figure 4-87 Drafting Settings dialog box

Snap and Grid Tab

This tab allows the user to specify **Snap** and **Grid** settings. Figures 4-88 and 4-89 explain the functions of this tab.

Snap On
Turns Snap mode on or off. You can also turn Snap mode on or off by clicking Snap on the status bar, by pressing F9.

Snap Spacing
Creates a rectangular grid of snap locations that restricts cursor movement to the settings assigned in these boxes. Sptecifies the snap spacing in the Y and X directions. The value must be a positive real number.

Polar Spacing
Polar Distance sets the snap increment distance when PolarSnap is selected under Snap Type & Style. If this value is 0, the PolarSnap distance assumes the value for Snap X Spacing. The Polar Distance setting is used in conjunction with polar tracking and/or object snap tracking. If neither tracking feature is enabled, the Polar Distance setting has no effect.

Grid On
Turns the grid on or off. You can also turn grid mode on or off by clicking Grid on the status bar, by pressing F7.

Grid Spacing
Controls the display of a grid that reflects the drawing's limits. Specifies the grid spacing in the X and Y directions.

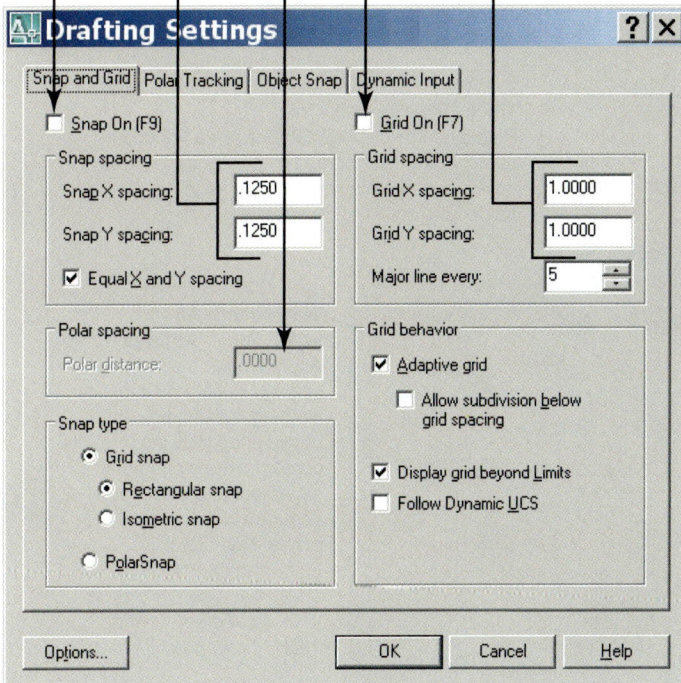

Figure 4-88   Snap and Grid tab

Polar Tracking Tab

Polar Tracking is a drawing tool that displays temporary alignment paths defined by user-specified polar angles. This tab allows the user to define the settings that are enabled when **Polar Tracking** is **On** (see Figure 4-90).

Grid Snap
Sets the snap type to Grid. When you specify points, the cursor snaps along vertical or horizontal grid points.

Rectangular Snap
Sets the snap style to standard rectangular snap mode. When the snap type is set to Grid snap and Snap mode is on, the cursor snaps to a rectangular snap grid.

Polar Snap
Sets the snap type to Polar. When Snap mode is on and you specify points with polar tracking turned on, the cursor snaps along polar alignment angles set on the Polar Tracking tab relative to the starting polar tracking point.

Adaptive Grid
Limits the density of the grid when zoomed out.

Allow Subdivision Below Grid Spacing
Generates additional, more closely spaced grid lines when zoomed in. The frequency of these grid lines is determined by the frequency of the major grid lines.

Display Grid Beyond Limits
Displays the grid beyond the area specified by the **LIMITS** comand.

Isometric Snap
Sets the snap type to Isometric snap mode. When the snap type is set to Grid snap and Snap mode is on, the cursor snaps to an isometric snap grid.

Follow Dynamic UCS
Changes the grid plane to follow the XY plane of the dynamic UCS.

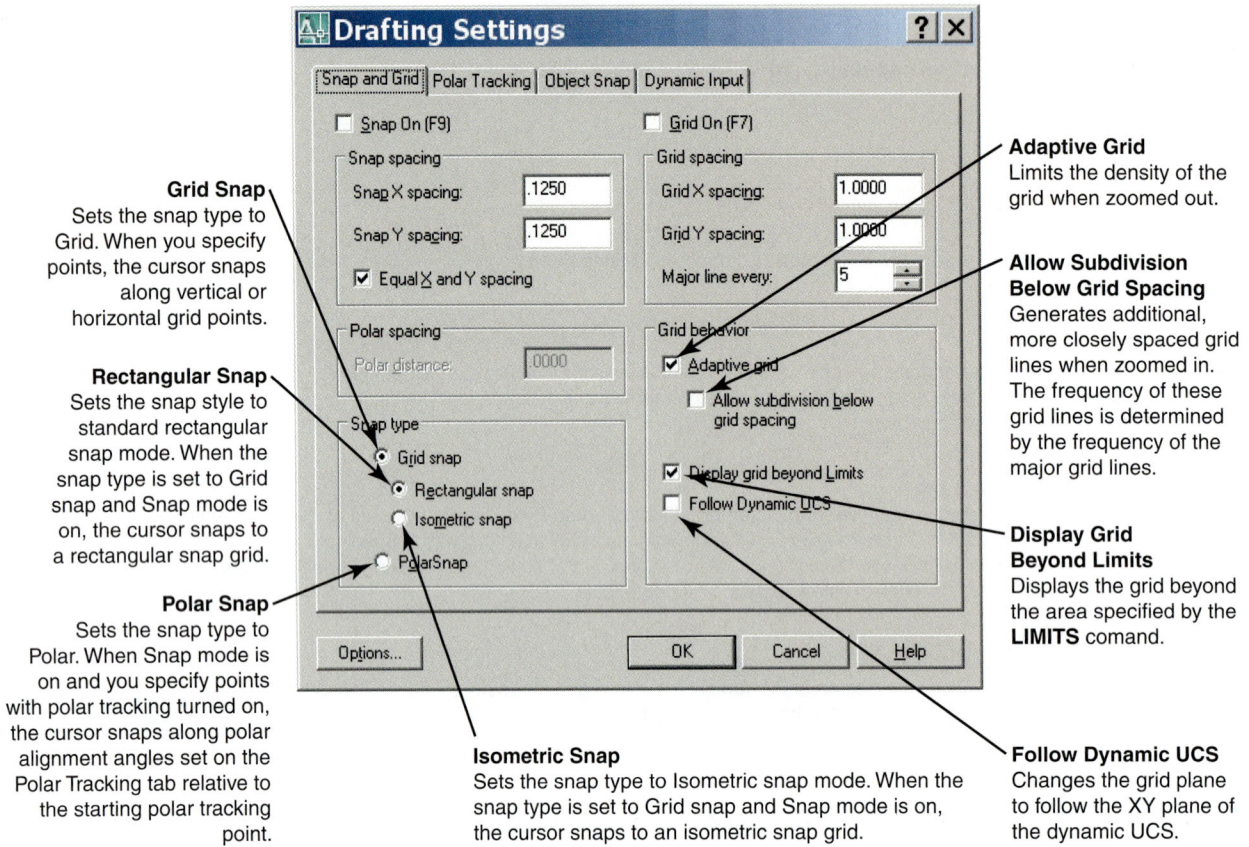

Figure 4-89   Snap and Grid tab

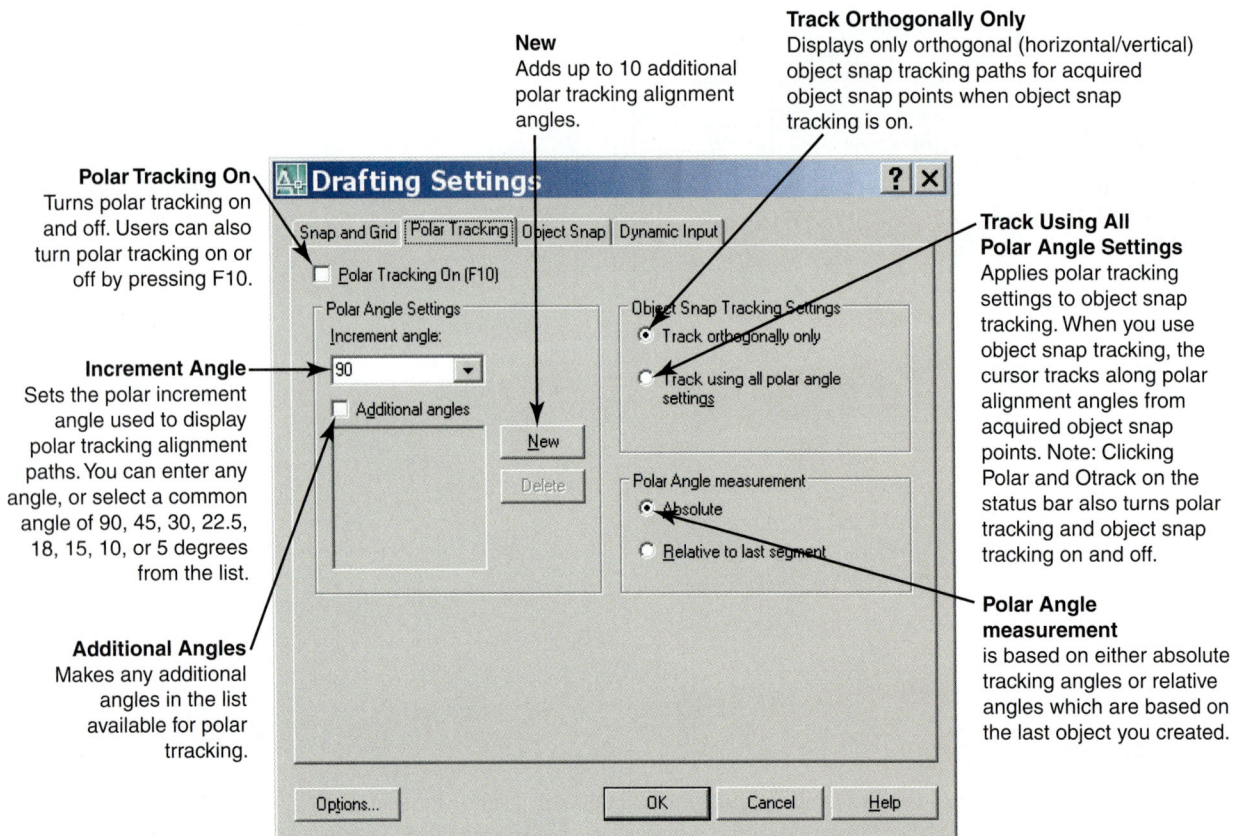

New
Adds up to 10 additional polar tracking alignment angles.

Track Orthogonally Only
Displays only orthogonal (horizontal/vertical) object snap tracking paths for acquired object snap points when object snap tracking is on.

Polar Tracking On
Turns polar tracking on and off. Users can also turn polar tracking on or off by pressing F10.

Increment Angle
Sets the polar increment angle used to display polar tracking alignment paths. You can enter any angle, or select a common angle of 90, 45, 30, 22.5, 18, 15, 10, or 5 degrees from the list.

Additional Angles
Makes any additional angles in the list available for polar trracking.

Track Using All Polar Angle Settings
Applies polar tracking settings to object snap tracking. When you use object snap tracking, the cursor tracks along polar alignment angles from acquired object snap points. Note: Clicking Polar and Otrack on the status bar also turns polar tracking and object snap tracking on and off.

Polar Angle measurement
is based on either absolute tracking angles or relative angles which are based on the last object you created.

Figure 4-90   Polar Tracking tab

Dynamic Input Tab

The **Dynamic Input** tab controls pointer input, dimension input, dynamic prompting, and the appearance of drafting tooltips (see Figure 4-91).

Displays a dimension with tooltips for distance value and angle value when a command prompts for a second point or a distance. The values in the dimension tooltips change as you move the cursor. You can enter values in the tooltip instead of on the command line.

Turns pointer input on or off.

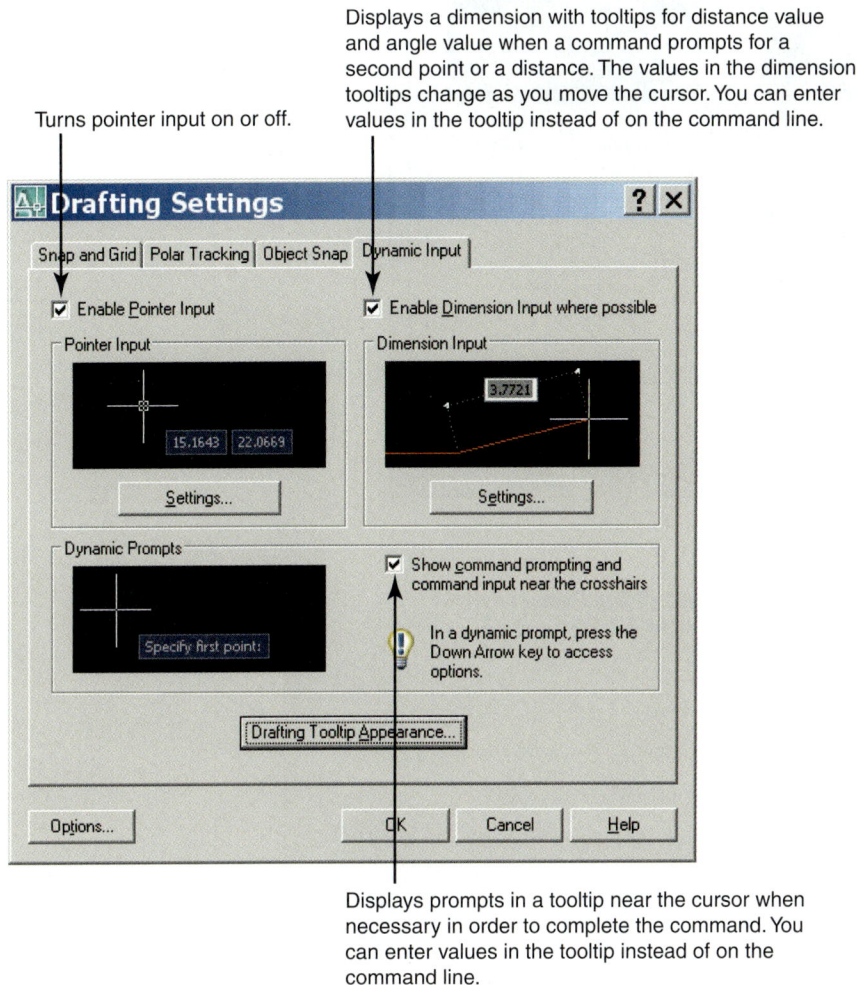

Displays prompts in a tooltip near the cursor when necessary in order to complete the command. You can enter values in the tooltip instead of on the command line.

Figure 4-91   Dynamic Input tab options

Dynamic Input Tab Settings.   The **Dynamic Input** tab allows the user to define pointer input settings and **dimension input settings**. Selecting the **Settings...** button opens the dialog boxes shown in Figure 4-92.

Object Snap Tab

Object Snap settings (osnaps) allow a drafter to snap exactly to a point on an object when prompted to select a point; for example, the **endpoint** or **midpoint** of a line, or the exact **center** of a circle or arc. **Osnaps** can be selected from the **Object Snap** toolbar or from the **Object Snap** tab of the **Drafting Settings** dialog box (see Figure 4-93). By checking the box next to an **Object Snap** in this tab, you can set *running object snaps* which can be used for repeated precise placements. For example, with the **Endpoint** box on this tab checked, you can use the line command to "snap" to the exact endpoint of another line. Do this by moving the cursor near the endpoint of the existing line and left-click when the object snap marker is displayed. If multiple object snaps are on, the user can press <**Tab**> to cycle through the available snap options before selecting the desired point.

Object Snap settings: A technique used in the creation and editing of AutoCAD drawings that allows the user to snap to exact points on an object. Common object snap settings include snap to endpoint, snap to midpoint, snap to intersection, snap to center, and snap to quadrant.

running object snaps: Setting an Object Snap mode so it continues for subsequent selections.

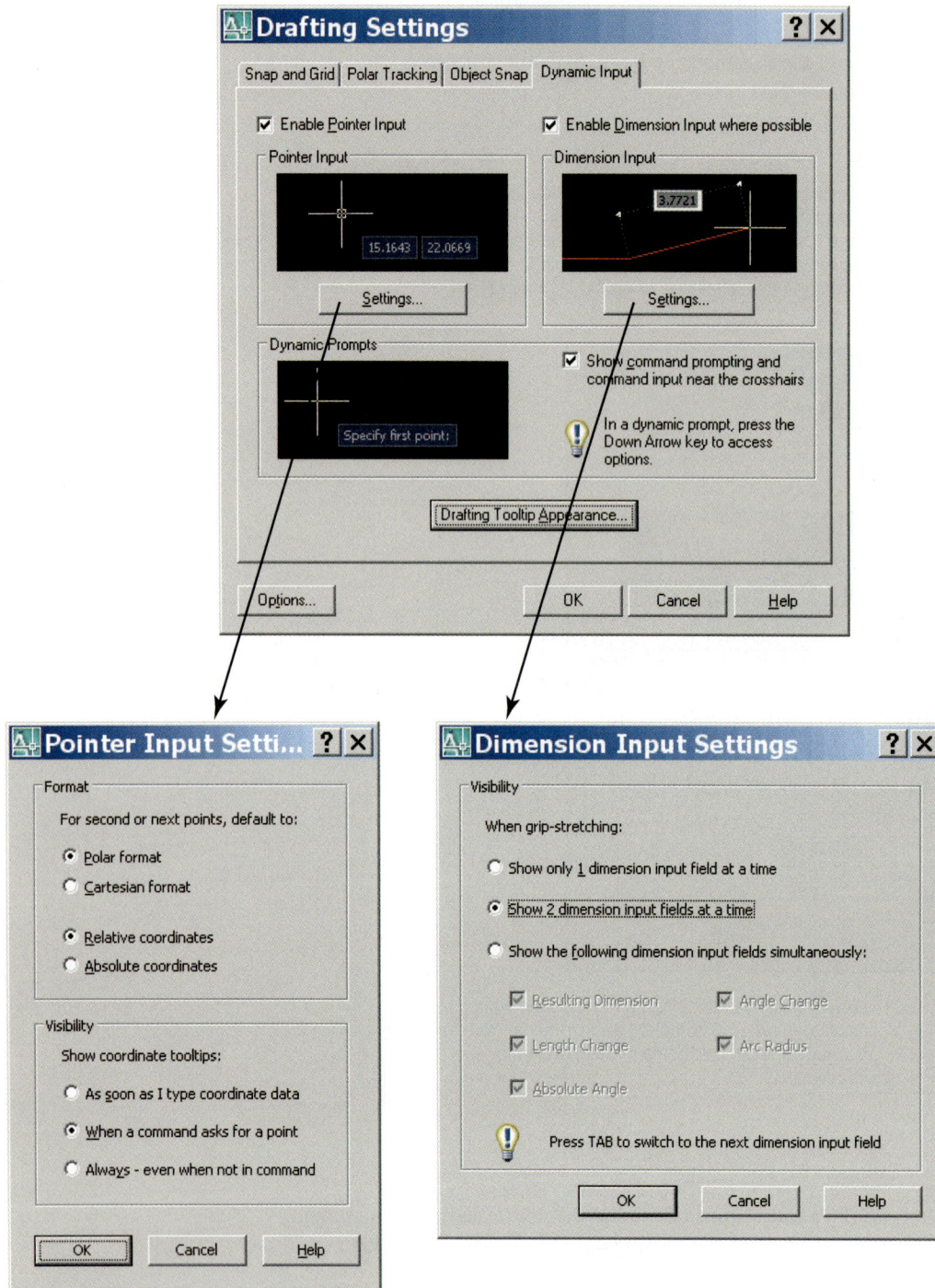

Figure 4-92   Dynamic Input tab settings

OBJECT SNAP TOOLBAR

Open the **Object Snap** toolbar by moving the cursor onto any existing toolbar or icon on the AutoCAD screen, right-clicking, and selecting the **Object Snap** toolbar with the left-click mouse button. Follow the numbered steps shown in Figures 4-94 through 4-103 to see how **Object Snaps** are used to simplify drawing. Video tutorials for the commands of the **Object Snap** toolbar are located on the Student Resource CD in the AutoCAD Tutorial Videos folder.

Turn Running
Object Snaps
On or Off

Turn Object
Snap Tracking
On or Off

Drafting Settings   ? ×

Snap and Grid | Polar Tracking | Object Snap | Dynamic Input |

☑ Object Snap On (F3)   ☑ Object Snap Tracking On (F11)

Object Snap modes

☐ ☑ Endpoint   ⅁ ☐ Insertion   Select All

△ ☐ Midpoint   ⊾ ☐ Perpendicular   Clear All

○ ☑ Center   ⊙ ☐ Tangent

⊠ ☐ Node   ⊠ ☐ Nearest

◇ ☐ Quadrant   ⊠ ☐ Apparent Intersection

× ☑ Intersection   ⫽ ☐ Parallel

--- ☑ Extension

💡 To track from an Osnap point, pause over the point while in a
command. A tracking vector appears when you move the cursor.
To stop tracking, pause over the point again.

Options...   OK   Cancel   Help

Set any Running
Objects Snaps that
you may need for
the drawing.

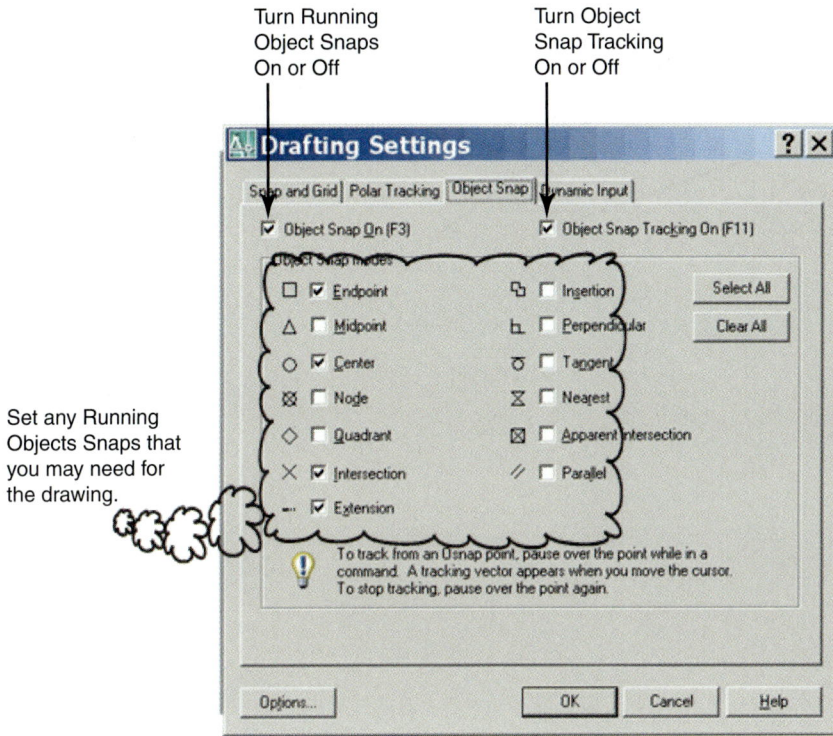

This dialog box may also be accessed by
selecting the **Object Snap Settings button**
on the Object Snap toolbar.

Figure 4-93 Objects
Snaps tab

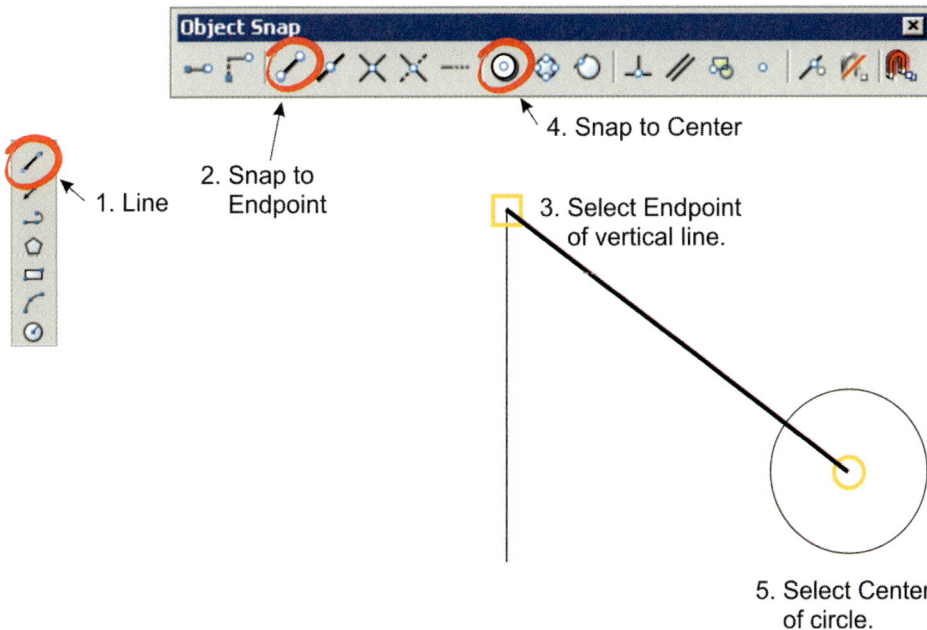

Object Snap   ×

4. Snap to Center

2. Snap to
Endpoint

1. Line

3. Select Endpoint
of vertical line.

5. Select Center
of circle.

Figure 4-94 Steps in
drawing a line from an
endpoint to a center point

127

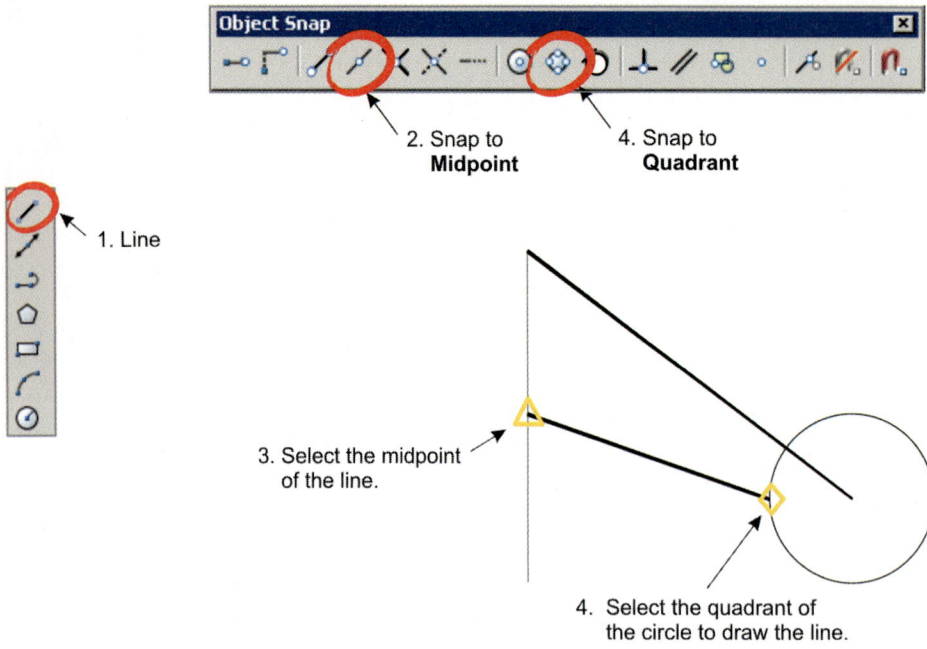

Figure 4-95   Steps in drawing a line from a midpoint to a quadrant

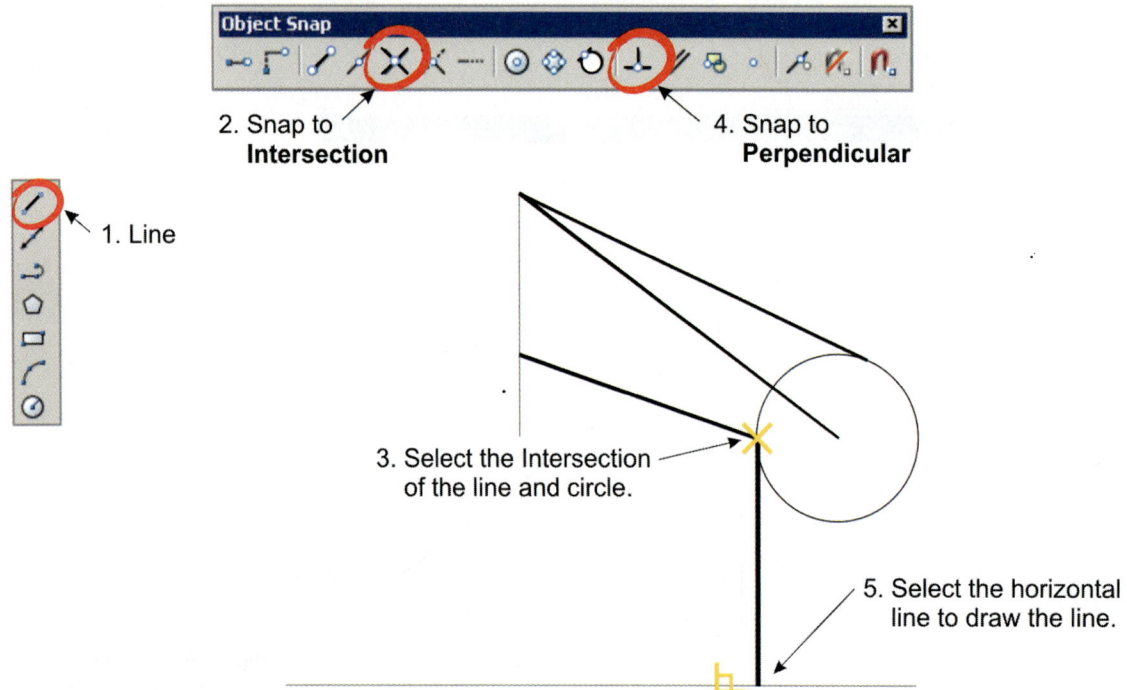

Figure 4-96   Steps in drawing a line from an intersection that is perpendicular to another line

Object Snap

2. Snap to
Endpoint

4. Snap to
Tangent

1. Line

3. Select the
Endpoint of
the line.

5. Select the
tangent of the
circle to draw
the line.

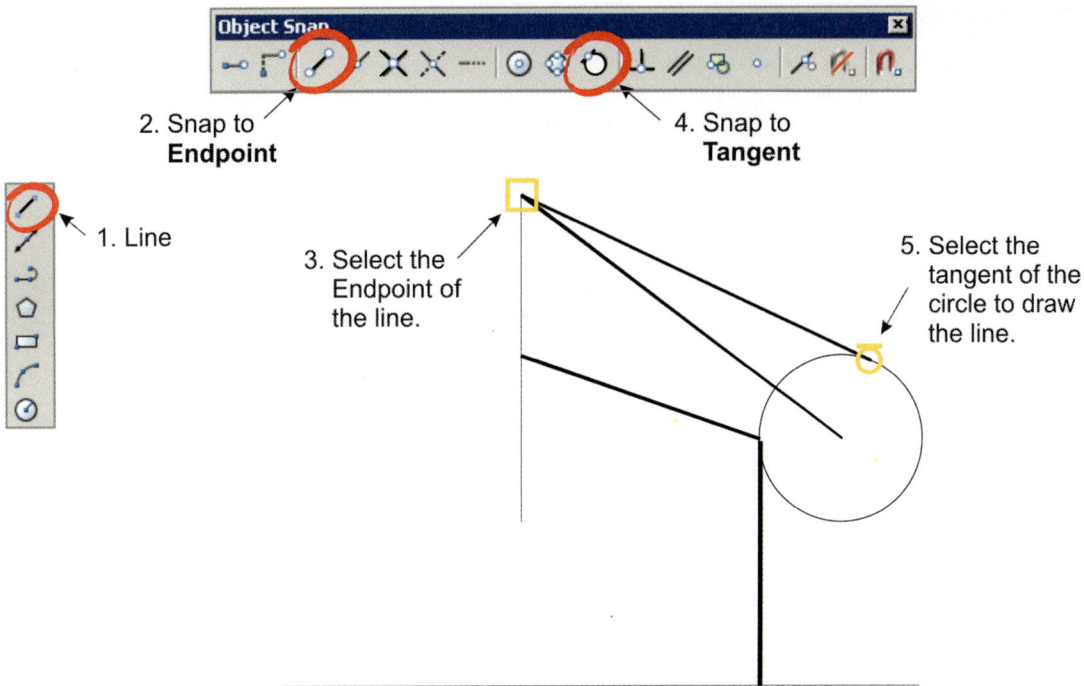

Figure 4-97   Steps in drawing a line from an endpoint to a tangency point on a circle

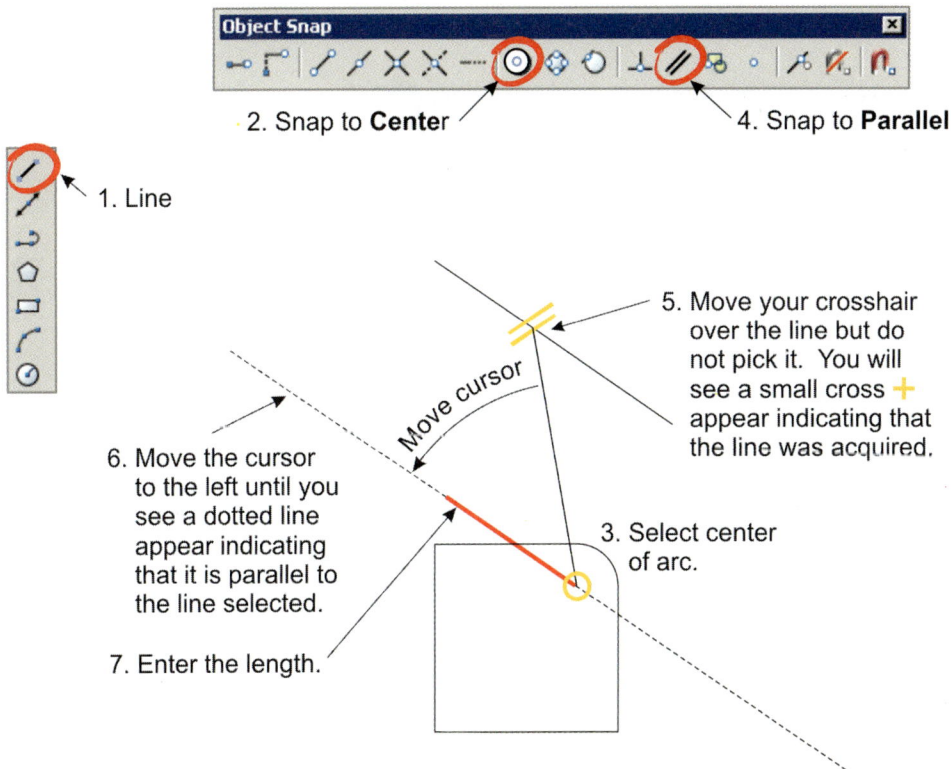

Object Snap

2. Snap to **Center**

4. Snap to **Parallel**

1. Line

5. Move your crosshair
over the line but do
not pick it. You will
see a small cross +
appear indicating that
the line was acquired.

Move cursor

6. Move the cursor
to the left until you
see a dotted line
appear indicating
that it is parallel to
the line selected.

3. Select center
of arc.

7. Enter the length.

Figure 4-98   Steps in drawing a line parallel to another line using the Parallel Osnap option

Object Snap

2. Snap to **Extension**

3. Acquire (do not pick)
the end of the line.

1. Circle

4. Move the cursor when
indicating the extended
dotted line and enter the
distance required. The
end of the extension is
the center of the circle.
Enter the radius or diameter.

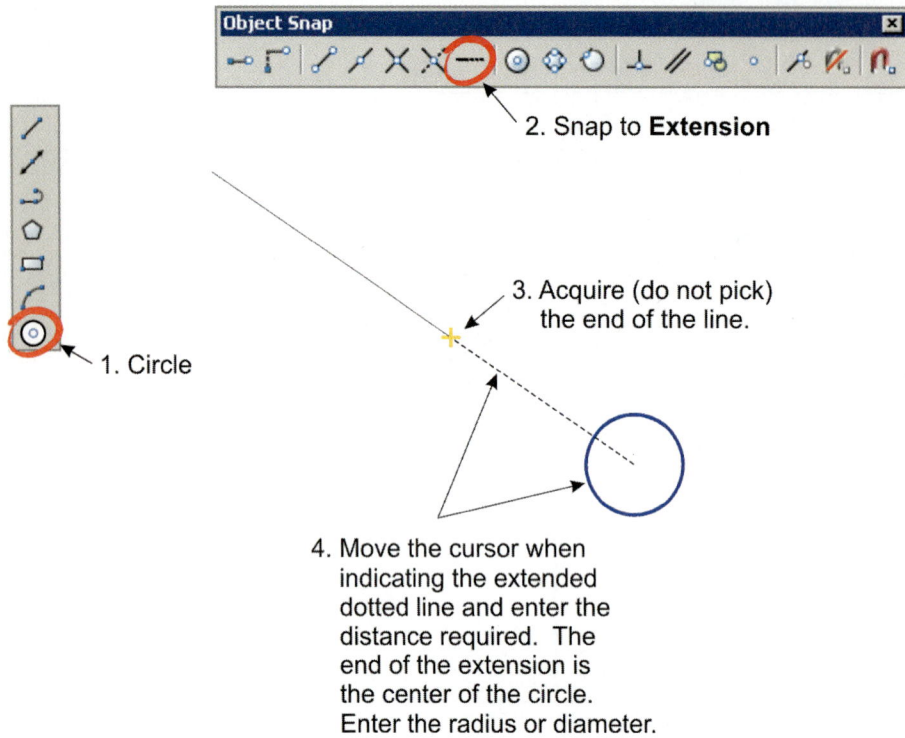

Figure 4-99   Steps in defining the center of a circle using Extension Osnap option

Note:
For Figure 4-100, Polar Tracking, Otrack Snap Tracking, and
Osnap mode must be turned
on; and the Snap to Endpoint
and Snap to Intersection object
snaps must be selected when
using this technique.

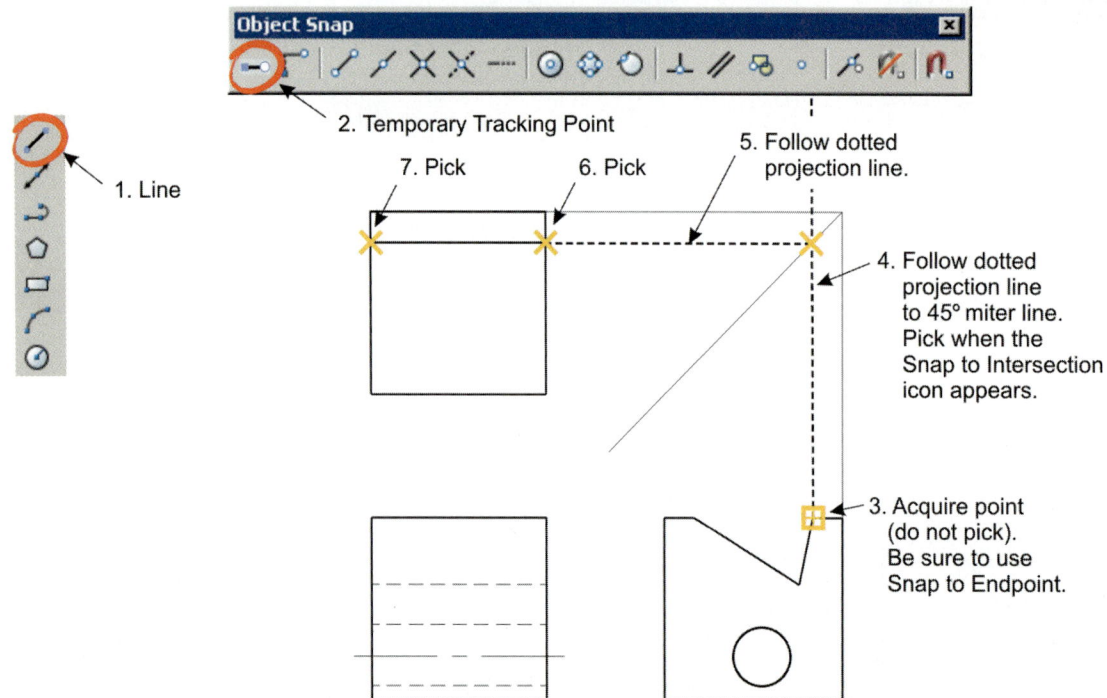

Object Snap

2. Temporary Tracking Point

1. Line

7. Pick   6. Pick

5. Follow dotted
projection line.

4. Follow dotted
projection line
to 45° miter line.
Pick when the
Snap to Intersection
icon appears.

3. Acquire point
(do not pick).
Be sure to use
Snap to Endpoint.

Figure 4-100   Steps in using Temporary Tracking Point to project information through a miter line

Object Snap ✕

2. Snap From

1. Line

4. Type @0,1.162
and enter.

1.162

5. Select Snap to
Perpendicular to
draw the line.

3. Command:_Line Specify first point:_from Base point: **Select
the bottom corner of the
object.**

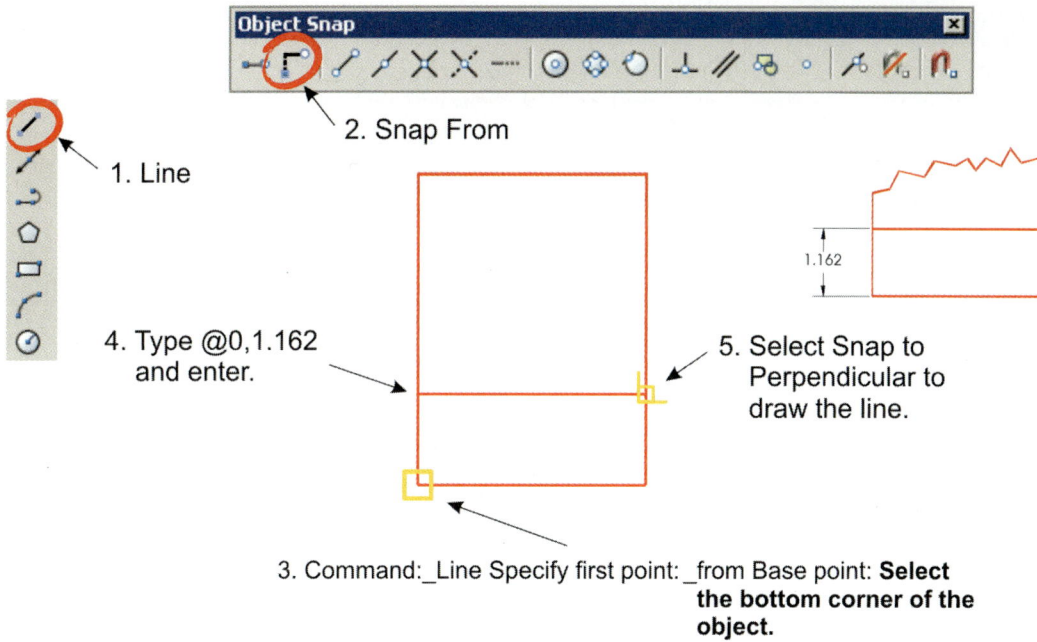

Figure 4-101   Steps in drawing a line using the Snap From Osnap option

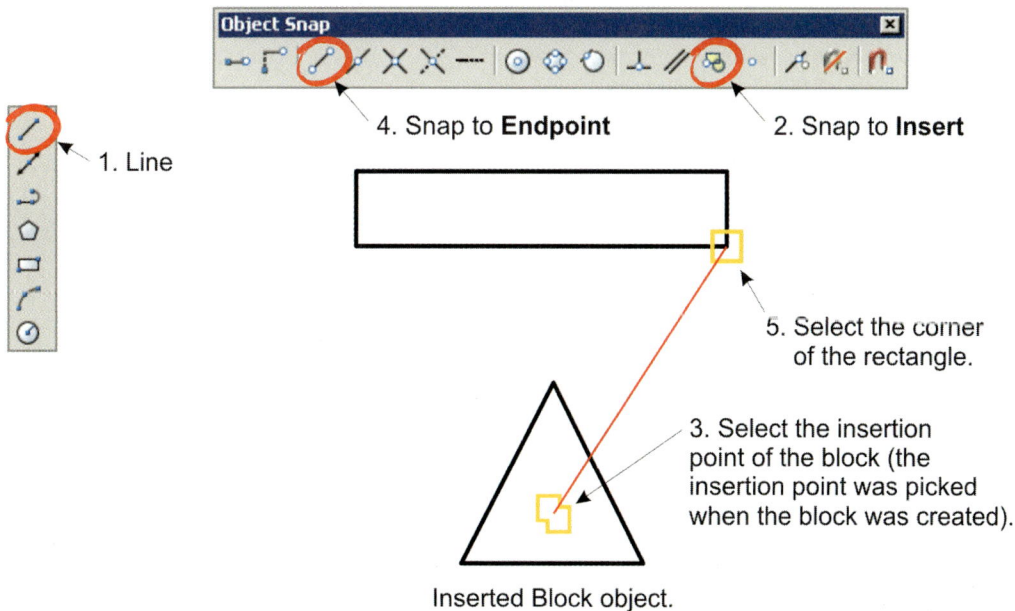

Object Snap ✕

4. Snap to **Endpoint**   2. Snap to **Insert**

1. Line

5. Select the corner
of the rectangle.

3. Select the insertion
point of the block (the
insertion point was picked
when the block was created).

Inserted Block object.

Figure 4-102   Steps in drawing a line from the insertion point of a block using the Snap to Insert option

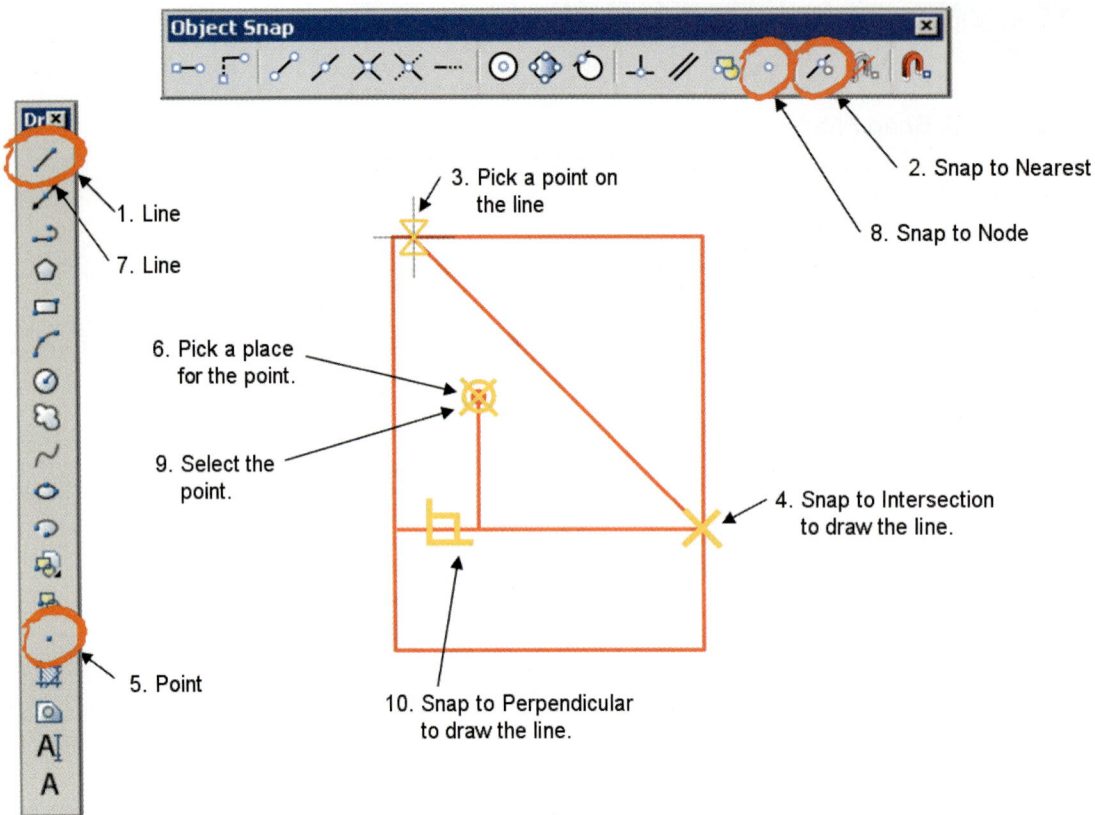

Figure 4-103   Drawing lines using the Snap to Nearest and Snap to Node options

Exercise 4-2:

Directions

1. Begin a new AutoCAD drawing (default to the **acad.dwt** template).
2. Set the upper right **Limit** to **22,17** and the grid spacing to **1.00**. Make the following layers: **Visible**, **Hidden**, and **Center**. Assign a color to each layer and set the **Hidden** layer's linetype to **Hidden** and the **Center** layer's line type to **Center**.
3. Set running osnaps for **endpoint**, **quadrant**, **tangency**, **perpendicular**, and **intersection**.
4. Draw the Front, Top and Right-side views of Exercise 2-1 located in Unit 2. Begin the views at the absolute coordinates shown in Figure 4-104(a). Count the grids to determine the size of the object's features.
5. **Save** the drawing as Figure 4-104(a).

Figure 4-104(a)
Exercise 4-2

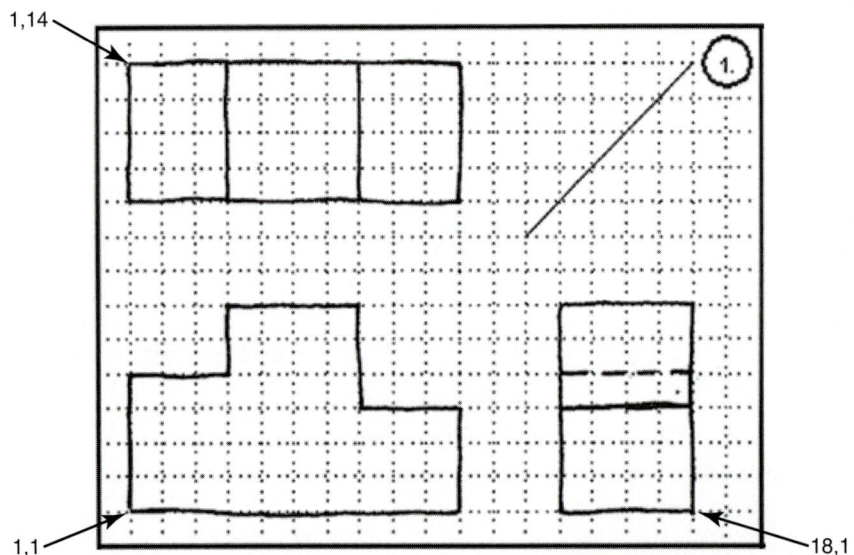

1,14

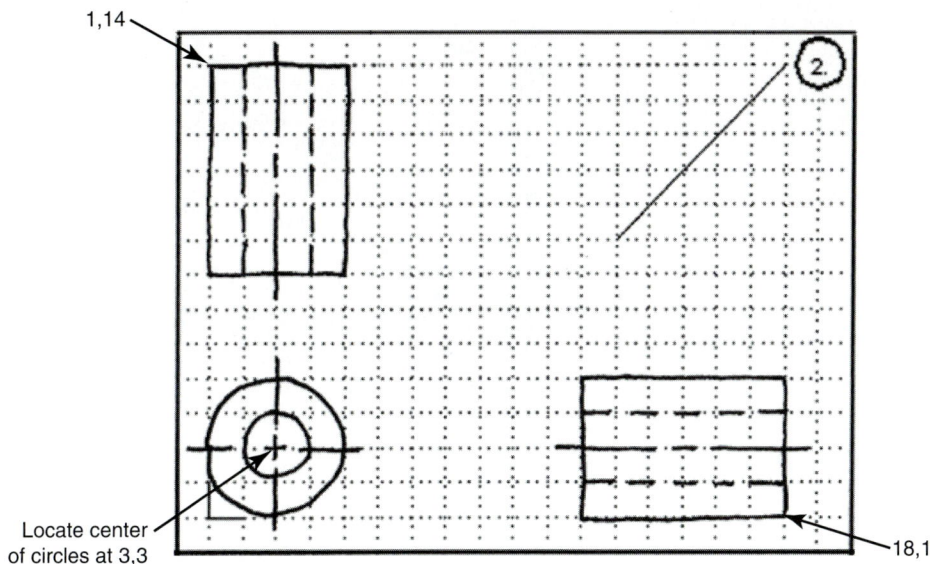

Locate center
of circles at 3,3

18,1

Figure 4-104(b)
Exercise 4-2 (continued)

1,14

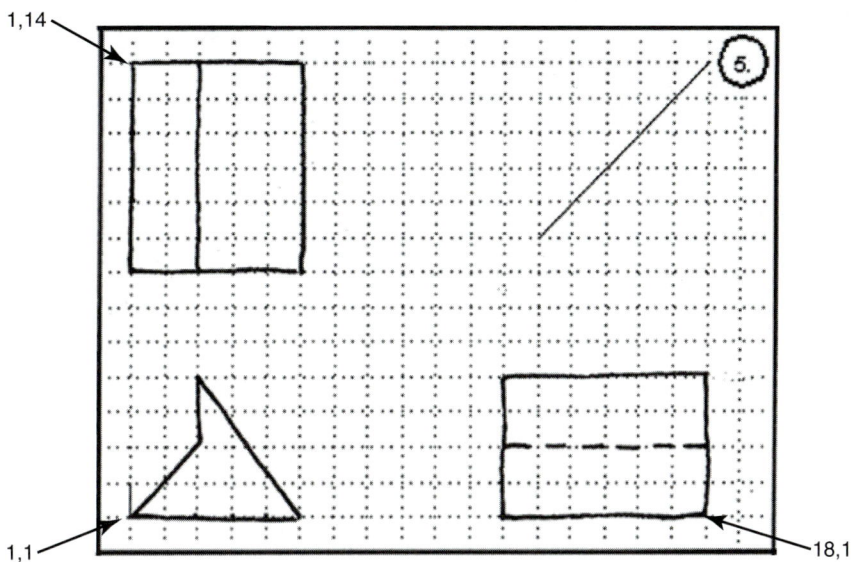

1,1

18,1

Figure 4-104(c)
Exercise 4-2 (continued)

6. Repeat the directions above and draw the Front, Top and Right-side views of Problems 2 and 5 from the Multiview Sketching Exercise 2-1 in Unit 2. See Figures 4-104(b) and 4-104(c). **Save** the drawings as Figure 4-104(b) and Figure 4-104(c) respectively.

PROPERTIES Command

The **PROPERTIES** command is used to display, or change, the **_properties_** of an object or group of objects, such as color, lineweight, layer, linetype, linetype scale, etc. This is a very useful feature in AutoCAD. Begin this command by choosing the **PROPERTIES** icon from the **Standard** toolbar (see Figure 4-105). Next, select an object, and its properties will be displayed in the **Properties Palette**. In the example in Figure 4-106, the circle has been selected, and its properties are displayed in the **Properties Palette**.

To change the properties of the circle, select the field next to the property to be changed. For example, the circle shown is on the **VIS** layer. By selecting the field next to the **Layer** property, a list of layer options appears. By selecting a different layer, the circle can be moved to the selected layer. The circle's other properties, such as **Linetype scale**, can be changed in a similar fashion.

Other information about the circle, such as its diameter, circumference, area, and center location, can be found in the **Properties Palette** under the **Geometry** heading.

properties: In an AutoCAD drawing, the properties of an object include its color, lineweight, layer, linetype, linetype scale, etc.

Figure 4-105   Properties icon

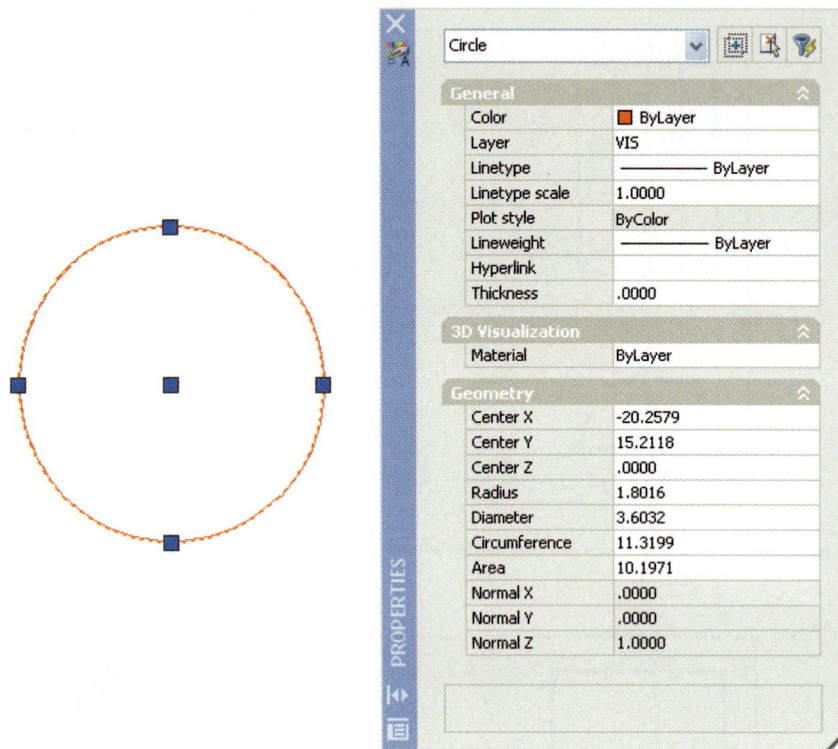

Figure 4-106 Properties dialog box displaying properties of a circle

INQUIRY TOOLBAR

Inquiry commands: Commands used to display information about AutoCAD entities such as distance between two points, area of a closed figure like a rectangle or circle, or the volume of a 3D object. These commands are located on the **Inquiry** toolbar.

The *Inquiry commands* on this toolbar are used to display information about AutoCAD entities, such as distance between two points, area of a closed figure like a rectangle or circle, or the volume of a 3D object. The icons on this toolbar are identified in Figure 4-107.

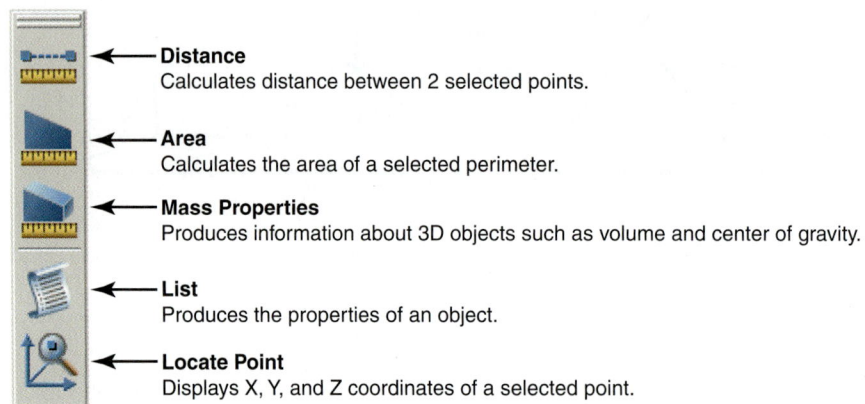

Distance
Calculates distance between 2 selected points.

Area
Calculates the area of a selected perimeter.

Mass Properties
Produces information about 3D objects such as volume and center of gravity.

List
Produces the properties of an object.

Locate Point
Displays X, Y, and Z coordinates of a selected point.

Figure 4-107 The Inquiry toolbar

PLOTTING WITH AUTOCAD 2009

Select the **PLOT** icon from the **Standard** toolbar (see Figure 4-108), and the **Plot** dialog box shown in Figure 4-109 will open.

The options in the **Plot** dialog box allow you to do the following:

- Select the name of the **Printer/plotter** where you want to send your plot.

- Change the **Plot style table** which controls color and pen width.

Note:
Under **Plot style table**, select **none** for color printing, or **monochrome** for black and white prints.

Figure 4-108 Plot icon

- Select the **Paper size**.
- Define the **Plot area**. Available choices are **Display**, **Extents**, **Window**, **View**, **Limits**, or **Layout**.
- Define the **Plot scale**.
- **Preview...** the plot before printing.

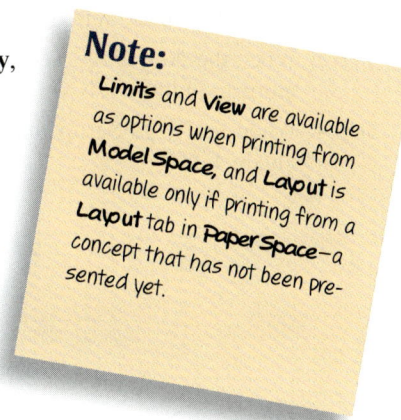

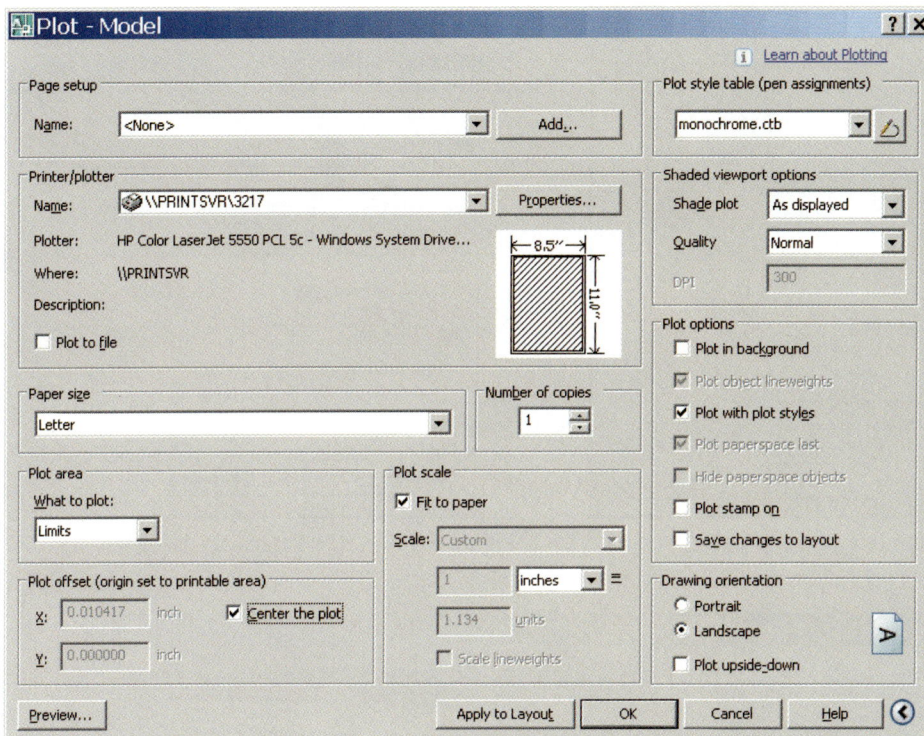

Note:

Limits and *View* are available as options when printing from *Model Space,* and *Layout* is available only if printing from a *Layout* tab in *Paper Space*—a concept that has not been presented yet.

Figure 4-109   Plot dialog box

TIP

Fit To Paper is an option, but the object will not be drawn to scale if this is selected. Select in the window next to **Scale:** to find a list of available plot scales.

Creating a Page Setup for Plotting

Creating named **Page Setups** for an AutoCAD project streamlines the process of plotting drawings. A **Page Setup** is a named set of predefined plot settings. When you wish to print the project, you simply select the name of the **Page Setup** from the list in the **Plot** dialog box rather than resetting the plot settings each time.

Steps in Creating a Page Setup

Step 1.   Select the **File** pull-down menu from the Menu Bar and pick **Page Setup Manager**. The **Page Setup Manager** dialog box opens as shown in Figure 4-110.

Step 2. Select the **New...** button from the dialog box as shown in Figure 4-110.

Step 3. When the **New Page Setup** dialog box opens, type Cottage into the **New page setup name:** window as shown in Figure 4-111 and click **OK**.

Step 4. When the **Page Setup** dialog box opens, set the desired printer settings as shown in Figure 4-112.

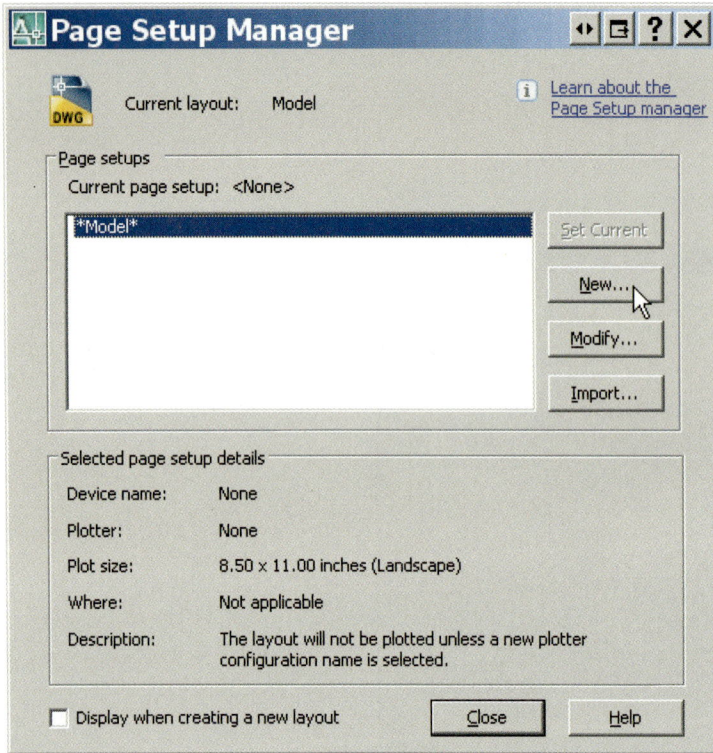

Figure 4-110   Page Setup Manager dialog box

Figure 4-111   New Page Setup dialog box

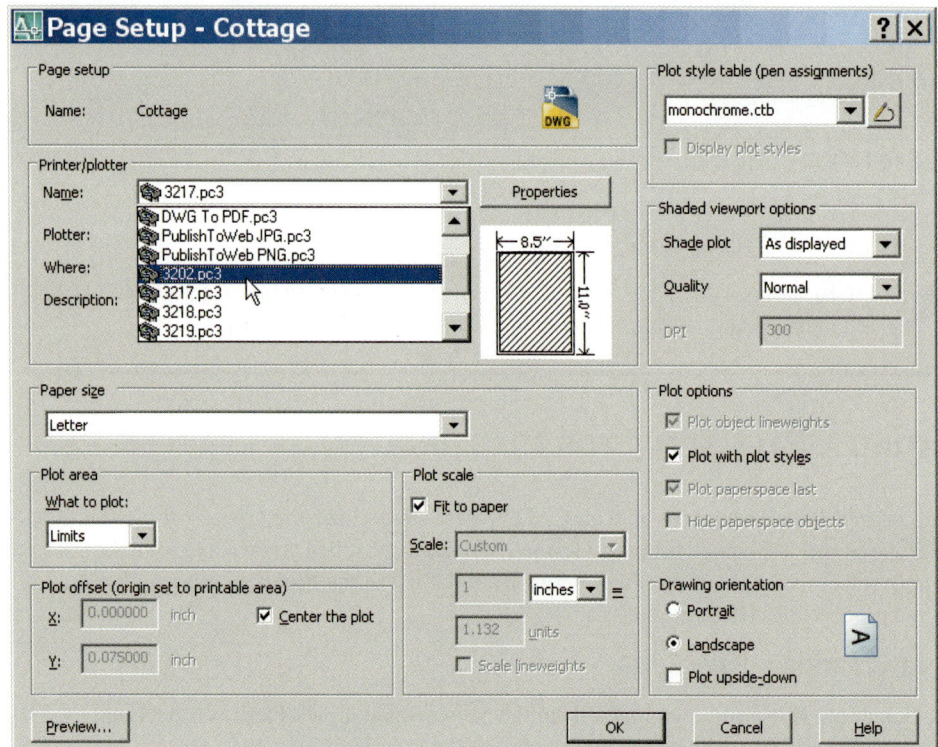

Figure 4-112   Page Setup dialog box

When the desired settings are in place, click **OK** to return to the **Page Setup Manager** dialog box and then click **Close** to return to your drawing.

Step 5. To plot the **Page Setup**, select the **PLOT** icon from the Standard toolbar (see Figure 4-108) and when the **Plot** dialog box opens, select Cottage in the **Name:** window as shown in Figure 4-113 and click **OK** to print the cottage project. **Note:** It is usually a good practice to **Preview** the plot before selecting **OK** to confirm that the printer settings will result in the desired plot.

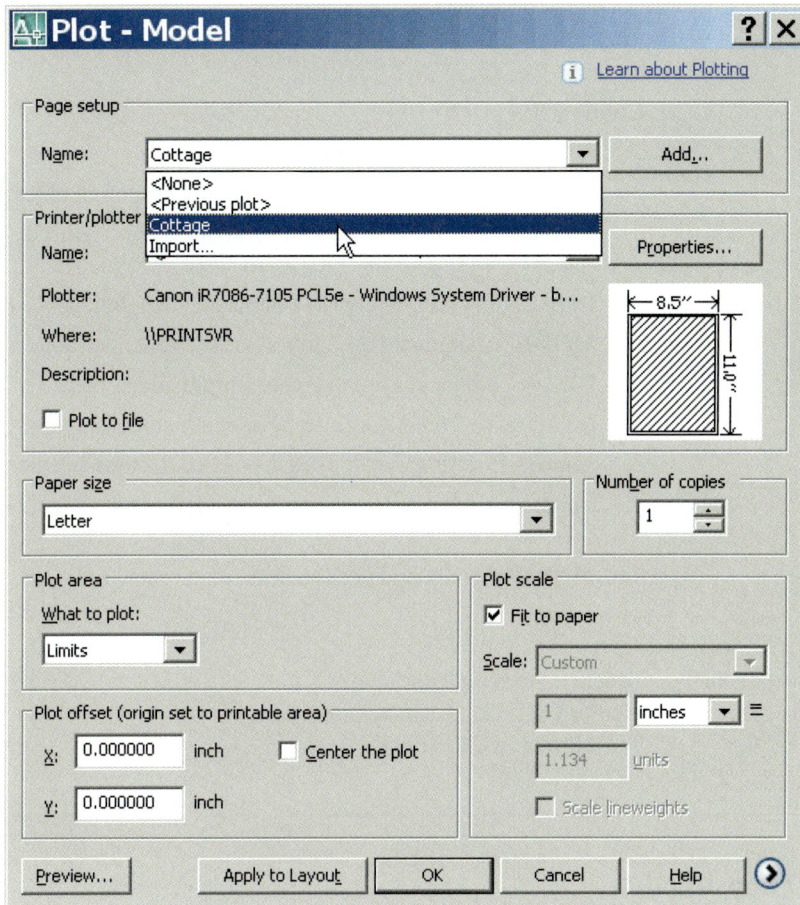

Figure 4-113   Plot dialog box

It is sometimes necessary to adjust the settings in the **Page Setup** dialog box through trial and error until the desired plot settings are in place. Selecting the **Preview...** button allows you to see the settings and adjust accordingly.

SUMMARY

With each release, computer aided design tools become more powerful, allowing CAD users to accomplish higher-level design and drafting tasks. However, as CAD software evolves, it also drives changes in the how technical drawings are created, viewed, shared, and managed. Persons employed in the CAD field will need to be life-long learners to keep up with the dynamic technology of this field.

UNIT TEST QUESTIONS

Short Answer

1. How is coordinate **0,0** represented in the graphics window of an AutoCAD drawing?

2. What is the direction (North, South, East, or West) of the default **Base Angle** setting in AutoCAD?

3. What setting should be defined before the limits of an AutoCAD drawing are set?

4. Which of the following is *not* an AutoCAD **Object Snap** setting: Node, Tangent, Quadrant, or Startpoint?

5. What is the name of the tool on the **Inquiry** toolbar that is used to calculate the distance between two points?

Matching

Column A

a. Polygon

b. Chamfer

c. Polar

d. Limits

e. Absolute

f. Scale

g. Rectangle

h. Decimal

i. Fillet

j. Relative

Column B

1. A type of coordinate that is relative to **0,0**

2. **MODIFY** command used to change the size of an object

3. A type of coordinate that is relative to the last point defined

4. **Units** setting commonly used in mechanical drawings

5. **MODIFY** command used to round the corners of objects

6. **DRAW** command used to create a multisided object

7. A type of coordinate defined by a length and an angle

8. AutoCAD setting that defines the size of the drawing area

9. Command that requires the drafter to define two diagonal points

10. **MODIFY** command used to bevel the corners of objects

Internet Resources

Autodesk (AutoCAD, Revit, Inventor): *www.autodesk.com*
Bentley (Microstation): *www.bentley.com*
Cadalyst: *www.cadalyst.com*

Cadence: *www.cadence.com*
Mentor Graphics: *www.mentor.com*
PTC (ProE): *www.ptc.com*

UNIT EXERCISES

Exercise 4-3: Geoquick CAD Construction

Directions

1. Select **FILE** and **OPEN**.
2. Select the **Student Resource CD**.
3. Select the **Prototype Drawings** folder.
4. Select the **Daily Work Prototype** drawing (open as "read-only").
5. **SAVE AS** to your **Home** directory renaming the drawing as **GEOQUIK**.
6. Set the drawing units to **decimal** and the limits to **0,0** for the lower left and **24,18** for the upper right. In this project, you will use the following steps to construct the object shown in Figure 4-114.

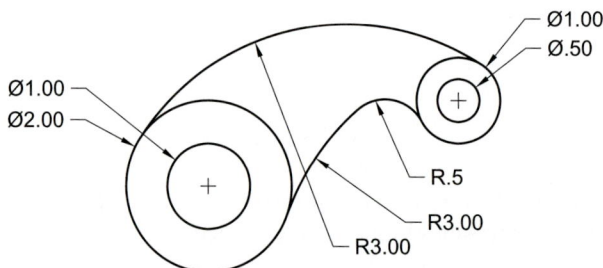

Figure 4-114   Geoquik exercise with dimensions

Step 1. Draw a **2″** diameter circle with a center point at absolute coordinate **4,5**. Draw another **1″** diameter circle at the same center point (see Figure 4-115). Draw a second set of concentric circles at absolute coordinate **7,6**. One circle will have a radius of **.5** and the other **.25** (see Figure 4-115).

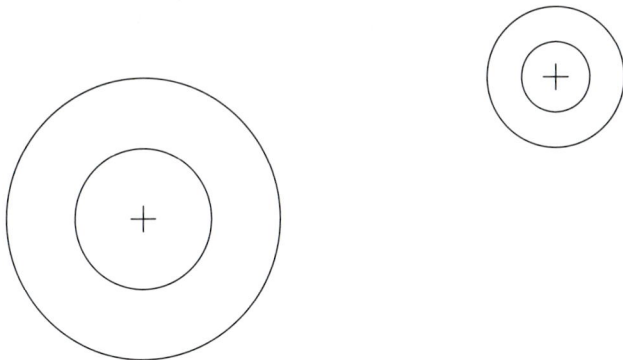

Figure 4-115   Circles created in Step 1

Step 2. From the **Draw** pull-down menu, select the **CIRCLE** command and the **Tangent**, **Tangent**, **Radius** option, or select the **CIRCLE** icon from the **Draw** toolbar and type **TTR** and press <**Enter**>. When prompted for the first and second tangency points, pick on the circles in the approximate locations shown in Figure 4-116. When prompted for the radius, type **3**. Your drawing should resemble Figure 4-117 when this step is completed.

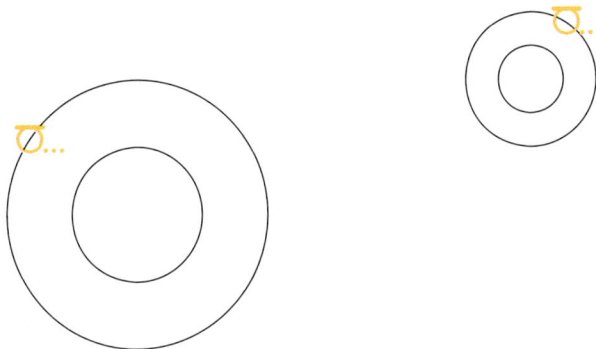

Figure 4-116   Location of tangency points for a circle drawn with tangent, tangent, radius option

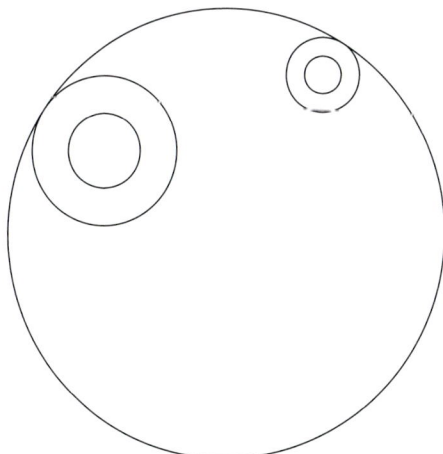

Figure 4-117   Geoquik drawing following completion of Step 2

Step 3. Draw a second **3″** circle by choosing **CIRCLE** and **Tangent**, **Tangent**, **Radius** again. Pick the first and second tangent locations as shown in Figure 4-118. When prompted for radius, type **3**. Your drawing should look like Figure 4-119 when you are finished.

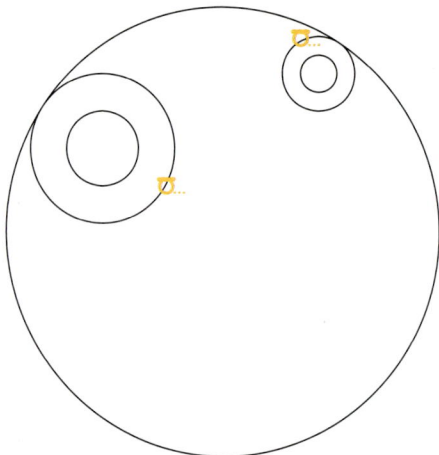

Figure 4-118 Location of tangency points for a circle drawn with tangent, tangent, radius option

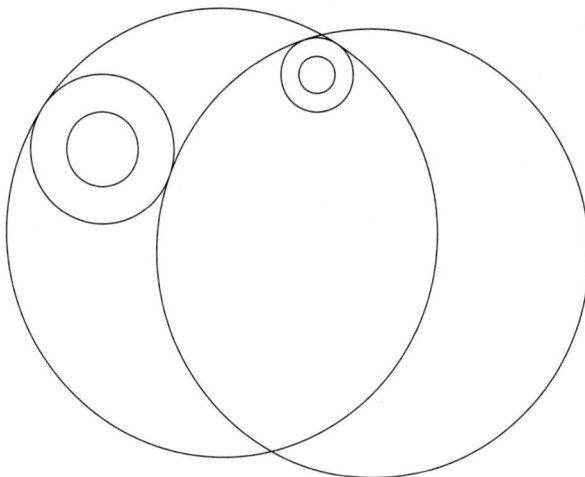

Figure 4-119 Geoquik Drawing following completion of Step 3

Step 4. Choose **CIRCLE** and the **Tangent**, **Tangent**, **Radius** option again. Pick the first and second tangent locations as shown in Figure 4-120. When prompted for radius, type **.5**. When you are finished, your drawing should look like the one in Figure 4-121.

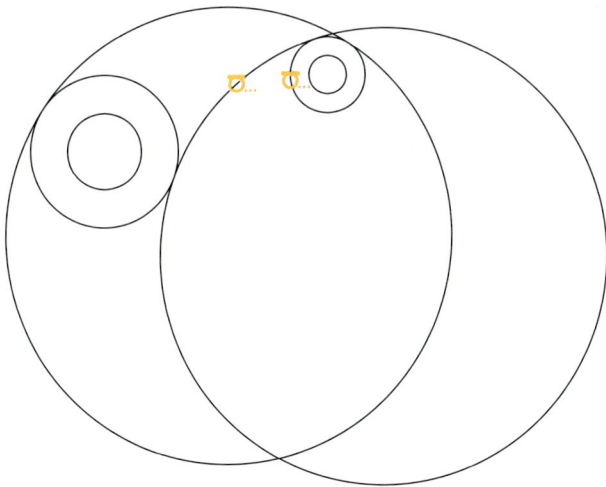

Figure 4-120   Location of tangency points for a circle drawn with tangent, tangent, radius option

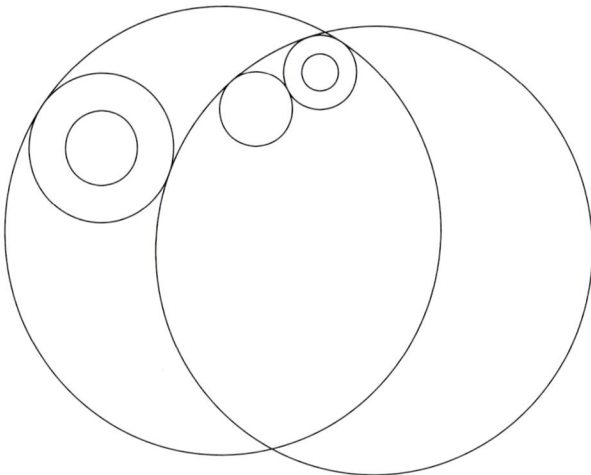

Figure 4-121   Geoquick drawing following completion of Step 4

Step 5. Select the **TRIM** icon from the **Modify** toolbar and press **<Enter>**.

TIP   Remember that pressing **<Enter>** will make every line an edge that can be trimmed to.

Select the circles in the locations shown in Figure 4-122 and begin to trim away the unnecessary construction lines. Continue to trim lines as shown in Figure 4-123.

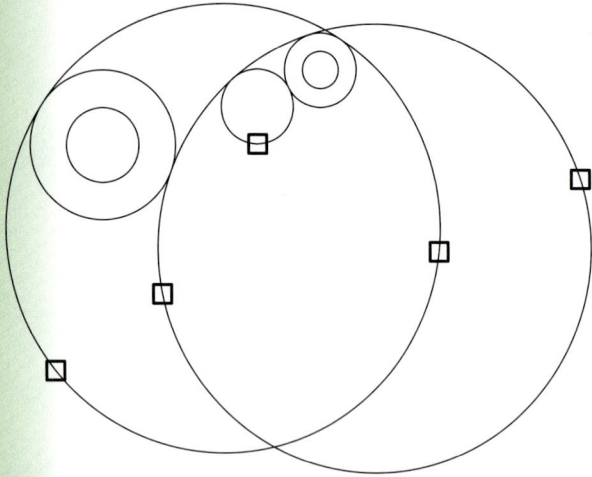

Figure 4-122   Location of points for Trim command

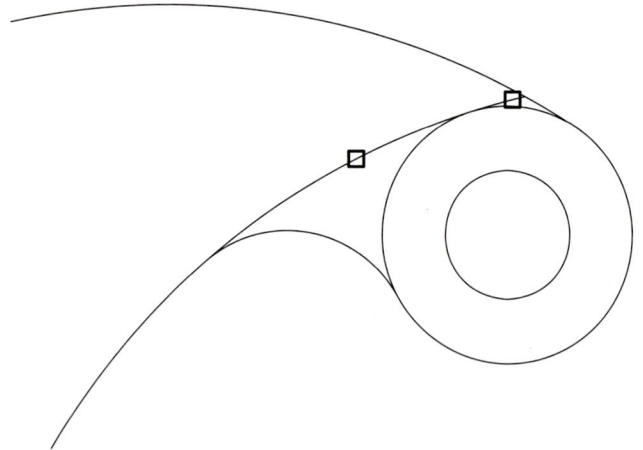

Figure 4-123   Location of points for Trim command

After trimming all the unnecessary lines, your completed drawing should look like Figure 4-124.

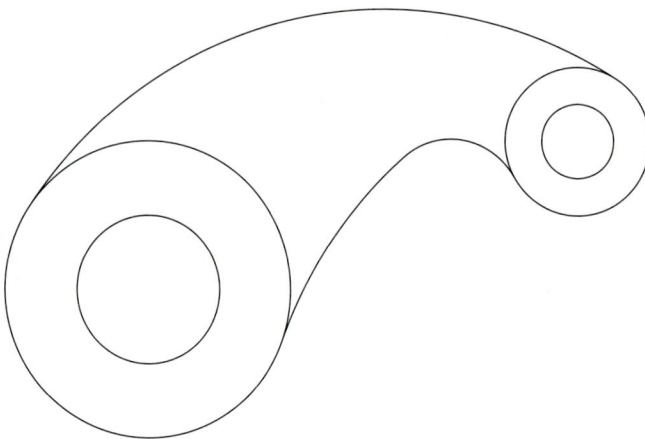

Figure 4-124   Geoquik drawing following completion of Step 5

Step 6. Save and print the drawing as instructed by your teacher.

Exercise 4-4: Cam Construction

Directions

1. Select **FILE** and **OPEN** the **GEOQUIK** drawing you created earlier.
2. In this exercise, you will follow the steps on the following pages to construct the object shown in Figure 4-125.

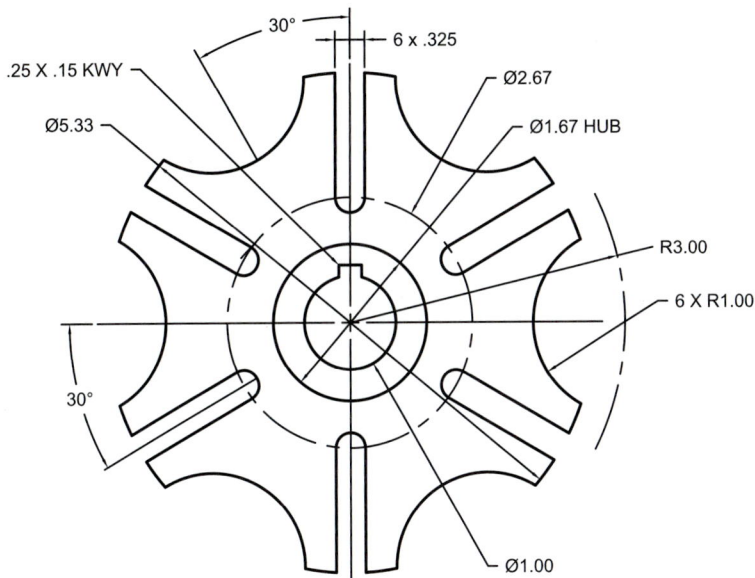

Figure 4-125   Cam

Constructing the Cam

Step 1. Draw two concentric circles with a center point at absolute coordinate **14,11**. Draw one circle with a **6″** diameter and the other with a **5.33″** diameter. Add a center line to the circles. See Figure 4-126.

Step 2. Draw a circle **2″** in diameter in the position shown in Figure 4-127.

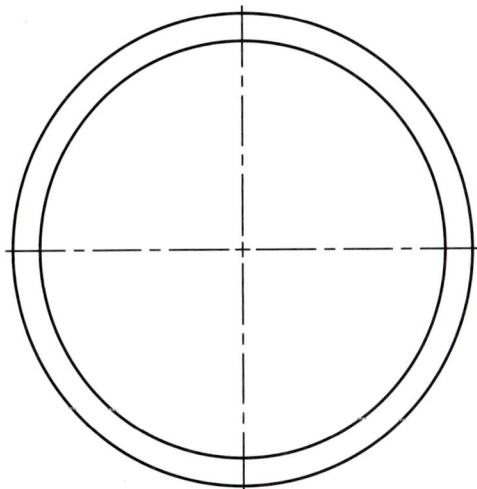

Figure 4-126   Concentric circles

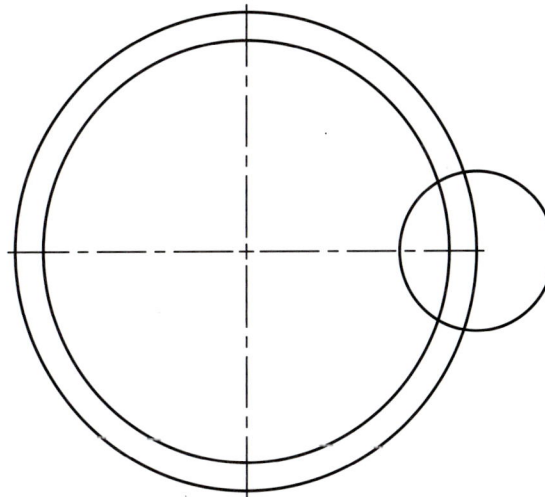

Figure 4-127   Adding 2″ circle

Step 3. Use the **ARRAY** command's **Polar** option to array **6** circles around the center of the cam to fill an angle of **360°** as shown in Figure 4-128.

Step 4. Erase the larger diameter circle and trim the object to create the shape shown in Figure 4-129.

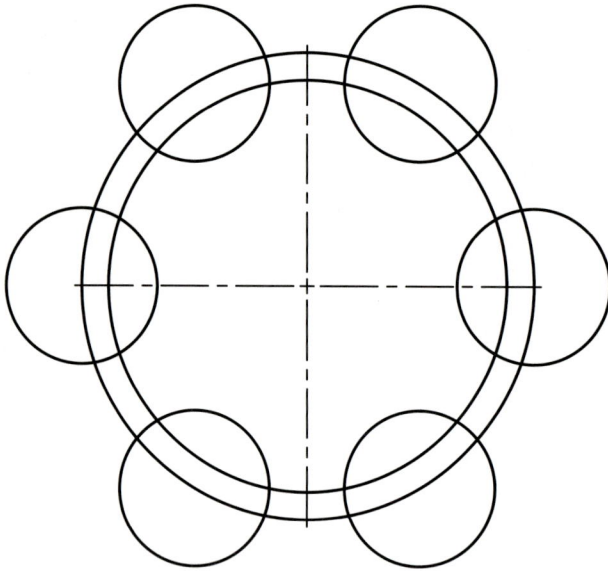

Figure 4-128   Polar array

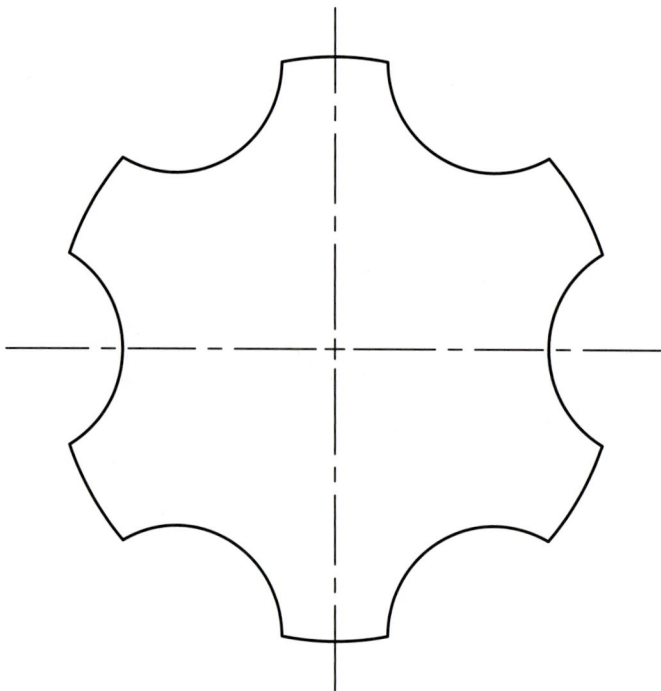

Figure 4-129   Erase and trim

Step 5. To create the slot, first draw a circle with a diameter of **2.67″**. Using the point where the vertical center line and this circle intersect as the center point, draw another circle with a diameter of **.325** as shown in Figure 4-130. Then draw two lines vertical lines (turn **Ortho On**) starting from the quadrants on each side of the small circle as shown in Figure 4-130.

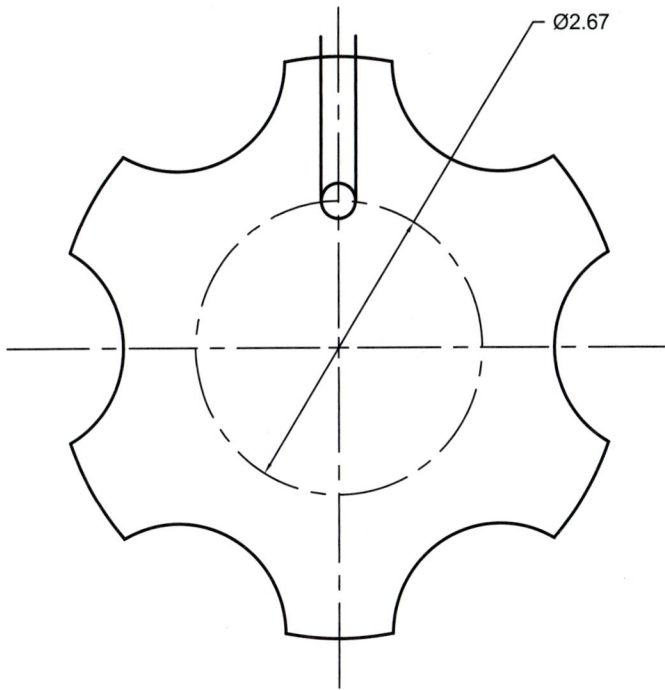

Figure 4-130   Creating a slot

Step 6. Use **Polar Array** to array the slot created in Step 5. Pick the center of the cam for the center of the array, and rotate **6** objects to fill an angle of **360°**, rotating the objects as they are arrayed. Your Cam drawing should look like the object shown in Figure 4-131.

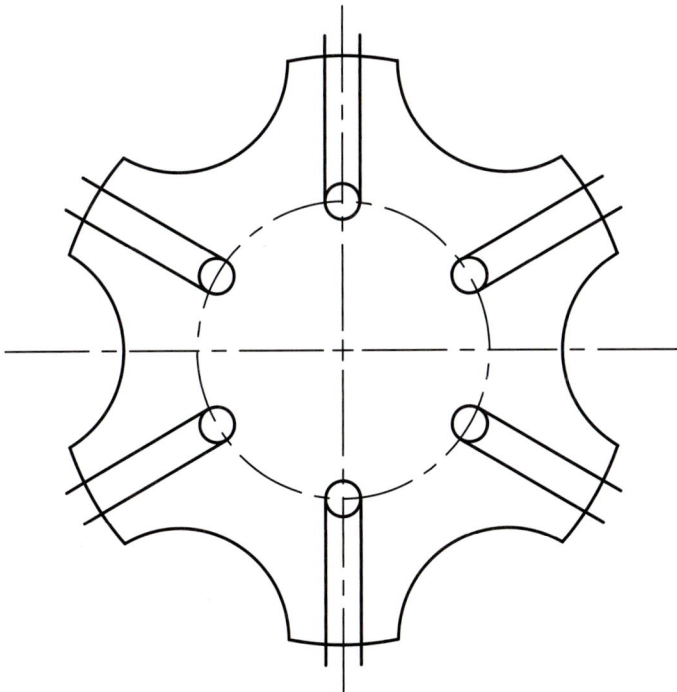

Figure 4-131   Polar array of slot

Step 7. Trim the slots as shown in Figure 4-132.

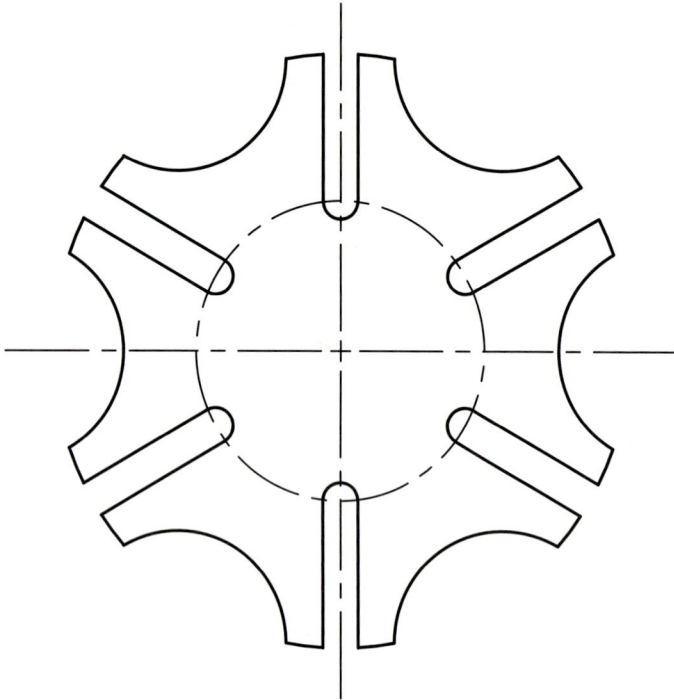

Figure 4-132   Trimmed slots

Step 8. Add the hub and keyway to the cam. Draw two concentric circles, one with a diameter of **1.67″** and another with a diameter of **1.00″** to create the hub. Add a keyway to the top of the **1.00″** circle. The keyway is **.25″** wide and **.15″** tall. Trim the top of the **1.00** circle to open the keyway. See Figure 4-133.

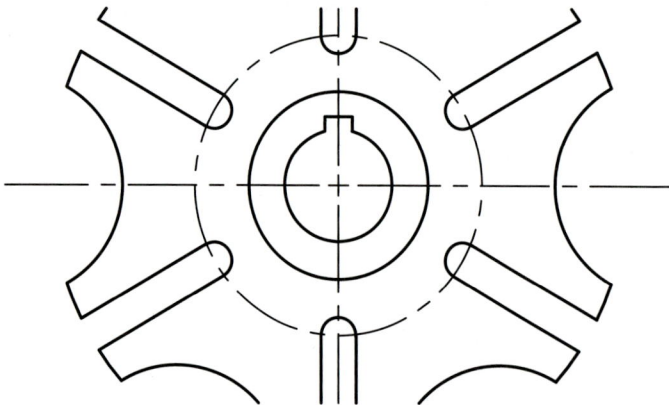

Figure 4-133   Adding the keyway

This completes the cam construction (see Figure 4-134). Follow your instructor's directions to print the drawing. Be sure to save the drawing file before you close AutoCAD.

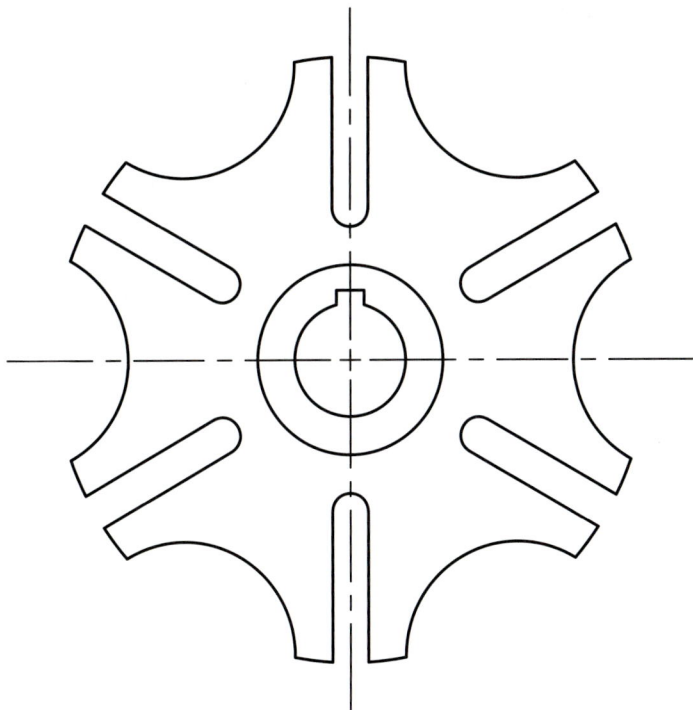

Figure 4-134   Completed Cam drawing

UNIT PROJECTS

Project 4-1: Guest Cottage

In this project, you will draw the floor plan of the guest cottage shown in the designer's sketch in Figure 4-135. In this project, you will be required to use many of the **Draw** and **Modify** tools that you learned about earlier in this unit. Your instructor will guide you through this project, so do not hesitate to ask for help if you get stuck.

Directions

1. Select **FILE** and **OPEN**.
2. Select the **Student Resource CD**.
3. Open the **Prototype Drawings** folder.
4. Open the **Cottage Prototype** drawing (open as "read-only").
5. **SAVE AS** to your **Home** directory renaming the drawing as **GUEST COTTAGE**.
6. Set **Units** to **Architectural** and precision to **1/16″**.

> **TIP**
>
> When Architectural units are in effect, values will default to inches unless you enter a foot mark (′). For example, to enter the length of a line 5 feet 4″ long, type **5′-4** (you do not need to type an inch mark after the 4 because AutoCAD defaults to inches).

7. Set the upper right **Limits** to **48′,38′** and the **Grid** to **4′**.

8. Make the following layers: Floor Plan, Windows, Doors, Labels, Text, Hatch, Clothes Rod, and Dimensions. Assign a color to each layer and set the linetype of the Clothes Rod layer to Dashed.

- In the **Text Style** dialog box, set *Stylus BT* as the font for the **Standard** text style.

- Set the following running object snaps: **Midpoint**, **Intersection**, **Endpoint**, and **Perpendicular**.

- Follow Steps 1 through 12 to create the floor plan.

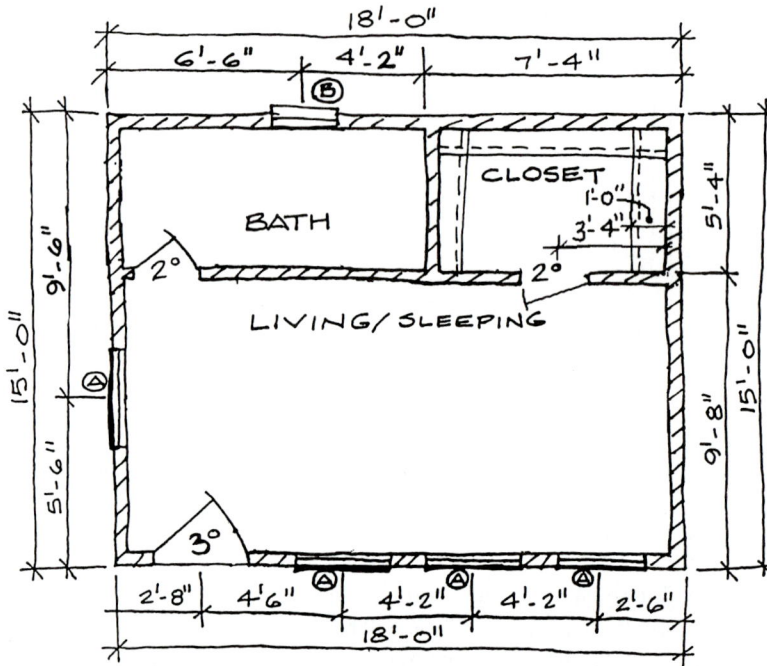

Figure 4-135 Designer's sketch of the guest cottage

Drawing the Cottage Floor Plan

Step 1. To draw the exterior walls, set Floor Plan as the current layer. Use the **RECTANGLE** command to draw a rectangle **18′ × 15′**. Begin the rectangle at absolute coordinate **10′,12′**. See Figure 4-136. Use the **OFFSET** command to offset the rectangle **4″** to the inside. The two rectangles represent the exterior walls of the guest cottage. **EXPLODE** both rectangles so that they can be edited.

Figure 4-136   Exterior walls

Step 2. To draw the interior walls, use the **OFFSET** command with the dimensions shown in Figure 4-137 to place the lines for the interior walls. Be careful to offset the wall to the side shown in the example.

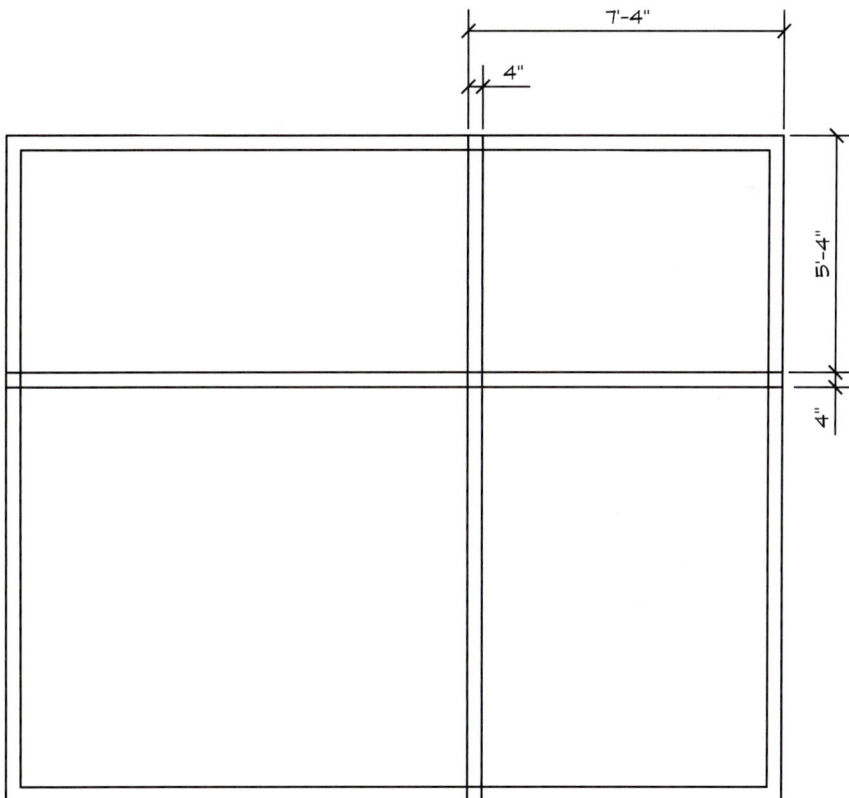

Figure 4-137   Interior walls

Use the **TRIM** command to trim the offset lines as shown in Figure 4-138.

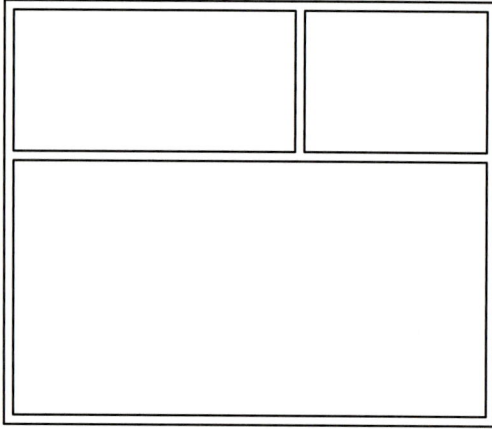

Figure 4-138   Trimmed wall openings

Step 3. To create the front door opening, **OFFSET** the left outside wall line **2′8″** to the right to locate the center of the front door opening. Next, offset this line **1′6″** to both its right and left sides as shown in Figure 4-139 to define the edges of an opening **3′-0″** wide.

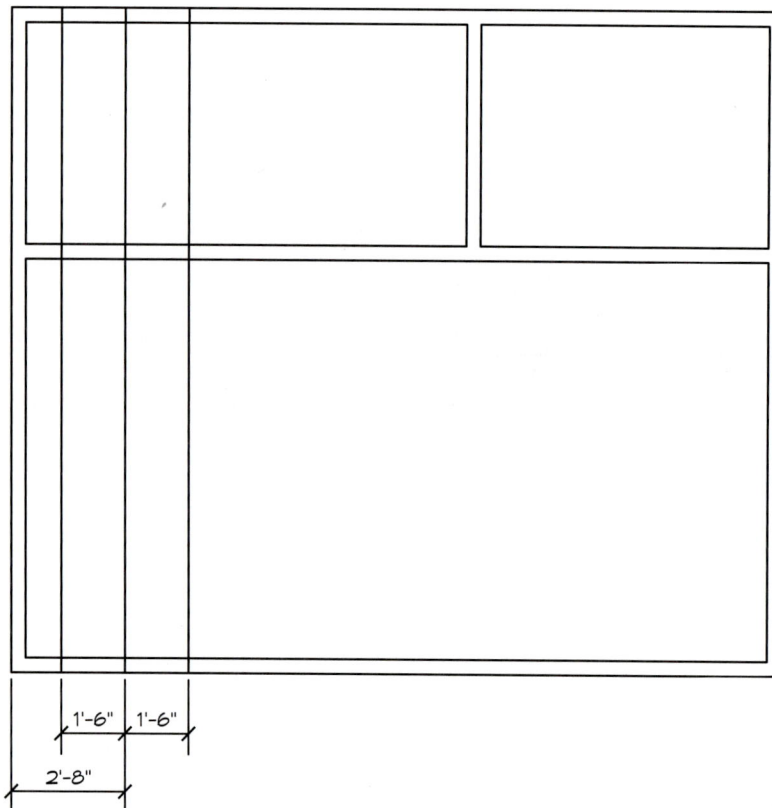

Figure 4-139   Offsetting walls to locate the front door

Trim the offset lines as needed to create the front door opening as shown in Figure 4-140.

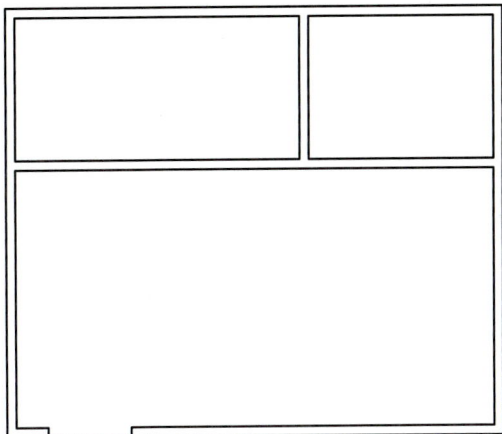

Figure 4-140   Trimming the front door opening

Step 4. To create the interior door openings, **OFFSET** the walls as
        shown in Figure 4-141 to create the two **2'-0"** wide openings
        for the interior doors to the Bath and Closet.

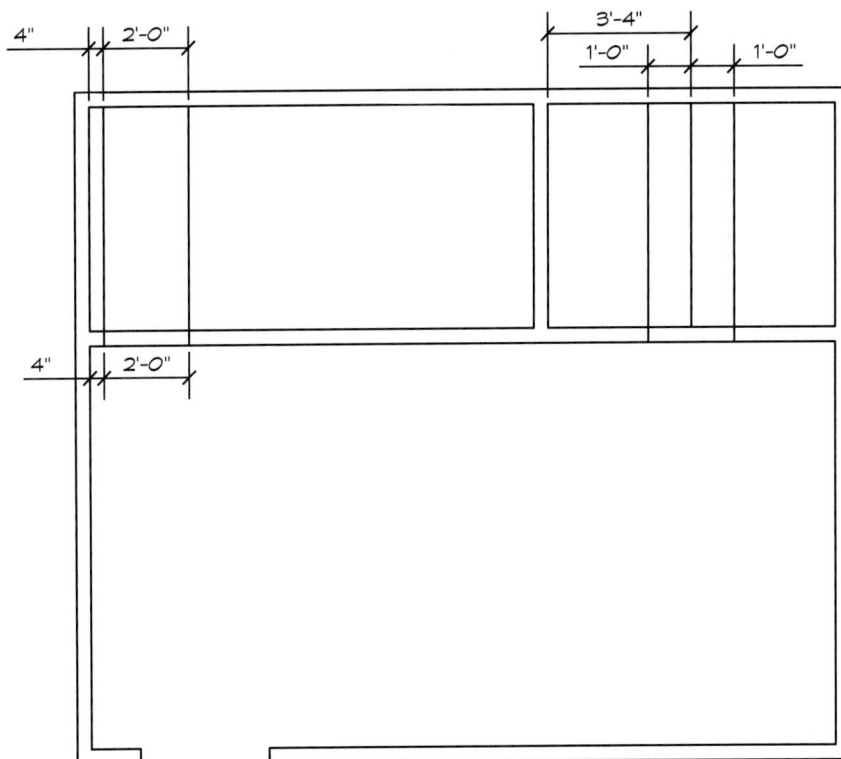

Figure 4-141   Constructing interior door openings

Trim the offset lines to create the door openings as shown
in Figure 4-142.

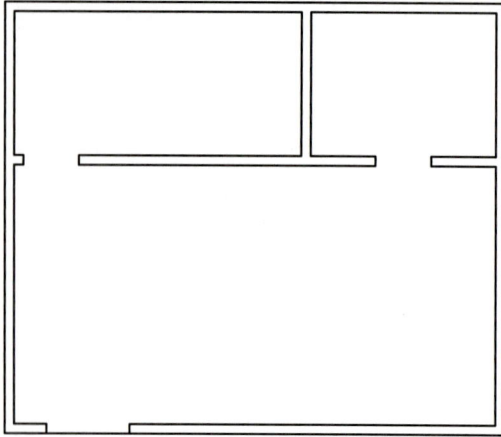

Figure 4-142   Trimming interior door openings

Step 5. To create and locate the windows, set Windows as the current layer. Next, **OFFSET** the walls as shown in Figure 4-143(a) to locate the centers of the windows.

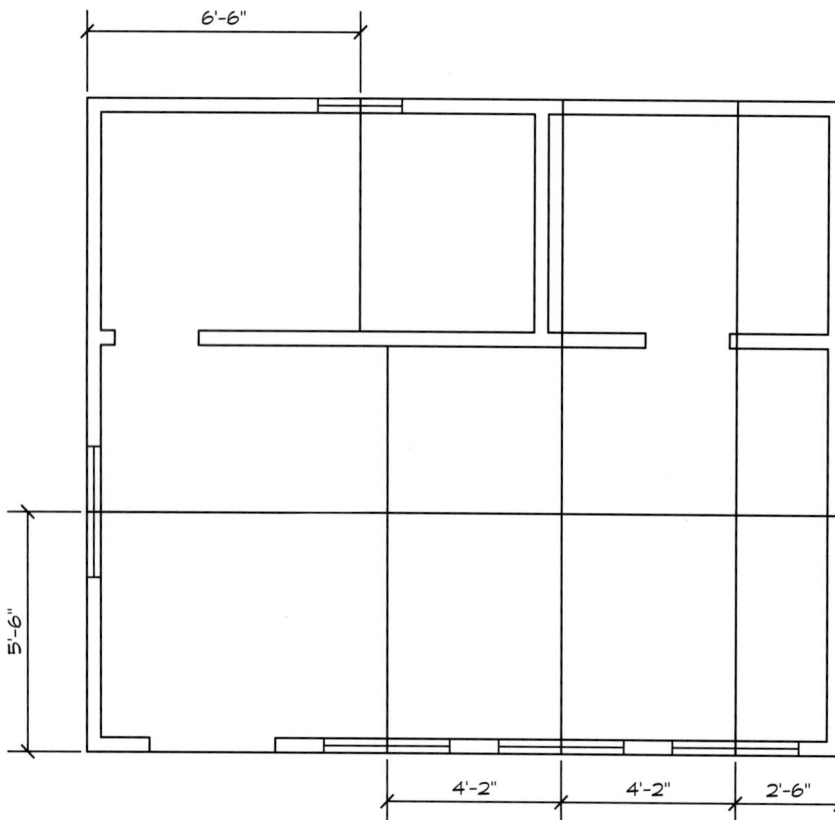

Figure 4-143(a)   Locating the centers of the windows

To make the windows, use the **RECTANGLE** command to draw 2 rectangles: one **36″ × 4″**, as shown in Figure 4-143(b), and another **24″ × 4″**. Draw a horizontal line through each rectangle from the midpoints of each vertical side as shown in Figures 4-143(c) and 4-143(d). Use **COPY** to make three copies of the larger window and then use the

MOVE command to place these copies in their positions along the front wall of the cottage as in Figure 4-143(a).

Figure 4-143(b) Drawing a 36″ × 4″ rectangle

Figure 4-143(c) Horizontal line drawn from midpoint to midpoint of the vertical sides of the rectangle

Figure 4-143(d) 24″ × 4″ window

> **TIP**
>
> Use the **Object Snap** settings to move the window from its midpoint to the intersection of the offset line and the outside edge of the wall to facilitate the exact placement of the windows.

Next, move the smaller window into position along the back wall. Finally, **ROTATE** the remaining window **90°** and move it to its place in the left side wall.

Step 6. To create and locate the doors, set Doors as the current layer and use **RECTANGLE** to draw the three doors. Begin each rectangle at the corner of the door opening where the hinge would be located, as shown in Figure 4-144(a) and use relative coordinates to define the door. The doors are drawn 1″ thick. For example, in Figure 4-144(a) the rectangle for the **3′-0″** front door starts at the top of the left inside corner of the door opening and then is drawn at a distance of **1″** along the X-axis and **36″** along the Y-axis. The other two interior doors are **1″ × 24″** rectangles and are placed as in Figure 4-144(a).

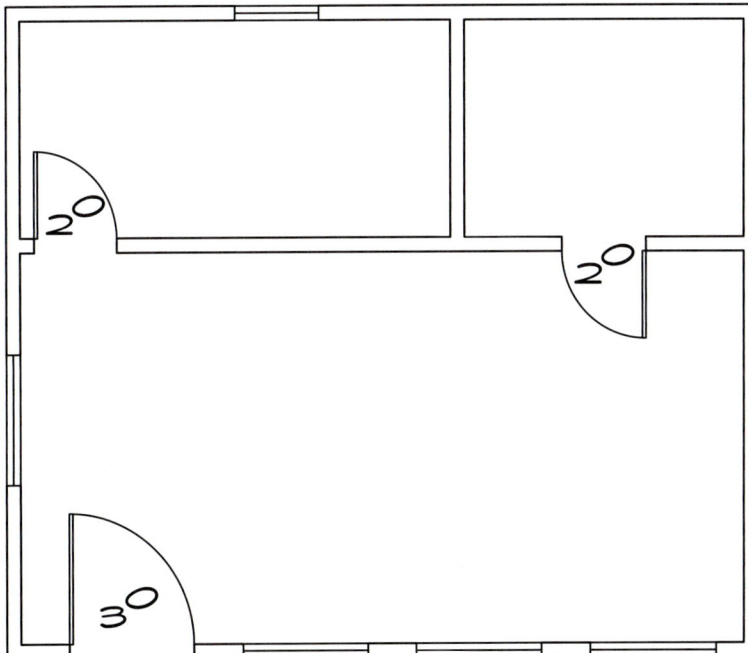

Figure 4-144(a) Creating the doors

Use the **Start**, **Center**, **End** option of the **ARC** command to draw the door swings. For example to draw the door swing for the front door, the start point will be the top corner of the door opening that is opposite the corner where the door is attached, the center of the arc will be the corner of the door opening where the door is attached, the end of the arc will be the top right corner of the door. See Figure 4-144(b).

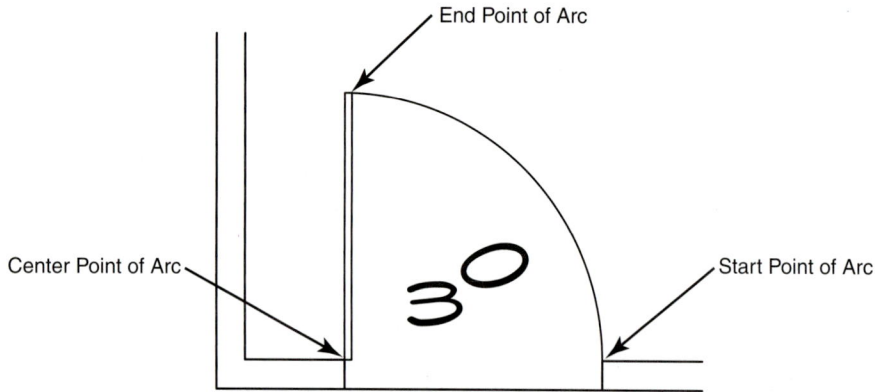

Figure 4-144(b)   Drawing door swings with the ARC command

Step 7.   To draw the closet shelves and clothes rods, set the Floor Plan layer current and add shelves to the closet by offsetting lines **12″** from the closet's inside walls. See Figure 4-145. Set the Clothes Rod layer current and add clothes rods to the closet by offsetting lines **10″** from the closet's inside walls. See Figure 4-145.

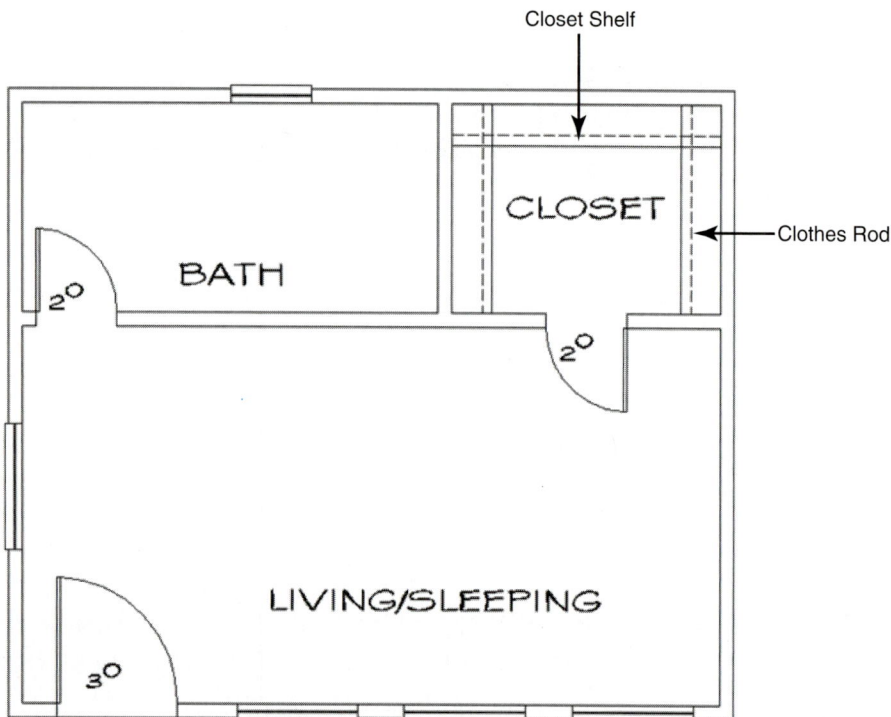

Figure 4-145   Closet shelves and clothes rods

Step 8. To add room labels to the floor plan, set the Text layer current and use the **MTEXT** command to add the room names shown in Figure 4-145. The text height for room names should be **6″**. For door call-outs, both numbers should be **4″**.

TIP

Place each door call-out number separately using the **DTEXT** command.

Step 9. Window marks are used to identify each window in the floor plan. Create the window mark by drawing a circle that is **9″** in diameter and centering the letter **A** or **B** inside of the circle (see Figure 4-146). Use **4″** text height for the letters. Place window marks next to each window as shown in the designer's sketch in Figure 4-135.

Figure 4-146   Window mark

Step 10. Complete the window schedule by using **Text Edit** to change the place-holders in the schedule to the values shown in Figure 4-147. In the Window Schedule, the $3^0 \times 3^0$ label refers to a window that is **3′-0″** wide by **3′-0″** tall. The *S.H.* labeled in the *TYPE* column indicates that the window type is Single Hung.

WINDOW SCHEDULE

MARK	SIZE	TYPE	QTY
A	$3^0 \times 3^0$	S.H.	4
B	$2^0 \times 2^0$	S.H.	1

Figure 4-147   Window schedule

Step 11. To Hatch the walls, set the Hatch layer current and use the **HATCH** command to place the pattern inside the walls. Select the **Net** pattern and set the scale to **10**, then use the **Pick Points** option to select inside the walls. After hatching, the plan should look like the one shown in Figure 4-148.

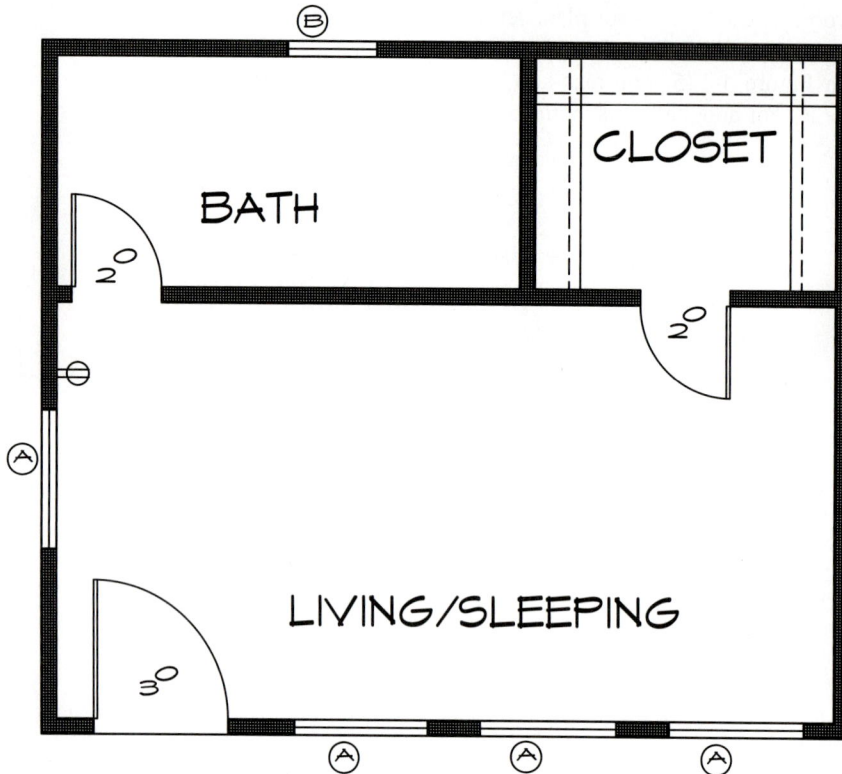

Figure 4-148   Walls with hatch pattern applied

Step 12. Follow the steps presented earlier in this unit to create a
Page Setup for the Cottage.

Plot the project by selecting the *Cottage* page setup
from the **Plot** dialog box. Be sure to save the drawing file
before closing AutoCAD.

Project 4-2: Bracket

Draw the Front, Top, and Side views of the Bracket shown in
the designer's sketch in Figure 4-149.

Note:
Your instructor may give you
different plotter settings than
the ones shown in the figures.

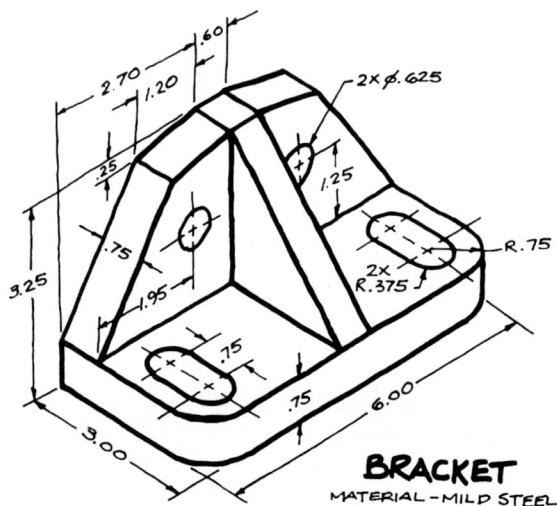

Figure 4-149   Designer's sketch of bracket

Directions

1. Select **FILE** and **OPEN**.
2. Select the **Student Resource CD**.
3. Open the **Prototype Drawings** folder.
4. Open the **Daily Work Prototype** drawing (open as "read-only").
5. **SAVE AS** to your **Home** directory and rename the new drawing **BRACKET**.
6. Set the drawing environment settings and create the layers specified below.
   a. Draw visible lines on the Visible layer.
   b. Draw hidden lines on the Hidden layer.
   c. Draw center lines on the Center layer.
   d. Place text on the Text layer.

Drawing Environment Settings

Units: Decimal

Units Precision: 0.000

Limits: 0,0 – 24,18

Text Style Font = Arial

Layer Settings:

Name	Color	Linetype	Lineweight
Visible	Red	Continuous	.50 mm
Hidden	Blue	Hidden	Default
Center	Green	Center	Default
Text	Green	Continuous	Default

Follow your instructor's directions to print the drawing. Be sure to save the drawing file when you close AutoCAD.

Project 4-3: Shaft Guide (SI)

Draw the Front, Top, and Side views of the shaft guide shown in the designer's sketch in Figure 4-150. The units of measurement provided on the designer's sketch are in millimeters.

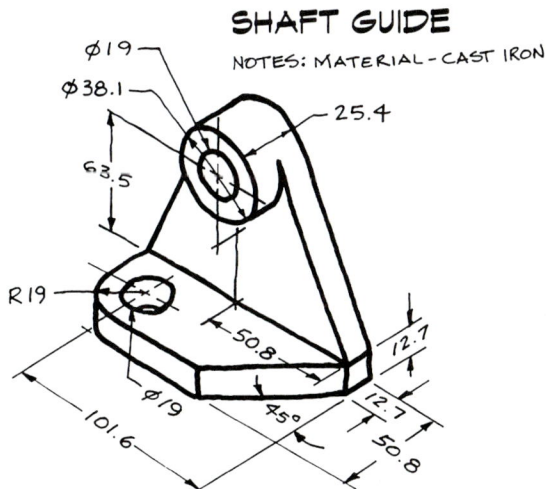

Figure 4-150   Designer's sketch of shaft guide

Directions

1. Select **FILE** and **OPEN**.
2. Select the **Student Resource CD**.
3. Open the **Prototype Drawings** folder.
4. Open the **Daily Work Metric Prototype** drawing (open as "read-only").
5. **SAVE AS** to your **Home** directory and rename the new drawing **SHAFT GUIDE**.
6. Set the drawing environment settings and create the layers specified below.
   a. Draw visible lines on the Visible layer.
   b. Draw hidden lines on the Hidden layer.
   c. Draw center lines on the Center layer.
   d. Place text on the Text layer.
7. Set **LTSCALE** to 25.4.

Drawing Environment Settings

Units: Decimal

Units Precision: 0.0

Limits: 0,0 – 594,420 (Metric A2)

Text Style Font = Arial

Layer Settings:

Name	Color	Linetype	Lineweight
Visible	Red	Continuous	.50 mm
Hidden	Blue	Hidden	Default
Center	Green	Center	Default
Text	Green	Continuous	Default

Follow your instructor's directions to print the drawing. Be sure to save the drawing file when you close AutoCAD.

Project 4-4: Tool Holder

Draw the Front, Top, and Side views of the tool holder shown in the designer's sketch in Figure 4-151.

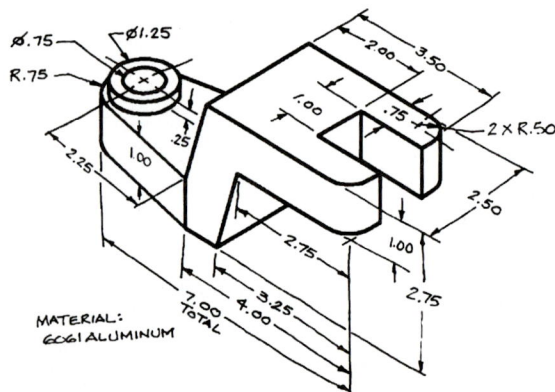

Figure 4-151   Designer's sketch of tool holder

Directions

1. Select **FILE** and **OPEN**.
2. Select the **Student Resource CD**.
3. Open the **Prototype Drawings** folder.
4. Open the **Daily Work Prototype** drawing (open as "read-only").
5. **SAVE AS** to your **Home** directory and rename the drawing **TOOL HOLDER**.
6. Set the drawing environment settings and create the layers specified below.
   a. Draw visible lines on the Visible layer.
   b. Draw hidden lines on the Hidden layer.
   c. Draw center lines on the Center layer.
   d. Place text on the Text layer.

Drawing Environment Settings

Units: Decimal

Units Precision: 0.000

Limits: 0,0 – 24,18

Text Style Font = Arial

Layer Settings:

Name	Color	Linetype	Lineweight
Visible	Red	Continuous	.50 mm
Hidden	Blue	Hidden	Default
Center	Green	Center	Default
Text	Green	Continuous	Default

Follow your instructor's directions to print the drawing. Be sure to save the drawing file when you close AutoCAD.

Project 4-5: Tool Slide

Draw the Front, Top, and Side views of the tool slide shown in the designer's sketch in Figure 4-152.

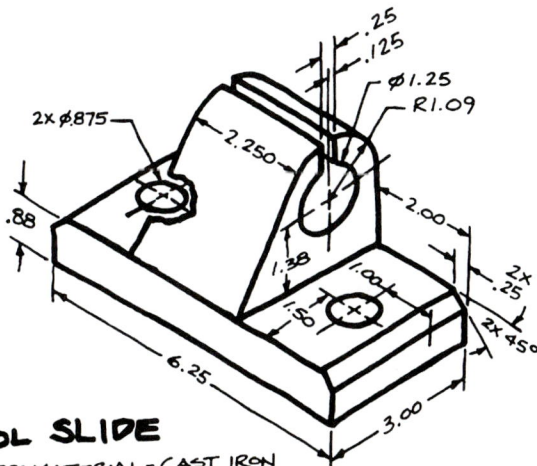

Figure 4-152   Designer's sketch of tool slide

Directions

1. Select **FILE** and **OPEN**.
2. Select the **Student Resource CD**.
3. Open the **Prototype Drawings** folder.
4. Open the **Daily Work Prototype** drawing (open as "read-only").
5. **SAVE AS** to your **Home** directory and rename the new drawing **TOOL SLIDE**.
6. Set the drawing environment settings and create the layers specified below.
   a. Draw visible lines on the Visible layer.
   b. Draw hidden lines on the Hidden layer.
   c. Draw center lines on the Center layer.
   d. Place text on the Text layer.

Drawing Environment Settings

Units: Decimal

Units Precision: 0.000

Limits: 0,0 – 24,18

Text Style Font = Arial

Layer Settings:

Name	Color	Linetype	Lineweight
Visible	Red	Continuous	.50 mm
Hidden	Blue	Hidden	Default
Center	Green	Center	Default
Text	Green	Continuous	Default

Follow your instructor's directions to print the drawing. Be sure to save the drawing file when you close AutoCAD.

OPTIONAL UNIT PROJECTS

Project 4-6: Offset Flange (SI)

Draw the Front and Side views of the offset flange shown in the designer's sketch in Figure 4-153. The units of measurement provided on the designer's sketch are in millimeters.

Directions

1. Select **FILE** and **OPEN**.
2. Select the **Student Resource CD**.
3. Open the **Prototype Drawings** folder.
4. Open the **Daily Work Metric Prototype** drawing (open as "read-only").
5. **SAVE AS** to your **Home** directory and rename the new drawing **OFFSET FLANGE**.
6. Set the drawing environment settings and create the layers specified below.
   a. Draw visible lines on the Visible layer.
   b. Draw hidden lines on the Hidden layer.

c. Draw center lines on the Center layer.

d. Place text on the Text layer.

7. Set **LTSCALE** to 25.4

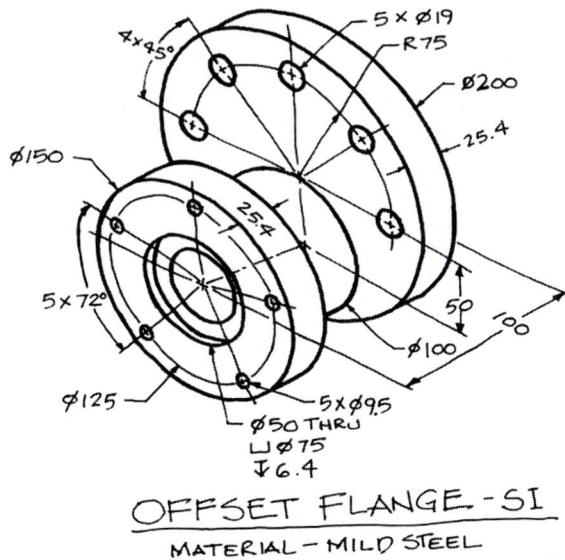

OFFSET FLANGE - SI

MATERIAL - MILD STEEL

Figure 4-153   Designer's sketch of the offset flange

Drawing Environment Settings

Units: Decimal

Units Precision: 0.0

Limits: 0,0 – 594,420 (Metric A2)

Text Style Font = Arial

Layer Settings:

Name	Color	Linetype	Lineweight
Visible	Red	Continuous	.50 mm
Hidden	Blue	Hidden	Default
Center	Green	Center	Default
Text	Green	Continuous	Default

Follow your instructor's directions to print the drawing. Be sure to save the drawing file when you close AutoCAD.

Project 4-7: Angle Stop (SI)

Draw the Front, Top, and Side views of the angle stop shown in the designer's sketch in Figure 4-154. The units of measurement provided on the designer's sketch are in millimeters.

Directions

1. Select **FILE** and **OPEN**.
2. Select the **Student Resource CD**.
3. Open the **Prototype Drawings** folder.
4. Open the **Daily Work Metric Prototype** drawing (open as "read-only").
5. **SAVE AS** to your **Home** directory and rename the new drawing **ANGLE STOP**.

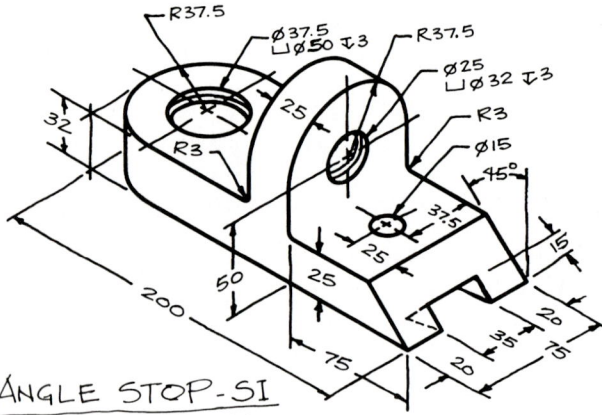

ANGLE STOP-SI

MATERIAL – ALUMINUM 6061

Figure 4-154 Designer's sketch of the angle stop

6. Set the drawing environment settings and create the layers specified below.
   a. Draw visible lines on the Visible layer.
   b. Draw hidden lines on the Hidden layer.
   c. Draw center lines on the Center layer.
   d. Place text on the Text layer.
7. Set **LTSCALE** to 25.4.

Drawing Environment Settings

Units: Decimal

Units Precision: 0.0

Limits: 0,0 – 594,420 (Metric A2)

Text Style Font = Arial

Layer Settings:

Name	Color	Linetype	Lineweight
Visible	Red	Continuous	.50 mm
Hidden	Blue	Hidden	Default
Center	Green	Center	Default
Text	Green	Continuous	Default

Follow your instructor's directions to print the drawing. Be sure to save the drawing file when you close AutoCAD.

Project 4-8: Swivel Stop

Draw the Front, Top, and Side views of the swivel stop shown in the designer's sketch in Figure 4-155.

Directions

1. Select **FILE** and **OPEN**.
2. Select the **Student Resource CD**.
3. Open the **Prototype Drawings** folder.
4. Open the **Daily Work Prototype** drawing (open as "read-only").
5. **SAVE AS** to your **Home** directory and rename the new drawing **SWIVEL STOP**.

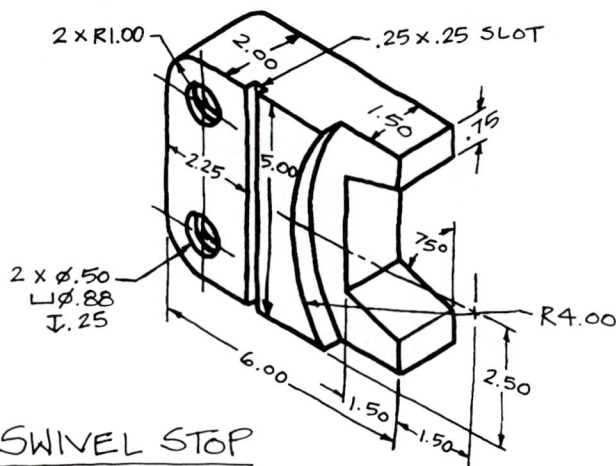

Figure 4-155 Designer's sketch of the swivel stop

6. Set the drawing environment settings and create the layers specified below.
   a. Draw visible lines on the Visible layer.
   b. Draw hidden lines on the Hidden layer.
   c. Draw center lines on the Center layer.
   d. Place text on the Text layer.

Drawing Environment Settings

Units: Decimal

Units Precision: 0.000

Limits: 0,0 – 24,18

Text Style Font = Arial

Layer Settings:

Name	Color	Linetype	Lineweight
Visible	Red	Continuous	.50 mm
Hidden	Blue	Hidden	Default
Center	Green	Center	Default
Text	Green	Continuous	Default

Follow your instructor's directions to print the drawing. Be sure to save the drawing file when you close AutoCAD.

Project 4-9: Alignment Guide

Draw the Front, Top, and Side views of the alignment guide shown in the designer's sketch in Figure 4-156.

Directions

1. Select **FILE** and **OPEN**.
2. Select the **Student Resource CD**.
3. Open the **Prototype Drawings** folder.
4. Open the **Daily Work Prototype** drawing (open as "read-only").
5. **SAVE AS** to your **Home** directory and rename the new drawing **ALIGNMENT GUIDE**.

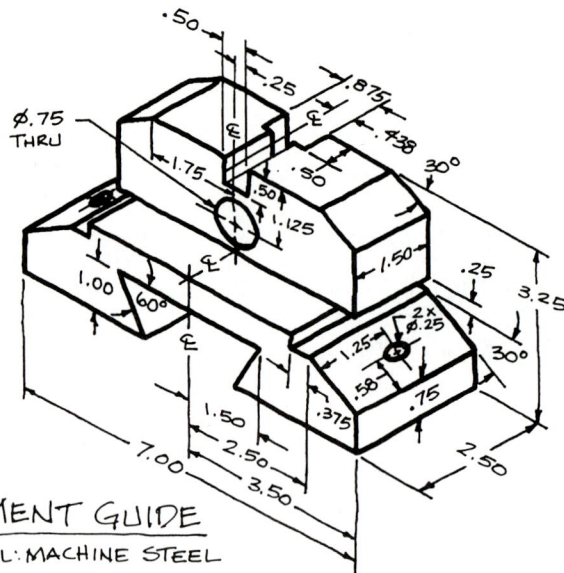

Figure 4-156   Designer's sketch of the alignment guide

6. Set the drawing environment settings and create the layers specified below.
   a. Draw visible lines on the Visible layer.
   b. Draw hidden lines on the Hidden layer.
   c. Draw center lines on the Center layer.
   d. Place text on the Text layer.

Drawing Environment Settings

Units: Decimal

Units Precision: 0.000

Limits: 0,0 – 24,18

Text Style Font = Arial

Layer Settings:

Name	Color	Linetype	Lineweight
Visible	Red	Continuous	.50 mm
Hidden	Blue	Hidden	Default
Center	Green	Center	Default
Text	Green	Continuous	Default

Follow your instructor's directions to print the drawing. Be sure to save the drawing file when you close AutoCAD.

Project 4-10: Flange #1105

Draw the Front, Top, and Side views of flange #1105 shown in the designer's sketch in Figure 4-157.

Directions

1. Select **FILE** and **OPEN**.
2. Select the **Student Resource CD**.
3. Open the **Prototype Drawings** folder.

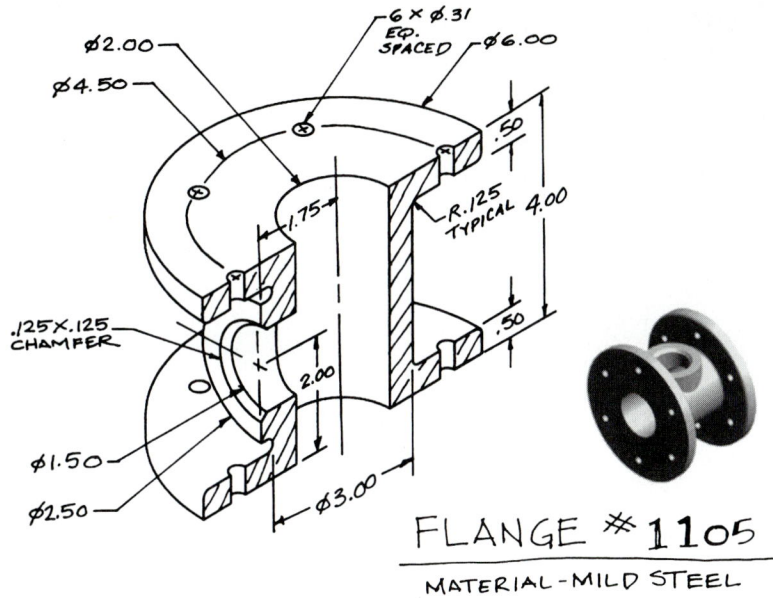

Figure 4-157   Designer's sketch of flange #1105

4. Open the **Daily Work Prototype** drawing (open as "read-only").
5. **SAVE AS** to your **Home** directory and rename the new drawing **FLANGE 1105**.
6. Set the drawing environment settings and create the layers specified below.
   a. Draw visible lines on the Visible layer.
   b. Draw hidden lines on the Hidden layer.
   c. Draw center lines on the Center layer.
   d. Place text on the Text layer.

Drawing Environment Settings

Units: Decimal

Units Precision: 0.000

Limits: 0,0 – 24,18

Text Style Font = Arial

Layer Settings:

Name	Color	Linetype	Lineweight
Visible	Red	Continuous	.50 mm
Hidden	Blue	Hidden	Default
Center	Green	Center	Default
Text	Green	Continuous	Default

Follow your instructor's directions to print the drawing. Be sure to save the drawing file when you close AutoCAD.

Dimensioning Mechanical Drawings

5

Unit Objectives

- Explain what dimensions are.
- Describe how dimensions are determined by designers.
- Describe the importance of tolerances to the dimensioning process.
- Calculate a fit between two simple parts.
- Describe how ASME and ISO standards affect the creation of mechanical drawings.
- Apply *ASME Y14.5M-1994* standards when dimensioning machine parts.
- List the "do's and don'ts" of dimensioning mechanical drawings.
- Describe the role of drafters in the dimensioning process.
- Use the commands on AutoCAD's **Dimension** toolbar.
- Create and modify mechanical dimension styles with AutoCAD's **Dimension Style Manager**.
- Add dimensions to mechanical engineering drawings.

INTRODUCTION

After the necessary views of a machine part have been drawn, the next step in the creation of a technical drawing is to add *dimensions* to the views. Dimensions communicate important information about the size and location of the features of an object, for example, the diameter of a hole and the location of its center point on a machine part.

Dimensional information may also include notes that provide other necessary information needed during the manufacture of a machine part. For example, the material the part is manufactured from or special processes performed on the part (heat treating, polishing, etc.) during manufacture can be labeled as notes in the field of the drawing.

Drafters play an important role in ensuring that the dimensions and notes on the designer's sketch are accurately represented on the finished mechanical drawing; otherwise, the part may not be manufactured as it was designed.

This unit presents the theory and practice of dimensioning drawings prepared for the mechanical engineering field and the use of AutoCAD's dimensioning tools and settings.

DIMENSION STANDARDS FOR MECHANICAL DRAWINGS

Dimension standards define the rules and guidelines for the preparation of technical drawings. In the United States, the industry-wide standard for dimensioning machine parts is published by the American Society of Mechanical Engineers (ASME) and is titled *ASME Y14.5M-1994 Dimensioning and Tolerancing*. According to ASME's website, "This standard establishes uniform practices for stating and interpreting dimensioning, tolerancing, and related requirements for use on engineering drawings."

---

dimensions: Annotations that are added to a technical drawing specifying the size and location of the features of an object. There are two types of dimensions: *size* and *location*. For example, the diameter of a hole is a size dimension, whereas the dimensions that indicate the placement of the center of the hole are location dimensions. Dimensional information may include notes concerning the material the object is manufactured from, special processes performed on the object (heat treating, polishing, etc.), and any other information needed during the manufacture of a part or the construction of a building.

dimension standards: Dimensioning rules that have been created to standardize dimensioning styles and techniques. Dimensioning standards for mechanical drawings have been defined by the American Society of Mechanical Engineers (ASME) and the International Standards Organization (ISO). Dimensioning standards for construction documents are defined by the United States National CAD Standard.

American National Standards Institute (ANSI): The national organization for the development of standards in the United States. ANSI represents the United States as a member of the International Standards Organization.

International Standards Organization (ISO): The international organization for the development of standards including technical drawing and dimensioning standards. ANSI represents the United States as a member of ISO. The ISO dimensioning standard is almost identical to the ASME dimensioning standard.

ASME also publishes other standards concerning engineering drawing practices. These standards are available for purchase from the ASME. In the United States, standards of this type are created under the aegis of the *American National Standards Institute (ANSI)*.

The *International Standards Organization (ISO)* also publishes drawing standards for the preparation of technical drawings including dimensioning. ANSI represents the United States as its delegate to ISO. The ISO dimensioning standard is very similar to the ASME dimensioning standard.

> **TIP**   To find out more about ANSI/ASME and ISO standards, see Appendices A and B.

HOW DIMENSIONS ARE CALCULATED FOR MECHANICAL DRAWINGS—

Mechanical engineers and designers are responsible for calculating the dimensions of an object. They determine the dimensional values by carefully considering the form, fit, and function of the object they are designing. For example, the material the part is to be manufactured from, how the object fits with other parts in an assembly, and the role it plays in the overall design are all factors that may affect the size and location of the features of a part. Sometimes, the dimensions define the aesthetic, rather than the functional, qualities of the finished object. The aesthetic qualities of an object refer more to the object's appearance than to its function.

Because the designer's dimensions are carefully calculated, it is very important that they are faithfully reproduced by the drafter during the preparation of an engineering drawing.

In order to appreciate how crucial it is that the designer's dimensions are accurately portrayed in a mechanical drawing, drafters must have an understanding of a very important concept of mechanical design: *tolerances*.

tolerances: The total permissible variation in the size and/or shape of the object's features as defined by applying tolerances to the nominal size dimension. The difference between the upper and lower size limits of the feature.

TOLERANCES

Manufacturing a machine part to an extreme degree of accuracy, is difficult *and* expensive; therefore, designers must decide how much the size and location of a part's features can deviate from the dimensions specified on the drawing and still perform their design function. This allowable variation in the location, or size, of a feature is called the *tolerance*. Once the acceptable tolerance for a feature has been determined, it is noted on the dimensioned drawing for the part.

On engineering drawings, the dimension to which the tolerances are applied is referred to as the *nominal size* of the feature.

> **Note:**
> To underscore the importance of tolerances in the dimensioning process, the ASME Y14.5M Standard states as one of its fundamental rules, "Each dimension shall have a tolerance except those dimensions specifically identified as reference, maximum, minimum or stock (commercial stock) size."

nominal size: A dimension that describes the general size of a feature. Tolerances are applied to this dimension.

For example, a designer may decide the allowable diameter of a hole is 1.00″, with a tolerance of plus or minus .01″. In this example, 1.00″ in diameter is the nominal size of the hole. By applying a tolerance of plus or minus .01″ to the nominal size of the hole, we can calculate that the hole could range in diameter from .99″ to 1.01″ and still fall within the acceptable size specified by the designer. The tolerance is the allowable difference between the hole's minimum and maximum size limits, so in this case, the tolerance for this hole would be .02″ (1.01 − .99 = .02).

The primary reason that designers calculate and specify tolerances on mechanical drawings is to control the size and location of the features of the part during the manufacturing process. After manufacture, the part is measured by quality control inspectors to verify that the features are within the allowable limits of size as defined by the tolerances on the drawing. Parts whose features measure within the allowable limits pass inspection. Parts whose features measure outside the limits are rejected.

Specifying tolerances on drawings has advantages for designers and manufacturers alike. The advantage for designers is that they can be confident that if the parts they design are manufactured within the tolerances specified on the drawing, their designs will fit and function as they intended. Manufacturers like toleranced parts because they can be confident that as long as the parts they make measure within the tolerances specified on the drawing, the client will purchase the parts.

Displaying Tolerances on Mechanical Drawings

Figures 5-1 and 5-2 show two examples of how tolerances may be specified for dimensions on a technical drawing. In Figure 5-1, the tolerance is shown beside the nominal dimension and noted with a plus/minus symbol ($\pm$). In Figure 5-2, the tolerance has been both added to and subtracted from the nominal size, and the allowable size limits are actually noted in the dimension.

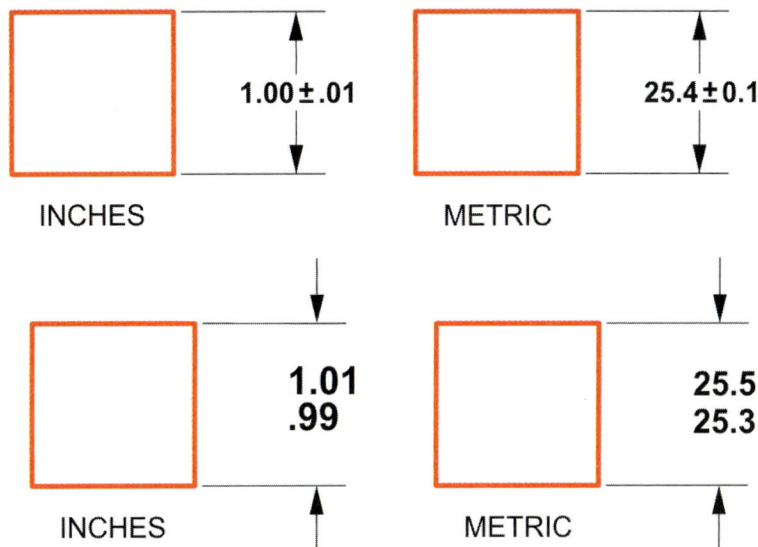

INCHES   1.00±.01

METRIC   25.4±0.1

Figure 5-1   Plus-minus tolerance of .01″ (left) and 0.1mm (right)

INCHES   1.01 / .99

METRIC   25.5 / 25.3

Figure 5-2   Limit tolerance

Another method of specifying tolerances is to add notations in the field of the drawing or in the drawing's title block. Examples 1, 2, and 3 show three different ways to depict tolerances with notations and include an interpretation of the specified tolerance.

Example 1.   *General tolerances* may be labeled in the title block or given as notes in the field of the drawing.

- **.X ± .05**   Dimensions noted with one decimal place of precision on the drawing will have a tolerance of plus or minus .05″.

- **.XX ± .02**   Dimensions noted with two decimal places of precision on the drawing will have a tolerance of plus or minus .02″.

- **.XXX ± .003**   Dimensions noted with three decimal places of precision on the drawing will have a tolerance of plus or minus .003″.

Example 2.   **Decimal Dimensions to Be ± .005″**   A general tolerance of .005″ will apply to all dimensions labeled in decimal units.

Example 3.   **Angular Tolerances ± 1 Degree**   Dimensions on the drawing labeled as angles will have a tolerance of plus or minus 1°. For example, an angle labeled 30° on the drawing could measure between 29° and 31° on the manufactured part.

Interpreting Tolerances on Technical Drawings

Figures 5-3 and 5-4 show examples of a machine part in which the dimensions *do not* include tolerances.

In Figure 5-3, the width of the slot is dimensioned with a ***continuous dimension***. A continuous dimension is referenced from termination of the dimension that preceded it. The width of the slot in Figure 5-3 is 1.00″.

In Figure 5-4, the width of the slot is dimensioned with ***baseline dimensions***. Baseline dimensions are referenced from a common geometric feature known as a ***datum***. In Figure 5-4, the datum feature is the left edge of the part. The width of the slot can be determined by calculating

continuous dimensions: A dimensioning technique in which a linear dimension is placed using the second extension line origin of a selected dimension as its first extension line origin. This technique is also called *chain dimensioning*.

baseline dimensions: A group of linear dimensions that are referenced from the same datum or baseline.

datum: A theoretically perfect feature (plane, axis, or point) from which dimensions are referenced.

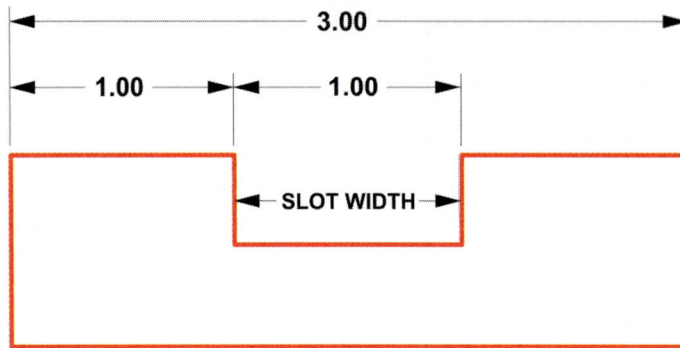

Figure 5-3 Part with slot dimensioned with a continuous dimension

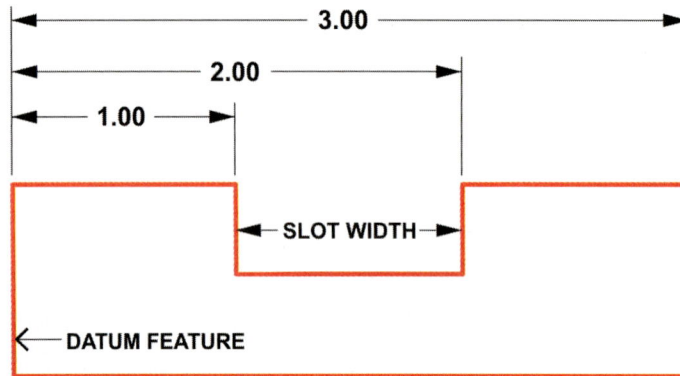

Figure 5-4 Part with slot dimensioned with a baseline dimension

the difference between locations of the right and left sides of the slot. In this case, the width of the slot would equal 1.00 (2.00 minus 1.00).

Because no tolerances are specified for the dimensions in Figures 5-3 and 5-4, the width of the slot will be 1.00″ whether it is defined by a single continuous dimension or by two baseline dimensions.

Figure 5-5 shows a view of a part that includes dimensions *and* tolerances. The overall length of this part is labeled 3.00″ plus or minus .01″. This means that when the part is manufactured, its overall length must measure between 2.99″ and 3.01″ in order to pass a quality control inspection.

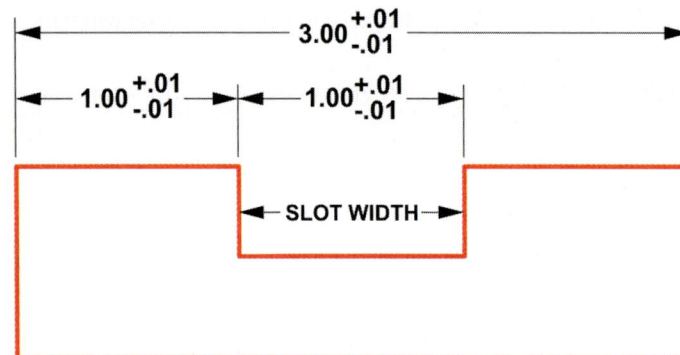

Figure 5-5 Part with slot dimensioned with a toleranced continuous dimension

The left side of the slot in Figure 5-5 is located 1.00″ plus or minus .01″ from the left edge of the part. This means that the location of the slot's left side must be measured between .99″ and 1.01″ from the left side of the part.

The width of the slot is dimensioned 1.0″ plus or minus .01″, so as long as the width of the slot on the part measures between .99″ and 1.01″, it will pass inspection.

Comparison of Continuous Dimensioning and Baseline Dimensioning.   Figure 5-6 shows a part that is dimensioned with three baseline dimensions include plus or minus tolerances. Here, the left edge of the part serves as the datum feature for each of the three dimensions.

The overall length of this part is labeled 3.00″ plus or minus .01″ so when this part is manufactured, its overall length must measure between 2.99″ and 3.01″.

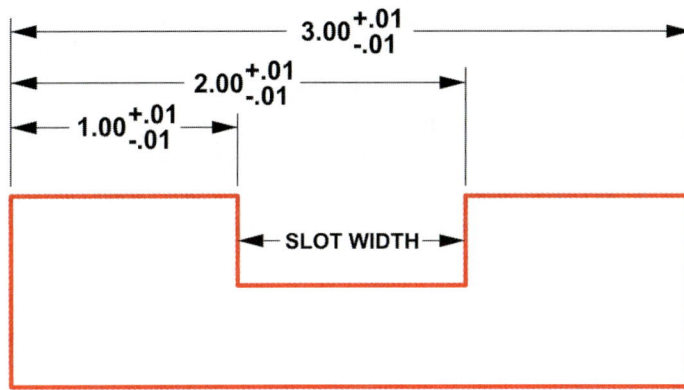

Figure 5-6 Part with baseline dimensions

The width of the slot in Figure 5-6 is defined by two baseline dimensions instead of one continuous dimension as in Figure 5-5. The location of the left side of the slot in Figure 5-6 must measure between .99″ and 1.01″ from the datum edge. The location of the slot's right side must measure between 1.99″ and 2.01″ from the datum.

If the left edge of the slot is located .99″ from the datum and the right edge is located 2.01″ from the datum, the width of the slot could be 1.02″.

If the slot's sides are located 1.01″ and 1.99″, respectively, from the datum edge, the slot's width would measure .98″.

Using baseline dimensions to define the slot could result in the slot measuring between .98″ and 1.02″ in width.

Dimensioning the slot with continuous dimensions as shown in Figure 5-5 resulted in the slot measuring between .99″ and 1.01″ wide.

As you can see from the comparison of the widths of the slots in Figures 5-5 and 5-6, the method used to dimension the slot (either baseline or continuous) affects its size, after tolerances are factored in. In this case, the slot's width defined with baseline dimensions may vary by as much as .04″ (.98″ to 1.02″), whereas dimensioning the slot with continuous dimensions results in the slot's width varying by .02″ (.99″ to 1.01″).

The point of this comparison is not that either dimensioning technique—continuous or baseline—is inherently better or worse than the other, but rather that the method used to define the widths of the slot in Figures 5-5 and 5-6 affects the slot's size after tolerances are applied. The designer chooses the dimensioning technique that will ensure that the part's features are manufactured within size limits that will allow the part to function in the application for which it was intended.

This comparison was intended to help you understand the importance of applying the same dimensioning method (continuous or baseline) to the creation of an engineering drawing as is shown on the design input. Mechanical drafters who do not follow the same dimensioning method as the designer run the risk of inadvertently changing the intended size, or location, of a feature.

Tolerancing Terminology

Following are terms that designers frequently use when discussing tolerances. As a drafter-in-training, you should become familiar with each term so that you can communicate effectively with designers and engineers.

- **Feature:** A geometric element that is added to the base part such as a slot, surface, or hole.
- **Nominal size:** A dimension used to describe the general size of the feature. Tolerances are applied to this dimension.
- **Tolerance:** The total permissible variation in a dimension value; the difference between the upper and lower size limits of a feature.
- **Limits:** The maximum and minimum sizes of a feature as defined by the toleranced dimension. For example, a hole dimensioned with a diameter of .50″, with a tolerance of ±.02″, has an upper limit of .52″ and a lower limit of .48″.
- **Allowance:** The minimum clearance, or maximum interference, between mating parts.

features: Geometric elements that are added to a base part. Features include holes, slots, arcs, fillets, rounds, angled planes, counterbored holes, and countersunk holes. Features are located on the object with location dimensions and are described with size dimensions.

limits: The maximum and minimum sizes of a feature as defined by its tolerances. For example, a feature with a nominal dimension of .50, with a tolerance of ±.02, has an upper limit of .52 and a lower limit of .48.

allowance: The minimum clearance, or maximum interference, between parts.

actual size: The measured size of a finished part. This size determines whether the part passes a quality control inspection.

reference dimension: A dimension that is included on a technical drawing for information only and is not necessary to manufacture, or inspect, the part. No tolerances are applied to reference dimensions. Reference dimensions are enclosed in parentheses.

Maximum Material Condition (MMC): The condition of a part when it contains the greatest amount of material. The MMC of an external feature, such as a shaft, is the upper limit. The MMC of an internal feature, such as a hole, is the lower limit.

Least Material Condition (LMC): The condition of a part when it contains the least amount of material. The LMC of an external feature, such as a shaft, is the lower limit of size defined by the tolerance. The LMC of an internal feature, such as a hole, is the upper limit of size defined by the tolerance.

- **Datum:** A theoretically perfect element (an edge, plane, axis, point, or other geometric feature) from which dimensional information is referenced. Finished parts are inspected by measuring from the datum geometry identified on the drawing.
- **Actual size:** The measured size of a feature of a manufactured part. This size determines whether the part passes a quality control inspection.
- **Reference dimension:** A dimension without a tolerance that is provided only for information purposes. It is not used for manufacture or inspection of the part. A reference dimension is enclosed in parentheses.
- **Maximum Material Condition (MMC):** The condition of a feature when it contains the greatest amount of material. The MMC of an external feature, such as a shaft, is the upper limit of size. The MMC of an internal feature, such as a hole, is the lower limit of size.
- **Least Material Condition (LMC):** The condition of a feature when it contains the least amount of material. The LMC of an external feature is the lower limit. The LMC of an internal feature is the upper limit.

Interpreting Design Sketch 1

In the design sketch shown in Figure 5-7, the designer has provided the dimensions required to define the nominal sizes of the diameter (.98″) and the length (2.00″) of a cylinder. The designer has also specified a plus or minus tolerance of .01″.

Analyzing Design Sketch 1.   Applying the tolerance to the part's nominal dimensions would result in the cylinder's *diameter* ranging between a minimum of .97″ (.98″ minus .01″) and a maximum of .99″ (.98″ plus .01″). By applying the same sort of calculation to the *length* of the cylinder, we see that it could range between a minimum of 1.99″ (2.00″ minus .01″) and a maximum of 2.01″ (2.00″ plus .01″).

The designer has calculated the size of the cylinder so that it should be able to perform as intended as long as its features fall within the allowable size limits.

Calculating Maximum Material Condition (MMC) for Design Sketch 1.   Applying the plus tolerance to the cylinder's nominal dimensions will define the MMC of the cylinder's diameter and length. Using this calculation, the MMC for the cylinder's diameter would equal .99″ (.98″ plus .01″) and the MMC for the length of the cylinder would be 2.01″ (2.00″ plus .01″).

Calculating Least Material Condition (LMC) for Design Sketch 1.   Applying the minus tolerance to the cylinder's nominal dimensions will define the cylinder's LMC. Using this calculation, the LMC of the cylinder's diameter would equal .97″ (.98″ minus .01″), and the LMC for the length of the cylinder would be 1.99″ (2.00″ minus .01″).

Interpreting Design Sketch 2

In the design sketch shown in Figure 5-8, the designer has provided the coordinate (*X* and *Y*) dimensions required to locate the center of the hole from the left side and bottom edge of the

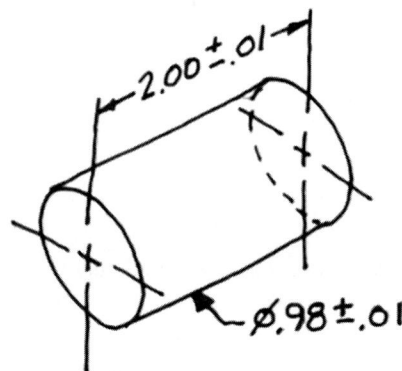

Figure 5-7   Design Sketch 1

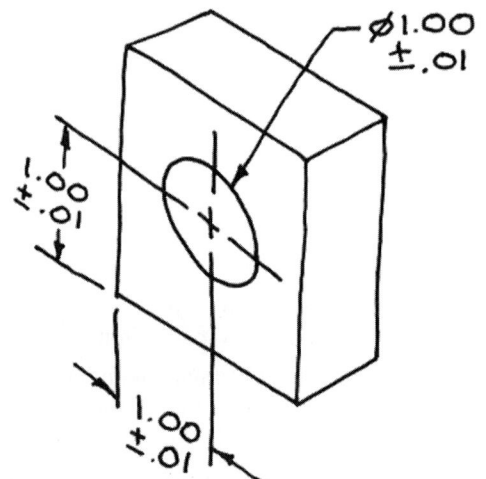

Figure 5-8   Design Sketch 2

object. These sides will be used as *datum features* for referencing dimensions on the drawing and later for inspecting the finished part.

The designer has also noted a tolerance range of plus or minus one hundredth of an inch (± .01″) that is to be applied to each nominal dimension.

Analyzing Design Sketch 2.   When this part is manufactured, the location of the center of the hole, as well as the diameter of the hole, must comply with the conditions noted in Figure 5-9. Otherwise, the part will be out of tolerance and may be rejected during a quality-assurance inspection check. When the part is inspected, the inspector will make measurements from the same datum features that are defined by the dimensions in the technical drawing.

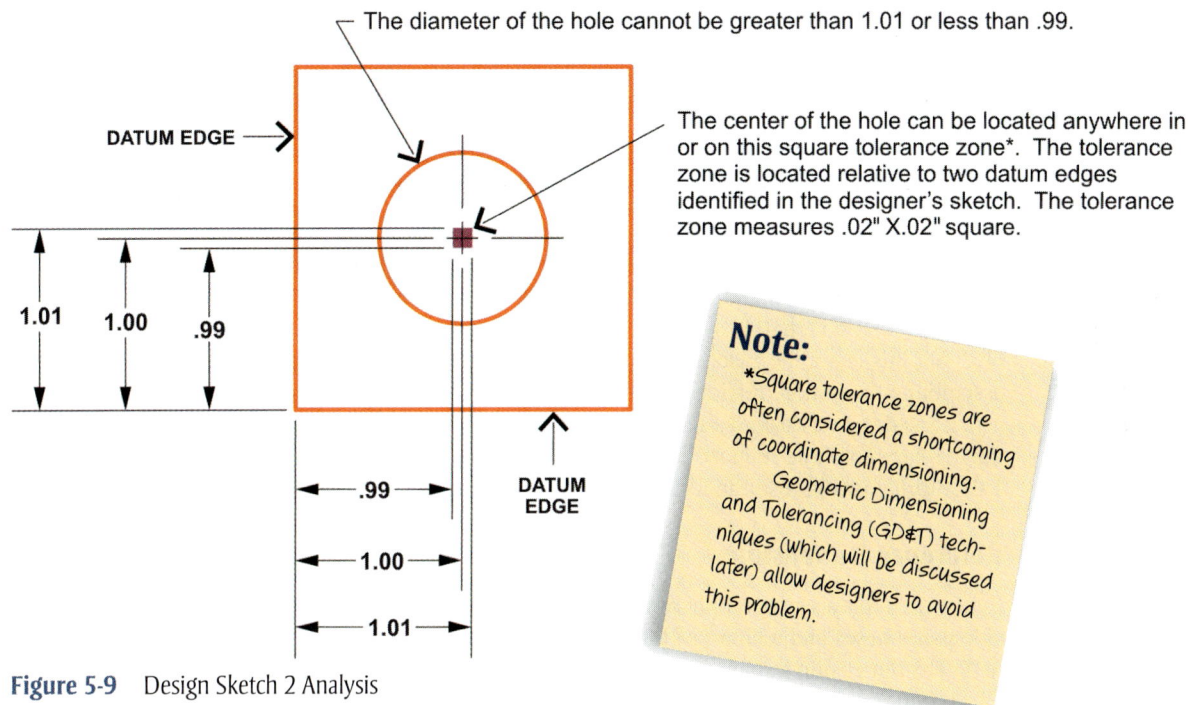

The diameter of the hole cannot be greater than 1.01 or less than .99.

DATUM EDGE →

The center of the hole can be located anywhere in or on this square tolerance zone*. The tolerance zone is located relative to two datum edges identified in the designer's sketch. The tolerance zone measures .02" X.02" square.

1.01   1.00   .99

.99

DATUM EDGE

1.00

1.01

Note:

*Square tolerance zones are often considered a shortcoming of coordinate dimensioning. Geometric Dimensioning and Tolerancing (GD&T) techniques (which will be discussed later) allow designers to avoid this problem.

Figure 5-9   Design Sketch 2 Analysis

Calculating Maximum Material Condition (MMC) for the Hole in Design Sketch 2.   Subtracting the *minus* tolerance from the hole's nominal dimension will define its MMC. The nominal size of the hole is 1.00″ in diameter and the minus tolerance is .01″, so in this case, the MMC for the diameter of the hole equals .99″.

Calculating Least Material Condition (LMC) for the Hole in Design Sketch 2.   Adding the plus tolerance to the hole's nominal dimension will define its LMC. The hole's nominal diameter is 1.00″ and the plus tolerance is .01″, so in this case, the LMC for the hole's diameter equals 1.01″.

Calculating the Fit Between the Parts in Design Sketches 1 and 2

Suppose the designer of the parts in Design Sketches 1 and 2 (see Figure 5-10) had intended that they be assembled after they were manufactured. During the design of each part, the designer would have needed to analyze the possible limits of size of the cylinder in Part 1 and the hole in Part 2 to see if an interference could exist that would prevent the parts from assembling.

Best-Case Scenario for Assembly.   If the cylinder in Design Sketch 1 is manufactured at its LMC, or smallest allowable diameter (.97″), and the hole in Design Sketch 2 is manufactured at

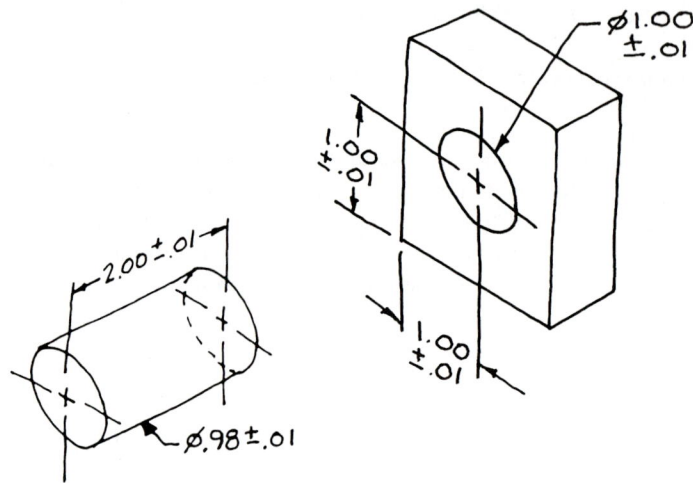

Figure 5-10   Assembling
Parts 1 and 2

its LMC, or largest allowable diameter (1.01″), the clearance between the two at LMC would be .04″. With a clearance of .04″, these parts can be easily assembled.

Worst-Case Scenario for Assembly.   The worst-case scenario for assembly exists if the cylinder in Design Sketch 1 is manufactured at its MMC, or largest allowable diameter (.99″), and the hole in Design Sketch 2 is manufactured at its MMC, or smallest allowable diameter (.99″). At first it may seem that both parts having the same diameter would cause them to interfere when assembled, but this is not the case. The resulting fit would, however, be very tight, which would make assembling them more difficult.

If the designer desired a looser fit at MMC of both parts, the nominal diameter specified for the cylinder could be reduced to .97″, or the nominal diameter of the hole could be enlarged to 1.01″. Either would result in a clearance fit between the parts at MMC.

Reference Dimensions

Figure 5-11 shows an example of a machine part with an overall dimension of 3.00″ labeled inside parentheses and a chain of three continuous dimensions, each labeled 1.00″ plus or minus .01″. The dimension in parentheses is a reference dimension. A reference dimension is an untoleranced dimension that is provided only for informational purposes and is not used to manufacture, or inspect, the part.

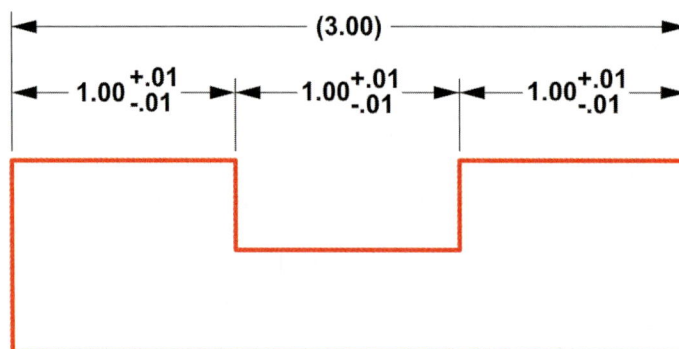

Figure 5-11   Reference
dimension

Because the reference dimension is not used to make the part in Figure 5-11, the overall length of the part will be a product of the cumulative effects of the three dimensions labeled 1.00″ plus or minus their tolerances. This could result in the overall length of the part ranging between 2.97″ (.99″ times 3) to 3.03″ (1.01 times 3). This phenomenon, where the sizes of toleranced features have a cumulative effect on the overall length, is called a *tolerance accumulation.*

In Figure 5-11, if the overall length dimension had been labeled 3.00″ plus or minus .01″ (instead of as a reference dimension), it would be impossible to reconcile its allowable limits (2.99″ to 3.01″) with the limits allowed by applying a tolerance to each of the dimensions labeled 1.00″ (2.97″ to 3.03″).

On the other hand, if the chain of 1.00″ dimensions were broken by removing one of them, the overall dimension would no longer be a reference dimension and would need to have a tolerance.

Confirming the Tolerances of Manufactured Parts

It is important to confirm that after the part is manufactured, it falls within the allowable size limits defined by the dimensions and tolerances noted on the drawing. This step in the manufacturing cycle is performed during a *quality control inspection*. Quality control (QC) inspectors use precise measuring (metrology) equipment to determine the actual size of the part. QC inspectors compare the actual size of the part to the dimensions noted on the technical drawing. Parts that measure within the allowable size limits will pass the QC inspection, whereas parts that measure outside the limits will be rejected.

Tolerance Costs

Designers must consider cost when determining the tolerances for a feature because as tolerances become tighter, the cost of manufacturing a part may increase. One reason is that as tolerance allowances become stricter, it may take longer to manufacture the part. Another reason is that due to tighter tolerances, fewer parts may pass a quality control inspection. The designer walks a fine line between the desired accuracy of the part and the cost of manufacturing the part within the budget constraints of the project.

Although tight tolerances usually add to the cost of a project, a type of tolerancing known as *Geometric Dimensioning and Tolerancing (GD&T)*, may actually lower the costs of producing a part. GD&T tolerances control the *form* (flatness, straightness, circularity and cylindricity), *orientation* (perpendicularity, angularity, parallelism), or *position* of a part's features. By using GD&T, the odds that parts will pass a quality control inspection rise, and fewer rejected parts results in lower production costs. The ASME Y14.5M-1994 standard covers the application of GD&T to technical drawings.

> **quality control inspection:** A step in the manufacturing cycle performed by a quality control (QC) inspector using precise measuring equipment to determine the actual size of the part. The QC inspector compares the actual size of the part to the dimensions noted on the technical drawing. Parts that measure within the allowable size limits will pass the QC inspection, whereas parts that measure outside the limits will be rejected.

> **Geometric Dimensioning and Tolerancing (GD&T):** A dimensioning technique that is used to control the *form* (flatness, straightness, circularity, and cylindricity), *orientation* (perpendicularity, angularity, and parallelism), or *position* of a part's features.

TIP   For more information on Geometric Dimensioning and Tolerancing, see Appendix D.

ADDING DIMENSIONS TO MECHANICAL DRAWINGS

An understanding of the following concepts is necessary in order to dimension an object:

- Every part has an overall X, Y and Z dimensional value. The X, Y and Z values correspond to the width (X), height (Y), and depth (Z) of the part in its multiview representation.

- There are two types of dimensions: size and location. For example, the overall height of an object is a size dimension, and a dimension that indicates the placement of the center of a hole is a location dimension. The dimension for the diameter of the hole would be a size dimension. Features are located on the object with location dimensions and described with size dimensions (see Figure 5-12).

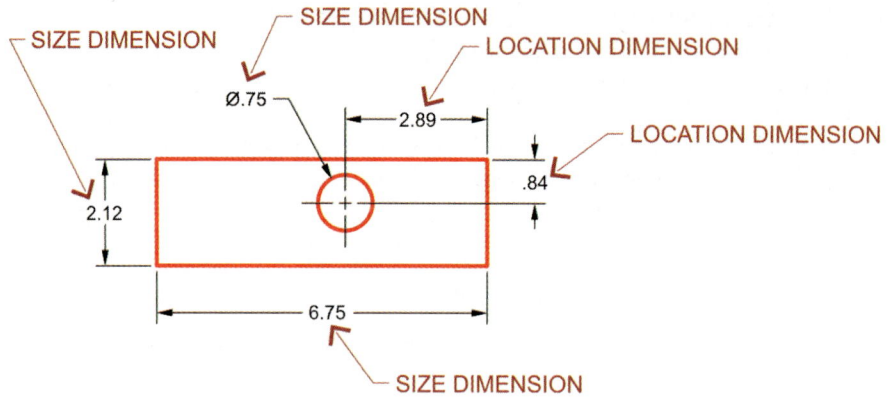

Figure 5-12 Size and location dimensions

Figure 5-13 shows the terminology used to refer to the elements of dimensions. Drafters must be familiar with this terminology in order to set the desired values for these components when creating an AutoCAD dimension style.

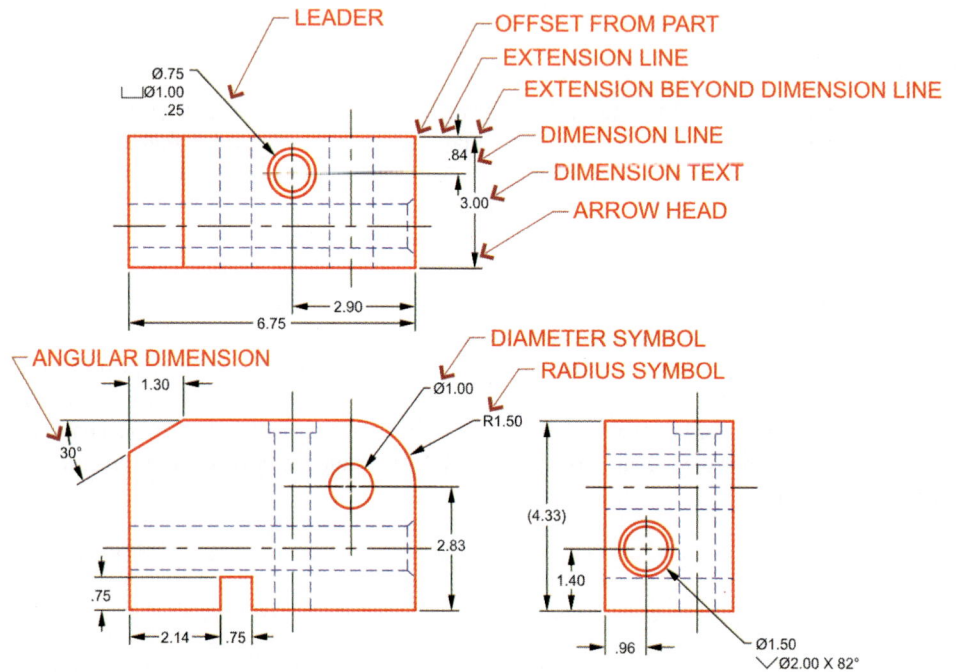

Figure 5-13 Dimensioning terminology

Guidelines for Dimension Placement

After drawing the necessary multiviews of an object, drafters place dimensions on the views. The dimensions should be exactly as they are given on the designer's sketch. For example, if a dimension on the sketch is to be shown to two decimal places of precision, it should be exactly the same on the drawing. Dimensions should also reference the same datum geometry as the designer's sketch.

Drafters are responsible for choosing the best placement of the dimensions on the multiviews. Drafters should use the following guidelines when determining the best location and placement of dimensions on a drawing.

1. Place dimensions on the *profile* view of the feature.
2. Avoid dimensioning to hidden features or center lines of hidden holes.
3. Whenever possible, group dimensions and place them between the views of the object.
4. Avoid drawing extension lines through dimension lines.
5. Avoid placing dimensions on the part unless it is absolutely necessary.

Figure 5-14 provides an example of a multiview drawing in which the dimensioning guidelines just listed have been ignored. Can you find the mistakes? Compare this drawing to the drawing in Figure 5-15.

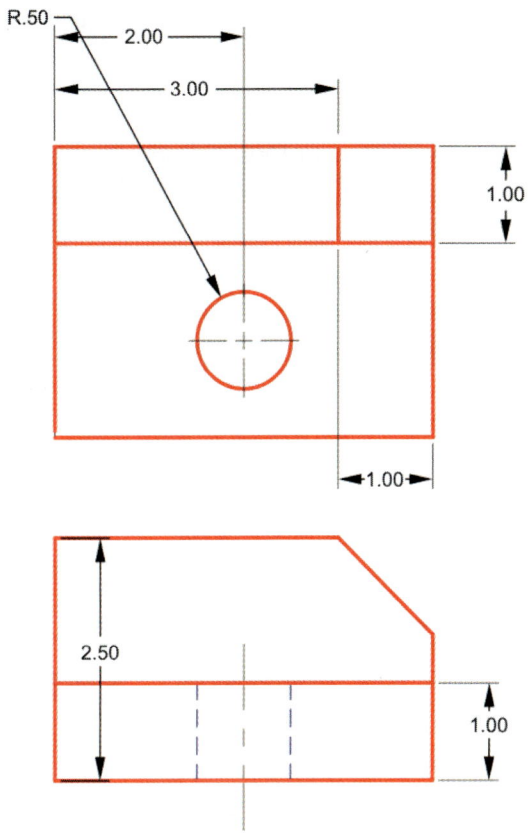

Figure 5-14   Poorly placed dimensions

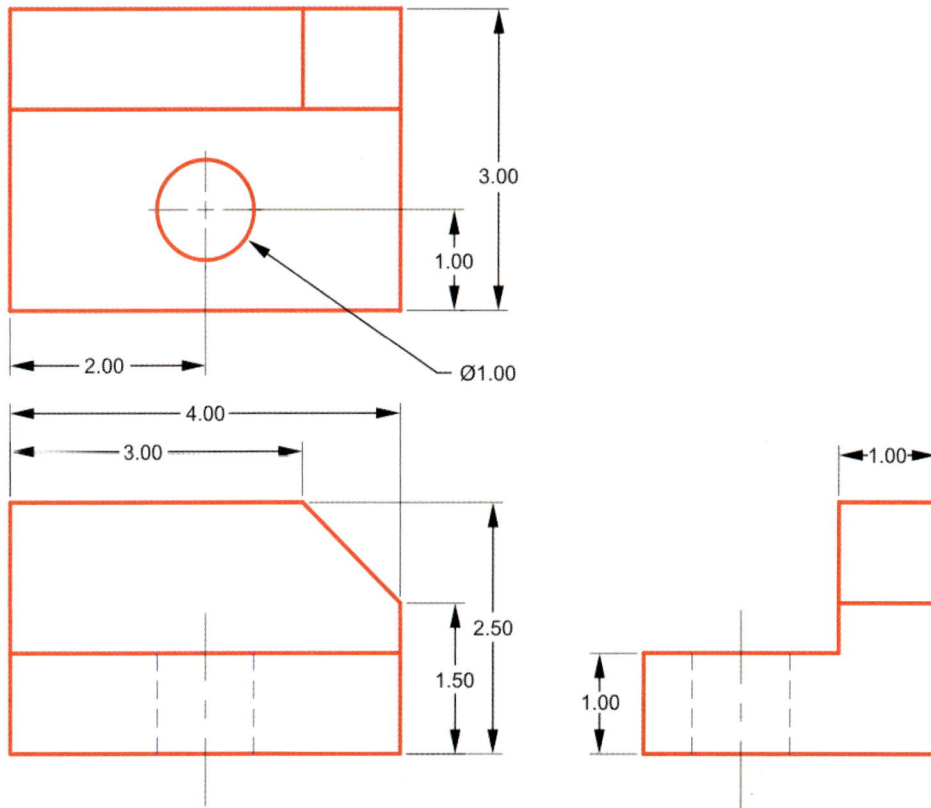

Figure 5-15   Dimensions placed following the recommended guidelines

Figure 5-16 shows the spacing of dimensions and dimensioning conventions defined by *ASME Y14.5M-1994*.

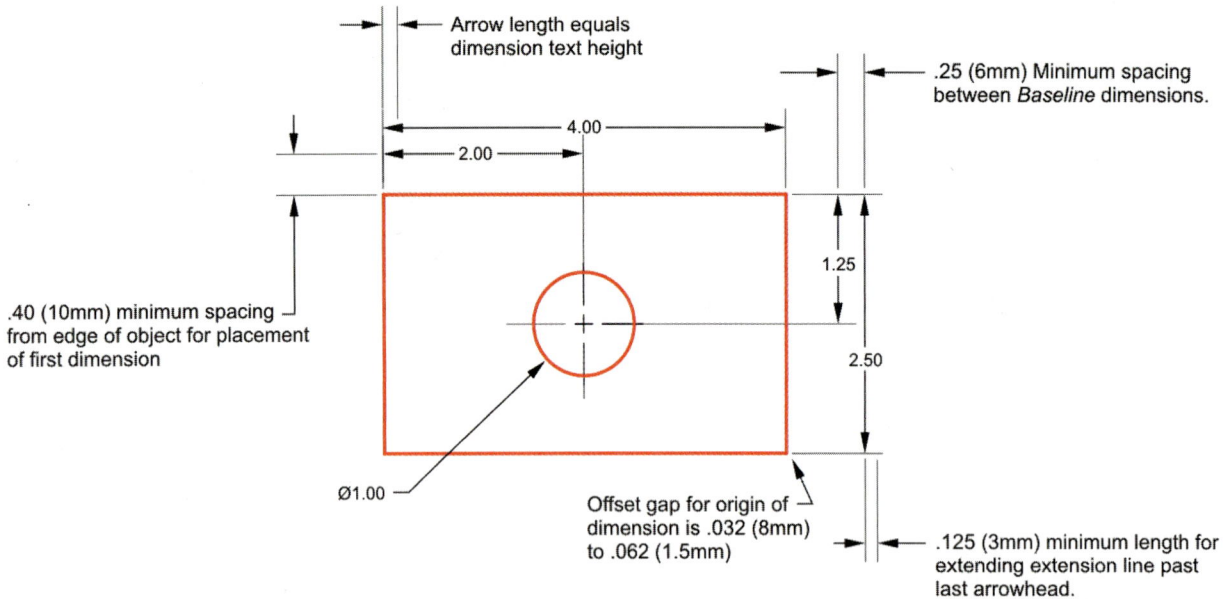

Figure 5-16   *ASME Y14.5M-1994* dimension spacing

> **TIP**
> The settings in the *ASME Y14.5M-1994* Standard are controlled in an AutoCAD drawing by the settings assigned in the **Dimension Style Manager** dialog box. This dialog box is covered in detail later in this unit.

Text Height and Style

With regard to placing text on a technical drawing, legibility is the primary concern. The ASME Standard governing text height and style on engineering drawings is *ASME Y14.2M Line Conventions and Lettering*. This Standard states that text used for titles and for denoting special characters, such as section view labels, should be no less than .24″ (6mm). All other characters should have a minimum text height of .12″ (3mm). Uppercase letters should be used unless lowercase letters are required. Single-stroke Gothic style letters are recommended. Gothic characters do not have serifs at the ends of the strokes.

Alignment of Dimension Text

unidirectional text: Text placed on a technical drawing that faces only the bottom of the sheet. This technique is required when the ASME text standard is applied to a drawing.

The ASME Standard for text states that text should face the bottom of the sheet. This is known as **unidirectional text**. An example of unidirectional text is shown in Figure 5-17.

Aligned text is aligned to dimension lines and may face the bottom and the right side of the sheet. This is allowed on mechanical drawings prepared following the ISO standard but is not al-

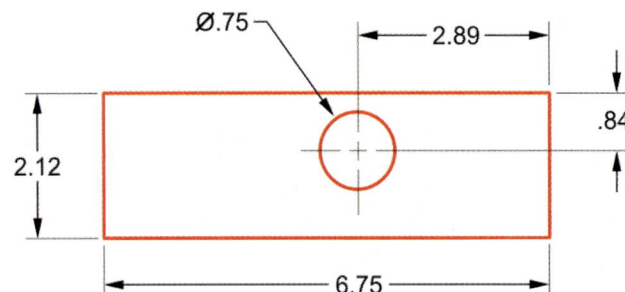

Figure 5-17   Unidirectional text

lowed on drawings employing the ASME dimensioning standard. An example of aligned text is shown in Figure 5-18.

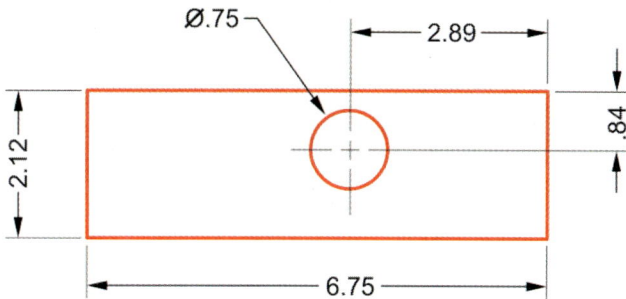

Figure 5-18   Aligned text

Notating Holes and Arcs

When dimensioning holes, provide the *X* and *Y* values to the center of the hole from the designer's datums and specify the hole's diameter. An arc should be described as a radius.

Figure 5-19 illustrates how leaders for small arcs and diameters are depicted. Figure 5-20 illustrates how leaders and notes are represented for large diameters and radii.

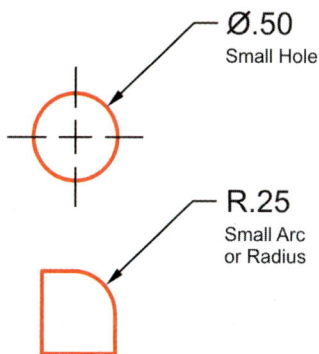

Figure 5-19   Dimensioning small holes and arcs

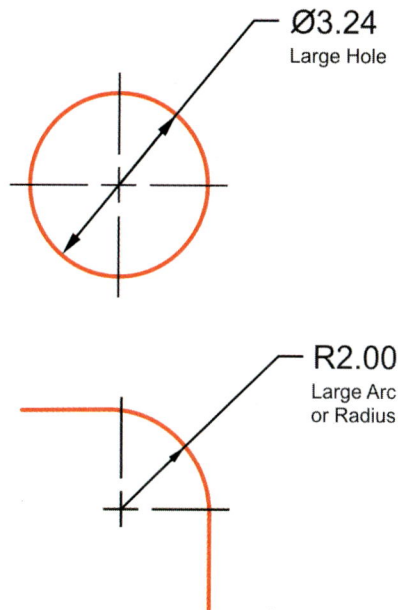

Figure 5-20   Dimensioning large holes and arcs

Dimensioning Cylindrical Shapes

ASME Y14.5M-1994 specifies that cylinders and other outside diameters should be dimensioned in their profile (or side) view. The dimension should be preceded by the diameter symbol. See Figure 5-21.

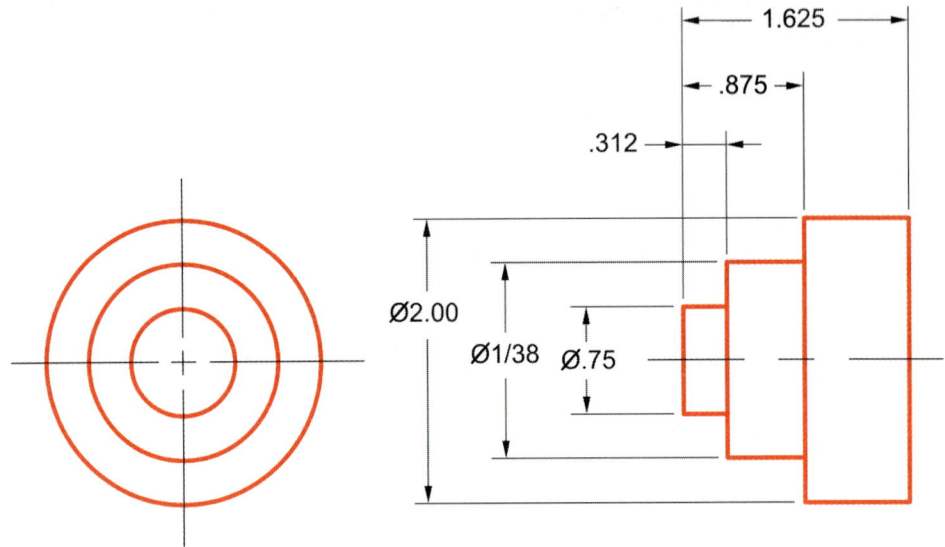

Figure 5-21 Dimensioning cylindrical shapes

Dimensioning Angles

An angle should also be dimensioned in its profile view, with the dimension value followed by the degree symbol (°). See Figure 5-22. When greater accuracy for noting angles is desired, angles may be specified in degrees, minutes ('), and seconds ("). A minute equals 1/60th of one degree, and a second equals 1/60th of a minute (see Figure 5-23).

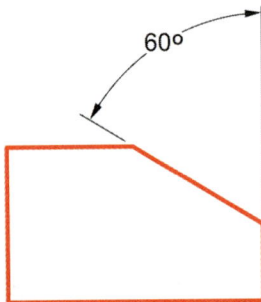

Figure 5-22 Dimensioning angles

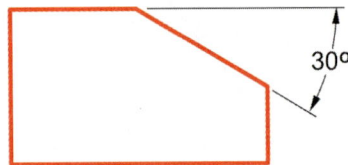

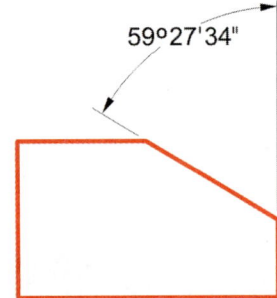

Figure 5-23 Specifying an angle in degrees, minutes, and seconds

Ordinate Dimensioning

In ordinate dimensioning, a **0,0** (zero X, zero Y) datum point is defined on the object and the location the object's features are located along the X- and Y-axes as referenced from the **0,0** datum point.

In the example in Figure 5-24, the **0,0** datum is at the lower left corner of the object, and the dimensions shown are all relative to this point. A table, like the one in Figure 5-25, is often placed on the drawing to describe hole diameters when ordinate dimensioning is employed.

Ordinate dimensioning is useful for dimensioning parts that are to be manufactured by **_Computer Aided Manufacturing (CAM)_** machinery, such as a CAM drill press. The operator of the CAM drill press mounts the material to be drilled on the bed of the press and programs the location of the **0,0** point on the material. The operator then programs the location of holes along the X and Y coordinates. The drill bit moves along the X- and Y-axes relative to the **0,0** datum to drill

Computer Aided Manufacturing (CAM): Manufacturing processes in which manufacturing equipment and processes are controlled by computer commands.

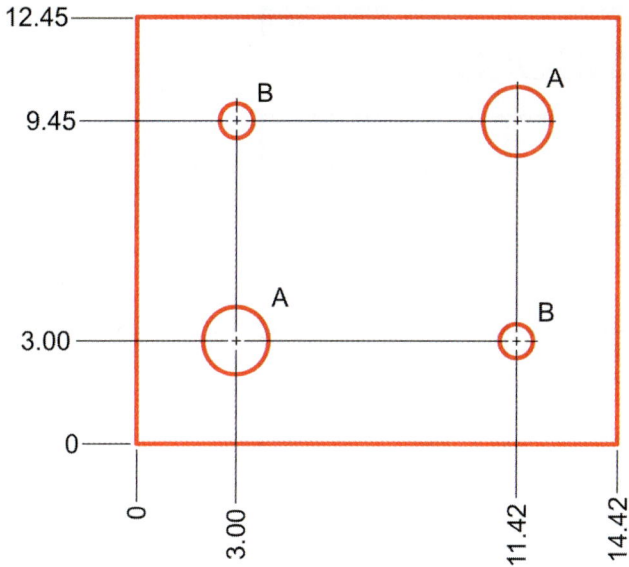

Figure 5-24   Ordinate dimensioning

Hole Table	
Mark	Diameter
A	Ø2.00
B	Ø1.00

Figure 5-25   Hole table

holes in the material (although in some machines, the drill bit remains stationary, and the table on which the material is mounted moves to the bit instead).

Notes for Drilling and Machining Operations

Notations for counterbored and countersunk holes are shown in Figure 5-26.

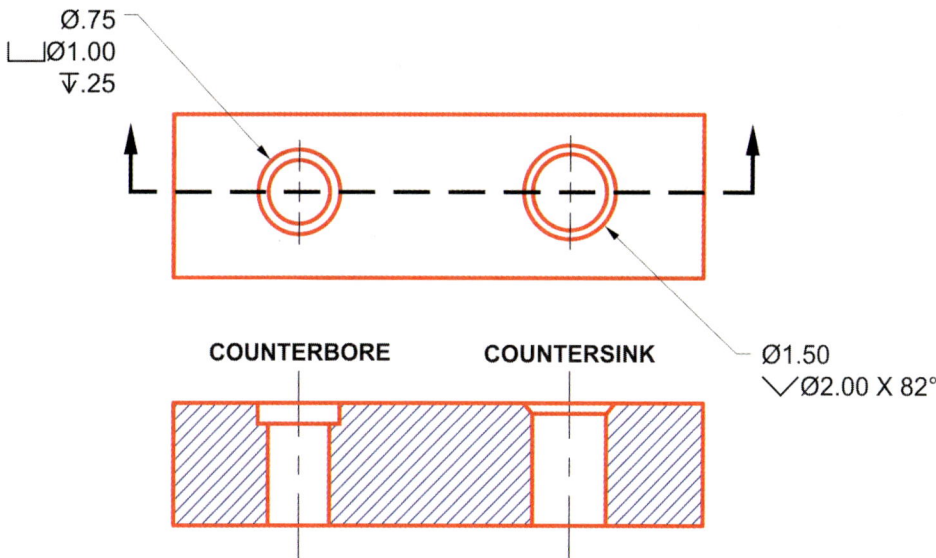

Figure 5-26   Notations for counterbored and countersunk holes.

Because the computer keyboard lacks the characters for geometric symbols such as counterbore, countersink, or depth, AutoCAD's Geometric Dimensioning and Tolerancing (GDT) font is often used to place these symbols. The alphabetical letter on the keyboard corresponding to the symbol in the GDT font is shown in Figure 5-27. For example, to place the counterbore symbol, the drafter types a lowercase **v** in the dimension text and then edits the text and changes the **v** to the GDT font. When the text edit is completed, the **v** will be replaced by the counterbore symbol.

GDT SYMBOLS		
Keyboard Character (lowercase)	GDT Symbol	Interpretation
V	⊔	Counterbored Hole
W	⌄	Countersunk Hole
X	⌄̲	Depth of Hole

Figure 5-27   Lowercase letters corresponding to symbols in AutoCAD's GDT font

DO'S AND DON'TS OF MECHANICAL DIMENSIONING (ASME Y14.5M-1994)

Do's

- Dimension to features in the profile (or most descriptive) view.
- Use the datums identified by the designer when placing dimensions.
- Use the same precision (number of decimal places) identified by the designer when specifying dimension values.
- Follow *ASME Y14.5M-1994* spacing guidelines and dimension settings (see Figure 5-16).
- Place dimensions applying to two views between the views, but project extension lines from only one of the views.
- A hole or cylinder (or other circular feature) should be described with a diameter dimension.
- An arc should be described with radius dimension.
- When dimensioning holes, provide the *X* and *Y* values of the center of the hole referenced from the designer's datum and specify the diameter and depth of the hole.
- Dimension cylinders and other outside diameters in their profile (side) view and include the diameter symbol.
- Include all notes required to manufacture the part (material, scale, etc.).
- Text for dimensions and notes should be single-stroke gothic in style and be fully legible when plotted or printed.
- When dimension values are in decimal inches, decimal points should be placed in line with the bottom of the dimension text.
- When dimension values are in millimeters, a zero precedes dimensions less than one millimeter in size.

Don'ts

- Do not dimension to hidden lines or hidden features such as the centers of hidden holes.
- Do not cross dimension (arrow) lines with extension lines.
- Do not over-dimension. Features should be dimensioned only once.
- Avoid placing dimensions on the object whenever possible.
- On drawings created with decimal inches, do not place a zero before the decimal point for a dimension with a value of less than one inch (suppress leading zero).

- When dimension values are in millimeters, do not place a decimal point, or a zero, after a dimension that is a whole number (suppress trailing zeros).
- Avoid unbroken "chains" of dimensions which may result in an unintended accumulation of tolerances by omitting one of the dimensions in the chain.

TIP If an unbroken dimension chain is specified on the designer's input, ask the designer if this is how it should be shown on the drawing.

ROLE OF DRAFTERS IN THE PREPARATION OF DIMENSIONED MECHANICAL DRAWINGS

The drafter's role in preparing a dimensioned drawing can be summarized by the following:

- In most cases, the drafter receives a design input—often a sketch—of the object to be drawn from a designer or engineer. The sketch will (or should) provide all of the dimensions and other information necessary to fabricate the object. The drafter should ask for clarification from the designer if he or she feels that the sketch is missing a dimension or contains incorrect or unclear dimensions.
- The drafter determines which multiviews are necessary to describe the features of the object, as well as the sheet format and layout. The drafter then draws the views of the object.
- The drafter dimensions the views following the dimensions defined on the designer's sketch.
- The drafter is responsible for ensuring that the *ASME Y14.5M-1994* (or other applicable) Standard is followed with regard to the placement, spacing, and style of dimensions.

CHECKING DIMENSIONS ON THE FINISHED DRAWING

When the drafter has finished dimensioning a drawing, it should be compared carefully with the designer's input. To assure accuracy and completeness, the drafter should ask the following questions about the finished drawing:

- Does the drawing provide the multiviews necessary to describe the object?
- Can each dimension and note included on the designer's input be accounted for on the final drawing?

TIP An effective strategy for checking the sketch against the drawing is to use a yellow marker to highlight the dimensions on the sketch *and* the drawing one-by-one until all are accounted for.

- Have all applicable drafting and dimensioning standards been followed?
- Could this object be manufactured using *only* the views, dimensions, and notes provided on the drawing? (Novice drafters may find this difficult to answer, but, as they gain experience with manufacturing processes and materials, it becomes easier.)

When the answers to all of the questions posed above are yes, the drawing is probably finished. However, in most offices, the final determination about whether a drawing is finished is made by an engineer, designer, or a ***checker***.

checker: An experienced designer/drafter with expertise in manufacturing, drafting techniques, and dimensioning conventions who is responsible for reviewing and approving the drawings prepared by other drafters.

Checkers are usually very experienced designer-drafters with expertise in manufacturing, drafting techniques, and dimensioning conventions. Often, there is an *Approved* box in the drawing's title block for the checker's initials. When a checker initials this box, it indicates that the drawing has passed the checker's review. It also means that the drafter is no longer the only one responsible for the accuracy of the drawing.

DIMENSIONING WITH AUTOCAD 2008

The AutoCAD 2008 **Dimension** toolbar is shown in Figure 5-28. Each *dimension command* icon on the toolbar is labeled with its function. An explanation of each icon is presented on the following pages. Video tutorials for the commands on the **Dimension** toolbar are located on the Student Resource CD in the AutoCAD Tutorial Videos folder.

Dimension commands: The commands used to dimension an AutoCAD drawing. These commands are located on the **Dimension** toolbar and include **LINEAR, BASELINE, CONTINUE, ANGULAR, DIAMETER**, and **RADIUS**.

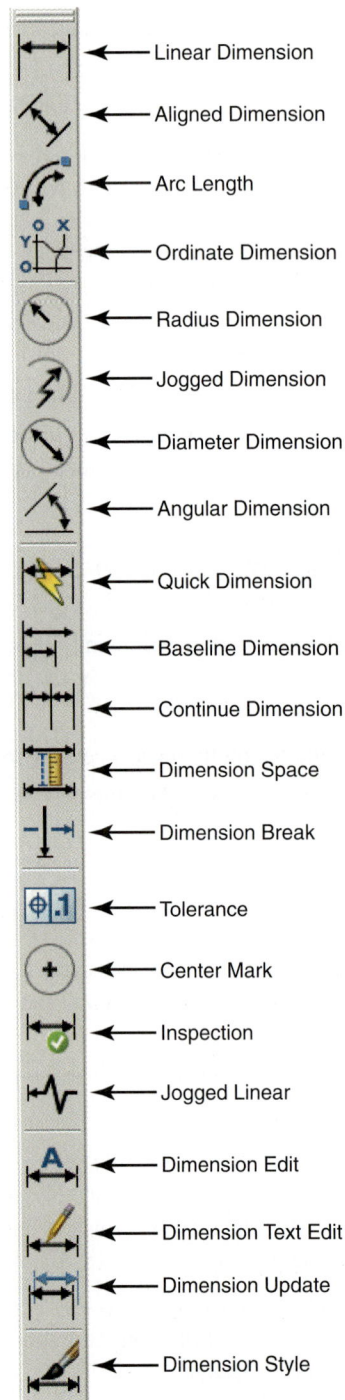

Linear Dimension
Aligned Dimension
Arc Length
Ordinate Dimension
Radius Dimension
Jogged Dimension
Diameter Dimension
Angular Dimension
Quick Dimension
Baseline Dimension
Continue Dimension
Dimension Space
Dimension Break
Tolerance
Center Mark
Inspection
Jogged Linear
Dimension Edit
Dimension Text Edit
Dimension Update
Dimension Style

Figure 5-28 AutoCAD 2008 Dimension toolbar

LINEAR DIMENSION Command

The icon for the **LINEAR DIMENSION** command is shown in Figure 5-29(a). This command displays the linear distance between two selected points. This option is used to dimension both vertical and horizontal features as shown in Figure 5-29(b).

Linear Dimension

Figure 5-29(a)   Linear Dimension icon

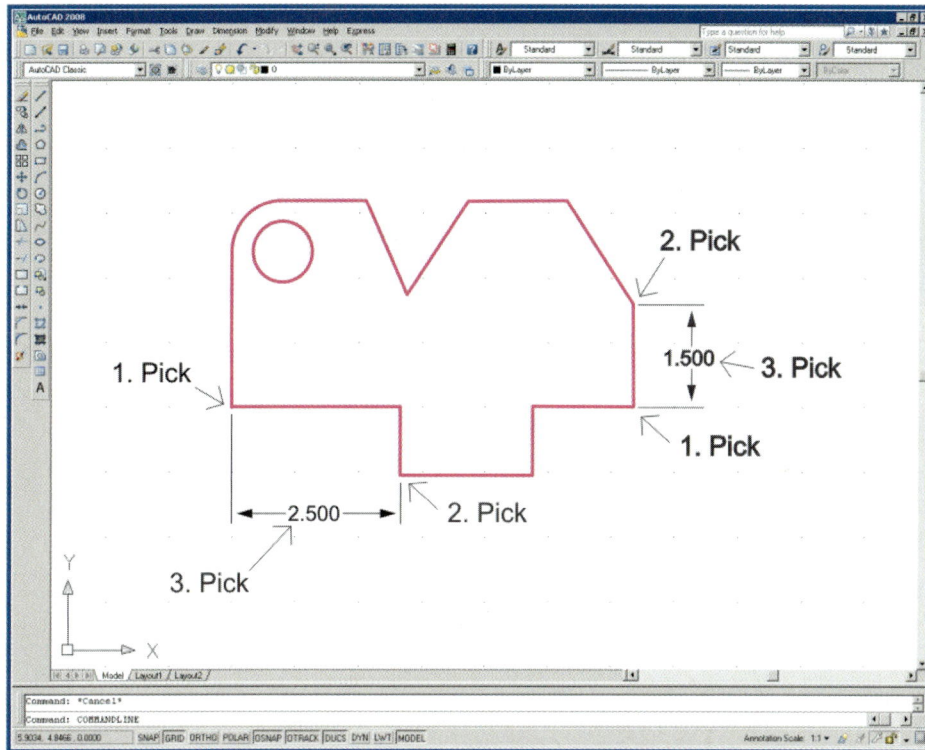

Figure 5-29(b)   LINEAR DIMENSION command

ALIGNED DIMENSION Command

The icon for the **ALIGNED DIMENSION** command is shown in Figure 5-30(a). This command is used to display the length of an angled line as shown in Figure 5-30(b).

Aligned Dimension

Figure 5-30(a)   Aligned Dimension icon

ARC LENGTH Command

The icon for the **ARC LENGTH** command is shown in Figure 5-31(a). This command is used to denote the length dimension of an arc or a polyline arc segment as shown in Figure 5-31(b).

ORDINATE DIMENSION Command

The icon for the **ORDINATE DIMENSION** command is shown in Figure 5-32(a). This command is used to denote distances along X- and Y-axes relative to a defined origin point (usually labeled **0,0**) as shown in Figure 5-32(b).

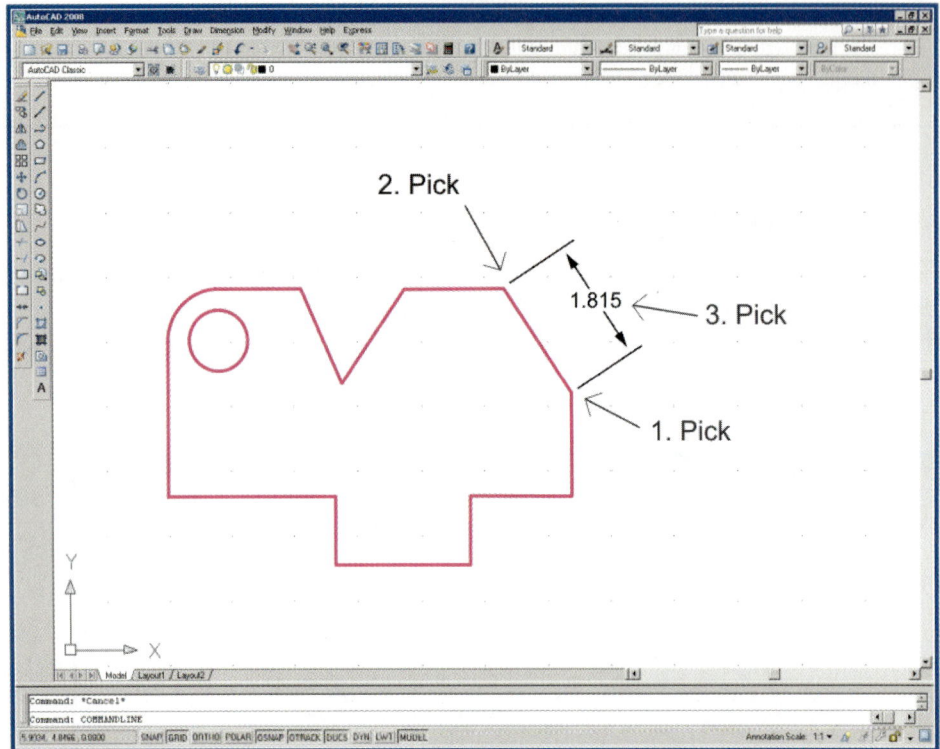

Figure 5-30(b)   ALIGNED DIMENSION command

Figure 5-31(a)   Arc Length icon

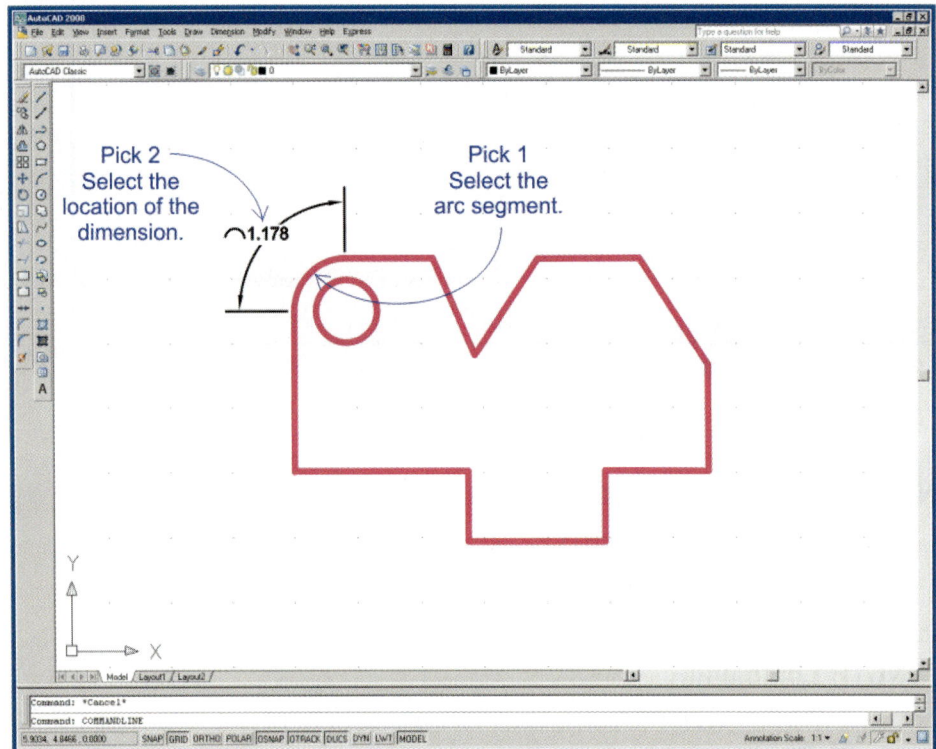

Figure 5-31(b)   ARC LENGTH command

RADIUS DIMENSION Command

The icon for the **RADIUS DIMENSION** command is shown in Figure 5-33(a). This command is used to denote the radius of an arc as shown in Figure 5-33(b).

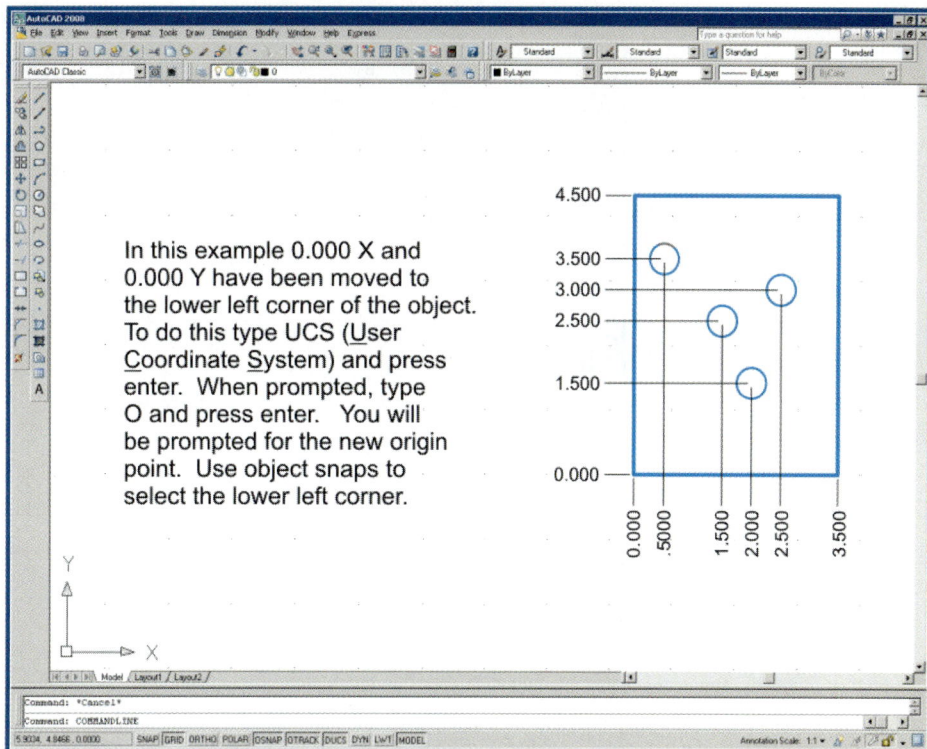

In this example 0.000 X and 0.000 Y have been moved to the lower left corner of the object. To do this type UCS (User Coordinate System) and press enter. When prompted, type O and press enter. You will be prompted for the new origin point. Use object snaps to select the lower left corner.

Ordinate Dimension

Figure 5-32(a)   Ordinate Dimension icon

Figure 5-32(b)   ORDINATE DIMENSION command

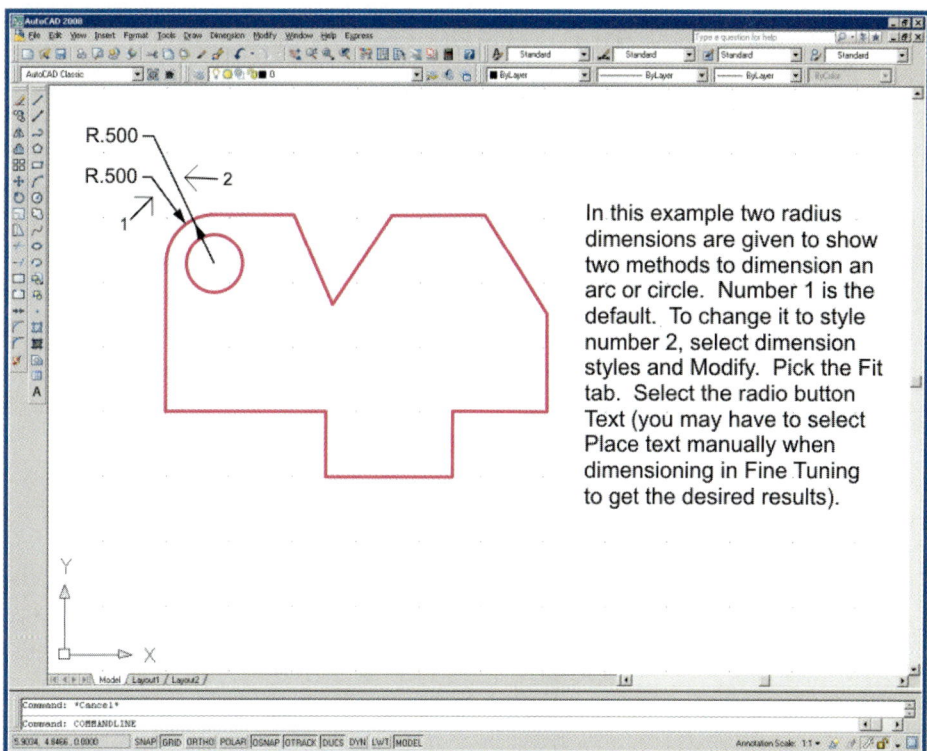

In this example two radius dimensions are given to show two methods to dimension an arc or circle. Number 1 is the default. To change it to style number 2, select dimension styles and Modify. Pick the Fit tab. Select the radio button Text (you may have to select Place text manually when dimensioning in Fine Tuning to get the desired results).

Radius Dimension

Figure 5-33(a)   Radius Dimension icon

Figure 5-33(b)   RADIUS DIMENSION command

JOGGED DIMENSION Command

The icon for the **JOGGED DIMENSION** command is shown in Figure 5-34(a). This command is used to create a jogged radius, or diameter, when dimensioning a large arc or circle whose center is outside of the drawing area as shown in Figure 5-34(b).

Jogged

Figure 5-34(a)   Jogged
Dimension icon

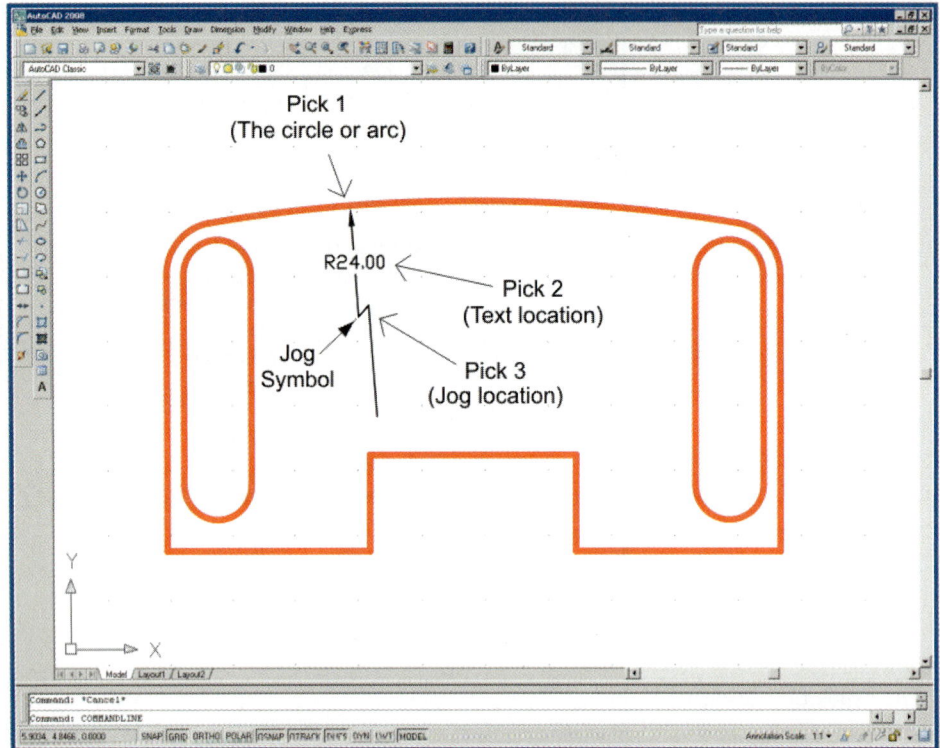

Figure 5-34(b)   JOGGED DIMENSION command

DIAMETER DIMENSION Command

The icon for the **DIAMETER DIMENSION** command is shown in Figure 5-35(a). This command is used to denote the diameter of a circle as shown in Figure 5-35(b).

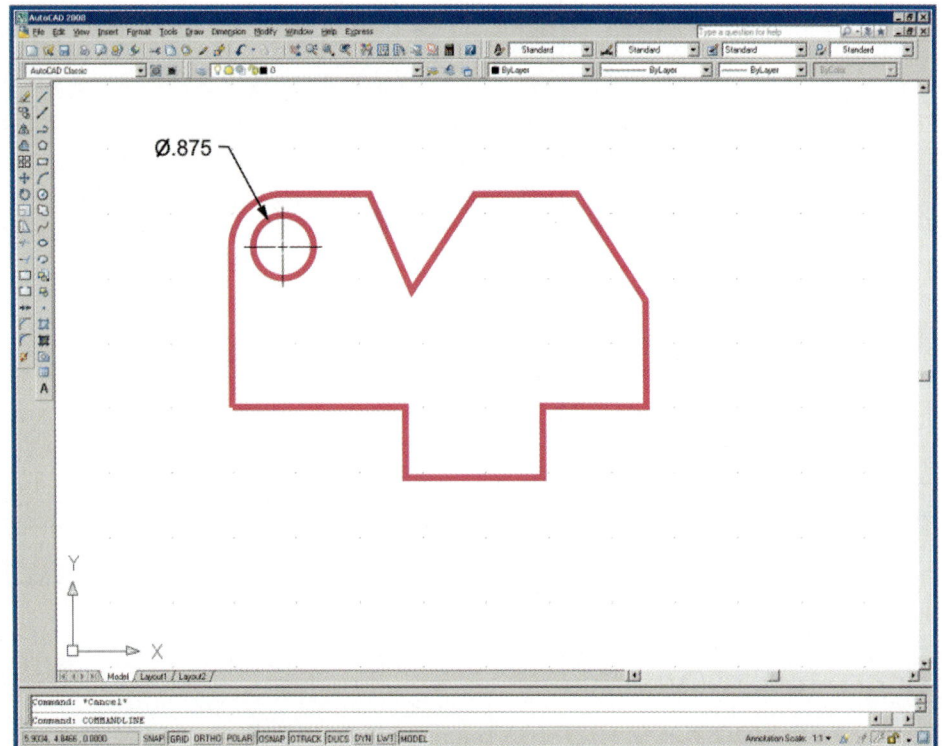

Diameter Dimension

Figure 5-35(a)   Diameter Dimension icon

Figure 5-35(b)   DIAMETER DIMENSION command

ANGULAR DIMENSION Command

The icon for the **ANGULAR DIMENSION** command is shown in Figure 5-36(a). This command is used to denote the angle between two features of an object as shown in Figure 5-36(b).

Figure 5-36(a)   Angular Dimension icon

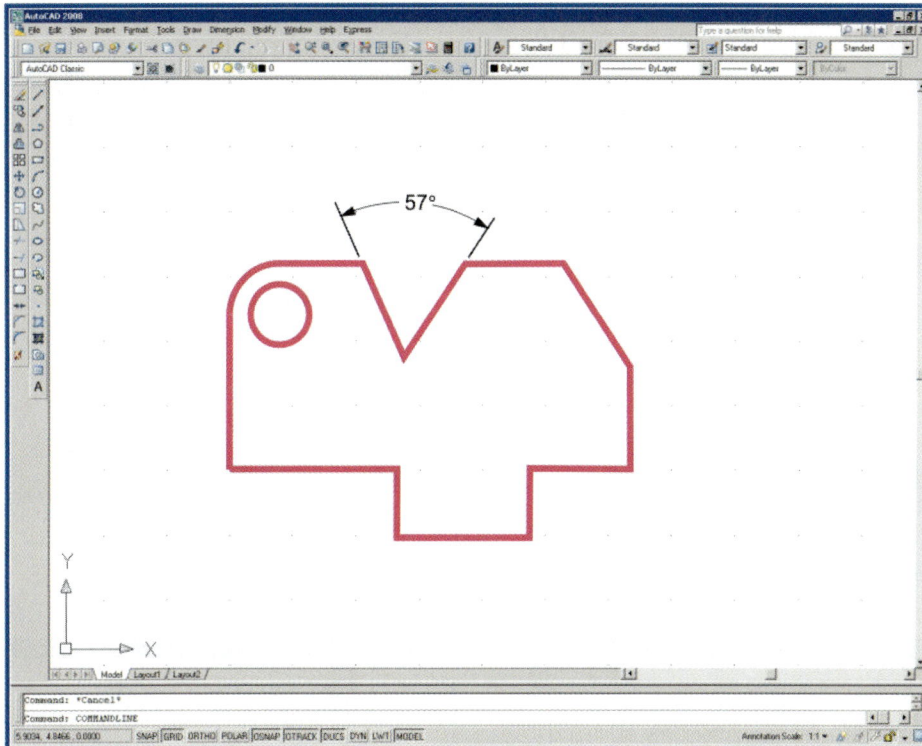

Figure 5-36(b)   ANGULAR DIMENSION command

QUICK DIMENSION Command

The icon for the **QUICK DIMENSION** command is shown in Figure 5-37(a). This command is used to create a group of dimensions quickly. You can either pick features on the object individually with the mouse or use a crossing window to select an area of the object to be dimensioned as shown in Figure 5-37(b). Several options are available: **Continue**, **Staggered**, **Baseline**, **Ordinate**, **Radius**, and **Diameter**.

Figure 5-37(a)   Quick Dimension icon

BASELINE DIMENSION Command

The icon for the **BASELINE DIMENSION** command is shown in Figure 5-38(a). This command is used to create a series of dimensions measured from the same baseline (datum). The first extension line of the first dimension placed (using the **LINEAR** command) defaults as the first extension line for dimensions placed after selecting **Baseline** dimension as shown in Figure 5-38(b).

CONTINUE DIMENSION Command

The icon for the **CONTINUE DIMENSION** command is shown in Figure 5-39(a). This command is used to create a string of continuous dimensions. The first dimension in the string is placed using the **Linear** dimension option. After selecting the **Continue** option, subsequent dimensions begin at the second extension line of the previously defined dimension as shown in Figure 5-39(b). This dimensioning method is also called *chain dimensioning*.

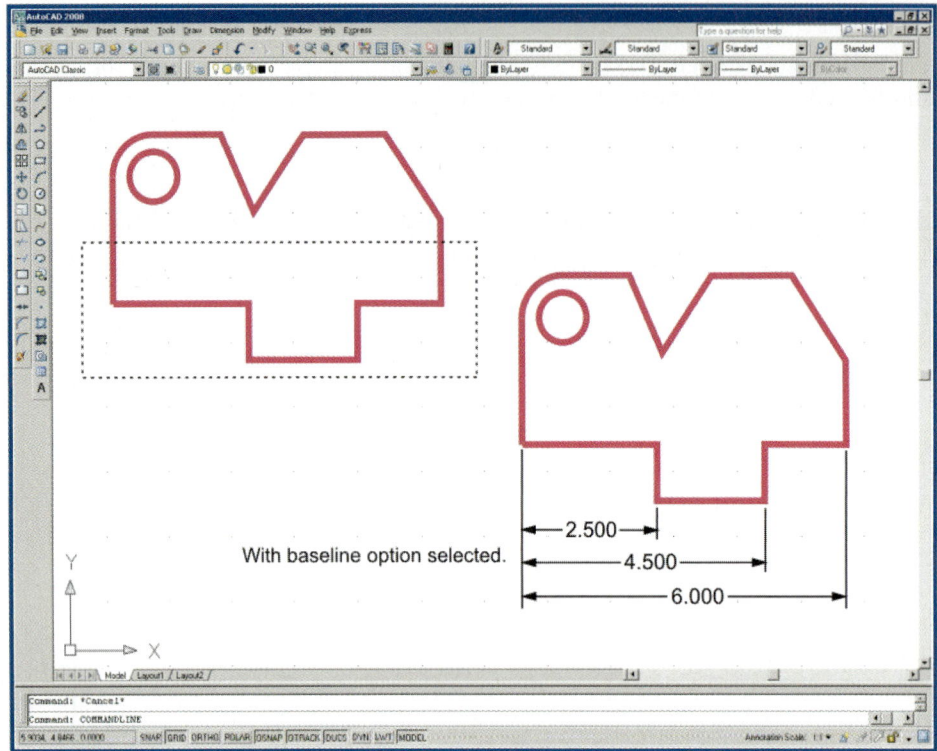

Figure 5-37(b)   QUICK DIMENSION command

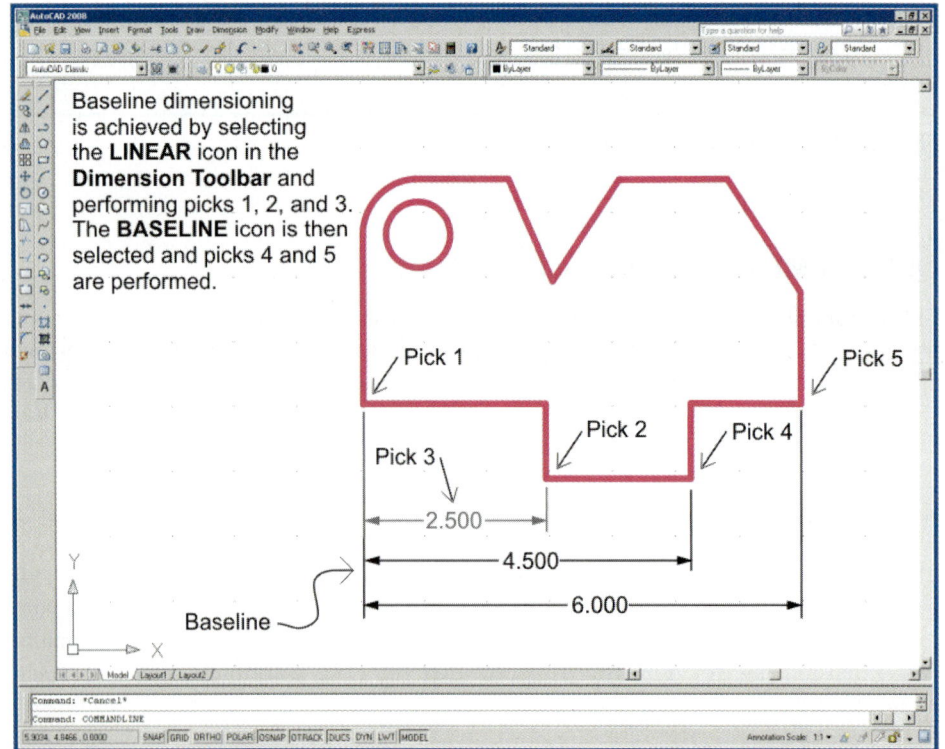

Baseline dimensioning
is achieved by selecting
the **LINEAR** icon in the
Dimension Toolbar and
performing picks 1, 2, and 3.
The **BASELINE** icon is then
selected and picks 4 and 5
are performed.

Baseline Dimension

Figure 5-38(a)   Baseline Dimension icon

Figure 5-38(b)   BASELINE DIMENSION command

DIMENSION SPACE Command

The icon for the **DIMENSION SPACE** command is shown in Figure 5-40(a). This command adjusts the space between parallel linear dimensions to match a defined distance as shown in Figure 5-40(b).

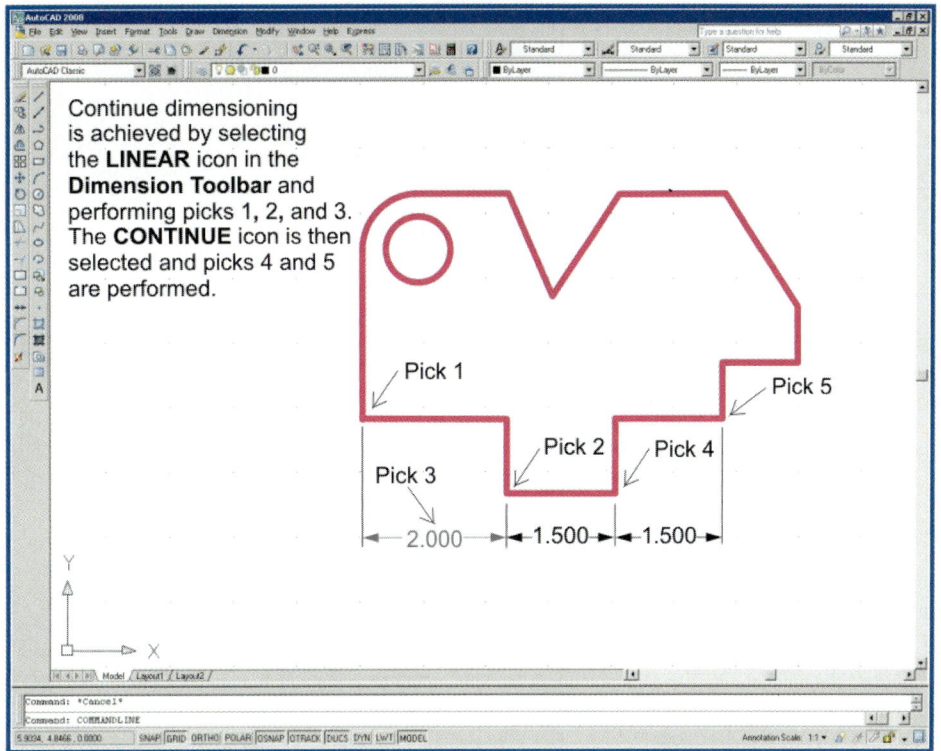

Continue dimensioning is achieved by selecting the **LINEAR** icon in the **Dimension Toolbar** and performing picks 1, 2, and 3. The **CONTINUE** icon is then selected and picks 4 and 5 are performed.

Pick 1
Pick 5
Pick 2   Pick 4
Pick 3
2.000   1.500   1.500

Continue Dimension

Figure 5-39(a)   Continue Dimension icon

Figure 5-39(b)   CONTINUE DIMENSION command

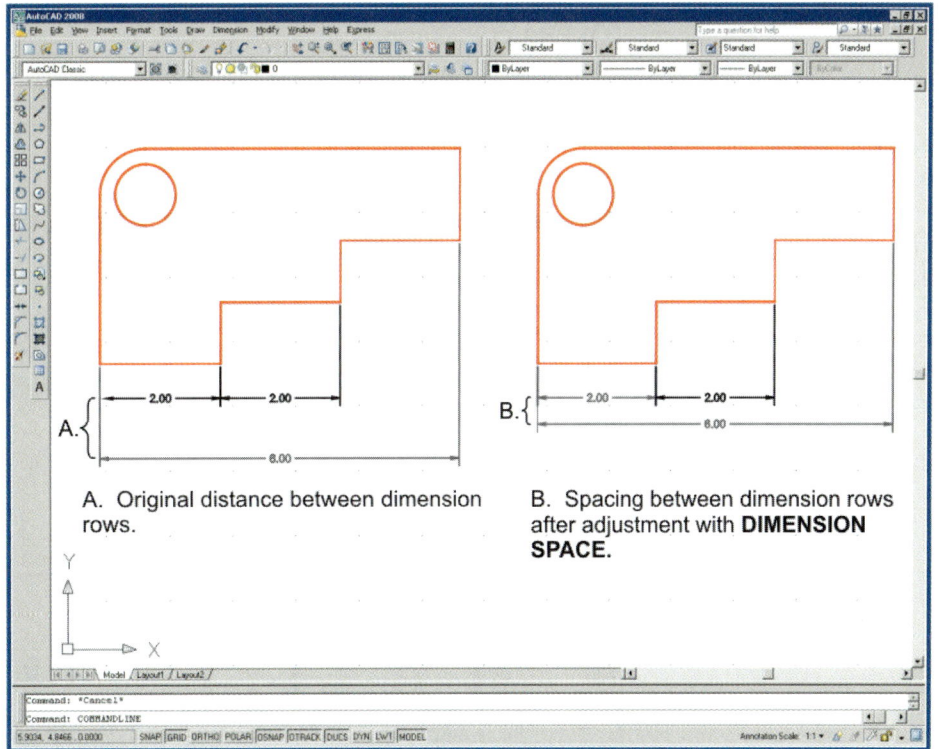

A.   2.00   2.00
6.00

B.   2.00   2.00
6.00

A. Original distance between dimension rows.

B. Spacing between dimension rows after adjustment with **DIMENSION SPACE**.

Dimension Space

Figure 5-40(a)   Dimension Space icon

Figure 5-40(b)   DIMENSION SPACE command

DIMENSION BREAK Command

The icon for the **DIMENSION BREAK** command is shown in Figure 5-41(a). This command is used to break dimension, or extension, lines where they overlap other lines as shown in Figure 5-41(b).

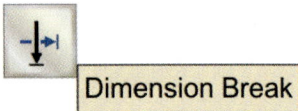

Figure 5-41(a) Dimension Break icon

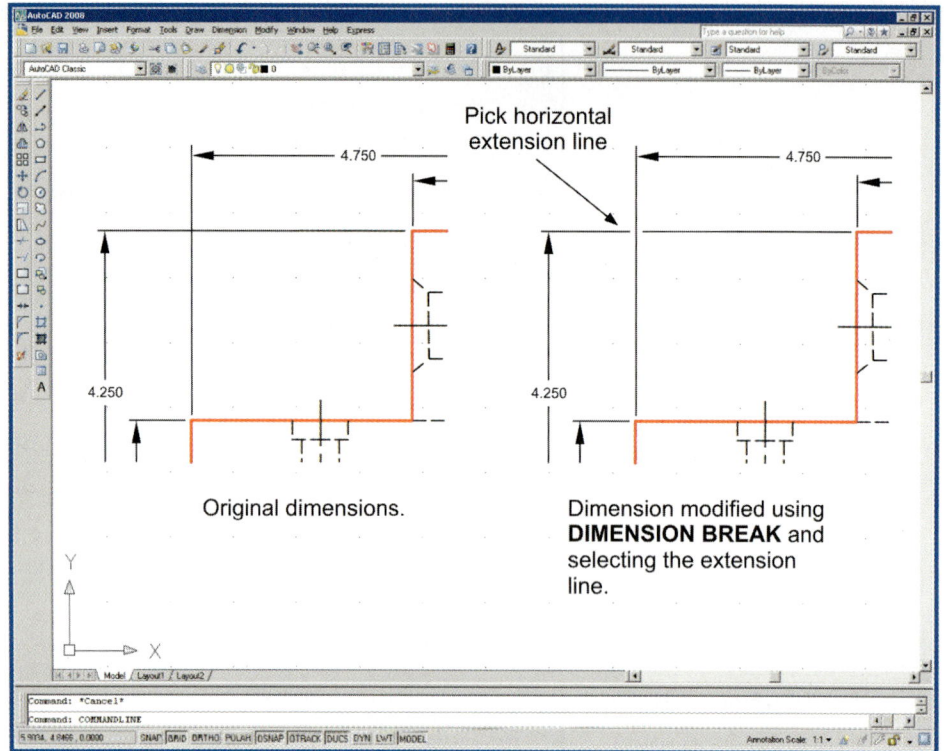

Figure 5-41(b) DIMENSION BREAK command

TOLERANCE Command

The icon for the **TOLERANCE** command is shown in Figure 5-42(a). This command is used to specify the symbols and values for Geometric Dimensioning and Tolerancing as shown in Figure 5-42(b).

Figure 5-42(a) Tolerance icon

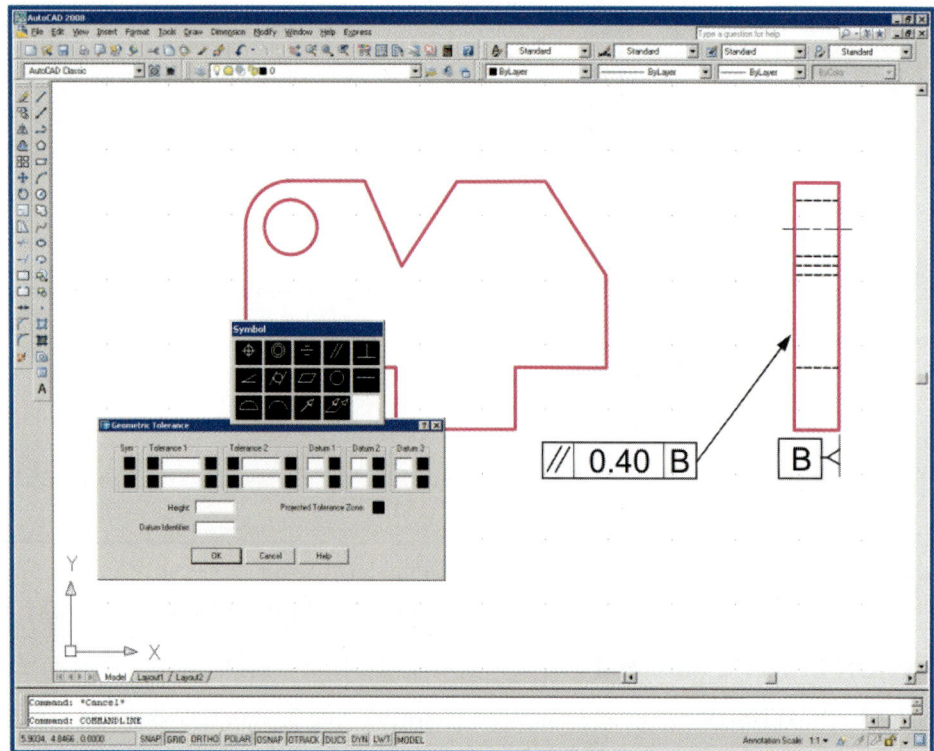

Figure 5-42(b) TOLERANCE command

CENTER MARK Command

The icon for the **CENTER MARK** command is shown in Figure 5-43(a). This command is used to create center marks, or center lines, on circles and arcs as shown in Figure 5-43(b). The option for center lines or center marks can be found in the **Dimension Styles** dialog box in the **Symbols and Arrows** tab.

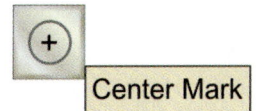

Figure 5-43(a)   Center Mark icon

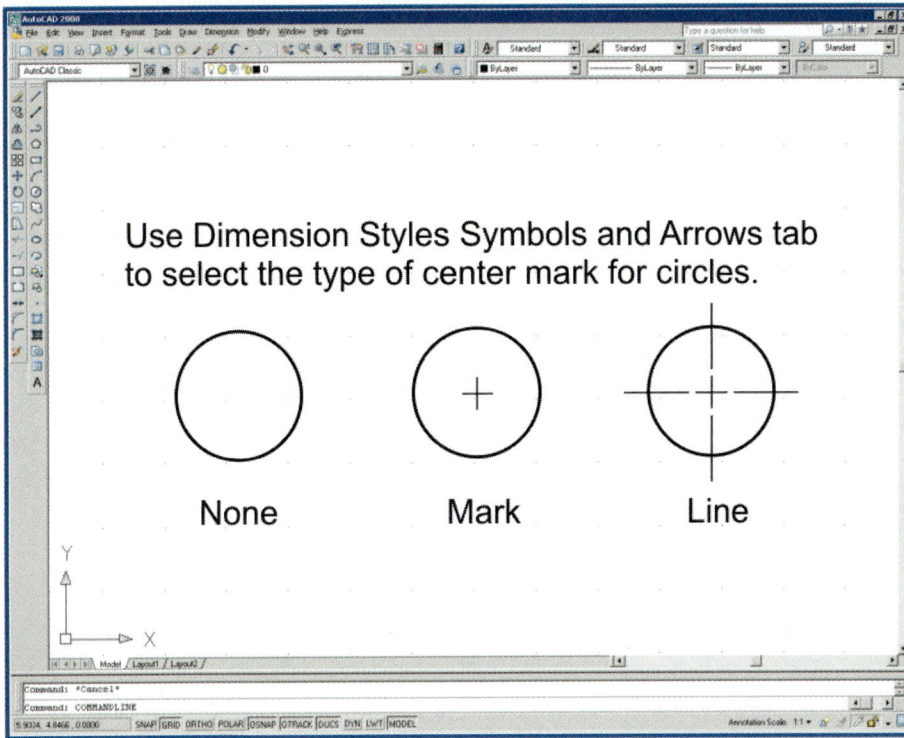

Figure 5-43(b)   CENTER MARK command

INSPECTION Command

The icon for the **INSPECTION** command is shown in Figure 5-44(a). This command creates a dimension inside a frame that is used to provide inspection information about the feature as shown in Figure 5-44(b).

Figure 5-44(a)   Inspection icon

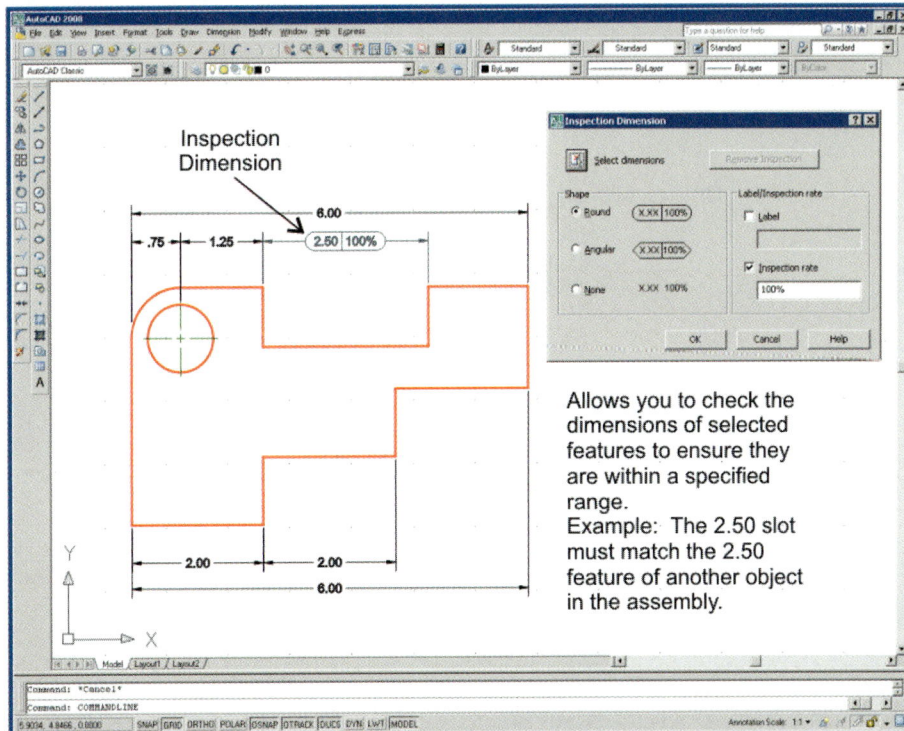

Figure 5-44(b)   INSPECTION command

Jogged Linear

Figure 5-45(a) Jogged
Linear icon

JOGGED LINEAR Command

The icon for the **JOGGED LINEAR** command is shown in Figure 5-45(a). This command is used to create a jog in a linear dimension line when the feature is not drawn full size as shown in Figure 5-45(b).

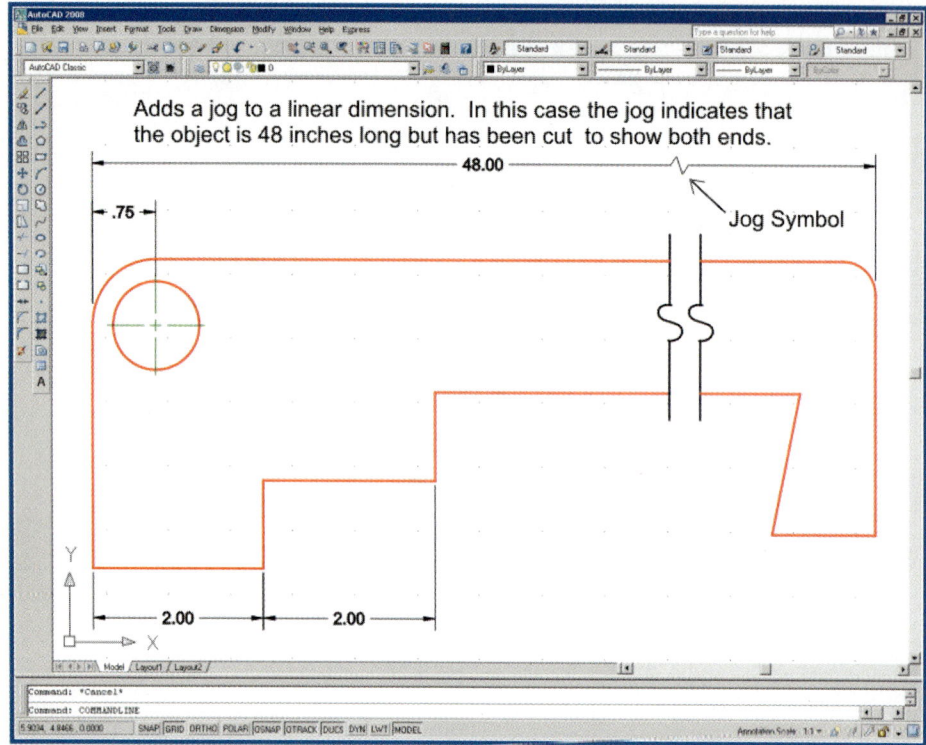

Figure 5-45(b) JOGGED
LINEAR command

Figure 5-46(a) Dimen-
sion Edit icon

DIMENSION EDIT Command

The icon for the **DIMENSION EDIT** command is shown in Figure 5-46(a). This command is used to edit existing dimensions as shown in Figure 5-46(b). Options in this command include **Home**, which changes rotated dimensions back to the default position; **New**, which changes di-

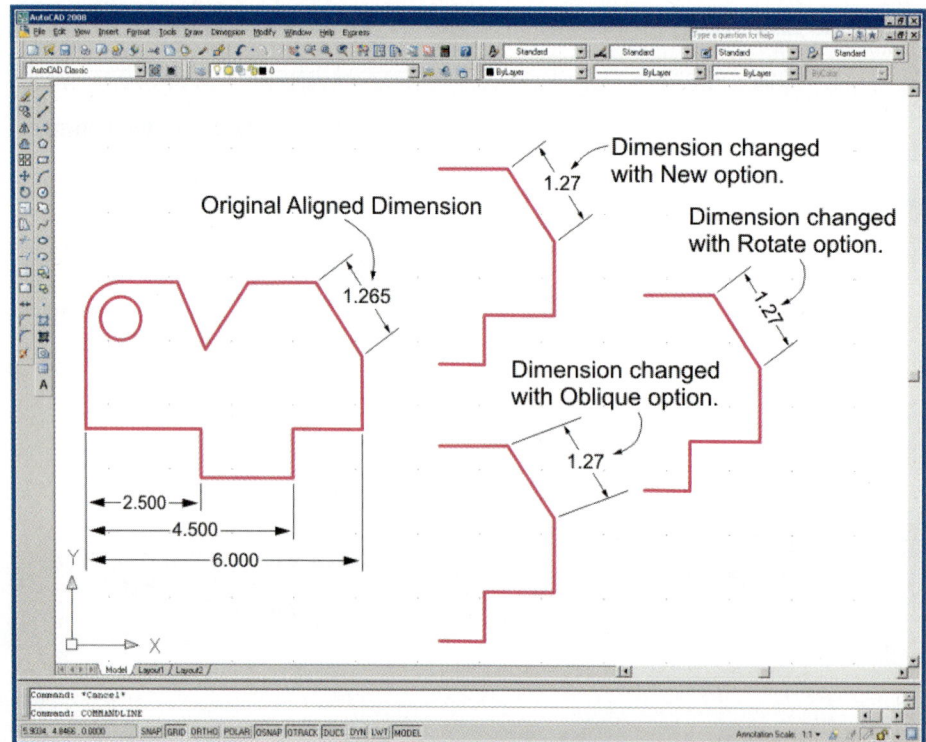

Figure 5-46(b) DIMEN-
SION EDIT command

mension text with the **Multiline Text Editor**; **Rotate**, which rotates dimension text; and **Oblique**, which changes extension lines to oblique angles.

DIMENSION TEXT EDIT Command

The icon for the **DIMENSION TEXT EDIT** command is shown in Figure 5-47(a). This command is used to move and rotate dimension text as shown in Figure 5-47(b).

Figure 5-47(a)   Dimension Text Edit icon

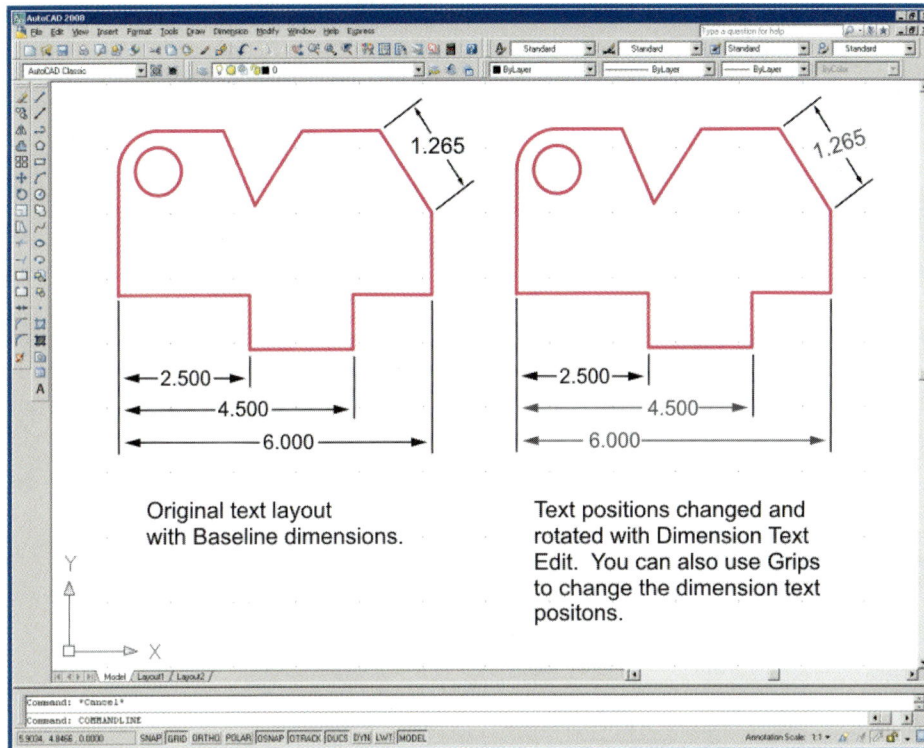

Figure 5-47(b)   DIMENSION TEXT EDIT command

DIMENSION UPDATE Command

The icon for the **DIMENSION UPDATE** command is shown in Figure 5-48(a). This command is used to apply redefined dimension style settings to existing dimensions as shown in Figure 5-48(b).

Figure 5-48(a)   Dimension Update icon

DIMENSION STYLE Command

The icon for the **DIMENSION STYLE** command is shown in Figure 5-49(a). A dimension style is a named set of values that define the appearance and format of dimensions, such as text height,

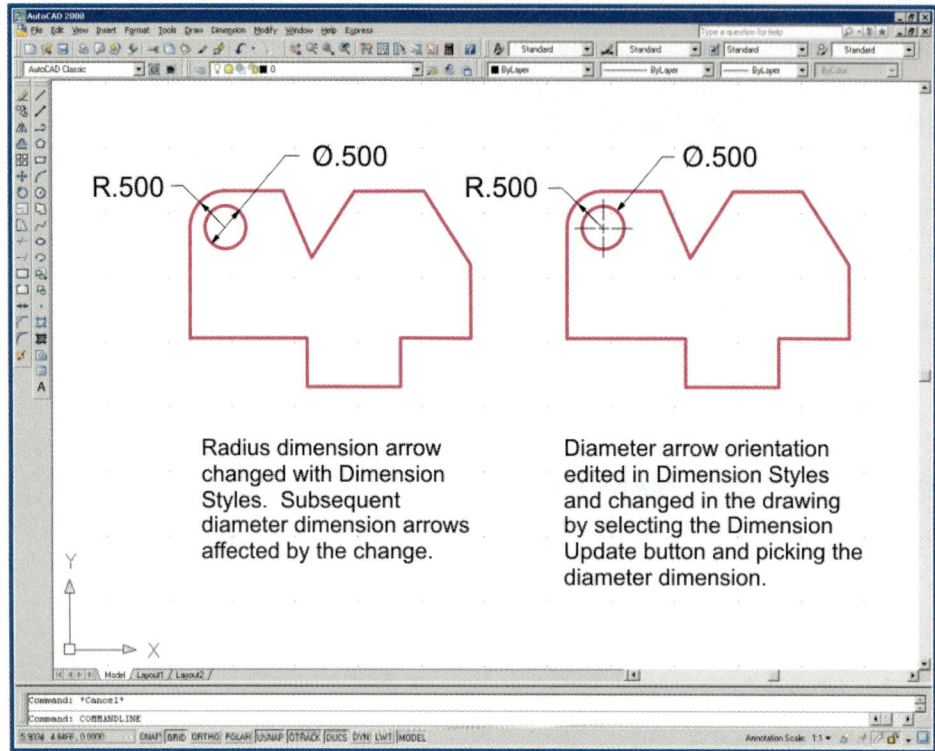

Figure 5-48(b) DIMEN-SION UPDATE command

Dimension Style

Figure 5-49(a) Dimen-sion Style icon

precision, and arrowhead length. By creating a dimension style, a drafter can set the dimension spacing and size values to match applicable dimension standards. The new dimension style becomes a part of the drawing file, but it is possible to insert a dimension style created in one drawing into another drawing. The **Dimension Style Manager** dialog box, shown in Figure 5-49(b), can be opened by selecting the **Dimension Style** icon or by typing **DIMSTYLE** and pressing <**Enter**>.

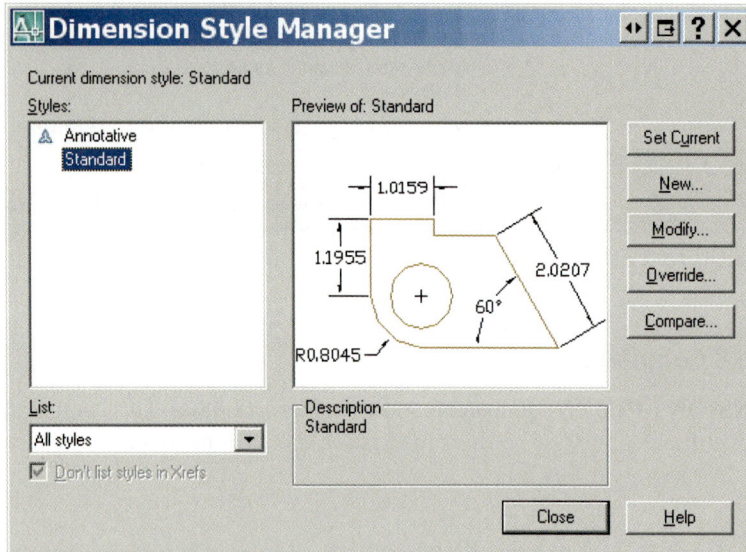

Figure 5-49(b) DIMEN-SION STYLE command

DEFINING DIMENSION SETTINGS WITH THE DIMENSION STYLE MANAGER

By choosing from the buttons on the right side of the **Dimension Style Manager** dialog box shown in Figure 5-50, it is possible to create a **new** dimension style, **modify** or **override** the current dimension style, set a different dimension style **current**, or **compare** the settings of two dimension styles.

Displays current dimension style.   Sets selected style current.

Dimension Style Manager   ? X

Current dimension style: Standard

Styles: Preview of: Standard

A Annotative   Set Current
Standard

Displays all dimension   1.0159   New...   Displays
styles in the drawing.
The current style is   1.1955   2.0207   Modify...
highlighted.
Annotative allows you   60°   Override...
to automate the process
of scaling annotations.   R0.8045   Compare...

List: Description
All styles   Standard

✓ Don't list styles in Xrefs

                                                          Close   Help

This Dialog Box

All Styles: Displays all dimension styles.   Modify Dimension Styles.
                                                          Makes changes to an existing
Styles in Use: Displays only styles   dimension style.
referenced in the drawing.
                                                          Set temporary overrides to
Does not display styles in   dimensions.
externally referenced drawings
under styles.   Displays Compare Dimension Styles
                                                          dialog box to compare two dimension
                                                          styles.

Create New Dimension Style   ? X

New Style Name: Copy of Standard   Name new style.

Start With: Standard   Sets the style you
                                                          want to start with.
Use for: All dimensions   Saves having to set
                                                          styles already created.

   Continue   Cancel   Help

Creates style that applies to specific
dimension types.

For example, the standard dimension
style color for text is white. However,
you want red for the diameter dimensions.
Select Continue - then the Text tab.
Change the color to red, then click
OK, and close.

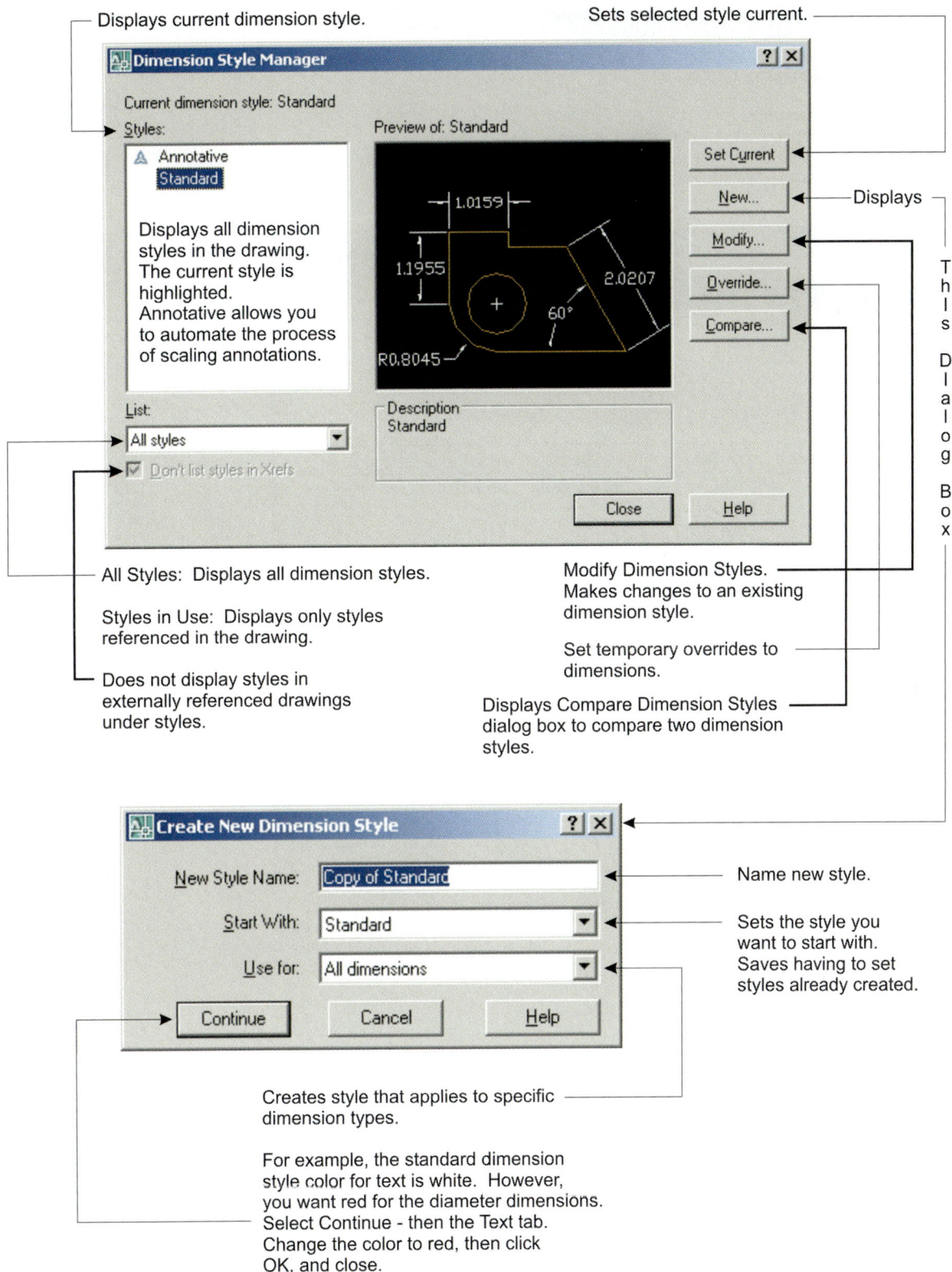

Figure 5-50   Dimension Style Manager dialog box

Select the **Compare** button on the **Dimension Style Manager** dialog box to open the
Compare Dimension Styles dialog box. Selecting two different dimension styles in the **Compare:**
and **With:** windows of this dialog box will show a comparison of the dimension style settings
assigned to each of the dimension styles. See Figure 5-51.

Displays and sets <u>first</u> dimension style for comparison.

Displays <u>second</u> dimension style for comparison.
If set to none, displays all settings for the style.

All Properties:
Description - dimension style property.
Variable - system variable that controls property.
Standard - system variable style properties.

Compare Dimension Styles   [?] [X]

Compare: Standard

With: Architectural

AutoCAD found 10 differences:

Description	Variable	Standard	Architectural
Arrow	DIMBLK	ClosedFilled	ArchTick
Dim line ext	DIMDLE	0"	0"
Length units	DIMLUNIT	2	4
Overall scale	DIMSCALE	1.0	20.0
Precision	DIMDEC	2	1
Text inside align	DIMTIH	On	Off
Text outside align	DIMTOH	On	Off
Text pos vert	DIMTAD	0	1
Text style	DIMTXSTY	Standard	ARCH
Tol precision	DIMTDEC	2	1

Close   Help

Prints results of comparison to
Windows Clipboard. You can then
paste results to word processors,
spreadsheets, etc.

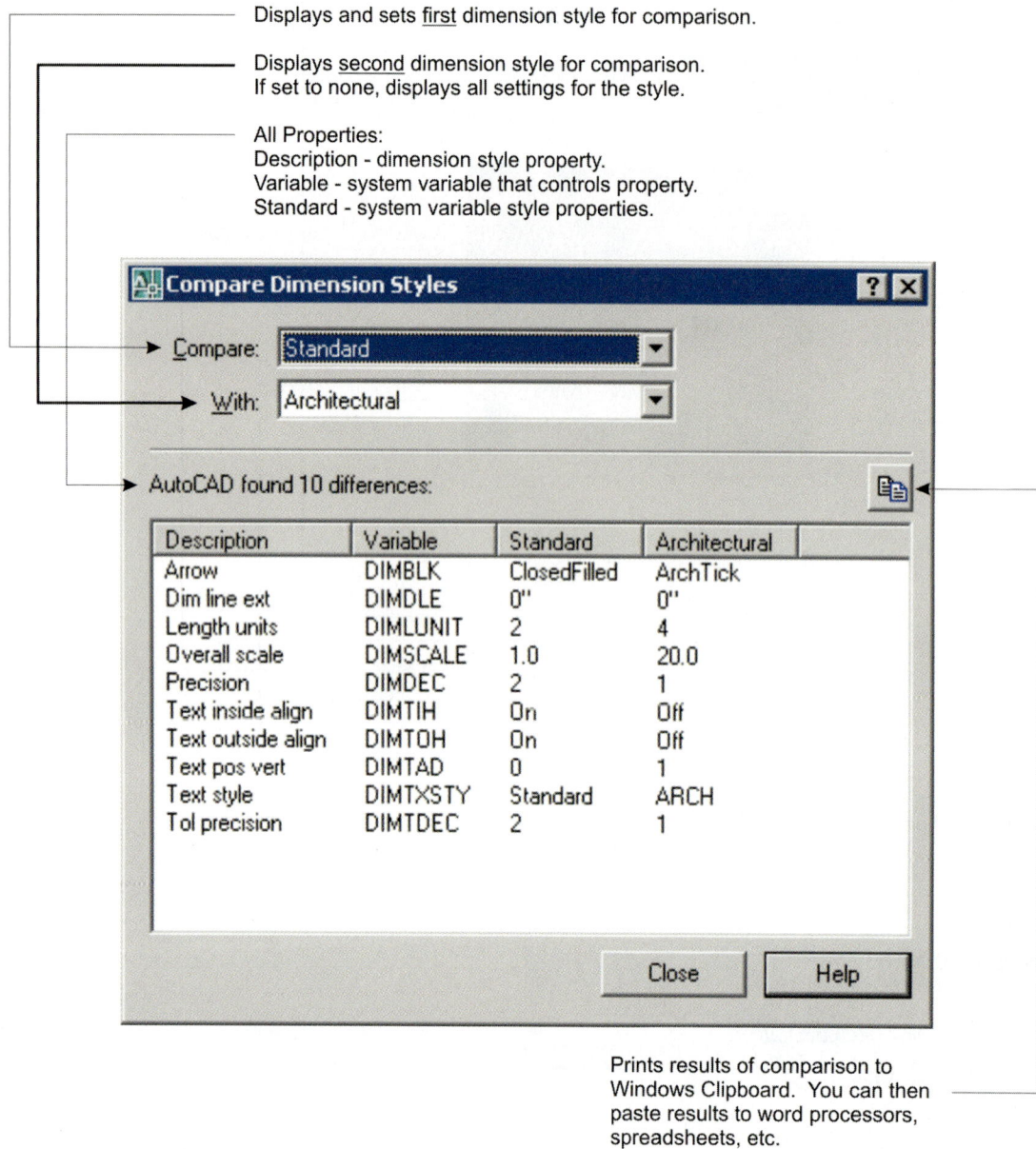

Figure 5-51   Compare Dimension Styles dialog box

TIP   A quick way to check a dimension's system variables is select the dimension, right-click your mouse, and select **Properties.** The **Properties** box wil open showing the dimension settings for the dimension. Many of these settings can be edited by changing the values shown in the **Properties** box.

Modifying a Dimension Style

By selecting the **Modify** button from the **Dimension Style Manager** dialog box, the **Modify Dimension Style** box will open as shown in Figure 5-52. This dialog box has seven tabs along the top edge: **Lines**, **Symbols and Arrows**, **Text**, **Fit**, **Primary Units**, **Alternate Units**, and **Tolerances**.

By selecting the appropriate tab, dimension settings such as text height, arrowhead style and size, and dimension spacing can be set for a drawing.

Lines Tab.   The **Lines** tab controls settings, such as the distance an extension line is offset from the object or extends past an arrowhead. It also controls the space between baseline dimensions. The values of this tab related to the settings of the ASME Y14.5M-1994 dimension standard are noted in Figure 5-52.

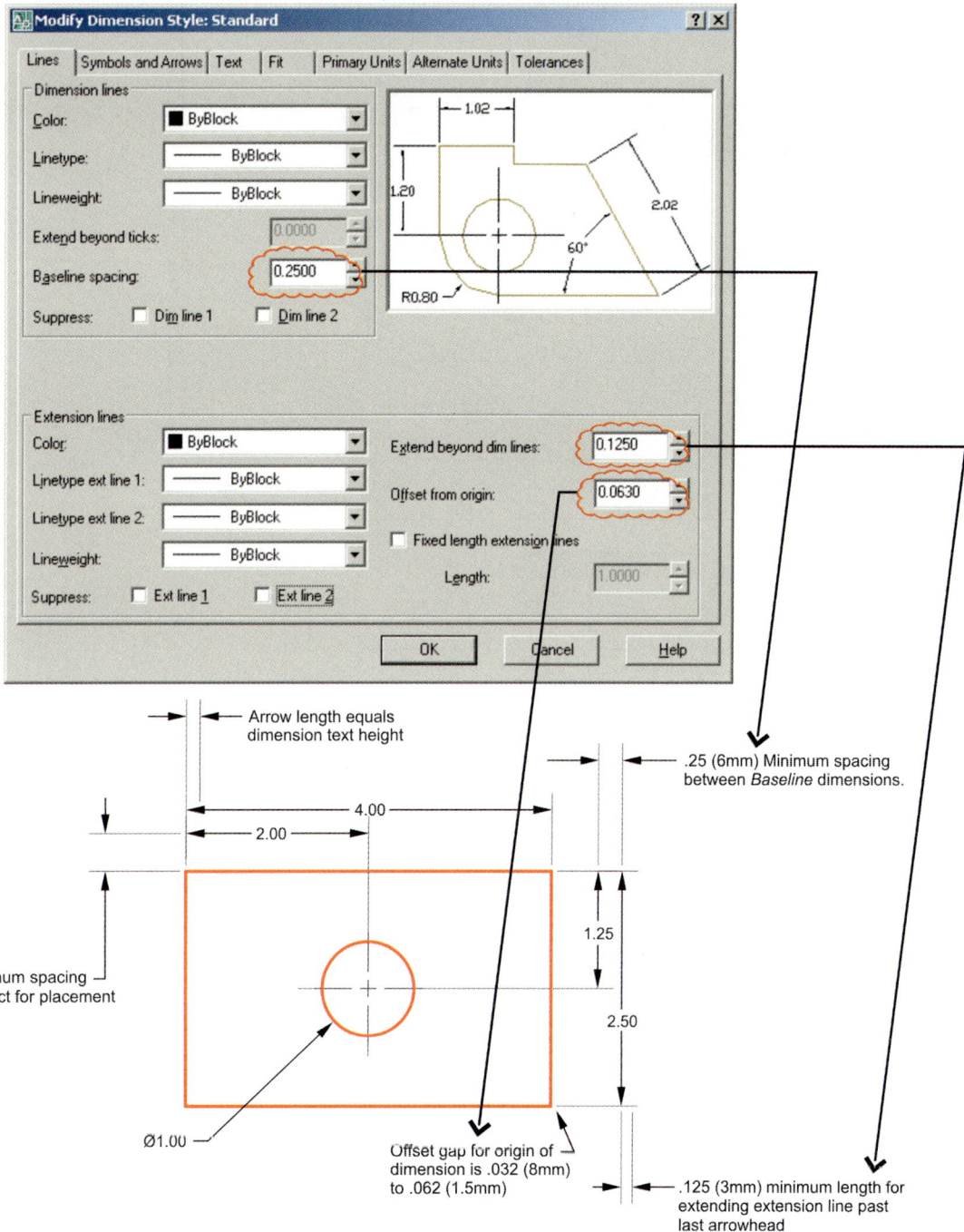

Figure 5-52   Relating the Lines tab to ASME Y14.5M-1994 dimension spacing

Symbols and Arrows Tab.   The **Symbols and Arrows** tab controls the size and type of arrowheads, including architectural tick marks, and the style of center marks used to dimension circles and arcs. The values of this tab related to the settings of the ASME Y14.5M-1994 dimension standard are noted in Figure 5-53.

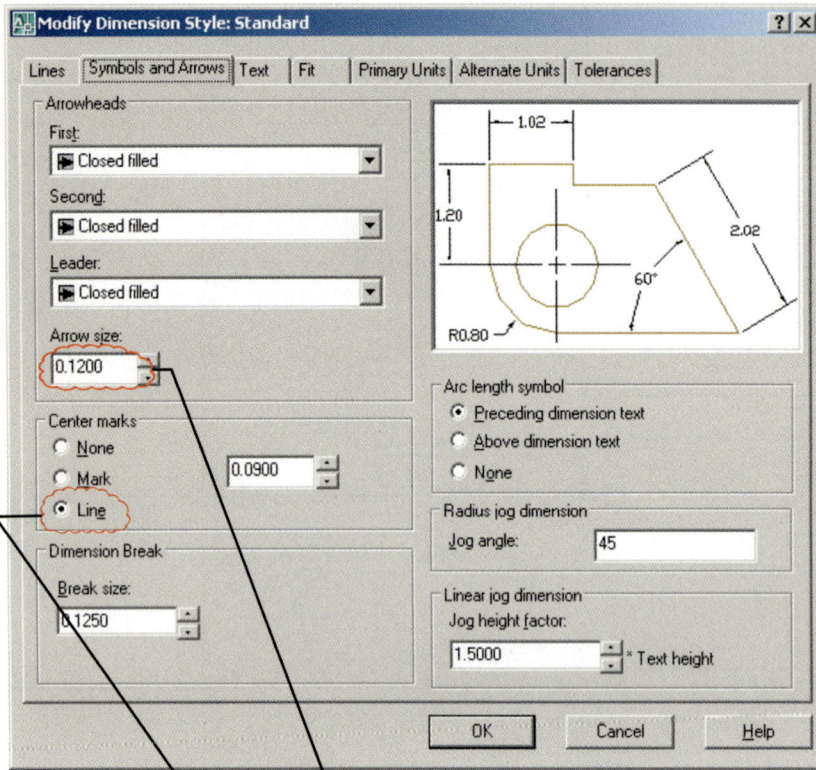

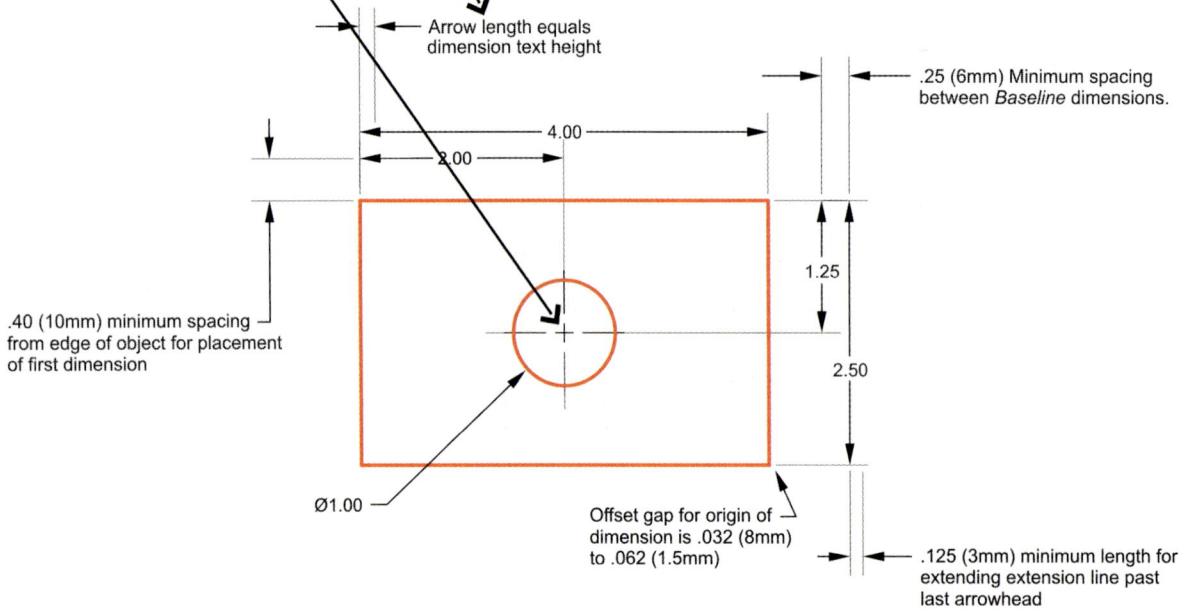

Figure 5-53   Relating the Symbols and Arrows tab to ASME Y14.5M-1994 dimension spacing

Text Tab.   The **Text** tab controls the text style, height, placement, and alignment of dimension text. Text style is based on the text properties defined in the **Text Style** dialog box. See Figure 5-54.

> **TIP**   Text height should not measure less than .12″ (3mm) to comply ASME standard for text height.

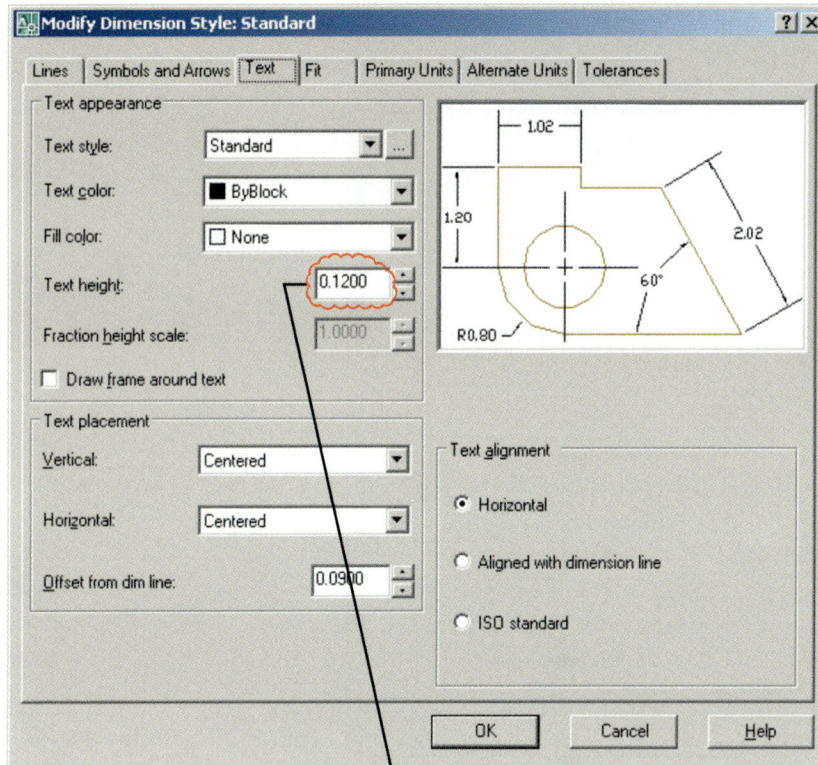

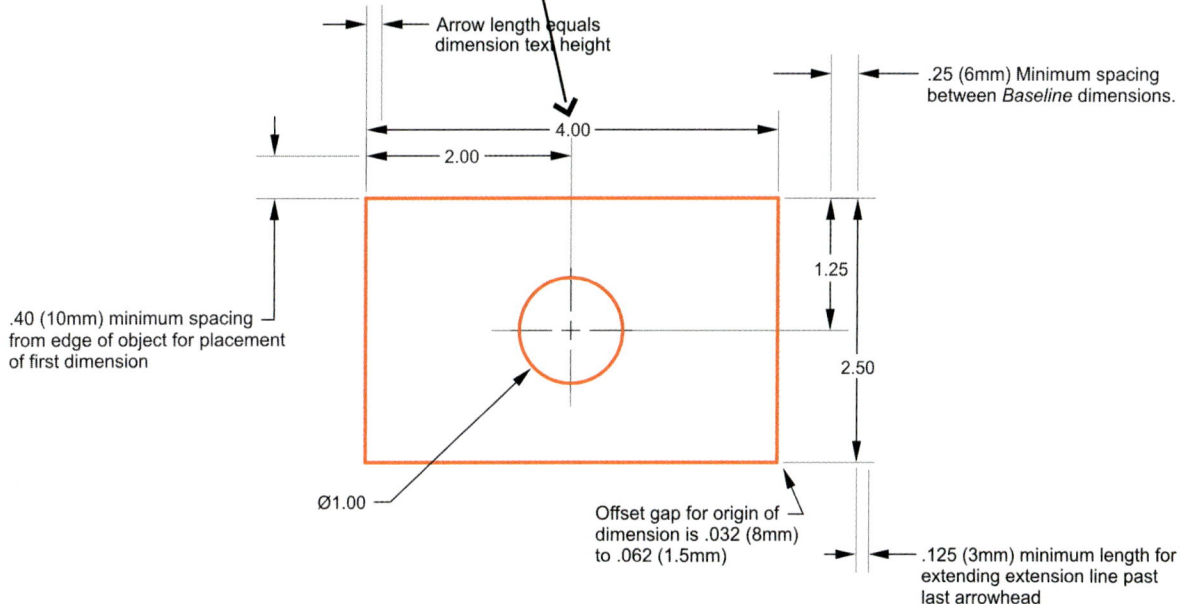

Figure 5-54   Relating the Text tab to ASME Y14.5M-1994 dimension spacing

Fit Tab.   The **Fit** tab controls the placement of text and the orientation of arrows on dimensions. Different combinations of settings from this tab can be used to force text between arrows or to force an arrow outside, etc. Study the examples in Figures 5-55(a)–(d) to see how different combinations result in different leader arrow placement.

Primary Units Tab.   The **Primary Units** tab controls the format of units (decimal, architectural, engineering, etc.) and the precision (the number of decimal places or fractional round-off) of dimensions. By selecting the **Leading** box in the **Zero suppression** area, the leading zero on decimal dimensions less than 1.00 unit in size will be suppressed. The setting in the **Scale factor:** box located in the **Measurement scale** area determines the value of the dimension. For example, setting the scale factor to **2.00** will make the value of the dimension placed on the drawing twice as large as the actual measurement. See Figure 5-56.

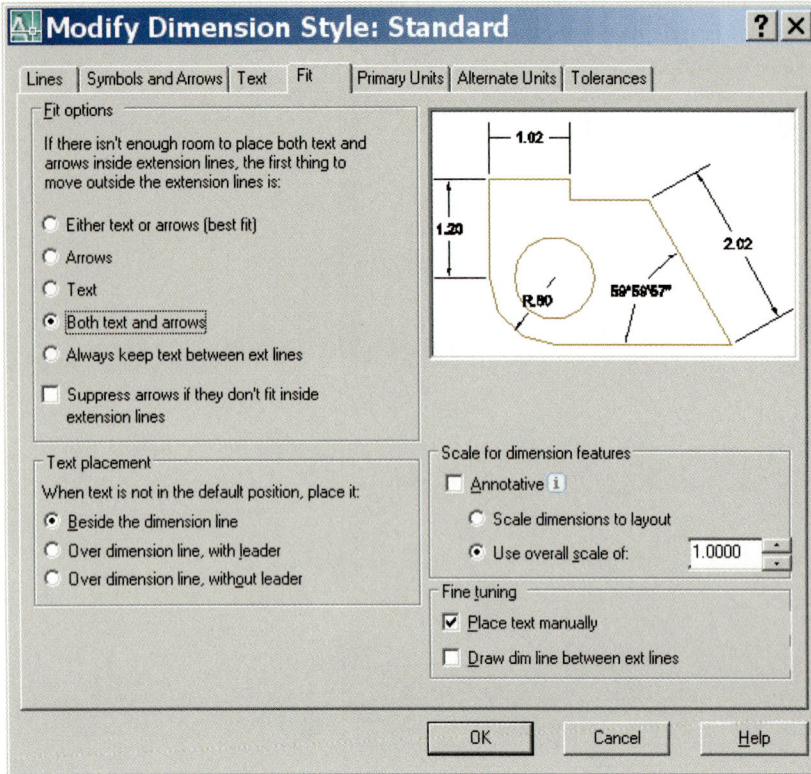

Figure 5-55(a)   Fit tab settings for controlling arrow orientation on small circles and radii

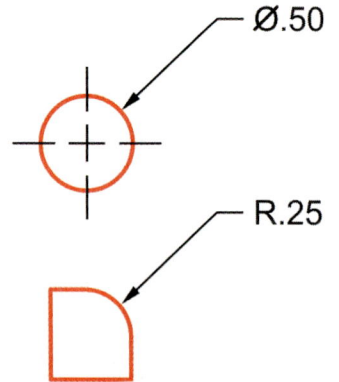

Figure 5-55(b)   Arrow orientations resulting from settings shown in Figure 5-55(a)

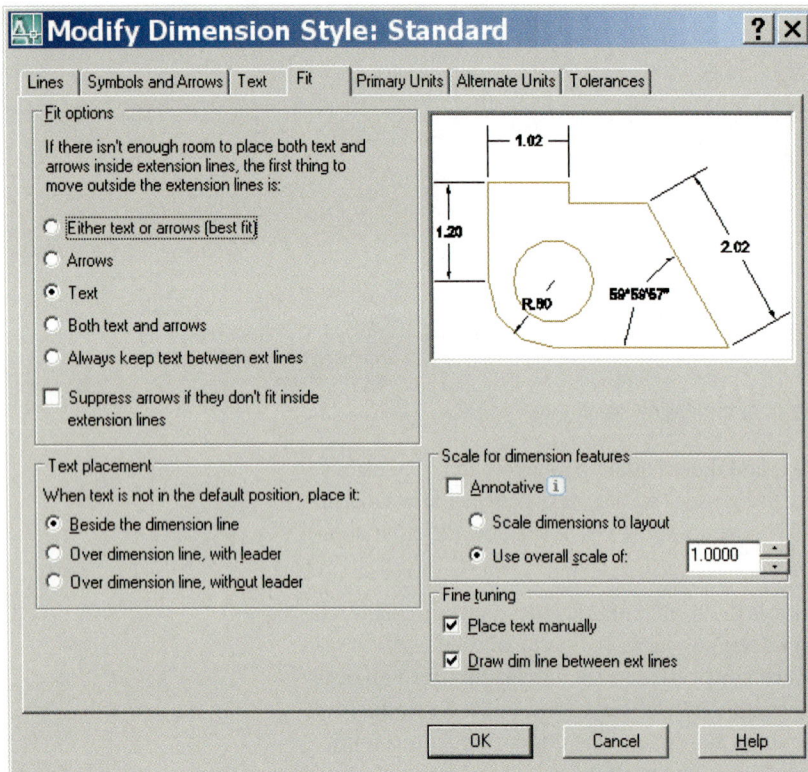

Figure 5-55(c)   Fit tab settings for controlling arrow orientation on large circles and radii

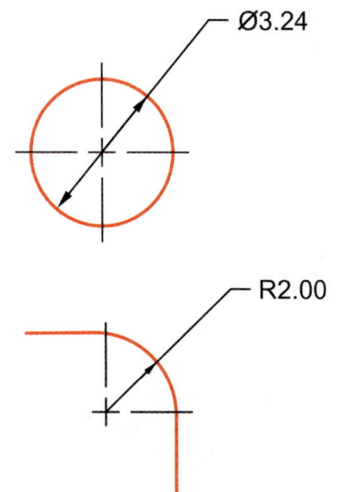

Figure 5-55(d)   Arrow orientations resulting from settings shown in Figure 5-55(c)

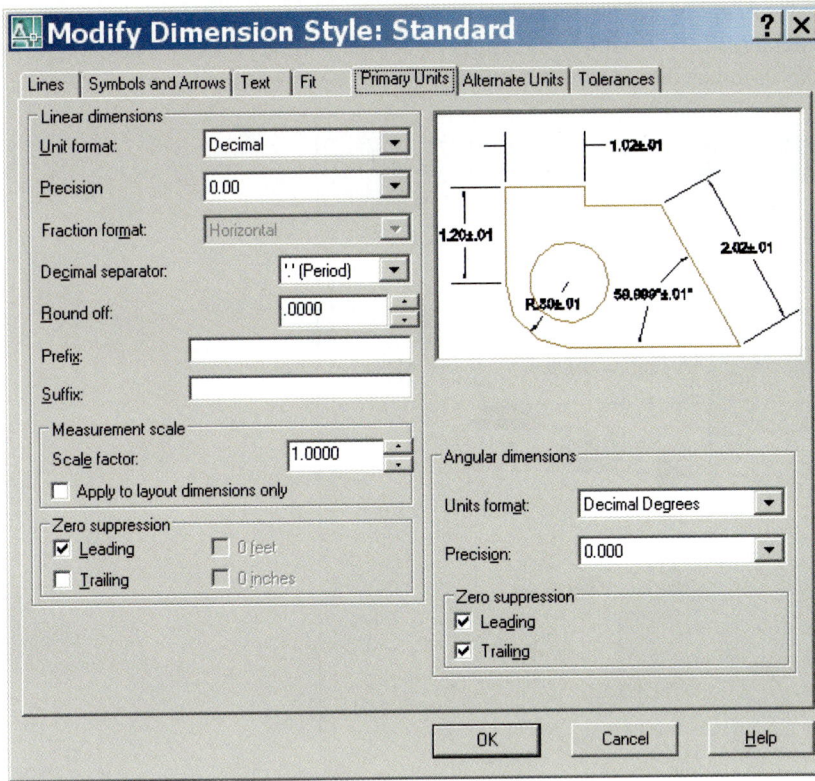

Figure 5-56   Primary Units tab

Alternate Units Tab.   The **Alternate Units** tab allows dual dimensions to be shown side by side on the drawing; for example, decimal units shown along side metric units. The alternate unit will be placed inside of brackets. See Figure 5-57.

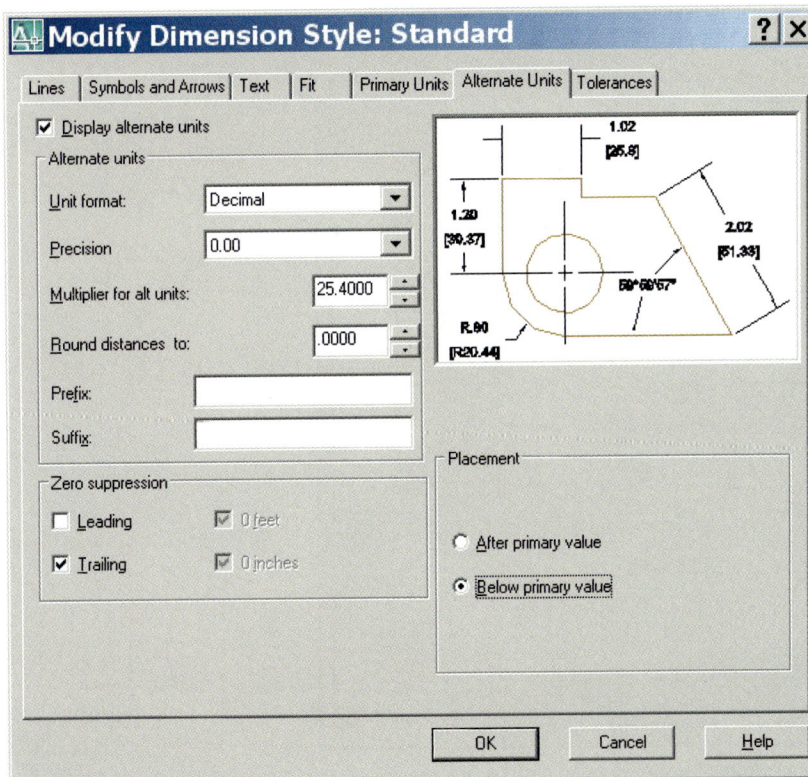

Figure 5-57   Alternate Units tab

Tolerances Tab.   The **Tolerances** *tab* allows tolerances to be incorporated into dimension text. Tolerances may be shown as **Limits** by applying the settings in Figures 5-58(a) and 5-58b, or as plus or minus dimensions by applying the settings shown in Figures 5-59(a) and 5-59(b).

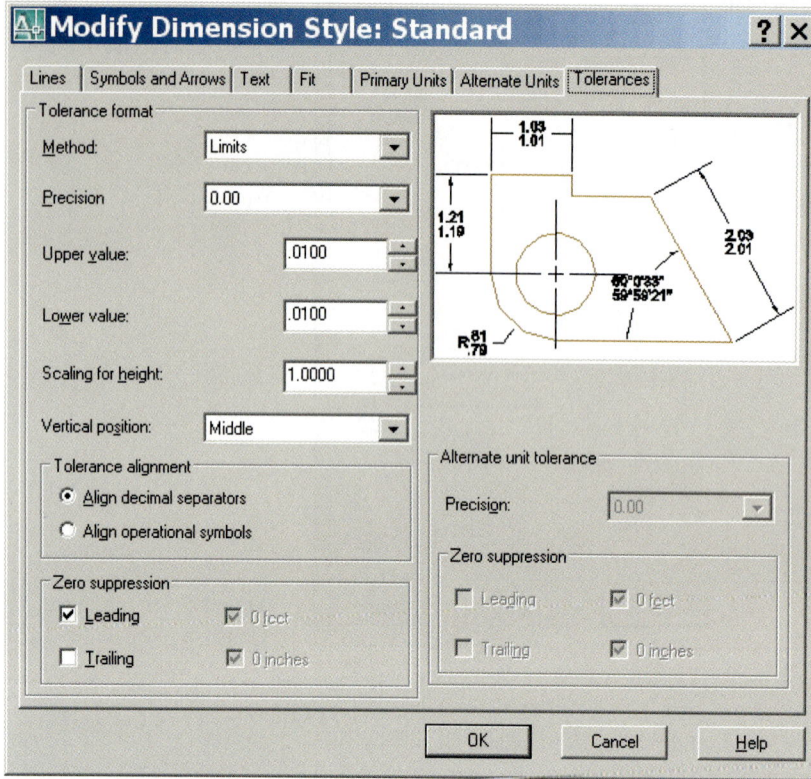

Figure 5-58(a)   Tolerances tab

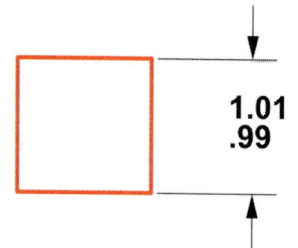

Figure 5-58(b)   Example of Limits dimensions

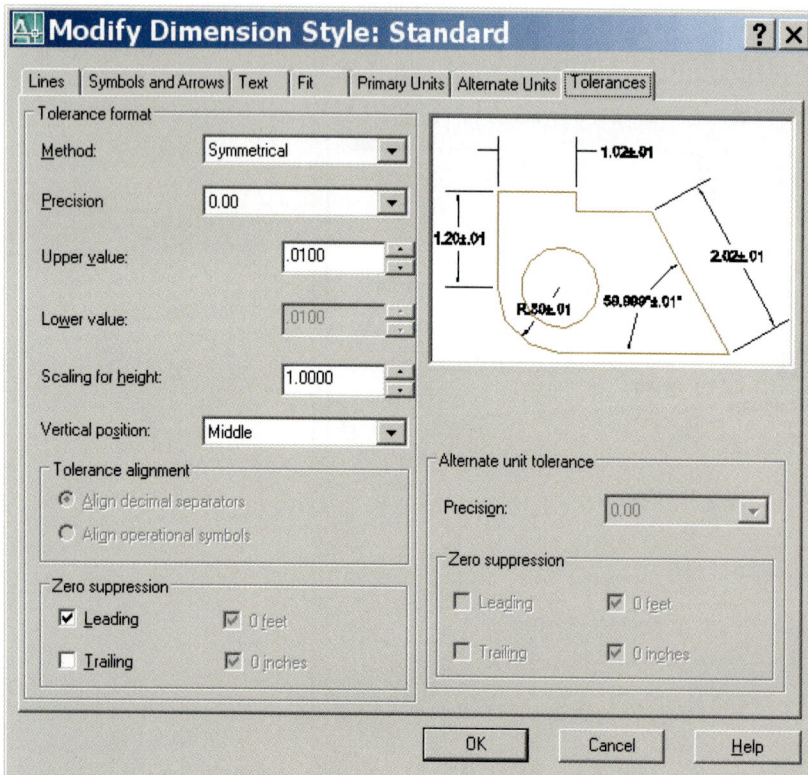

Figure 5-59(a)   The Tolerances tab settings for symmetrical dimensions

Figure 5-59(b)   Example of plus or minus (symmetrical) dimensions

CREATING A NEW DIMENSION STYLE

By selecting the **New. . .** button from the **Dimension Style Manager** dialog box, the **Create New Dimension Style** box will open. The first step in the process is to enter a name for the new dimension style in the **New Style Name:** box. In the **Start With:** box, an existing style can be selected that may already have some desired dimension settings in place so that the new style is not created from scratch. In the example in Figure 5-60, the new style name is **Arch 48**. Next, select **Continue** and choose the appropriate tab(s) from the **New Dimension Style** dialog box to change the desired dimension setting(s) for the new style. When the changes are made, click **OK**.

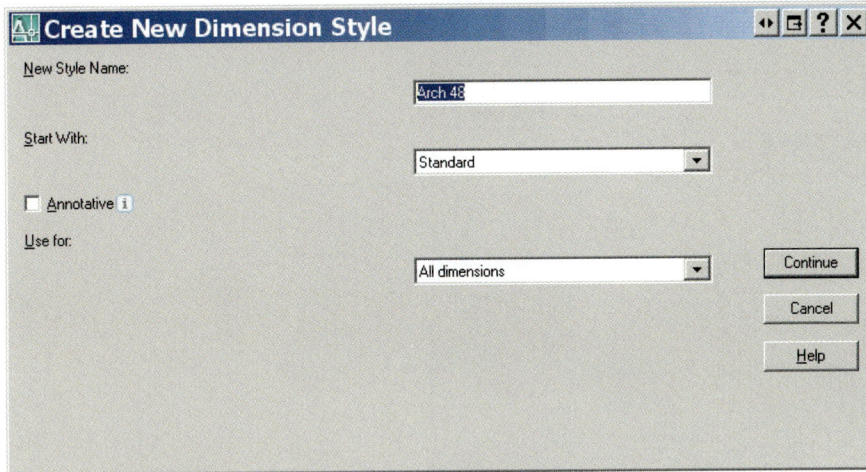

Figure 5-60   Create New Dimension Style dialog box

> **TIP**
>
> A new dimension style can be based on an existing style defined in the **Start With:** box.

SETTING THE NEW STYLE CURRENT

In the **Styles:** area of the **Dimension Style Manager** dialog box, select the new style's name, then select the **Set Current** button and click **Close**. The new dimension style settings will go into effect. See Figure 5-61.

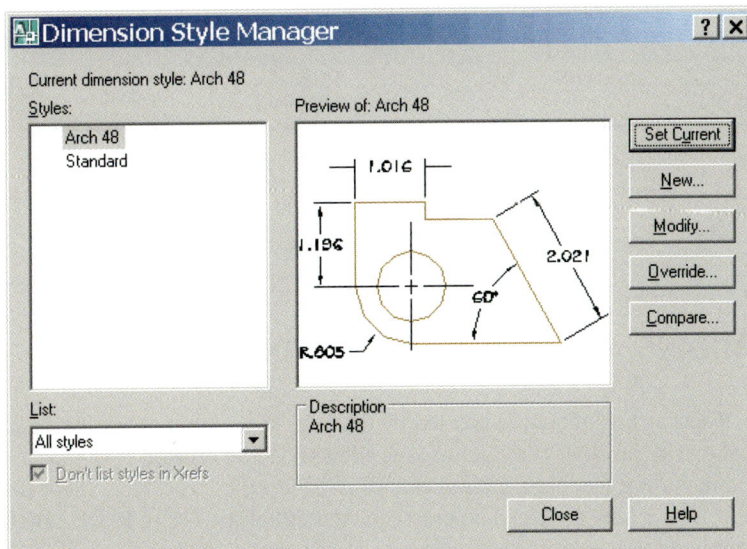

Figure 5-61   Setting a new dimension style current

OVERRIDING A DIMENSION SETTING

Sometimes it is necessary to have a different style setting apply to only a few dimensions. In a case like this, a dimension style override can be performed.

To perform a dimension override, first select the **Override. . .** button from the **Dimension Style Manager** dialog box. See Figure 5-62. Then, select the appropriate tab(s) and assign the new setting(s) and click **OK**. Then click **Close** to exit the **Dimension Style Manager**.

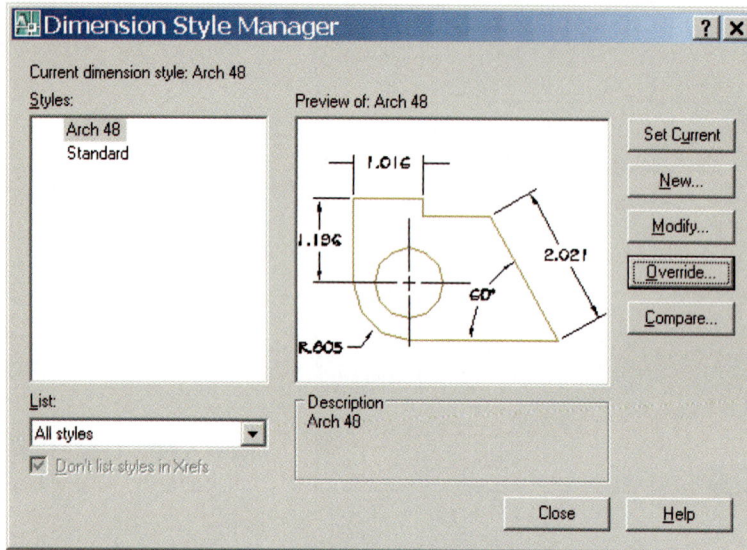

Figure 5-62   Override dimension style

Note:
The new settings will not go into effect until a **Dimension Update** is performed on the dimensions to be changed.

Performing a Dimension Update

Select the **Dimension Update** icon located on the **Dimension** toolbar and select the dimension(s) that you want the overridden setting(s) to apply to and press **<Enter>**. The dimension's style will update to reflect the new setting(s). See Figure 5-63.

Dimension Update Icon

Figure 5-63   Dimension Update icon

ADDING A LEADER TO A DRAWING

A leader is an annotation created by drawing a line (or a spline) with an arrowhead at one end and text at the other end. Figure 5-64 shows a leader created for a mechanical engineering drawing including the three components of a leader: the arrowhead, the leader line, and the landing line. This leader style can be created using either the **QUICK LEADER** or **MULTILEADER** command.

LEADER

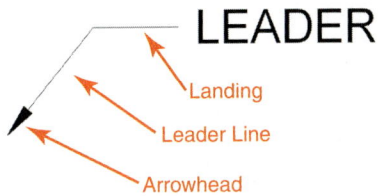

Figure 5-64 Mechanical leader style

> The landing line is also referred to as the *shoulder* of a leader.
> **TIP**

Quick Leader Command Tutorial (Mechanical Style Leader)

Step 1. Type **QLEADER** on the command line and press **<Enter>**.

> The alias for **QLEADER** is **LE**.
> **TIP**

Step 2. At the Specify first leader point prompt, select the point for the arrowhead to begin.

Step 3. At the Specify next point prompt, select the point for the Leader Line to end.

Step 4. At the Specify next point prompt, select the point where the Landing Line should end.

> Turn **Ortho On** to draw a horizontal landing line.
> **TIP**

Step 5. At the Specify text width prompt, either press **<Enter>**, or type in a width and press **<Enter>**.

Step 6. At the Enter first line of annotation text prompt, enter the text for the first line of the leader annotation and press **<Enter>**.

Step 7. At the Enter next line of annotation prompt, enter the text for the next line or press **<Enter>**.

MULTILEADER Command Tutorial

Step 1. Type **MLEADER** on the command line and press **<Enter>**, or select the **MULTILEADER** icon from the **Multileader** toolbar.

> The alias for **MLEADER** is **MLD**.
> **TIP**

Step 2. At the Specify first leader point prompt, select the point for the arrowhead to begin.

Step 3. At the Specify leader landing location prompt, select the point for the leader line to end.

Step 4. When the **Text Formatting** dialog box opens, enter the desired text and click **OK**.

SUMMARY

In every field of technical drawing, drafters are responsible for creating drawings that accurately reflect the designer's intentions.

In the mechanical engineering field, drafters must ensure that each dimension on the drawing reflects the exact dimensional value, including the exact degree of precision (the number of decimal places), as the designer's input. Drafters must also ensure that the dimensions on the drawing reference the same datum geometry that the designer specified in the input. If the drafter fails to portray the de-signer's input faithfully, the manufactured part might not fit, or function, as the designer intended. Also, parts that do not pass a quality control inspection may have to be scrapped, thus hampering the organization's ability to produce the product in the desired timeframe, or worse, prevent the organization from profitably manufacturing the product.

In every field of technical drawing, CAD drafters must be familiar with applicable dimensioning standards and appropriate dimension style settings.

UNIT TEST QUESTIONS

Short Answer

1. What is a tolerance?
2. What does ISO stand for?
3. What is a datum?
4. Who is responsible for calculating tolerances on an engineering drawing?
5. Is the diameter of a hole a size, or a location, dimension?
6. In the United States, what is the name of the organization that publishes a dimensioning standard for engineering drawings?
7. What does the term CAM stand for?
8. What is a reference dimension?
9. What is meant by nominal size?
10. Which method of labeling text complies with the ASME standard: unidirectional or aligned?

Matching

Column A

a. Primary Units

b. Lines

c. Text

d. Fit

e. Symbols and Arrows

Column B

1. The **Dimension Style Manager** tab where dimension arrow size is defined

2. The **Dimension Style Manager** tab where dimension height is defined

3. The **Dimension Style Manager** tab where dimension precision is defined

4. The **Dimension Style Manager** tab where the orientation of arrowheads on circles and arcs is defined

5. The **Dimension Style Manager** tab where center mark style is defined

UNIT PROJECTS

Dimensioning Project 5-1

Open the **Bracket** drawing that you created in Unit 4. Create a new layer named Dimensions and set it current. Add dimensions and notations to the views of the bracket (see Figure 5-65). Follow the rules of dimensioning outlined in this unit.

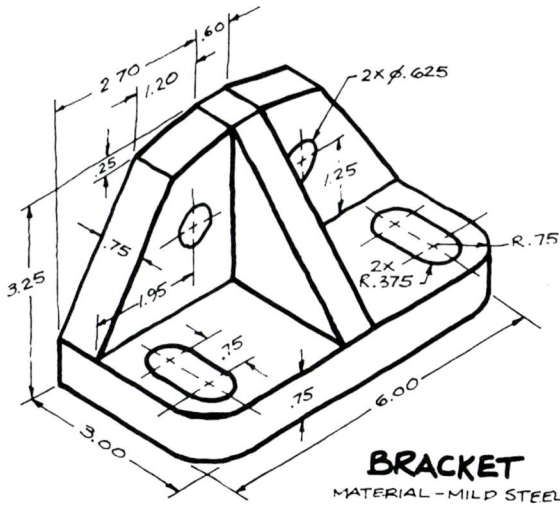

Figure 5-65 Designer's sketch of the bracket

Follow your instructor's directions to print the drawing when you are finished placing the dimensions.

In the **Dimension Style Manager** dialog box, set the following variables:

Text height:	.125
Arrow size:	.125
Center marks:	Line
Extend beyond dim lines:	.125
Precision:	Varies—match precision of dimensions on sketch
Zero suppression:	Leading
Offset from origin:	.062

Dimensioning Project 5-2

Open the **Shaft Guide** drawing that you created in Unit 4. Create a new layer named Dimensions and set it current. Add dimensions and notations to the views of the shaft guide (see Figure 5-66). Follow the rules of dimensioning outlined in this unit.

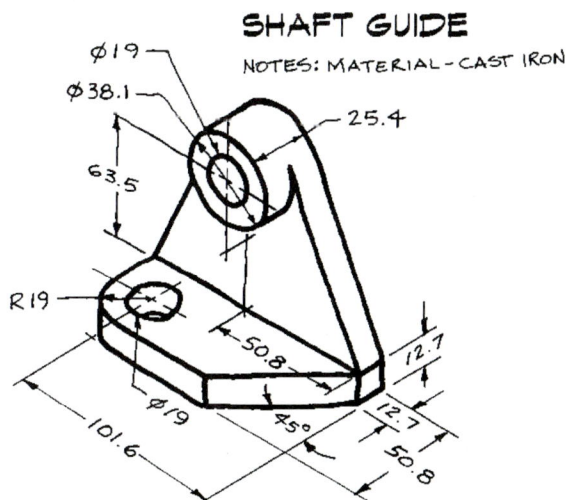

Figure 5-66 Designer's sketch of shaft guide

Follow your instructor's directions to print the drawing when you are finished placing the dimensions.

In the **Dimension Style Manager** dialog box, set the following variables (values are in millimeters):

Text height:	3
Arrow size:	3
Center marks:	Line
Center mark size:	2
Extend beyond dim lines:	3
Precision:	0.0
Zero suppression:	Trailing
Offset from origin:	1.5
Offset from dim line:	1.5

Dimensioning Project 5-3

Open the **Tool Holder** drawing that you created in Unit 4. Create a new layer named Dimensions and set it current. Add dimensions and notations to the views of the tool holder (see Figure 5-67). Follow the rules of dimensioning outlined in this unit.

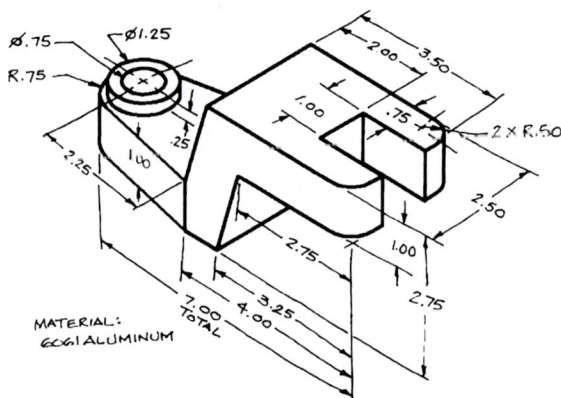

Figure 5-67   Designer's sketch of the tool holder

Follow your instructor's directions to print the drawing when you are finished placing the dimensions.

In the **Dimension Style Manager** dialog box, set the following variables:

Text height:	.125
Arrow size:	.125
Center marks:	Line
Extend beyond dim lines:	.125
Precision:	0.00
Zero suppression:	Leading
Offset from origin:	.062

Dimensioning Project 5-4

Open the **Tool Slide** drawing that you created in Unit 4. Create a new layer named Dimensions and set it current. Add dimensions and notations to the views of the tool slide (see Figure 5-68). Follow the rules of dimensioning outlined in this unit.

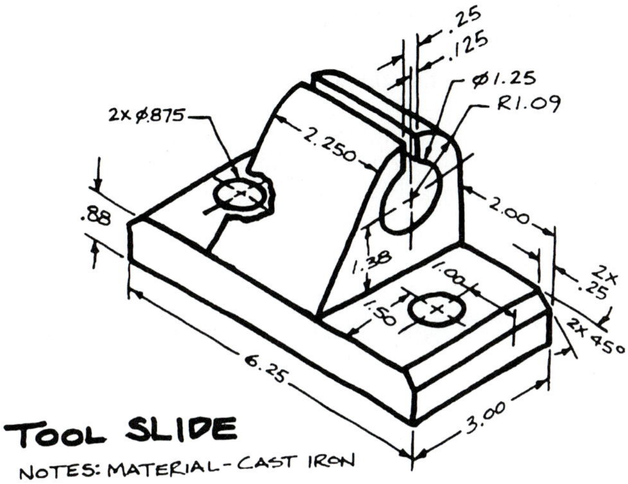

Figure 5-68   Designer's sketch of the tool slide

Follow your instructor's directions to print the drawing when you are finished placing the dimensions.

In the **Dimension Style Manager** dialog box, set the following variables:

Text height:	.125
Arrow size:	.125
Center marks:	Line
Extend beyond dim lines:	.125
Precision:	Varies—match precision of dimensions on sketch
Zero suppression:	Leading
Offset from origin:	.062

OPTIONAL UNIT PROJECTS

Dimensioning Project 5-5

Open the **Offset Flange** drawing you created in Unit 4. Create a new layer named Dimensions and set it current. Add dimensions and notations to the views of the offset flange (see Figure 5-69). Follow the rules of dimensioning outlined in this unit.

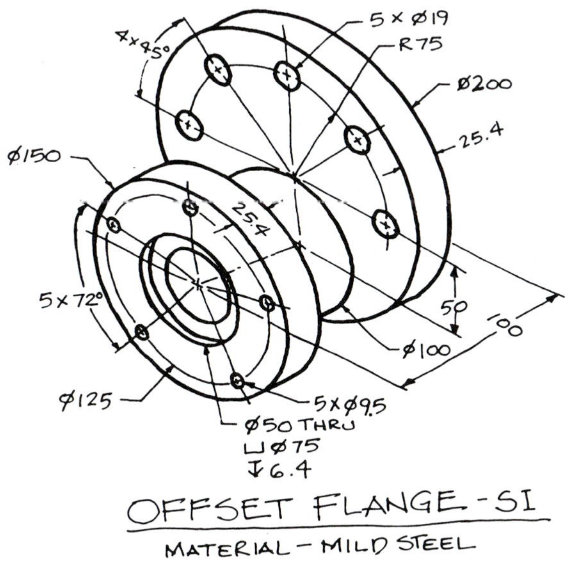

Figure 5-69   Designer's sketch of the offset flange

Follow your instructor's directions to print the drawing when you are finished placing the dimensions.

In the **Dimension Style Manager** dialog box, set the following variables (the values are in millimeters):

Text height:	3
Arrow size:	3
Center marks:	Line
Center mark size:	2
Extend beyond dim lines:	3
Precision:	0.0
Zero suppression:	Trailing
Offset from origin:	1.5
Offset from dim line	1.5

Dimensioning Project 5-6

Open the **Angle Stop** drawing you created in Unit 4. Create a new layer named Dimensions and set it current. Add dimensions and notations to the views of the angle stop (see Figure 5-70). Follow the rules of dimensioning outlined in this unit.

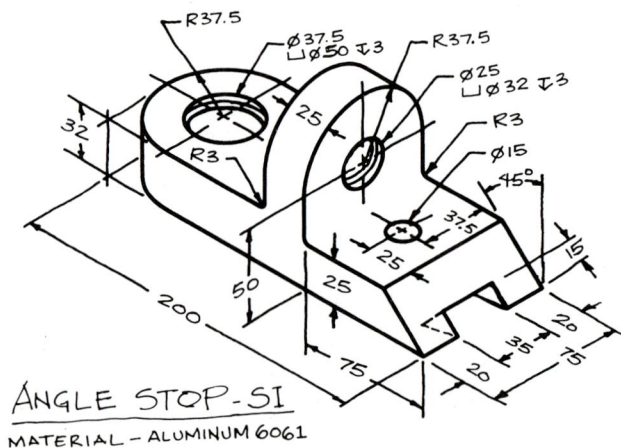

Figure 5-70   Designer's sketch of the angle stop

Follow your instructor's directions to print the drawing when you are finished placing the dimensions.

In the **Dimension Style Manager** dialog box, set the following variables (the values are in millimeters):

Text height:	3
Arrow size:	3
Center marks:	Line
Center mark size:	2
Extend beyond dim lines:	3
Precision:	0.0
Zero suppression:	Trailing
Offset from origin:	1.5
Offset from dim line	1.5

Dimensioning Project 5-7

Open the **Swivel Stop** drawing you created in Unit 4. Create a new layer named Dimensions and set it current. Add dimensions and notations to the views of the swivel stop (see Figure 5-71). Follow the rules of dimensioning outlined in this unit.

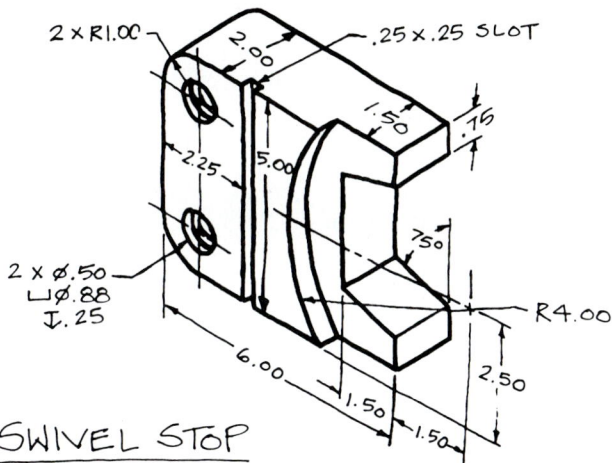

SWIVEL STOP
MATERIAL—MACHINE STEEL

Figure 5-71   Designer's sketch of the swivel stop

Follow your instructor's directions to print the drawing when you are finished placing the dimensions.

In the **Dimension Style Manager** dialog box, set the following variables:

Text height:	.125
Arrow size:	.125
Center marks:	Line
Extend beyond dim lines:	.125
Precision:	0.00
Zero suppression	Leading
Offset from origin:	.062

Dimensioning Project 5-8

Open the **Alignment Guide** drawing you created in Unit 4. Create a new layer named Dimensions and set it current. Add dimensions and notations to the views of the alignment guide (see Figure 5-72). Follow the rules of dimensioning outlined in this unit.

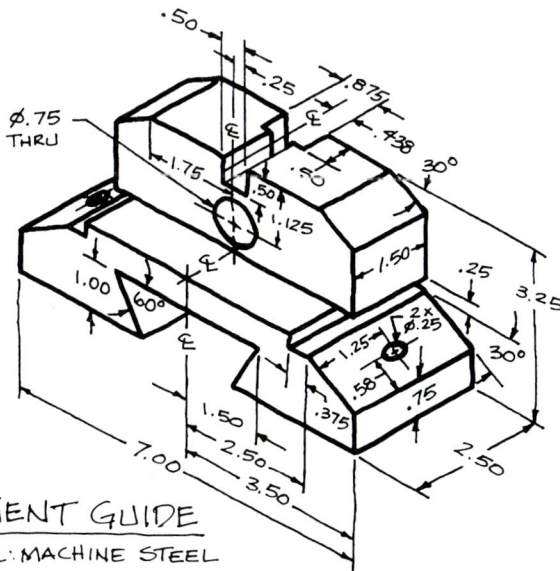

ALIGNMENT GUIDE
MATERIAL: MACHINE STEEL

Figure 5-72   Designer's sketch of the alignment guide

Follow your instructor's directions to print the drawing when you are finished placing the dimensions.

In the **Dimension Style Manager** dialog box, set the following variables:

Text height:	.125
Arrow size:	.125
Center marks:	Line
Extend beyond dim lines:	.125
Precision:	Varies—match precision of dimensions on sketch
Zero suppression:	Leading
Offset from origin:	.062

Dimensioning Project 5-9

Open the **Flange #1105** drawing you created in Unit 4. Create a new layer named Dimensions and set it current. Add dimensions and notations to the views of the flange #1105 (see Figure 5-73). Follow the rules of dimensioning outlined in this unit.

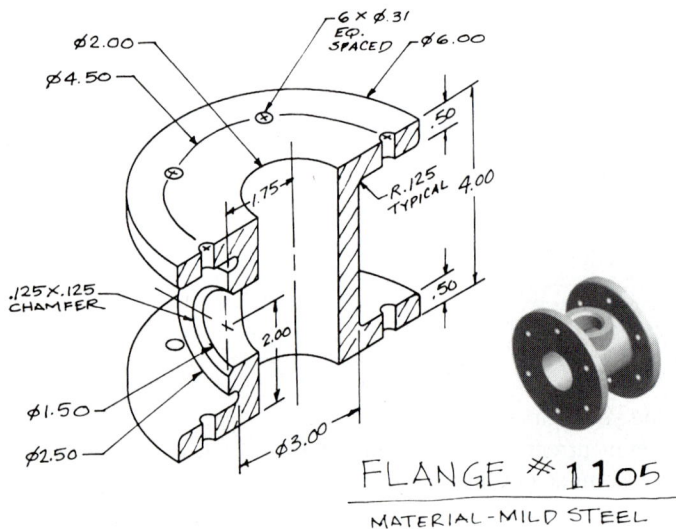

Figure 5-73 Designer's sketch of flange #1105

Follow your instructor's directions to print the drawing when you are finished placing the dimensions.

In the **Dimension Style Manager** dialog box, set the following variables:

Text height:	.125
Arrow size:	.125
Center marks:	Line
Extend beyond dim lines:	.125
Precision:	Varies—match precision of dimensions on sketch
Zero suppression:	Leading
Offset from origin:	.062

Dimensioning Architectural Drawings

Unit Objectives

- Explain how dimensions are determined on architectural drawings.
- Describe standards that affect the creation of architectural drawings.
- List the guidelines for adding dimensions to architectural drawings.
- Use the commands on AutoCAD's **Dimension** toolbar.
- Create and modify architectural dimension styles with AutoCAD's **Dimension Style Manager**.
- Add dimensions and notes to a floor plan.

INTRODUCTION

In preparing a set of architectural plans, drafters add dimensions to floor plans, elevations, and construction details. Drafters are responsible for accurately transferring dimensions from the designer's input to the finished drawing. A mistake on an architectural drawing could lead to a costly revision on a construction site, or even worse, the failure of a structural system.

This unit presents the theory and practice of dimensioning architectural drawings and the use of AutoCAD's dimensioning tools and settings for architectural design.

DIMENSIONING ARCHITECTURAL DRAWINGS

As with mechanical drawings, drawing and dimensioning standards apply to the creation of architectural drawings. Often the standard is described in an in-house drafting manual that has been developed by the designers, architects, and drafters of the firm. This manual is used to guide placement and spacing of dimensions, text height for dimensions and notations, and naming conventions for the title block and layers.

Increasingly, however, national standards are being adopted by architectural design firms, especially those that bid on publicly funded projects such as schools and government buildings. At present, the ***United States National CAD Standard (NCS)*** is gaining acceptance by the building design and construction industry. The NCS is being developed by experts from the fields of architecture, engineering, and the construction industry to standardize building design data and improve communication among owners, designers, and construction professionals. The NCS defines standards for drafting conventions, CAD layering, drawing sheets, schedules, drawing sets, terms and abbreviations, graphic symbols, notations, and plotting.

Advances in CAD modeling and linking of digital information are driving another paradigm in design, construction, and building management called ***Building Information Models (BIM)***. Projects incorporating BIM technologies will move away from 2D drawings that do not have intelligence built into them toward a linked database that all users of the system can tap into. This standard, entitled the *National Building Information Model Standard*^{TM} (NBIMS), is being developed by the National Institute for Building Sciences (NIBS). The NIBS website describes the mission

United States National CAD Standard (NCS): A standard developed for the preparation of technical drawings for the building design and construction industry.

Building Information Model (BIM): A system of modeling and linking building system information into a digital, and increasingly 3D, database.

of the NBIMS standard as an attempt to "improve the performance of facilities over their full life-cycle by fostering common and open standards and an integrated life-cycle information model for the A/E/C & FM (Architectural/Engineering/Construction & Facilities Management) industry."

BIM covers all aspects of the project, from information management to the design, construction, and operation of the facility. BIM is not intended to compete with the NCS standard.

> **TIP** To learn more about the United States National CAD Standard®, see Appendix C. To learn more about the National Building Information Model Standard™, visit the National Institute for Building Sciences website: www.nibs.org.

DETERMINING DIMENSIONS ON ARCHITECTURAL DRAWINGS

Architectural designers determine the dimensions and notes that will go into the design of a building and use this information to create a design input. Often, the design input comes to the drafter as a sketch showing placement of walls, doors, windows, and other features of the building. Drafters use the design input to draw the building's features and add the dimensions and notes necessary to build the project.

ARCHITECTURAL DRAFTING CONVENTIONS

Dimensioning conventions for architectural drawings differ from mechanical drawings. For example, the very small tolerances common on mechanical drawings are usually much less important in architectural drawings; as long as walls are placed within 1/8"–1/4" of their dimensioned location, no problems should result. Another difference is that instead of arrows, **tick marks** (short diagonal lines) are often used to show the termination of dimensions.

tick mark: A short diagonal line used to note the termination of dimensions in architectural drawings.

In the detail of the floor plan shown in Figure 6-1, note how dimensions are given from the outside edges of framed exterior walls to the centers or edges of interior walls; doors and windows are dimensioned to their centers; and, unlike mechanical drawings, unbroken chains of dimensions are allowed. Architectural style fonts are usually assigned for notes and dimension text.

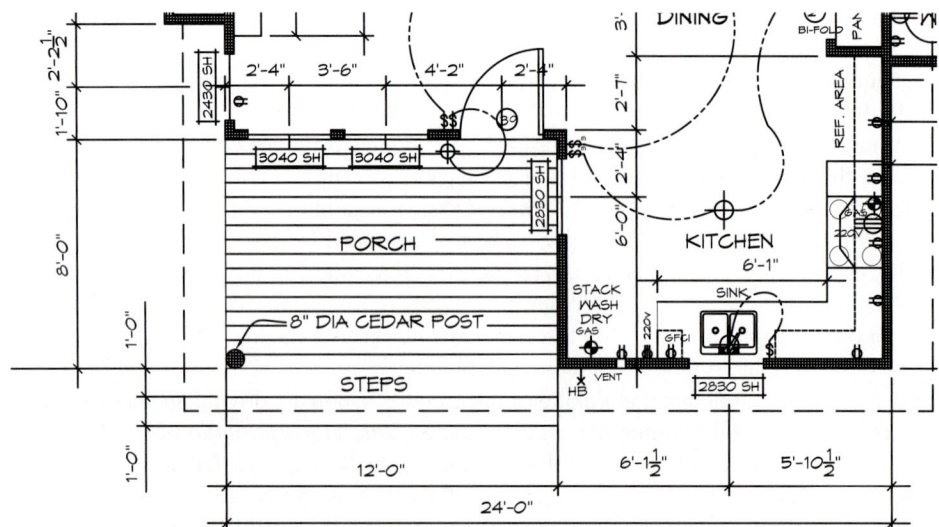

Figure 6-1   Detail from a floor plan

> **TIP**
>
> AutoCAD's architectural fonts include CityBlueprint, CountryBlueprint, and Stylus BT. If a desired font cannot be located in the **Text Style** dialog box, the drafter may need to ask the organization's computer systems administrator to install it.

Dimensions and notations are also added to the elevation drawings of the project. These dimensions may include floor heights, overhangs of eaves, roof pitches, and building materials. It is helpful for architectural designers and drafters to have an understanding of construction processes, particularly framing, to determine the best placement of dimensions.

ALIGNMENT OF DIMENSION TEXT

Aligned text is used on architectural drawings. Aligned text faces the bottom and the right side of the drawing sheet. Unlike mechanical drawing, the text is placed above the dimension line on architectural drawings. Examples of aligned text are shown on the dimensions in Figure 6-1.

ARCHITECTURAL DIMENSIONING GUIDELINES

Review the following architectural dimensioning guidelines:

- Dimensions are placed from the outside face of framing of exterior walls.
- The first line of dimensions locates interior walls and the centers of doors and windows.
- The second line of dimensions denotes distances between outside walls and interior walls.
- The third line of dimensions denotes the overall distance between outside walls.
- Interior dimensions usually locate interior walls and other features from edges of outside walls.
- Dimensions should be aligned with dimension lines and read from the bottom and right side of sheet.
- Dimensions should be centered above dimension lines.
- Dimensions should be spaced consistently from the outside edges of the building.
- Dimensions should be consistently spaced from each other.
- Dimensions may terminate in architectural tick marks, dots, or arrowheads.

ARCHITECTURAL DIMENSION SPACING

Minimum suggested spacing for architectural dimensions are shown in Figure 6-2. The spacing values shown apply to the plotted drawing.

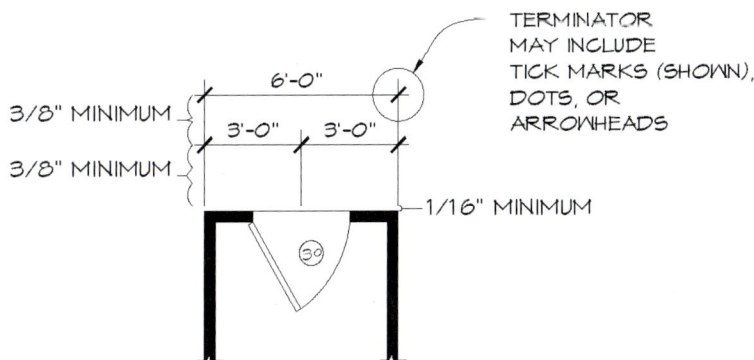

Figure 6-2 Minimum suggested dimension spacing for architectural drawings

> **TIP**
> A 3/8″ spacing would be equal to 18″ if measured in model space in an AutoCAD drawing at a scale of 1/4″ = 1′-0″.

ADDING A LEADER TO A DRAWING

Figure 6-3 shows an example of a leader that is appropriate for an architectural drawing. These leader styles can be created using either the **QUICK LEADER** or **MULTILEADER** command.

LEADER

Figure 6-3   Architectural leader style

> **TIP**
> The landing line is also referred to as the *shoulder* of a leader.

Quick Leader Command Tutorial

Step 1. Type **QLEADER** on the command line and press **<Enter>**.

Step 2. At the Specify first leader point prompt, type **S** (for **Settings**) and press **<Enter>**.

Step 3. When the **Leader Settings** dialog box opens, select the **Leader Line & Arrow** tab and choose the **Spline** option for the **Leader Line** setting and click **OK**.

Step 4. At the Specify first leader point prompt, select the point for the arrowhead to begin.

Step 5. At the Specify next point prompt, select the point for the leader line to end.

Step 6. At the Specify next point prompt, select the point where the Landing Line should end.

Step 7. At the Specify text width prompt, either press **<Enter>** or type in a width.

Step 8. At the Enter first line of annotation text prompt, enter the text for the first line of the leader annotation and press **<Enter>**.

Step 9. At the Enter next line of annotation prompt, enter the text for the next line or press **<Enter>**.

Multileader Command Tutorial

Step 1. Type **MLEADER** on the command line and press **<Enter>** or select the **MULTILEADER** icon from the **Multileader** toolbar.

Step 2. Select the **MULTILEADER STYLE** icon from the **Multileader** toolbar and click the **Modify** button.

Step 3. Under the **Leader Format** tab, change the **Type** from **Straight** to **Spline** and click **OK** and **Close**.

SUMMARY

Architectural drafters are responsible for creating dimensioned drawings that are used by construction professionals to locate walls, windows, doors, and other features of the building project. Drafters are responsible for ensuring that the dimensions on the drawing accurately reflect the dimensions provided on the designer's input. Drafters must be familiar with architectural drawing and dimensioning standards and the appropriate dimension style settings for the preparation of CAD drawings.

UNIT TEST QUESTIONS

Short Answer

1. Who is responsible for determining dimensions on architectural drawing?

2. In the United States, what is the name of the organization that publishes a drawing standard for architectural drawings?

3. How are the locations of doors and windows defined with dimensions on architectural drawings?

4. What termination symbol is often used on architectural dimensions?

5. Which direction(s) does aligned text face?

Matching

Column A

a. Primary Units

b. Lines

c. Text

d. Fit

e. Symbols and Arrows

Column B

1. The Dimension Style Manager tab where the size of tick marks (obliques) is defined

2. The Dimension Style Manager tab where the distance that dimension lines should extend beyond tick marks is defined

3. The Dimension Style Manager tab where the format for dimension unit is defined

4. The Dimension Style Manager tab where placing text above vertical dimension lines is defined

5. The Dimension Style Manager tab where the overall scale of dimension features is defined

UNIT PROJECT

Project 6-1: Dimensioning the Guest Cottage

In this project, you will open the **Guest Cottage** drawing that you created in Unit 4 and apply the dimensions shown in Figure 6-4. Follow the rules of dimensioning outlined earlier in this unit.

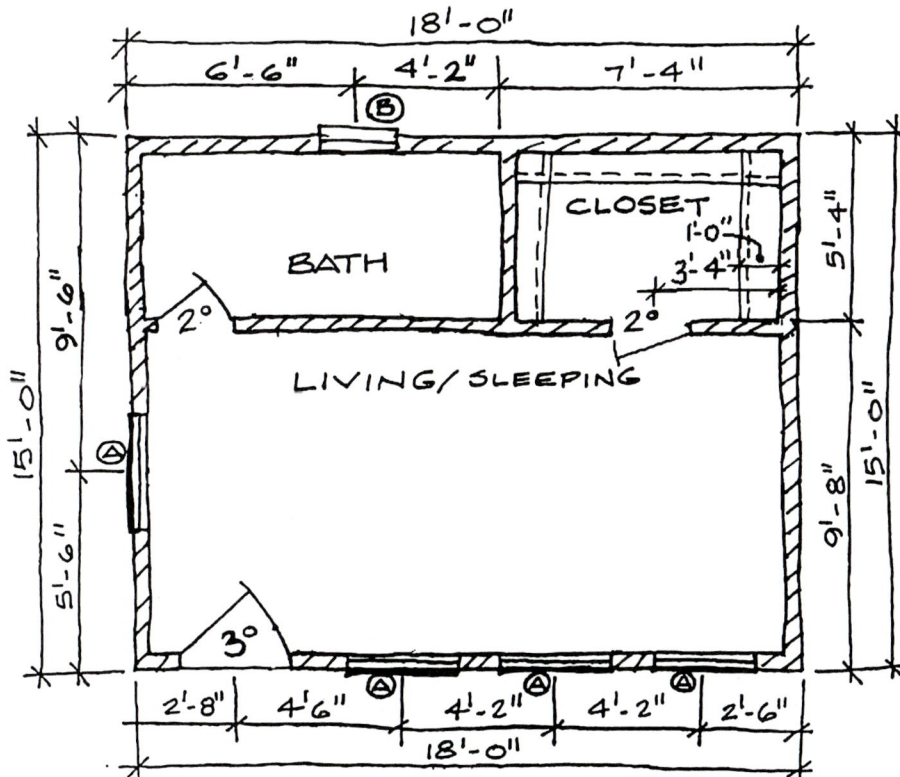

Figure 6-4 Designer's sketch of the guest cottage

Before dimensioning the cottage, create a new dimension style named **ARCH48** using the dimension style settings shown in Figures 6-5(a)–(e).

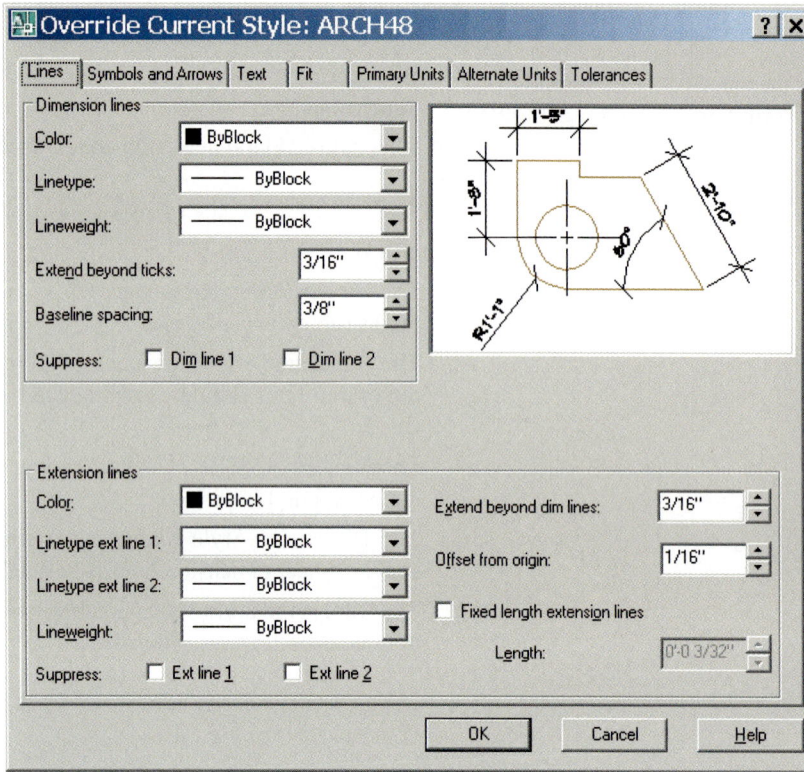

Figure 6-5(a)   Lines tab settings for cottage

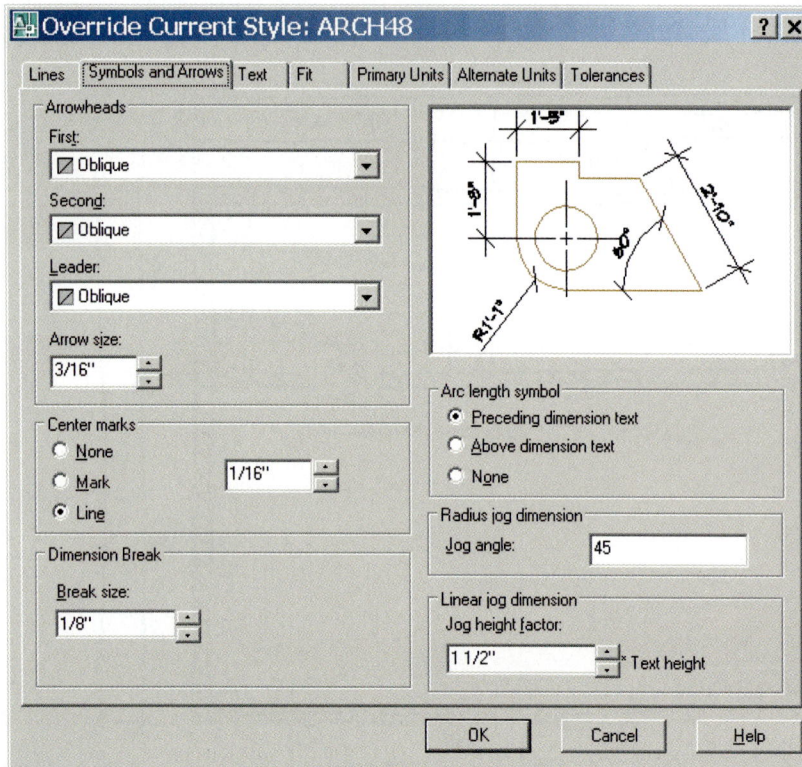

Figure 6-5(b)   Symbols and Arrows tab settings for cottage

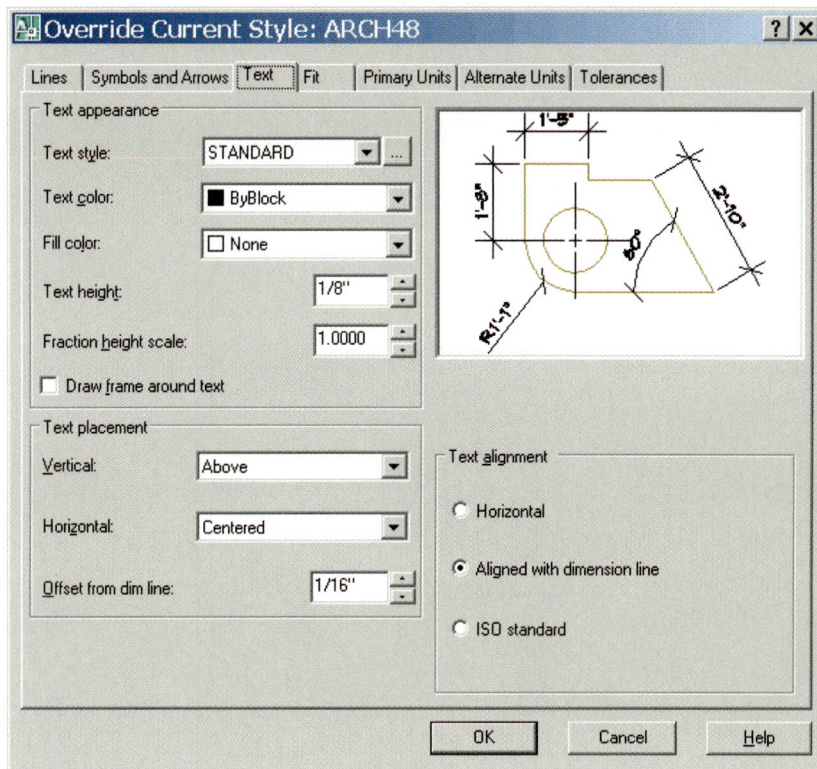

Figure 6-5(c)   Text tab settings for cottage

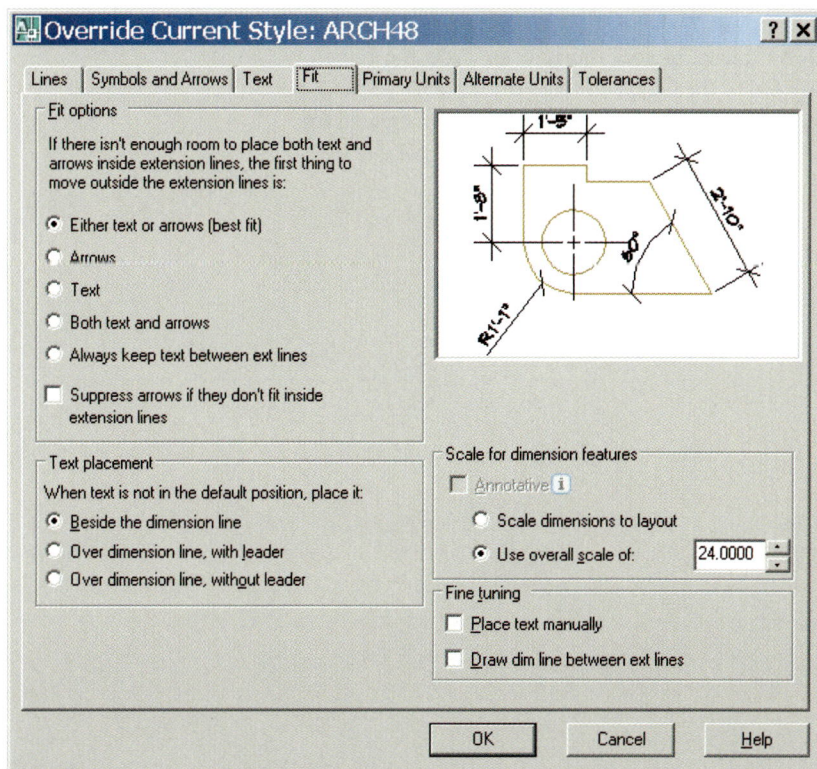

Figure 6-5(d)   Fit tab settings for cottage

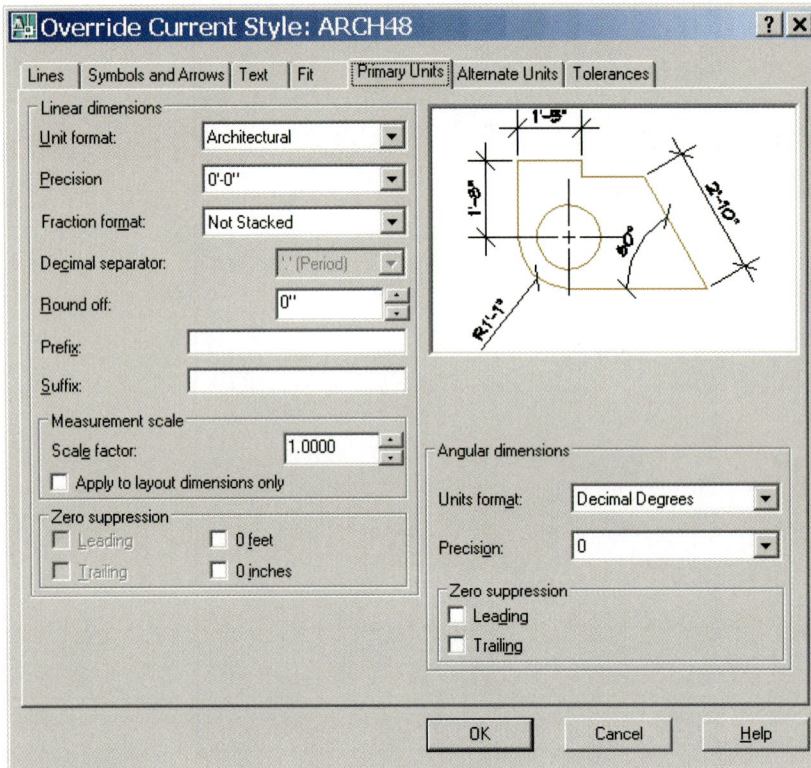

Figure 6-5(e)   Primary Units tab settings for cottage

Set the Dimensions layer current and begin the first row of dimensions by placing a **Linear Dimension**, then use the **Continue Dimension** option to finish the first row of dimensions. Place the first dimension row **2'-0"** from the outside wall of the cottage and space the second row of dimensions **1'-6"** from the first dimension line.

Set the **ARCH48** dimension style current and add the dimensions shown in Figure 6-4 to the Guest Cottage floor plan.

Plot the drawing using the **Page Setup** named *Cottage* created for this project in Unit 4. Save the drawing file before closing AutoCAD.

Isometric Drawings

7

Unit Objectives

- Define the term *isometric drawing*.
- Correctly orient lines, ellipses, fillets, and rounds in isometric drawings.
- Construct inclined planes in isometric drawings.
- Construct isometric drawings with AutoCAD.

INTRODUCTION

An ***isometric drawing*** is a type of drawing known as a ***pictorial drawing***. In a pictorial drawing, an object appears to be three-dimensional, that is, it appears to have width, height, and depth. However, unlike a 3D model, a pictorial drawing is constructed using 2D drawing techniques. The type of drawing discussed in this unit—isometric drawing—is generally used for drawing machine parts pictorially. Figure 7-1(a) shows an isometric drawing of a box. In this drawing, the box appears to have width, depth, and height, but this drawing was constructed using only X and Y coordinates, so it is considered a 2D drawing.

In the architectural field, pictorial drawings are created using ***perspective drawing*** techniques. In a perspective drawing, lines appear to recede toward a vanishing point, whereas in an isometric drawing, the receding lines are drawn parallel. Figure 7-1(b) shows an example of a perspective drawing. Study the differences in the two types of pictorial drawings—isometric and perspective—shown in Figures 7-1(a) and (b).

isometric drawing: A type of pictorial drawing in which receding lines are drawn at 30° relative to the horizon. Commonly used in the mechanical engineering field. See *pictorial drawing*.

pictorial drawing: A type of drawing in which an object appears to be three dimensional, that is, it appears to have width, height, and depth. But unlike an actual 3D model, a pictorial drawing is constructed using only X and Y coordinates. See *isometric drawing* and *perspective drawing*.

perspective drawing: A type of pictorial drawing in which receding lines appear to converge at a vanishing point. Commonly used in the architectural field. See *pictorial drawing*.

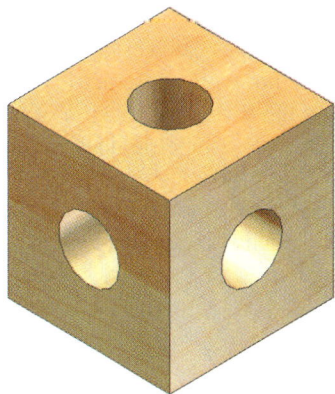

Figure 7-1(a)   Isometric drawing   **Figure 7-1(b)**   Perspective drawing

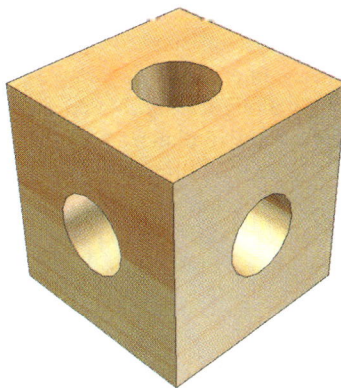

In many modern CAD applications, a drafter can construct a 3D model of an object and use it to generate a pictorial image instead of using isometric or perspective drawing techniques.

Even though 3D CAD models are used to generate pictorial images, designers and drafters often use the principles of isometric drawing to make freehand design sketches to facilitate communication quickly between other design team members or clients.

ORIENTATION OF LINES IN ISOMETRIC DRAWINGS

In an isometric drawing, an object's horizontal lines are drawn at 30° angles relative to the horizon, and its vertical lines are drawn at a 90° angle relative to the horizon (in other words, the object's vertical lines are drawn vertically) as shown in Figure 7-2. When isometric lines meet at their ends to define a regular (not inclined) plane, the plane is classified in AutoCAD terminology as either *isoright*, *isoleft*, or *isotop* as shown in Figure 7-2.

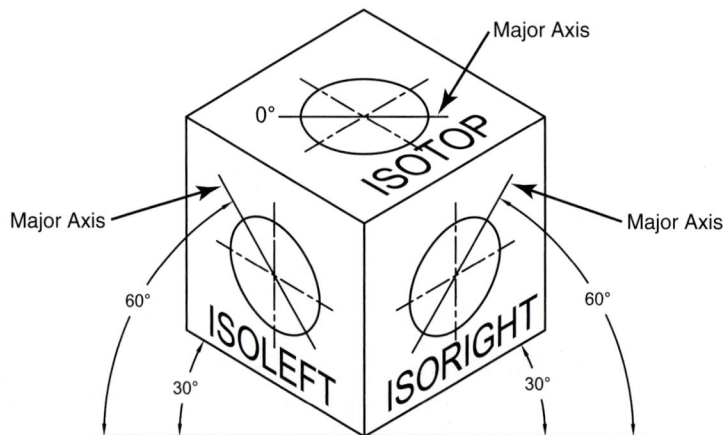

Figure 7-2 Orientation of isometric lines and ellipses

TIP Even though isometric planes are described in AutoCAD terms as *isoright*, *isoleft*, and *isotop*, drafters may also refer to these planes, as *right vertical*, *left vertical*, and *horizontal planes*, respectively.

ORIENTATION OF ELLIPSES IN ISOMETRIC DRAWINGS

Another characteristic of isometric drawings is that round shapes, such as circles and cylinders, appear as ellipses. In order for the drawing to look natural, however, an isometric ellipse must be aligned correctly along its *major axis*. Figure 7-2 shows the correct orientation of isometric ellipses on their respective planes. Note that although horizontal isometric lines are drawn at 30° angles, the major axes of ellipses located on isoright and isoleft planes are aligned along 60° angles. The major axis of an isometric ellipse on an isotop plane is aligned along a horizontal (zero degree) line. Study the differences in the orientation of the major axes of the isoright, isoleft, and isotop planes as shown in Figure 7-2.

CONSTRUCTING AN ISOMETRIC DRAWING USING THE BOUNDING BOX TECHNIQUE

One of the easiest ways for beginners to construct an isometric drawing is to start by creating an isometric *bounding box*. A bounding box is a box drawn along isometric axes that can completely enclose the object. The bounding box is constructed using the overall width, depth, and height dimensions of the object. After the bounding box is drawn, a drafter can reference measurements from its corners to locate the object's features.

The steps in creating an isometric drawing of the object shown in Figure 7-3 using a bounding box are shown in Figures 7-4 through 7-6. These steps illustrate how to locate the object's features by measuring from the corners of the bounding box.

Step 1. Construct an isometric bounding box using height, width, and depth dimensions taken from Figure 7-3. The completed bounding box is shown in Figure 7-4.

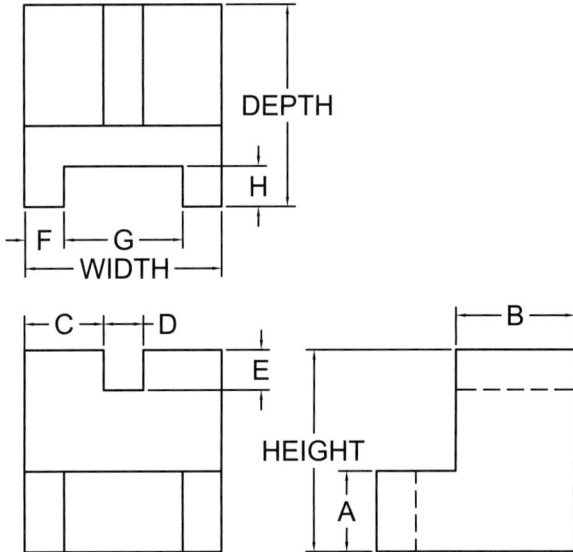

Figure 7-3   Multiview drawing of an object to be drawn isometrically

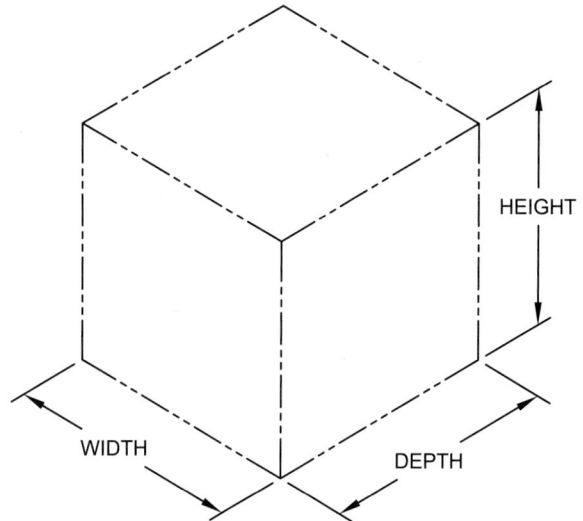

Figure 7-4   Isometric bounding box

Step 2. Transfer distances *A* and *B* from Figure 7-3 to the isometric drawing by measuring from the corners of the bounding box. Add lines as needed to construct the view shown in Figure 7-5.

Step 3. Transfer distances *C* through *H* from Figure 7-3 to locate the two slots. Trim and add lines as needed to complete the view as shown in Figure 7-6.

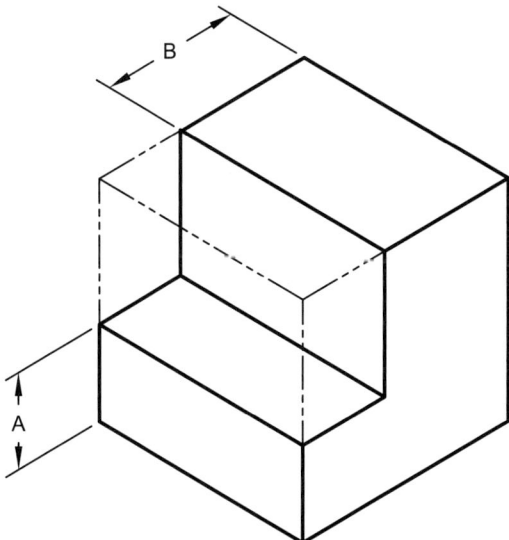

Figure 7-5   Transferring distances A and B

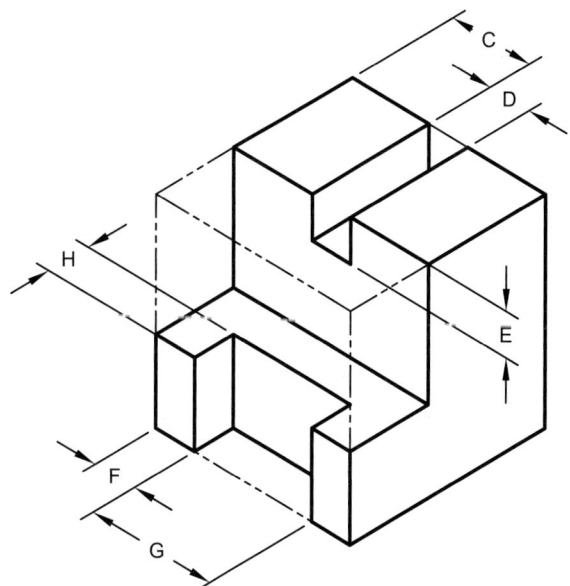

Figure 7-6   Transferring distances C through H to locate the slots

CONSTRUCTING INCLINED PLANES IN ISOMETRIC DRAWINGS

The angle of an inclined plane cannot be measured directly in an isometric drawing. The drafter must instead locate the start and end points of the corners of the inclined plane by using measurements taken from a multiview drawing and connecting the points to define the angled plane. The bounding box technique discussed earlier is also helpful for locating the corners of inclined planes.

The object shown in Figure 7-7 contains an inclined plane. The steps in creating an isometric drawing of this object, including the construction of the inclined plane, are illustrated in Figures 7-8 through 7-10.

Step 1.   Construct an isometric bounding box using height, width, and depth dimensions taken from Figure 7-7. The completed bounding box is shown in Figure 7-8.

Step 2.   Transfer distances A and B from Figure 7-7 by measuring from the corners of the bounding box. Add lines as needed to construct the view shown in Figure 7-9.

Step 3.   Connect the points that define the inclined plane to complete the view as shown in Figure 7-10.

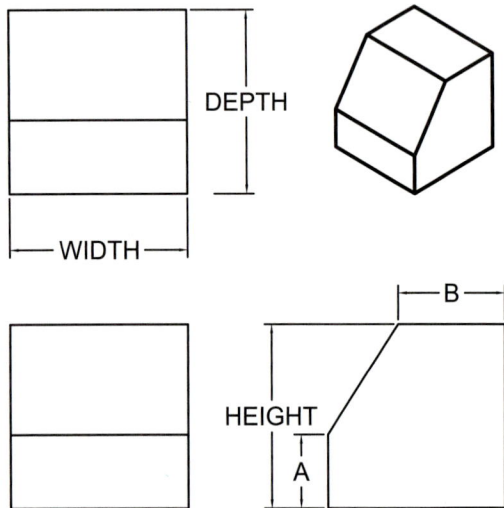

Figure 7-7   Multiview drawing of object to be drawn isometrically

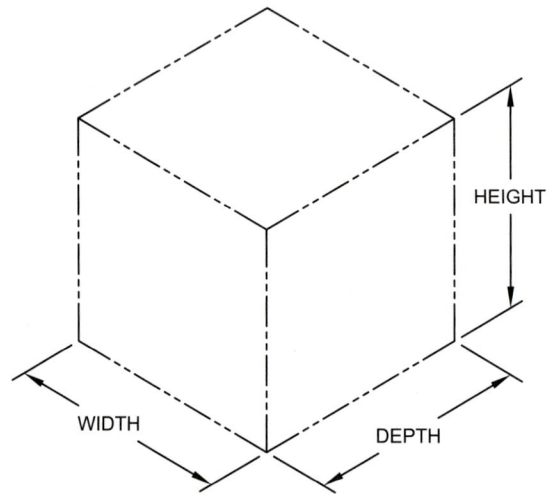

Figure 7-8   Isometric bounding box

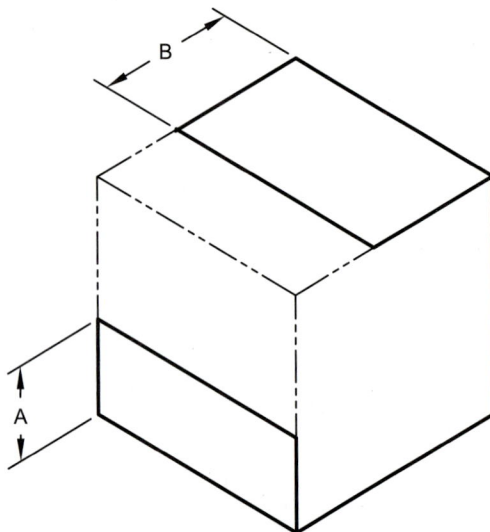

Figure 7-9   Transferring distances A and B

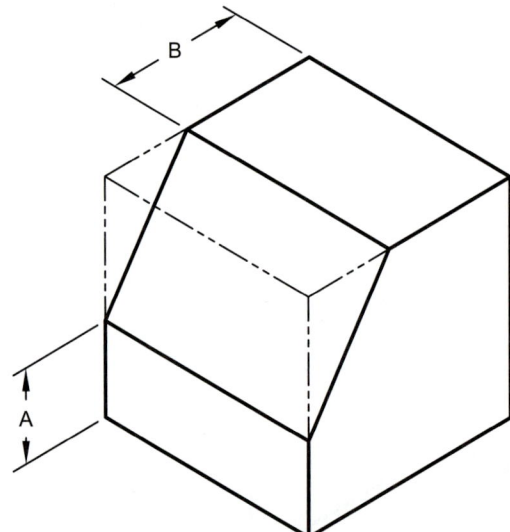

Figure 7-10   Connecting the points to create the inclined plane

CREATING ISOMETRIC DRAWINGS WITH AutoCAD

Begin the isometric drawing by going into the **Tools** pull-down menu and selecting **Drafting Settings**, then pick the **Snap and Grid** tab and check the button next to **Isometric snap** (see Figure 7-11) in the dialog box. This puts AutoCAD into the mode for isometric drawing.

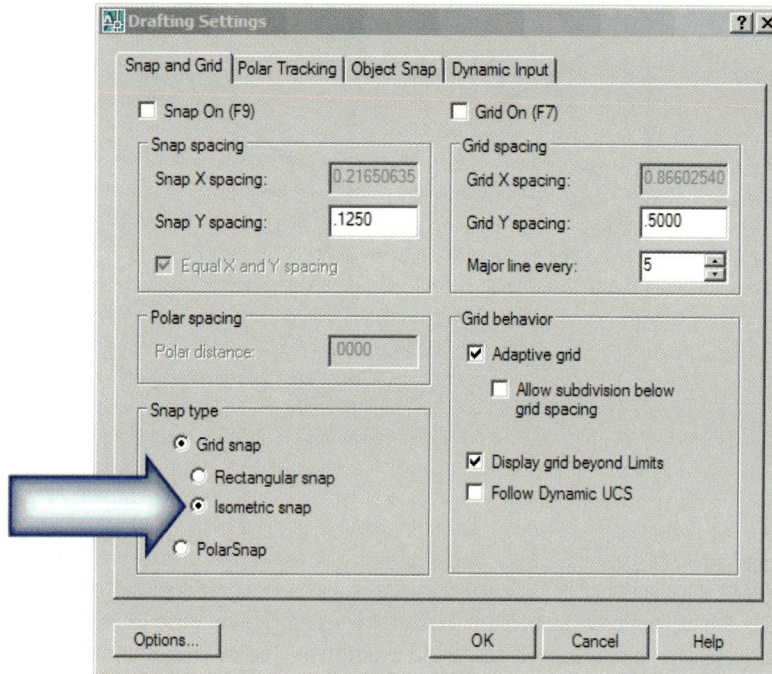

Figure 7-11   Setting isometric snap

After changing from the **Rectangular Snap** mode to the **Isometric Snap** mode, you will notice that the AutoCAD crosshair is oriented at an isometric angle. When **Ortho** is **On**, lines will be automatically oriented along isometric axes but can only be drawn along the axes of the visible crosshair. By changing the orientation of the crosshair, the drafter can draw along other isometric axes. The orientation of the crosshair can be changed by pressing the <Ctrl> and <E> keys simultaneously.

> **TIP**
>
> A quick way to change the orientation of the crosshair, is to press <**F5**>. You can toggle in this manner among the **Isotop**, **Isoleft**, and **Isoright** orientations.

In an isometric drawing, horizontal isometric lines are drawn at 30° relative to the horizon. When making an isometric drawing with AutoCAD, however, the angle of isometric lines must be converted to polar coordinates (where *East* equals zero degrees). In Figure 7-12, the possible isometric angles for the lines of the object are noted on each corner (this example depicts the angles with **Dynamic Input On**). The angle of each line is determined by the position of its start point and its direction relative to *East*. Note that the angle of the inclined plane does not fall on an isometric angle and, therefore, would need to be drawn using the technique described in Figures 7-8 through 7-10.

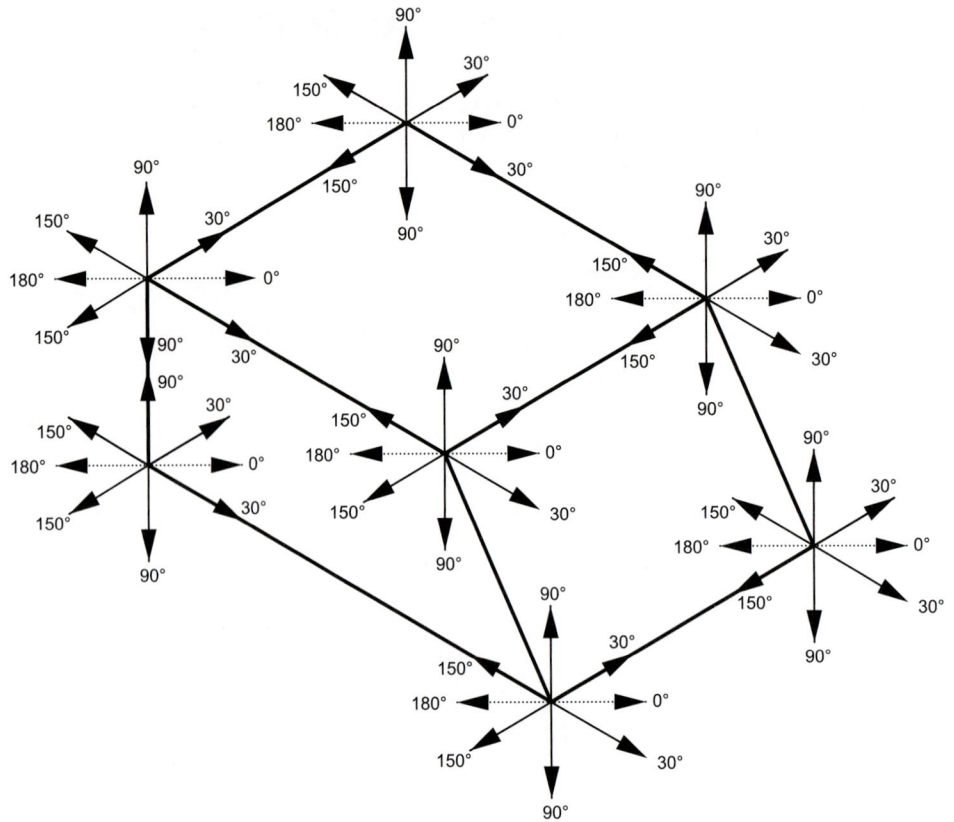

Figure 7-12 Isometric angles

An example of the principal isometric axes for isometric lines using AutoCAD polar coordinates is shown in Figure 7-13. Some examples of polar coordinates for isometric lines (assuming **Dynamic Input** is **On**) are 6 tab 30, 4 tab 90, 7 tab 150, and 6 tab 180.

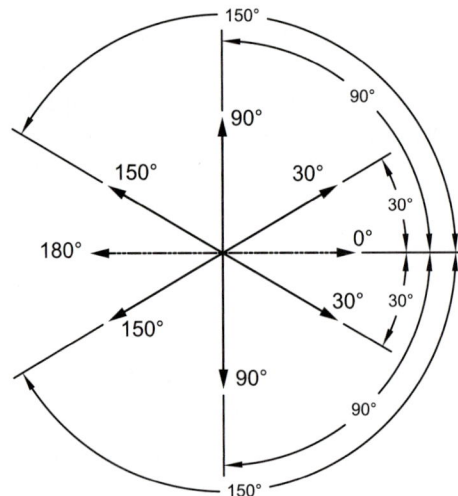

Figure 7-13 Isometric axes

DRAWING ISOMETRIC ELLIPSES WITH AUTOCAD

AutoCAD refers to isometric ellipses as *Isocircles*. To draw an isocircle, first determine the isometric plane that the ellipse is to be drawn on and toggle <F5> to the appropriate isoplane (isotop, isoright, or isoleft). Then select the **ELLIPSE** icon from the **Draw** toolbar and type **I** for **Isocircle** and press <Enter>. Next, specify the center of the ellipse and its radius and press <Enter> to complete the command. Figure 7-14 shows a box with three isometric ellipses. Each ellipse was drawn with a different isoplane setting.

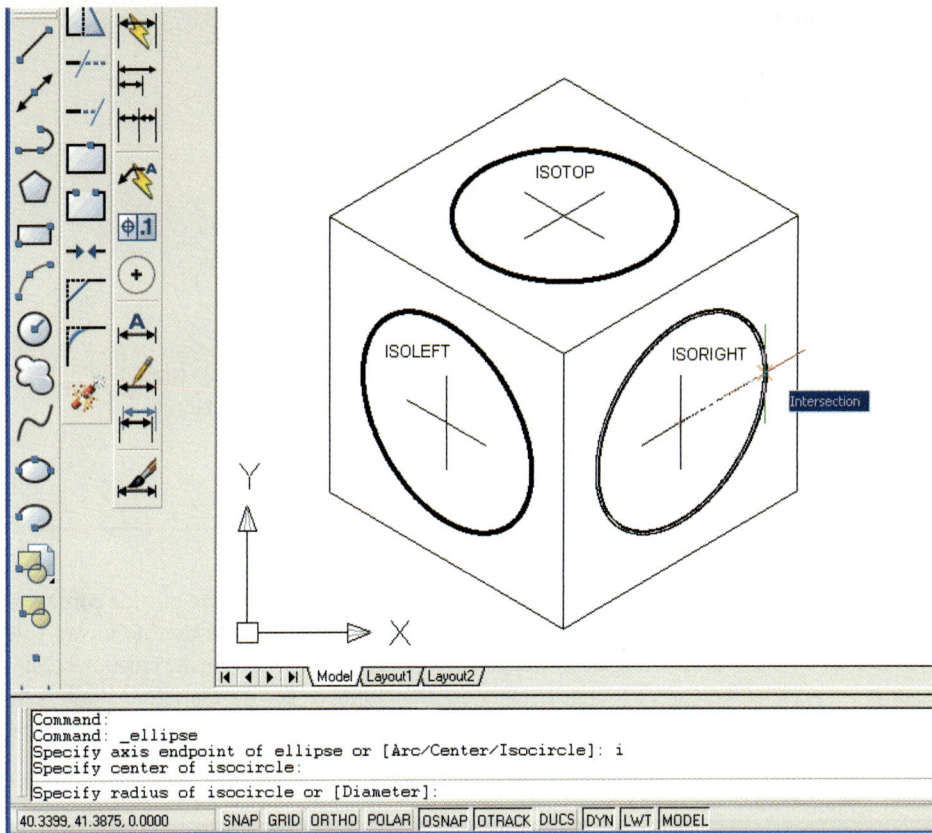

Figure 7-14 AutoCAD isocircles

The steps in constructing an isometric cylinder are shown in Figures 7-15(a)–(c).

Step 1. Draw an ellipse (isocircle) at the desired diameter and orientation of the cylinder. Copy the ellipse along an isometric angle the desired length of the cylinder as shown in Figure 7-15(a).

Step 2. Draw lines from the quadrants of the front ellipse to the corresponding quadrants of the back ellipse as shown in Figure 7-15(b).

Step 3. Trim the back ellipse to complete the cylinder as shown in Figure 7-15(c).

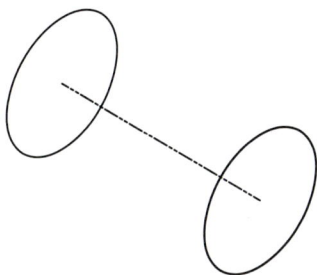

Figure 7-15(a) Copying an ellipse

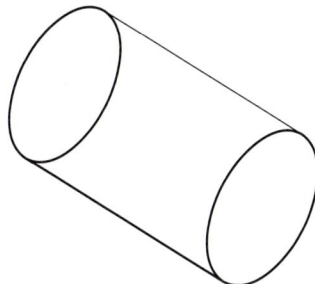

Figure 7-15(b) Drawing lines from quadrants of ellipses

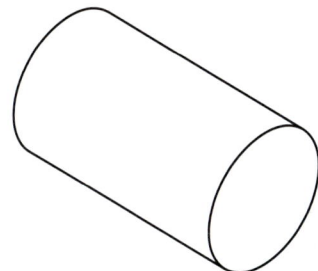

Figure 7-15(c) Completed cylinder

Possible orientations for **horizontal** and **vertical** cylinders are shown in Figure 7-16.

Isocircles can be used to create counterbored, countersunk, and through holes in isometric drawings as shown in Figure 7-17.

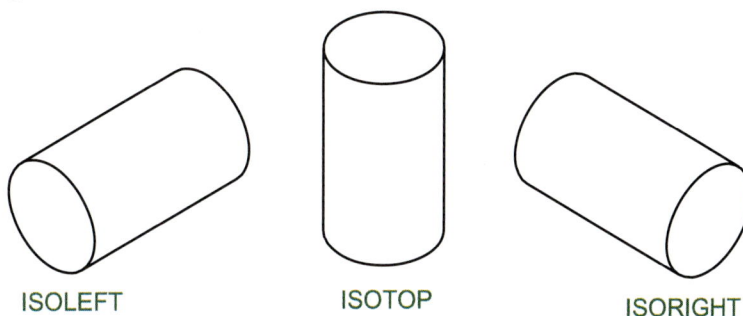

ISOLEFT

ISOTOP

ISORIGHT

Figure 7-16 Orienting isometric cylinders

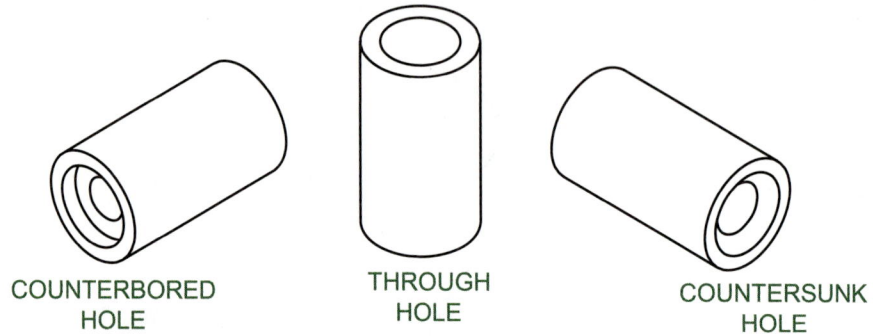

Figure 7-17 Orienting isometric cylinders

COUNTERBORED
HOLE

THROUGH
HOLE

COUNTERSUNK
HOLE

CONSTRUCTING ISOMETRIC ARCS AND RADII (FILLETS AND ROUNDS)

fillet: A rounded inside corner of a part.

round: A rounded outside corner of a part.

In technical drawing, the terms *fillet* and *round* refer to rounded inside and outside corners, respectively. On a multiview drawing, AutoCAD's **FILLET** command can be used to create fillets and rounds at an object's corners, but the **FILLET** command cannot be used in the creation of an isometric drawing because it creates a rounded, rather than an elliptical, corner. Therefore, drafters must use the steps illustrated in Figures 7-18 through 7-22 to add isometric fillets and rounds to an object.

Step 1. Copy edges *1* through *6* along isometric axes 1″ inside the edges of the part as shown in Figure 7-19.

Step 2. Using the intersections of the lines copied in Step 1 for center points, draw four **2″** diameter ellipses (isocircles) as shown in Figure 7-20.

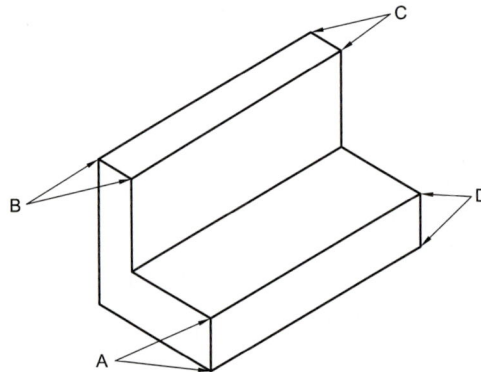

Figure 7-18 Add a 1″ radius to corners A, B, C, and D of this object

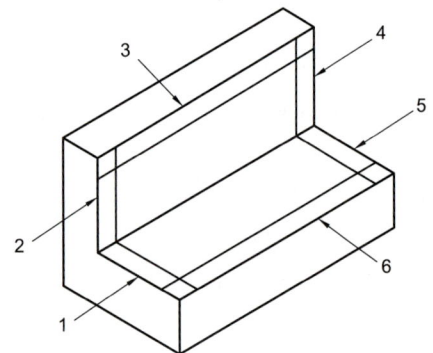

Figure 7-19 Copying edges 1–6 along isometric axes

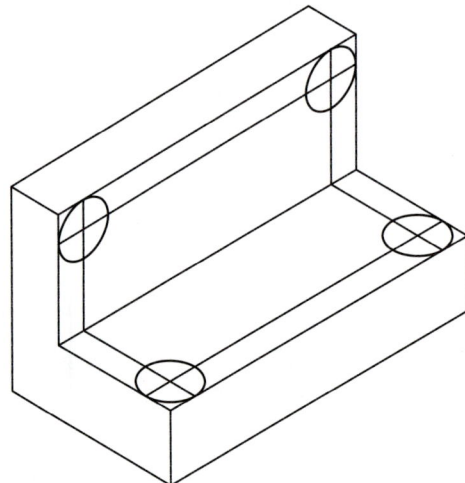

Figure 7-20 Adding ellipses at intersections of copied lines

Step 3. Trim the ellipses to create a round for each corner and copy the rounds along isometric axes to the corresponding corner as shown in Figure 7-21.

Step 4. Erase and edit construction lines as needed to complete the view as shown in Figure 7-22.

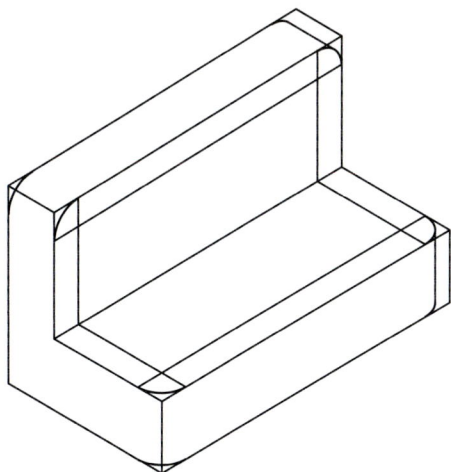

Figure 7-21   Trimming ellipses to create filleted corners

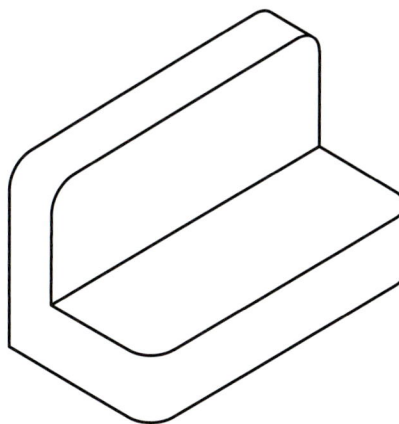

Figure 7-22   The finished isometric drawing

SUMMARY

Pictorial drawings appear to be three-dimensional but are constructed using only *X* and *Y* coordinates so they are actually a form of two-dimensional drawing. In order to create isometric drawings in AutoCAD, drafters must master the concept of polar coordinates, be able to orient correctly the major axes of isometric ellipses, and be able to use the techniques needed to construct inclined planes.

Pictorial views created from 3D CAD models are replacing isometric and perspective drawings in many fields of technical graphics. However, the ability to create freehand pictorial sketches quickly to facilitate communication of design ideas remains an important job skill for designers and drafters, and one that students seeking to be competitive in this field should strive to develop.

UNIT TEST QUESTIONS

Multiple Choice

1. In what field of technical drawing are isometric techniques most often used to create a pictorial image?

   a. Archeology field
   b. Architectural field
   c. Mechanical field
   d. Civil field

2. What is the angle of an object's horizontal lines in an isometric drawing?

   a. 60° angle
   b. 180° angle
   c. 30° angle
   d. None of the above

3. What is the angle of an object's vertical lines in an isometric drawing?

   a. 90° angle
   b. 45° angle

   c. 30° angle
   d. 60° angle

4. On what tab in the **Drafting Settings** dialog box is the **Rectangular snap** button located?

   a. **Snap and Grid** tab
   b. **Polar Tracking** tab
   c. **Dynamic Input** tab
   d. **Object Snap** tab

5. What is the angle of the major axis of an ellipse drawn on an isoright plane?

   a. 90° angle
   b. 45° angle
   c. 30° angle
   d. 60° angle

True or False

1. *True or False*: Angles of inclined planes can be measured directly in an isometric drawing.

2. *True or False*: **<F5>** and **<Ctrl-E>** are the two ways to change the orientation of the crosshair in an isometric drawing created with AutoCAD.

3. *True or False*: When drawing an isometric ellipse, **Isoplane** must be selected after beginning AutoCAD's **Ellipse** command.

4. *True or False*: It is not important for drafters and designers to understand isometric drawing techniques.

5. *True or False*: In an isometric drawing, the lines appear to recede toward a vanishing point.

UNIT PROJECTS

Project 7-1: Tee Connector Isometric

Directions

Create an isometric drawing of the tee connector shown in Figure 7-23.

1. Select **FILE** and **OPEN.**
2. Select the **Student Resource CD**.
3. Select the **Prototype Drawings** folder.
4. Select the **Daily Work Prototype** drawing (open as "read-only").
5. **SAVE AS** to your **Home** directory renaming the drawing as **TEE CONNECTOR**.

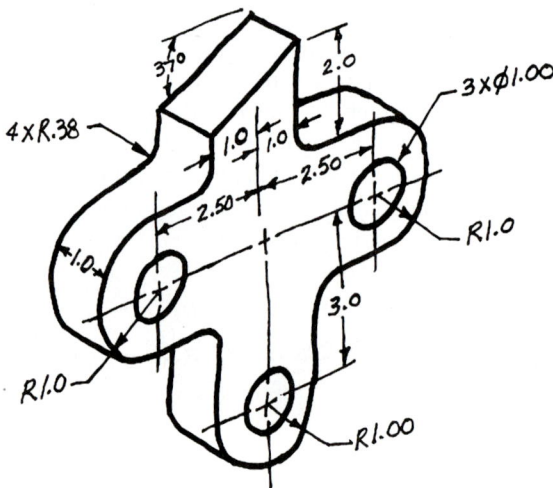

Figure 7-23　Designer's sketch of the tee connector

Steps in Constructing the Tee Connector

Step 1. Draw the horizontal and vertical isometric lines shown in Figure 7-24.

Step 2. Draw a **1″** diameter and **2″** diameter ellipse (isocircle) at the three endpoints as shown in Figure 7-25.

Step 3. Add the isometric lines shown in Figure 7-26.

TIP You may need to add construction lines from the center points of the isocircles drawn in Step 2 to find snap points tangent to the ellipses.

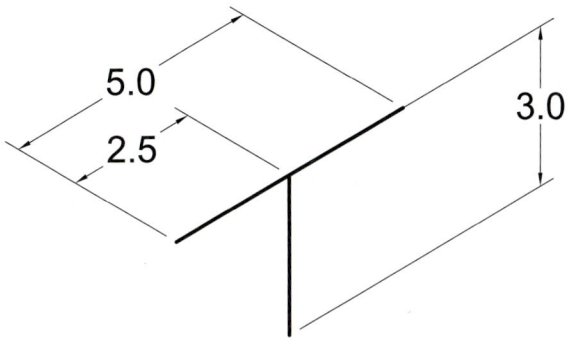

Figure 7-24   Isometric lines

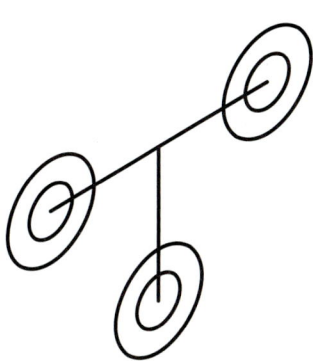

Figure 7-25   Isocircles

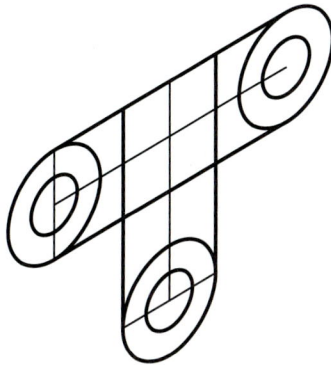

Figure 7-26   Adding tangency lines

Step 4. To construct the **37°** angle shown in the designer's sketch, you will need to create a orthographic front view of the top part of the object in order to find Distance **A** as shown in Figure 7-27(a). You will need to change the snap type in the **Drafting Settings** dialog box from **Isometric** to **Rectangular snap** to construct the view shown in Figure 7-27(a). Transfer distance **A** from the orthographic drawing to the isometric construction as shown in Figure 7-27(b), and set the **Snap type** in the **Drafting Settings** dialog box back to **Isometric** from **Rectangular snap**.

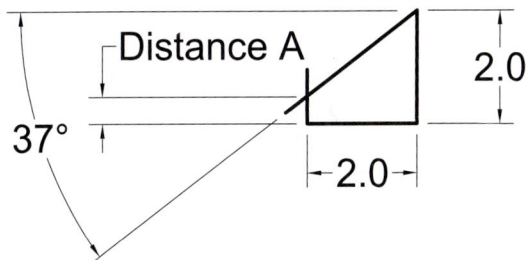

Figure 7-27(a)   Orthographic view drawn to determine distance A

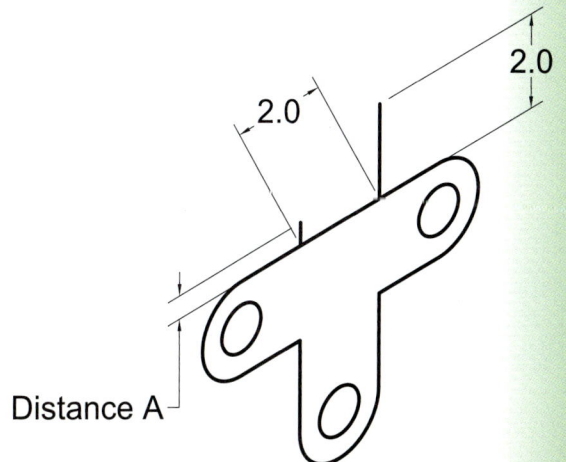

Figure 7-27(b)   Transferring distance A to the isometric drawing

Step 5. Connect the endpoint of the lines drawn in Step 4 and trim out any unnecessary construction lines. Following completion of this step, your drawing should resemble the object shown in Figure 7-28.

Step 6. With **Ortho On**, copy construction lines .**38″** from the inside corners of the object and draw .**76″** (.38″ × 2) diameter ellipses (isocircles) at the intersections of these lines as shown in Figure 7-29.

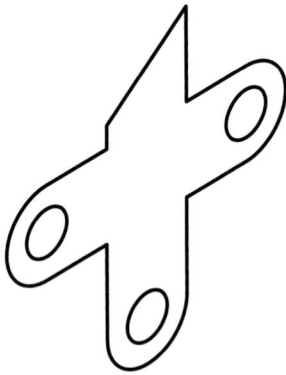

Figure 7-28   Front surface of tee connector

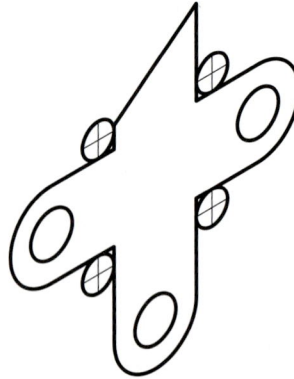

Figure 7-29   Adding fillets

Step 7. Trim the .**38″** diameter ellipses to create the fillets at the inside corners of the tee connector and then copy the object **1″** behind the first object at an angle of **150°** as shown in Figure 7-30.

Step 8. Trim any unnecessary lines and add isometric lines connecting the front edge of the tee connector to the back edge as shown in Figure 7-31 to complete the construction.

Step 9. Follow your instructor's directions to plot the drawing. Be sure to save the drawing file when you close AutoCAD.

Figure 7-30   Copy edges of front face back 1″ at a 150° angle

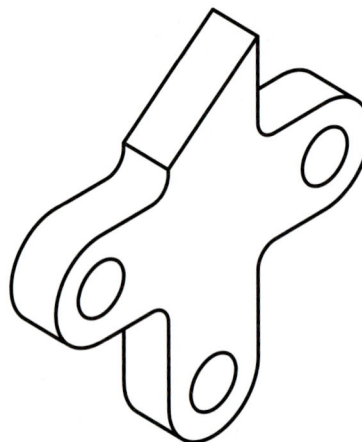

Figure 7-31   Finished project

Project 7-2: Tool Holder Isometric

Directions: Open the drawing titled **Tool Holder** that you created in Unit 4 and follow the steps outlined here to create an isometric view of the object shown in Figure 7-32 in the upper right corner of

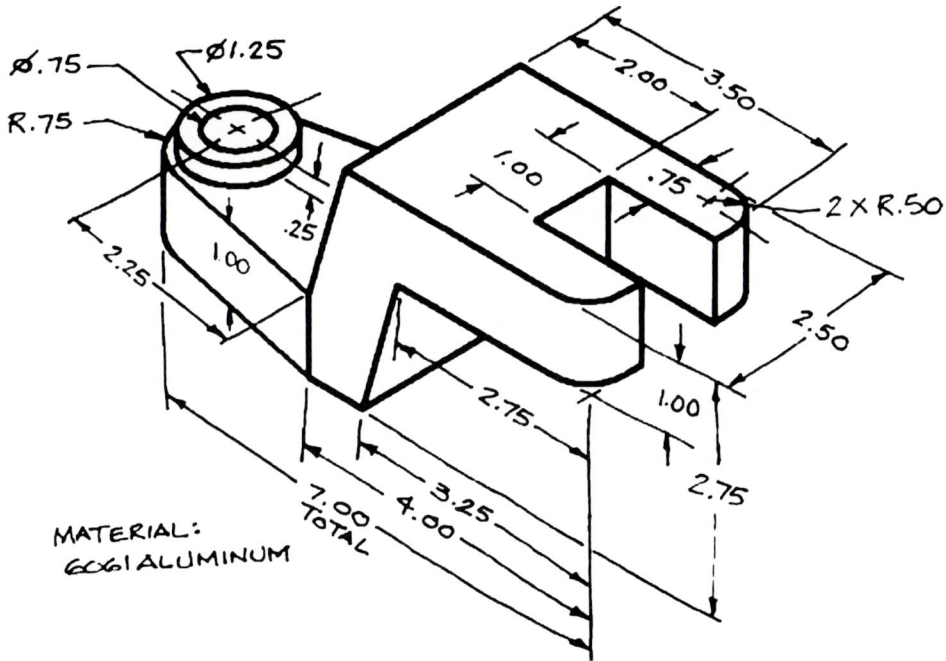

Figure 7-32   Designer's sketch of tool holder

the sheet. When you are finished, this drawing will have both the isometric drawing of the tool holder as well its dimensioned multi-views.

Steps in Constructing the Tool Holder Isometric Project

Step 1. Create a **7.00″ × 2.50″ × 2.75″** bounding box that can contain the tool holder as shown in Figure 7-33.

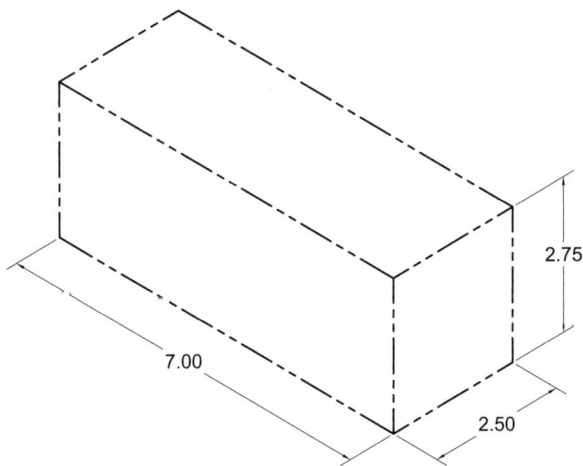

Figure 7-33   Bounding box for tool holder

Step 2. Construct the basic shape of the front of the tool holder within the bounding box and add fillets to the front corners using the dimensions shown in Figure 7-34.

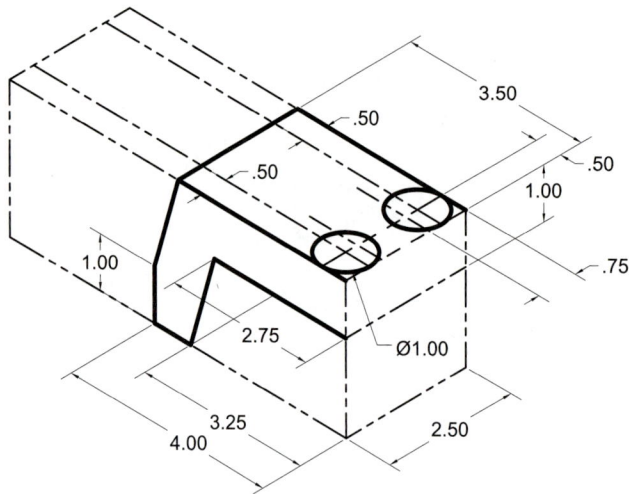

Figure 7-34   Constructing the front part of the tool holder

Step 3. Copy the filleted corners created in Step 2, down 1.00″ along the *Y* axis and add the slot using the dimensions shown in Figure 7-35.

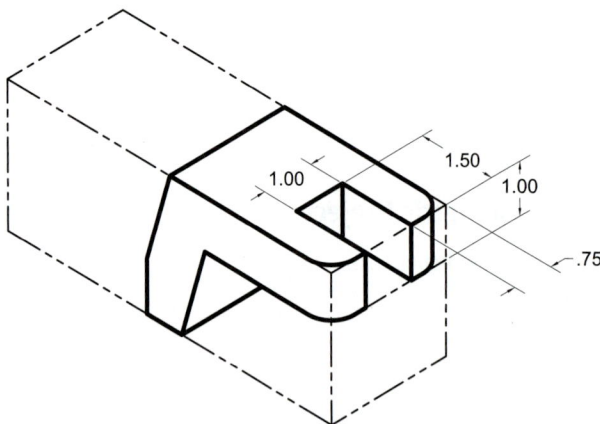

Figure 7-35   Completed construction of front part of tool holder

Step 4. Locate the two **1.50″** diameter ellipses at the back of the object and add tangency lines between the ellipses as shown in Figure 7-36. (The front part of the tool holder has not been shown in Figure 7-36 for clarity.)

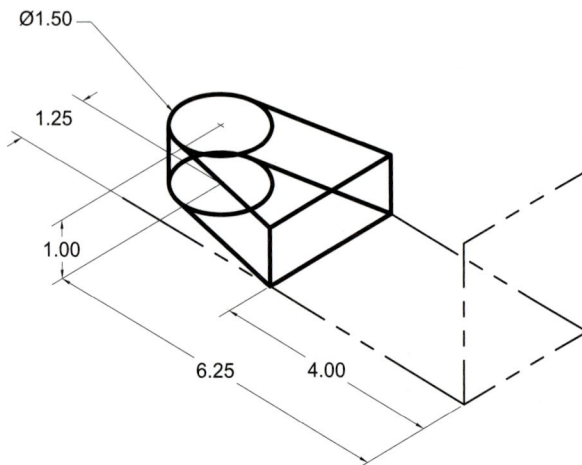

Figure 7-36   Constructing the back part of the tool holder

Step 5. Trim the lines drawn in Step 4 so that your drawing resembles the one shown in Figure 7-37.

Step 6. Add two **1.25″** diameter ellipses **.25″** apart to form the raised cylinder (also referred to as a *boss*) shown in Figure 7-38. Next, add a **.75″** diameter ellipse at the top of the boss as shown in Figure 7-38. Trim the lines as needed to complete the boss as shown in Figure 7-39.

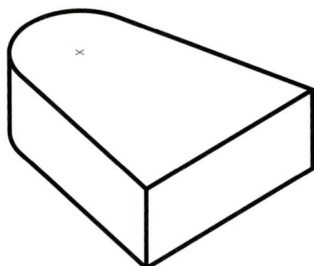

Figure 7-37   The back of the tool holder after ellipses are trimmed

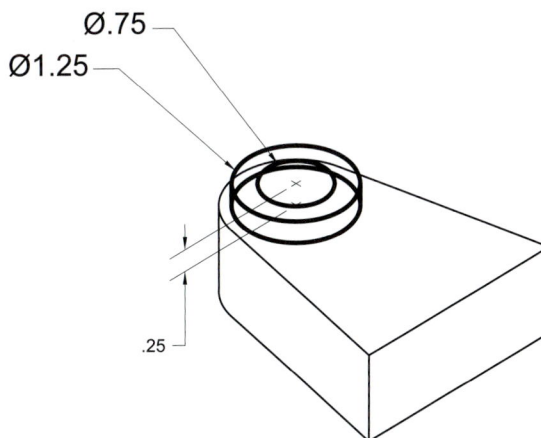

Figure 7-38   Constructing the boss

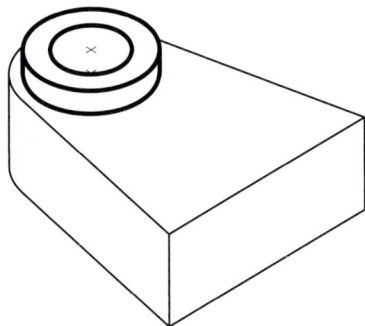

Figure 7-39   Completed construction of back part of tool holder

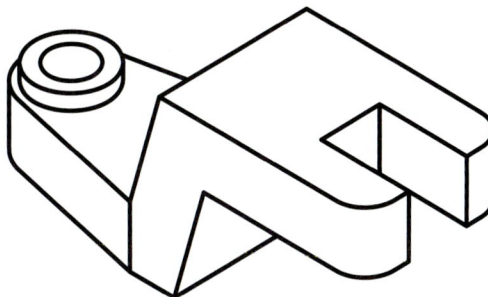

Figure 7-40   Completed isometric drawing of the tool holder

Step 7. Trim any unneeded lines to finish the isometric view of the tool holder. Your drawing should resemble the one shown in Figure 7-40.

Step 8. Follow your instructor's directions to plot the drawing. Be sure to save the drawing file when you close AutoCAD.

Sections

8

Unit Objectives

- Define what section views are.
- Describe how section views are used in technical drawings.
- Provide the names and descriptions of the different types of sections and the terminology associated with section views.
- Use AutoCAD to create section views, including proper placement of cutting plane lines and hatch patterns.

INTRODUCTION

A *section* view is a type of drawing in which part of an object's exterior is removed to reveal its interior features. For example, in mechanical engineering drawings, sections are used to show interior features of machine parts that would not be clearly represented by hidden lines. In architectural drawings, sections are used to reveal the interior details of walls, roofs, and foundations.

In creating a section view, an imaginary *cutting plane* is used to slice through the object to reveal its interior features. In mechanical drawings, the sectioned areas are usually filled with diagonal *section lines* (also called *cross-hatching*), which indicate where the cutting plane line passes through the part. In architectural drawings, the sectioned areas may be filled with hatch patterns that represent building materials such as concrete or insulation.

SECTIONS IN MECHANICAL DRAWINGS

In Figure 8-1, an object with five machined holes is shown as it would look if it were cut in half to show its inside detail. The heavy dashed line shown between the views is called a *cutting plane line*. The cutting plane line defines the line along which the object is cut. The arrows on the ends of the cutting plane line indicate the direction in which the viewer is observing the sectioned object. The diagonal lines drawn on the plane created by cutting the part in half are called *section lines*.

The principal reason for using section views in mechanical drawings is to better define the complex interior features of an object by replacing hidden lines with visible lines in a view. This allows the drafter to dimension to interior features without dimensioning to hidden lines.

In Figure 8-2, the object shown in Figure 8-1 has been drawn as a front and top view. The heavy dashed line running through the center of the top view is the *cutting plane line*. This line defines the line along which the object is cut. The upward-pointing arrows on the ends of the cutting plane line indicate the direction in which the viewer is observing the sectioned object. The diagonal lines shown on the front, or sectioned, view in Figure 8-1 indicate that the view is a section. The profile edges of the machined holes, that would have been represented by hidden lines in a regular multiview drawing, are represented by visible lines in the section view.

section: A drawing technique in which an object is drawn as if part of its exterior has been removed to reveal its interior features and details.

cutting plane: An imaginary plane that slices through an object to reveal its interior features in a section view.

section lines: Diagonal lines drawn that are placed on a section view to indicate the areas of the object that came in contact with the cutting plane line. On AutoCAD drawings, section lines are placed with the **HATCH** command.

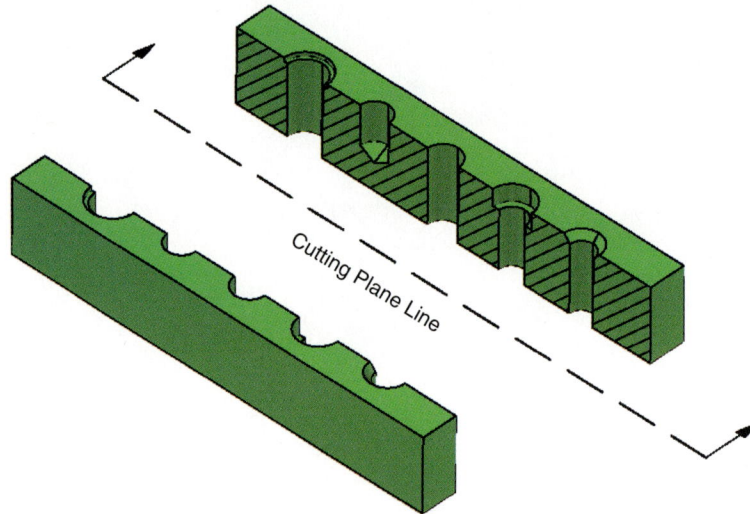

Figure 8-1 Section view of a 3D object

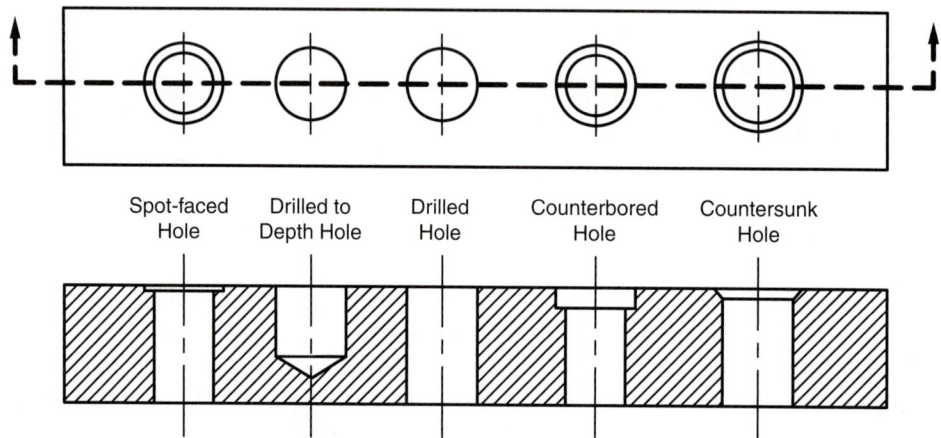

Figure 8-2 Top view and sectioned front view of an object

Spot-faced Hole   Drilled to Depth Hole   Drilled Hole   Counterbored Hole   Countersunk Hole

SECTIONS IN ARCHITECTURAL DRAWINGS

In architectural drawings, sections are often included on foundation plans to show details through beams and footings as shown in the designer's sketch in Figure 8-3. A CAD drafter might work from a sketch of this type to create a technical drawing.

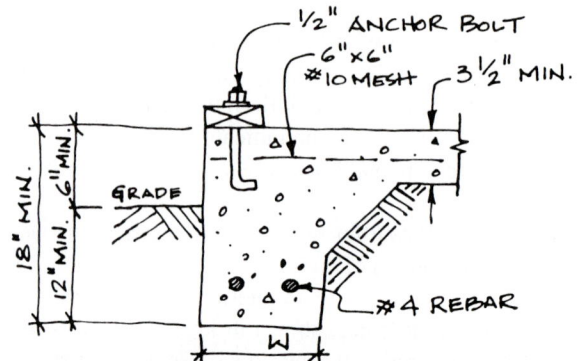

½" ANCHOR BOLT
6"x6" #10 MESH
3½" MIN.
GRADE
18" MIN.
12" MIN. 6" MIN.
#4 REBAR
W

MONOLITHIC SLAB DETAIL
WITH INTEGRAL FOOTING
1.) TWO #4 REBAR @ 4" O.C.
2.) 6"x6" - #10 MESH IN FLOOR SLAB
3.) ½" ANCHOR BOLTS @ 6'-0" O.C.
4.) FOOTING WIDTH, W = 10" MIN.

Figure 8-3 Architectural designer's sketch of a foundation section detail

Wall sections are prepared in architectural drawings to specify the composition of a wall as shown in Figure 8-4. Often, a wall section of a building must be included in a set of architectural plans before a building permit will be granted for a project. Study Figure 8-4 and note the hatch patterns used to represent the insulation and concrete as well as the earth around the foundation.

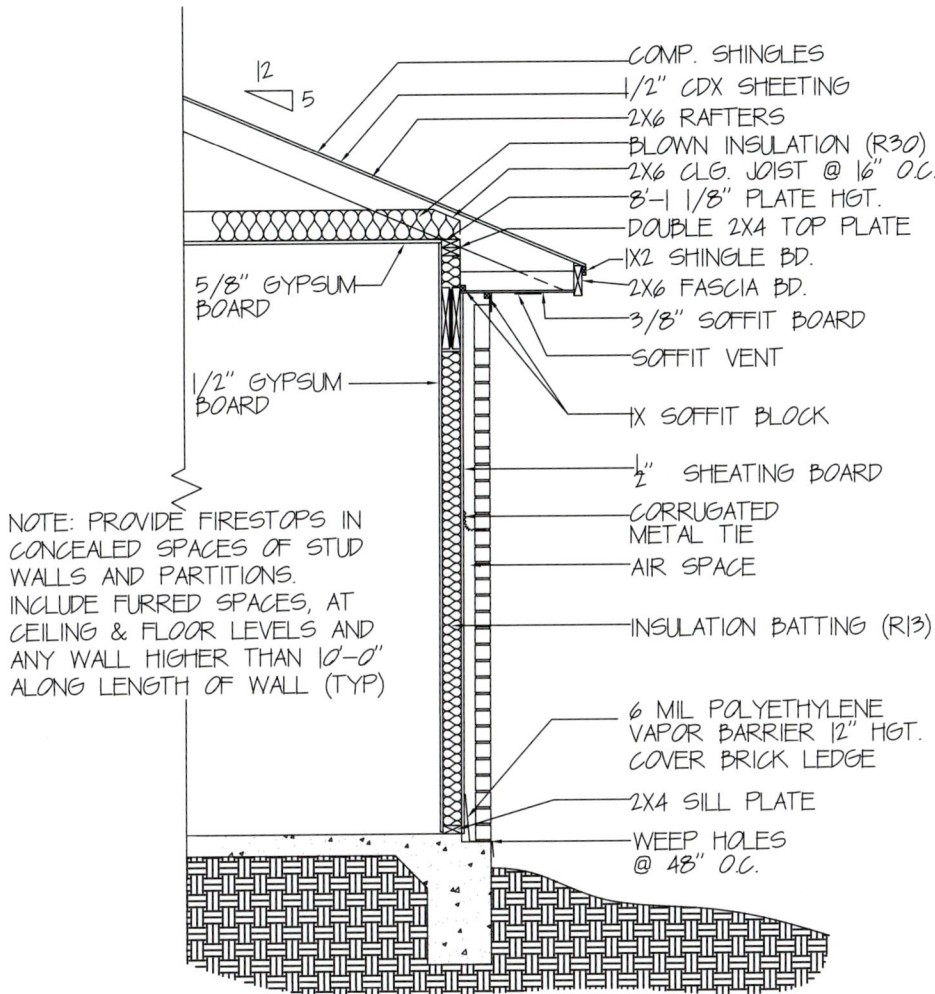

COMP. SHINGLES
1/2" CDX SHEETING
2X6 RAFTERS
BLOWN INSULATION (R30)
2X6 CLG. JOIST @ 16" O.C.
8'-1 1/8" PLATE HGT.
DOUBLE 2X4 TOP PLATE
1X2 SHINGLE BD.
2X6 FASCIA BD.
3/8" SOFFIT BOARD
SOFFIT VENT
1X SOFFIT BLOCK
1/2" SHEATING BOARD
CORRUGATED METAL TIE
AIR SPACE
INSULATION BATTING (R13)
6 MIL POLYETHYLENE VAPOR BARRIER 12" HGT. COVER BRICK LEDGE
2X4 SILL PLATE
WEEP HOLES @ 48" O.C.

5/8" GYPSUM BOARD

1/2" GYPSUM BOARD

NOTE: PROVIDE FIRESTOPS IN CONCEALED SPACES OF STUD WALLS AND PARTITIONS. INCLUDE FURRED SPACES, AT CEILING & FLOOR LEVELS AND ANY WALL HIGHER THAN 10'-0" ALONG LENGTH OF WALL (TYP)

ONE STORY DETAIL
1   3/8" = 1'-0"

Figure 8-4   Architectural wall section

TYPES OF SECTIONS

Common section view types are *full, half, broken-out, revolved, removed,* and *offset.* Drafters decide on the type of section view to draw based on which one most clearly represents the features the drafter is attempting to depict.

Full Sections

A *full section* shows the object as if it has been cut in half. In Figures 8-5 and 8-6, the front and side views of an apple are shown. The knife blade in the front view (Figure 8-5) represents the cutting plane line. In the side view (Figure 8-6), the apple appears as it would look if it were sliced in half along the cutting plane line. This view would represent a full section.

Figure 8-5   Knife blade represents the cutting plane line in Front view   **Figure 8-6**   Side view shown as a full section

In Figures 8-7 and 8-8, the front and side views of a machine part are shown. The thick dashed lines with the arrows pointing to the left in the front view represent the cutting plane line (see Figure 8-7). The view shown in Figure 8-8 represents a full section. This view is drawn as it would appear if the part of the object in the front view that lies behind the cutting plane line had been removed. The diagonal lines shown in the side view are the section lines.

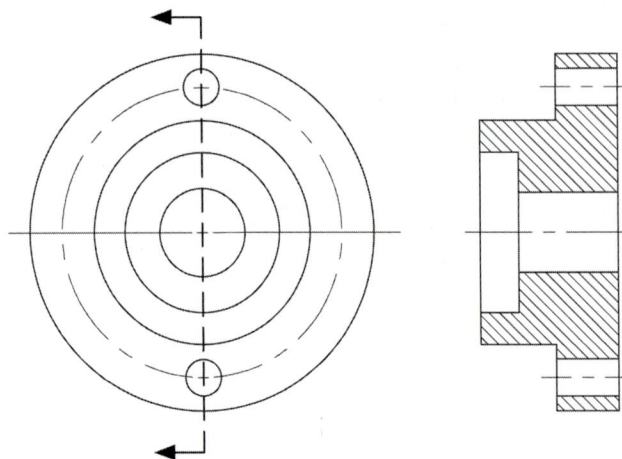

Figure 8-7   Front view with the cutting plane line for a full section   **Figure 8-8**   Side view drawn as a full section

> **TIP** Cutting plane lines take precedence over center lines. When a cutting plane line is used as the center line, show the cutting plane line only.

Half Sections

A *half section* shows the object as if one-fourth of it has been removed. In Figures 8-9 and 8-10, the top and front views of an apple are shown. The knife blades in the top view (Figure 8-9)

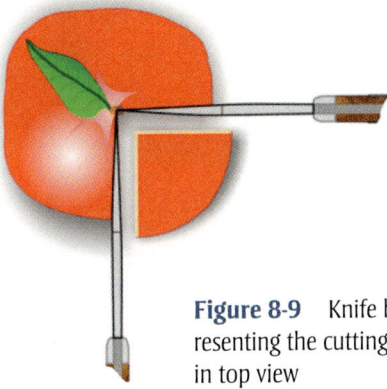

Figure 8-9   Knife blades representing the cutting plane line in top view

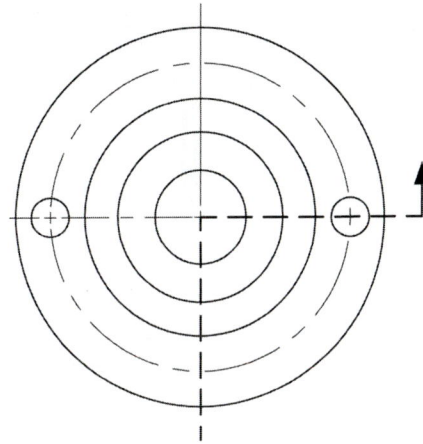

Figure 8-11   Top view with the cutting plane line for a half section

Figure 8-12   Front view drawn as a half section

Figure 8-10   Front view of an apple shown as a half section

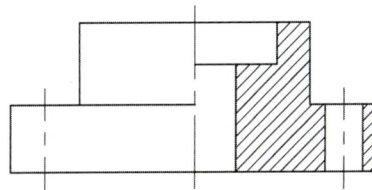

represent the cutting plane line. In the front view (Figure 8-10), the apple appears as it would look if the part of the apple framed by the knife blades in the top view had been removed. This view would represent a half section.

In Figures 8-11 and 8-12, the top and front views of a machine part are shown. The thick dashed line with the arrow pointing up in the top view represents the cutting plane line (see Figure 8-11). The view in Figure 8-12 would represent a half section. This view is drawn as it would appear if the part of the object in the top view framed by the cutting plane line had been removed.

TIP   In a half section, it is not necessary to draw hidden lines in the un-sectioned half of the view.

In Figures 8-13 and 8-14, the front and side views of an apple are shown. The knife blades in the front view (Figure 8-13) represent the cutting plane line. In the side view (Figure 8-14), the apple appears as it would look if the part of the apple inside the knife blades in the front view had been removed. This view would represent a half section.

In Figures 8-15 and 8-16, the front and side views of a machine part are shown. The thick dashed line with the arrow pointing toward the left in the front view represents the cutting plane line (see Figure 8-15). The view in Figure 8-16 would represent a half section. This view is drawn as it would appear if the part of the object in the front view framed by the cutting plane line had been removed.

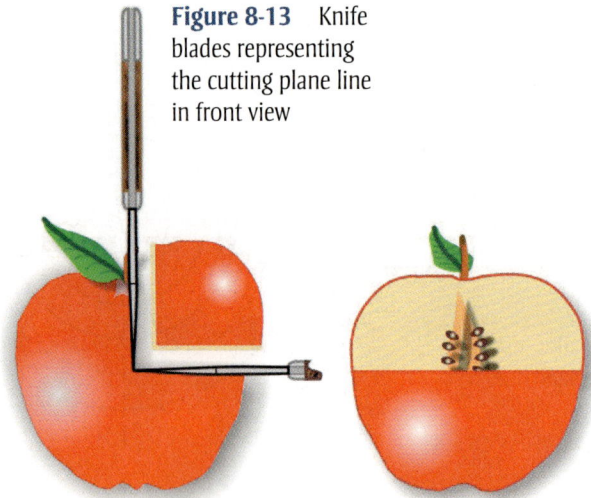

Figure 8-13   Knife blades representing the cutting plane line in front view

Figure 8-14   Side view of apple shown as a half section

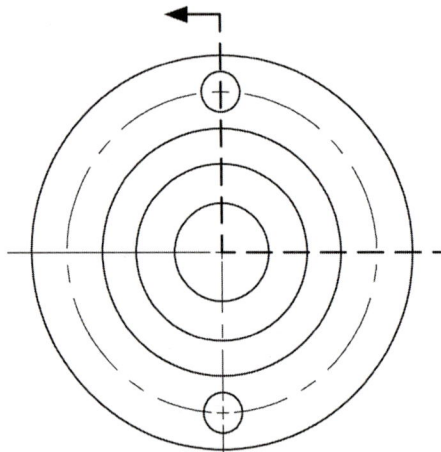

Figure 8-15   Front view with the cutting plane line for a half section

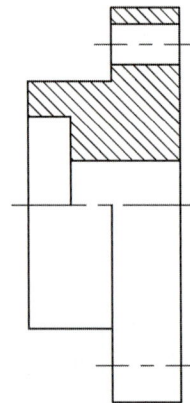

Figure 8-16   Side view drawn as a half section

Broken-Out Sections

broken-out section: A view of an object that shows only a small area of the object as a section.

A ***broken-out section*** is used when only a small portion of the object needs to be sectioned. Figure 8-17 shows the front and side views of an apple. In the side view, a piece of the apple has been broken-out to show the worm on the inside.

Figure 8-17   Front and side views of apple with a broken-out section shown in the side view

TIP

A cutting plane line is not necessary on a broken-out section.

Figure 8-18 shows the front and top views of a machine part. The front view is drawn as it would appear if a small part of the object has been removed, or broken-out, to reveal the desired interior feature. The broken line is drawn as a visible line, and diagonal section lines are added to the sectioned areas. The front view in Figure 8-18 represents a broken-out section.

Revolved Sections

In a *revolved section*, the cross-sectional view of the object is drawn on the object as shown in Figures 8-19 and 8-20. Drafters can place dimensions directly on the revolved section.

revolved section: A cross-sectional view of an object drawn on the object see Figures 8-19–8-20.

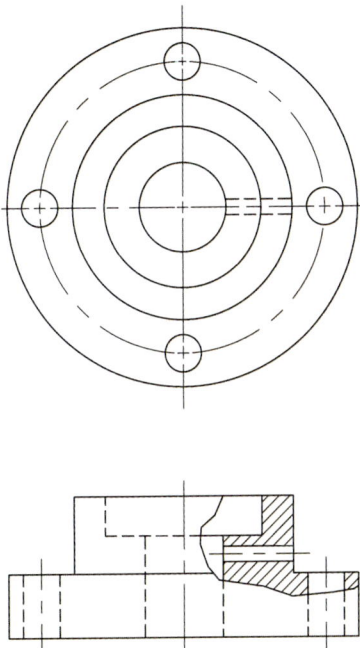

Figure 8-18   Front and top views with a broken-out section in the front view

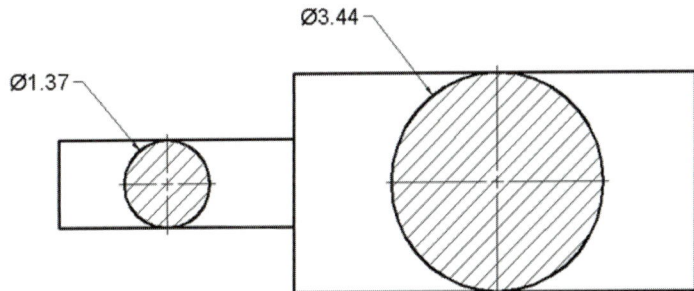

Figure 8-19   Revolved section

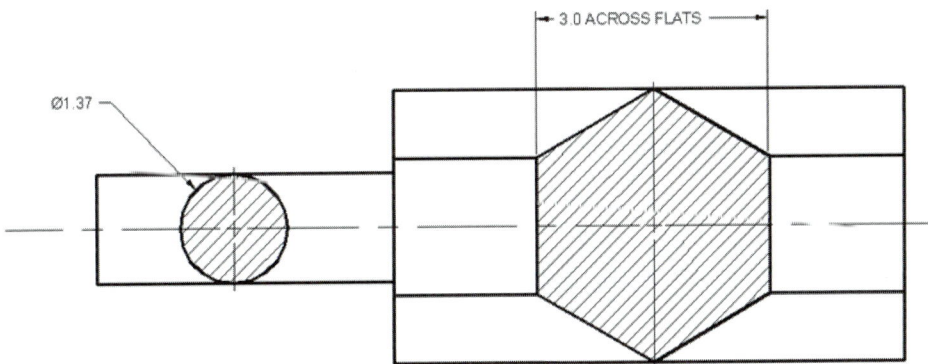

Figure 8-20   Revolved section

Removed Sections

In a *removed section*, the view created by the cutting plane line is not drawn in its normal "projected" position but is drawn somewhere else on the sheet. In Figure 8-21, two cutting planes

removed section: A section view that is not drawn in its normal projected position but somewhere else on the sheet. Requires labeling of both the cutting plane line and the view it references.

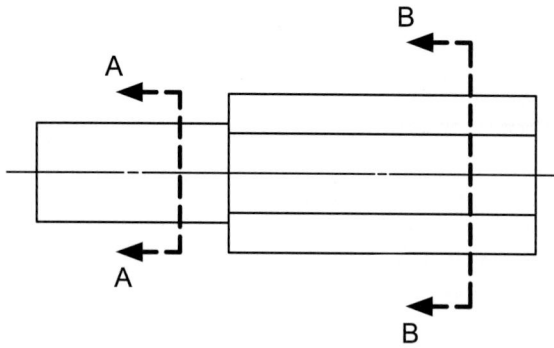

Figure 8-21   Cutting plane lines for removed Sections A-A and B-B

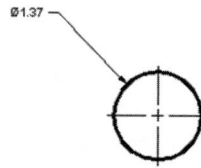

SECTION A-A

Figure 8-22   Removed Section A-A

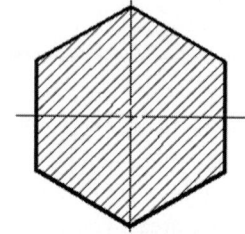

SECTION B-B

Figure 8-23   Removed Section B-B

lines are shown; one is labeled *A-A*, and the other is labeled *B-B*. Figure 8-22 shows the removed section of the object resulting by viewing the object through the cutting plane line labeled *A-A*. The section is labeled *SECTION A-A* to show its relationship to cutting plane *A-A*. Figure 8-23 shows the removed section resulting from viewing the object through the cutting plane labeled *B-B*. This view is labeled *SECTION B-B* to correspond to cutting plane *B-B*. Drafters must ensure that the cutting plane line *and* the resulting removed section view are both labeled alike in order to avoid confusion when using this technique.

Offset Sections

offset section: A section that includes features that would not lie along the path of a straight cutting-plane line. The cutting plane line is offset to take in these features.

Offset sections allow a drafter to create a section with features that would not lie on the path of a straight cutting plane line (see Figure 8-24). In an offset section, the cutting plane line is offset at 90° angles to allow it to pass through the features that the drafter would like to be shown in the resulting section view (see Figure 8-25).

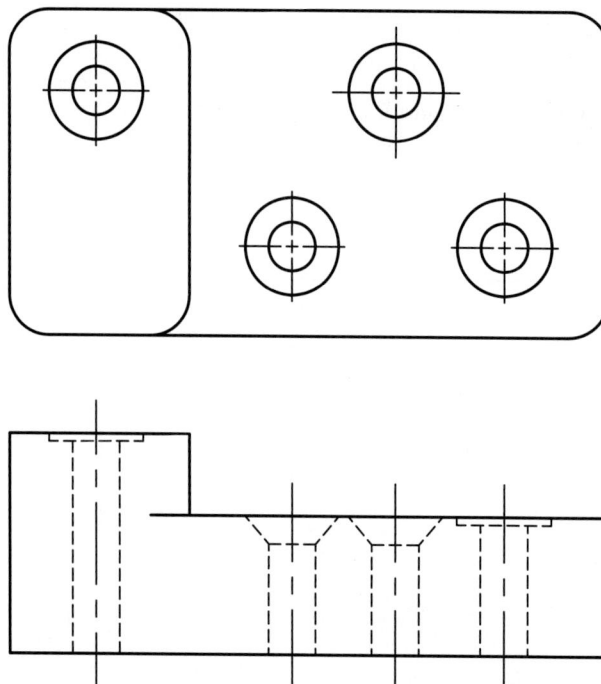

Figure 8-24   Front and top views of a mechanical part with hidden lines shown.

Note that no hidden lines
are shown for this hole in
the sectioned view.

Note that this "edge"
does not project as a
line in the sectioned
view.

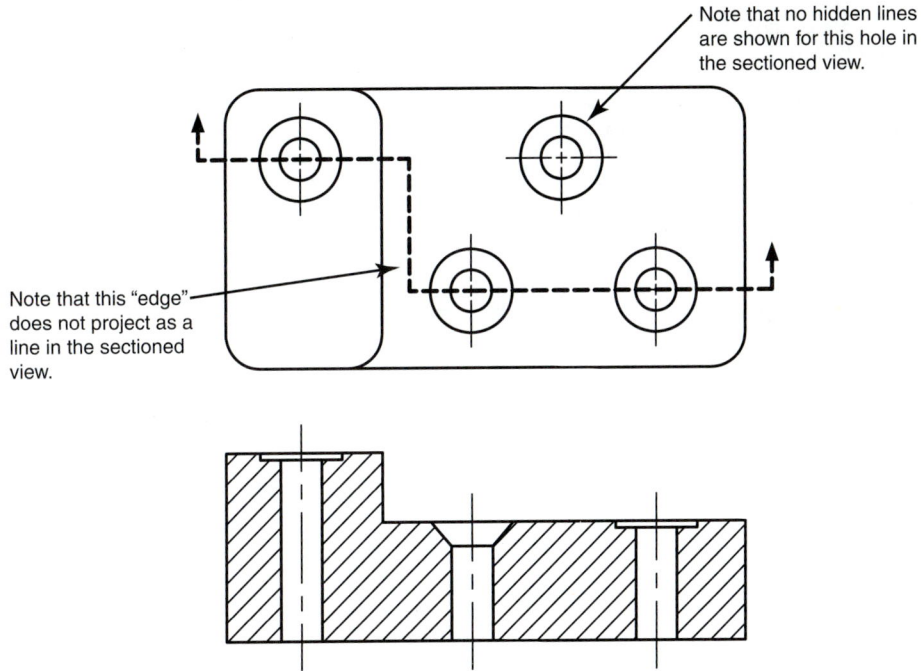

Figure 8-25   Offset section
view

STEPS IN CREATING A SECTION VIEW

The steps involved in creating a full section view of the object shown in Figure 8-26 are as follows:

Step 1. Determine the view to be sectioned and the location of the cutting plane line. In this case, the cutting plane line will be located on the front view (see Figure 8-27). The

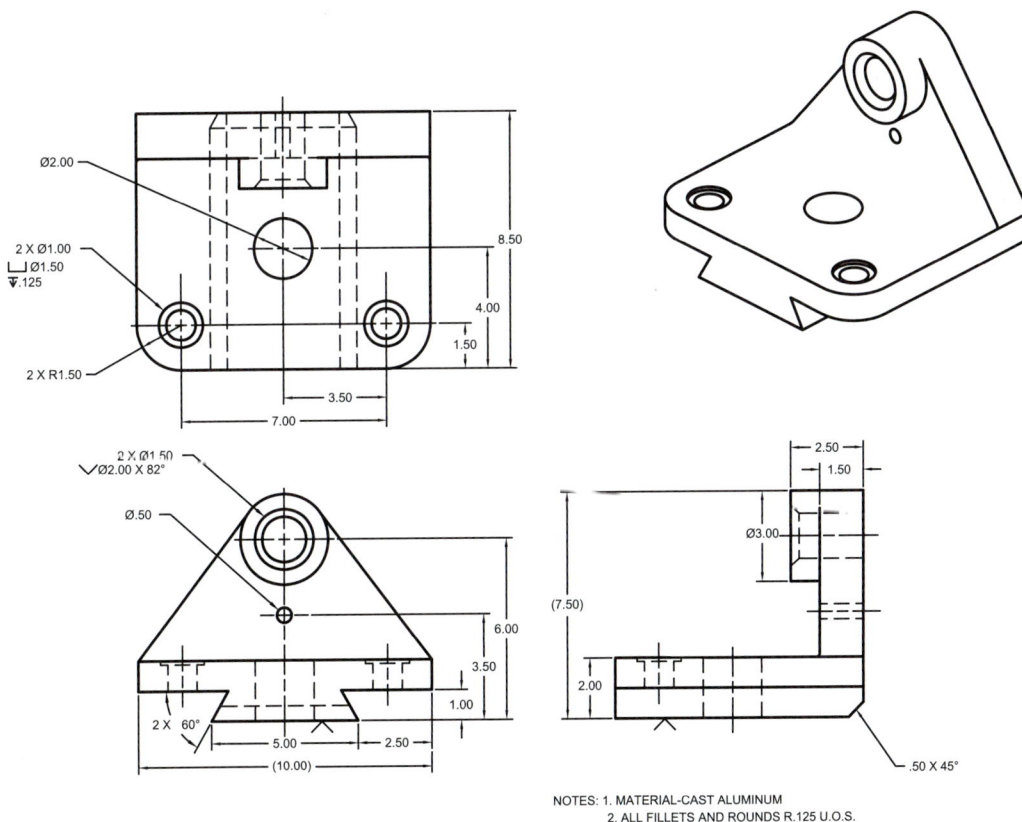

Ø2.00

2 X Ø1.00
⊔ Ø1.50
▼.125

2 X R1.50

8.50

4.00
1.50

3.50
7.00

2 X Ø1.50
∨ Ø2.00 X 82°

Ø.50

(7.50)

2 X 60°
5.00   2.50
(10.00)

6.00
3.50
1.00

2.50
1.50

Ø3.00

2.00

.50 X 45°

NOTES: 1. MATERIAL-CAST ALUMINUM
2. ALL FILLETS AND ROUNDS R.125 U.O.S.

Figure 8-26   Multiview drawing to be sectioned

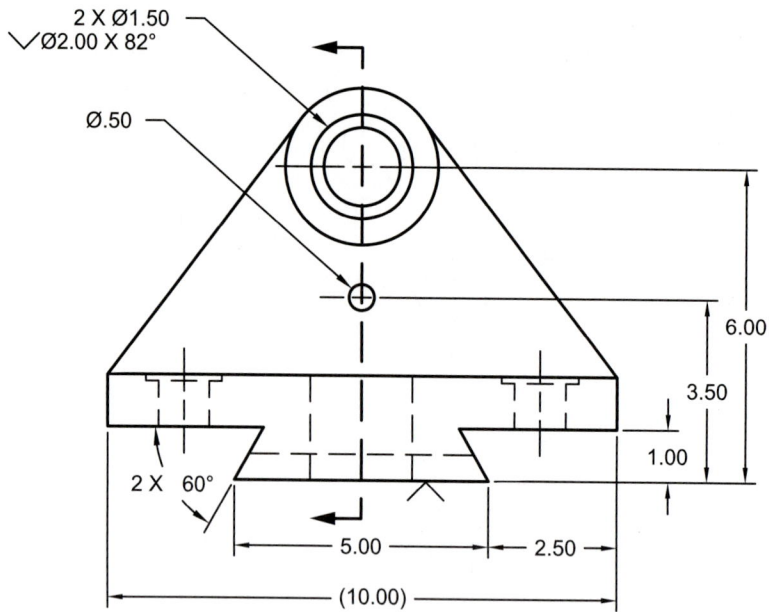

Figure 8-27 Placing the cutting plane line on the front view

placement of the cutting plane line on the object's front view, along with the direction of the cutting plane line's arrows, indicates that the right-side view of the object will be drawn as a full section.

> **TIP** The **POLYLINE** command can be used to make the large arrowheads needed at the ends of cutting plane lines. This is done by drawing a short polyline (about .25″ long) that has a starting width of 0 (zero) and an ending width of .10. See Figure 4-44(b).

Step 2. Study the right-side view (see Figure 8-28) and determine which of the object's hidden lines will become visible after it has been sectioned. Convert these hidden lines to visible lines as shown in Figure 8-29.

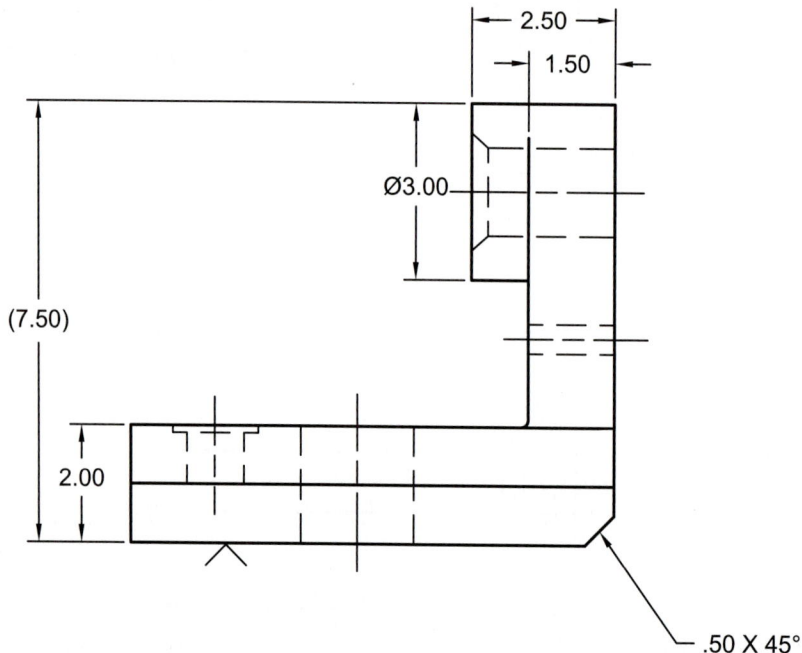

Figure 8-28 View showing hidden lines

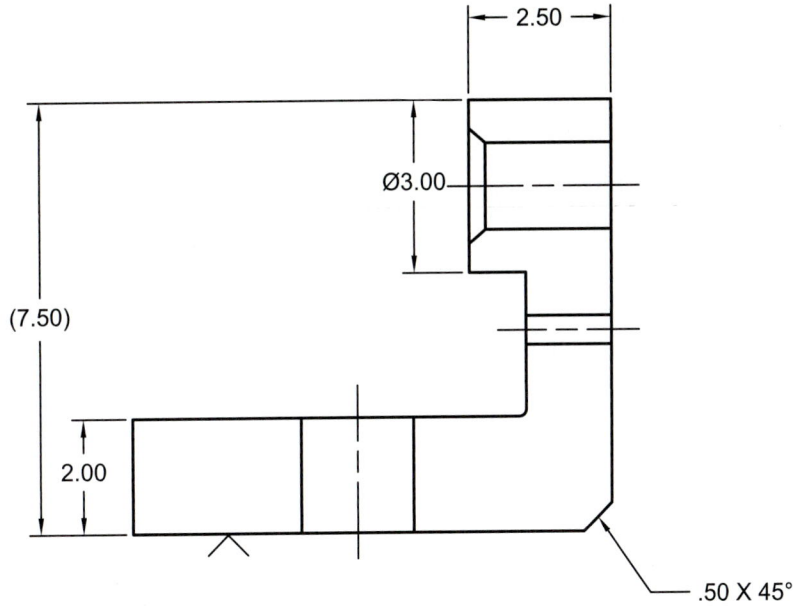

Figure 8-29   View showing visible lines

Step 3. Place section lines inside the areas created by the cutting plane line passing through the object. In an AutoCAD drawing, the **HATCH** command is used to place section lines, or other hatch patterns, into a sectioned area.

Using the HATCH Command

Select the **HATCH** command icon from the **Draw** toolbar as shown in Figure 8-30(a). The **Hatch and Gradient** dialog box shown in Figures 8-30(b) will open. The *placement, scale,* and *angle* of

Figure 8-30(a)   Hatch Icon

In an AutoCAD drawing, *placement, scale,* and *angle* of the Hatch pattern is controlled by the settings in the **Hatch and Gradient** Dialog Box.

Angle controls the angle of the lines in the hatch pattern. (Note: a hatch pattern that shows a default angle of **0** in this dialog box will be inserted at a **45** degree angle into the drawing.)

Scale controls the density of the hatch pattern. In metric and architectural drawings, the scale is often quite large.

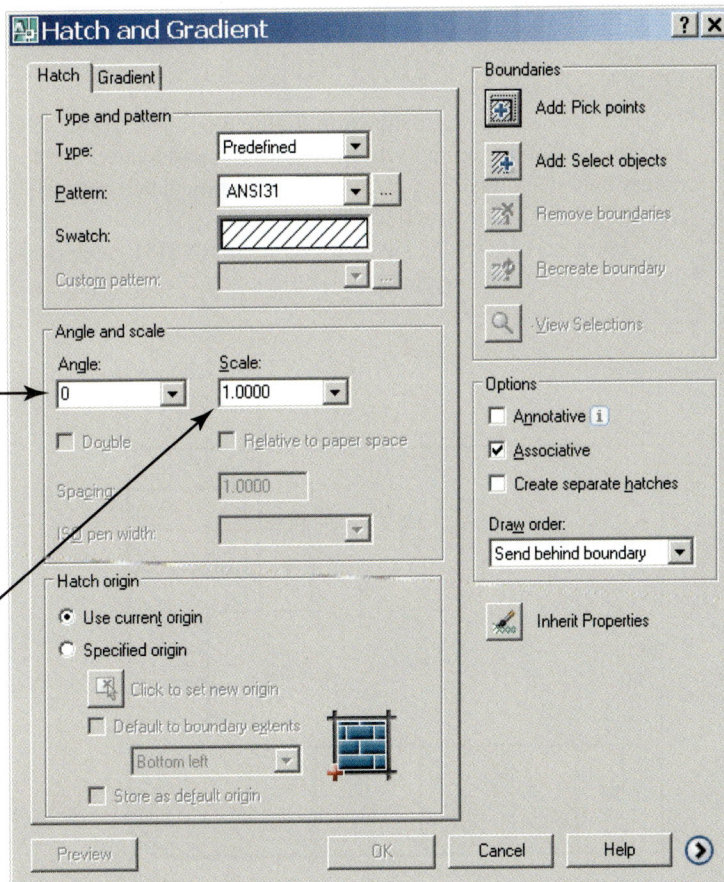

Figure 8-30(b)   Hatch and Gradient dialog box

the Hatch pattern is controlled by the settings in this dialog box. **Angle:** controls the angle of the lines in the hatch pattern.

> **TIP**
> A hatch pattern that shows a default angle of **0** in this dialog box will be inserted at a **45°** angle into the drawing.

Scale: controls the density of the hatch pattern. In metric and architectural drawings, the scale often has to be set quite large.

> **TIP**
> The scale for hatching in metric units is:
>
> | Decimal | Metric |
> |---------|--------|
> | .25 | 6.35 |
> | .50 | 12.7 |
> | .75 | 19.5 |
> | 1.0 | 25.4 |
> | 1.25 | 31.75 |
> | 1.50 | 38.1 |
> | 1.75 | 44.45 |
> | 2.00 | 50.8 |

To place the hatch pattern shown in Figure 8-31, you would select **ANSI31** for the type of pattern, and accept the default settings for **Angle:** and **Scale:**. Then, you would select the **Add: Pick Points** button and pick inside the areas where the hatch pattern should be applied. When you are finished, click **OK** or **Preview**.

Figure 8-31 shows the side view with the **ANSI31** hatch pattern applied. Note that no hidden lines are shown in this view.

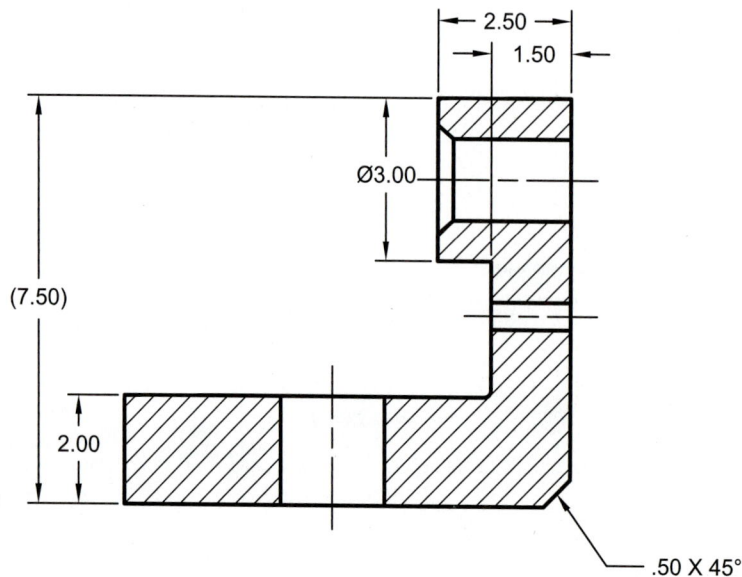

Figure 8-31   Sectioned side view

Figure 8-32 shows the finished drawing with the right-side view replaced with a full section view. The full section provides a clear portrayal of the part's interior features.

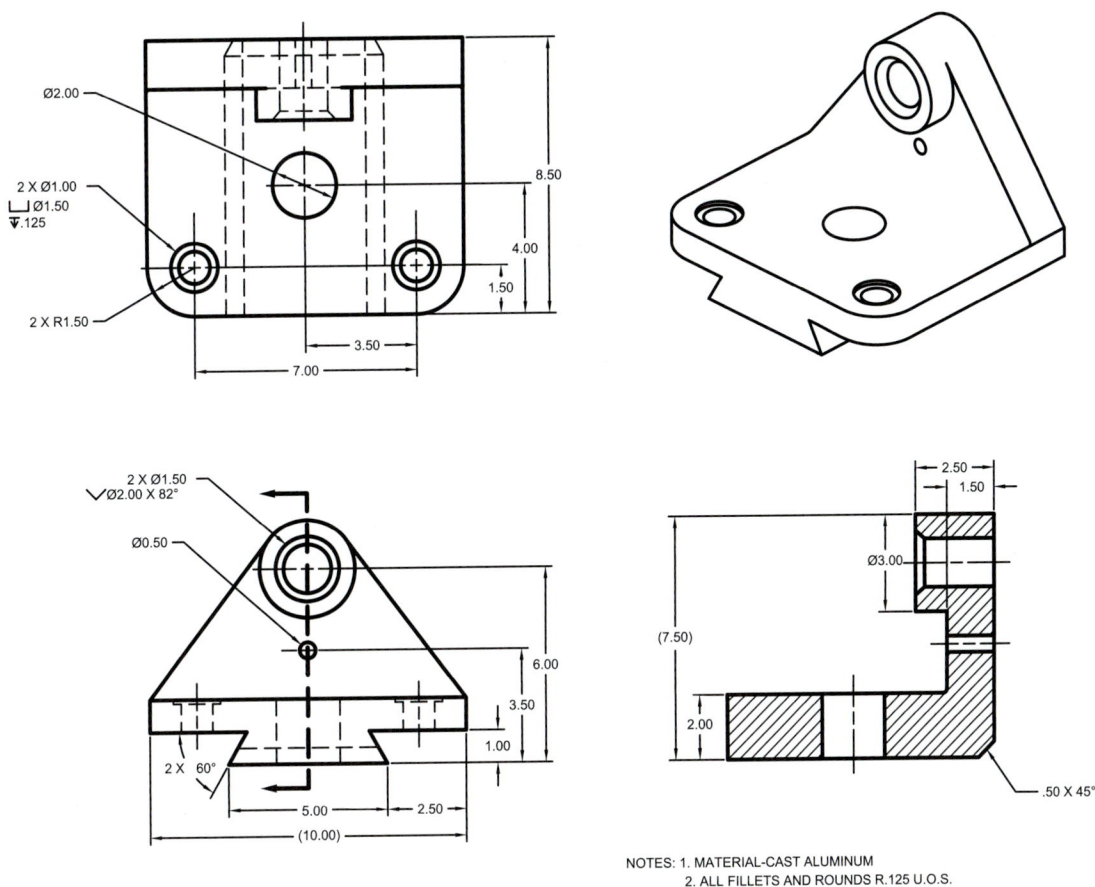

Figure 8-32   Drawing with sectioned side view shown

SUMMARY

Interpreting an object's complex interior features is often difficult, especially when the features are represented by hidden lines in a front, top, or side view. By creating a section view of the object, the drafter can clarify the interior details by replacing the hidden lines with visible lines. Because dimensioning to hidden lines should be avoided in technical drawings, creating a section view also facilitates the placement of dimensions by making complex interior features visible.

In order to create and interpret section views, drafters and designers must be familiar with the different types of sections and the CAD techniques used to create them.

Before drawing a section view of an object, a drafter must imagine what the object will look like after it has been sliced along a cutting plane line. The ability to visualize section views is a skill that can be developed and improved with practice.

UNIT TEST QUESTIONS

Multiple Choice

1. What are the diagonal lines in a section view called?

   a. Section lines
   b. Cutting plane lines
   c. Horizontal plane lines
   d. Straight plane lines

2. In an offset section, the cutting plane line is offset at what angle?

   a. 45° angle
   b. 35° angle
   c. 90° angle
   d. 25° angle

3. Where is a removed section placed on a drawing?

   a. On another sheet
   b. Somewhere else on the sheet
   c. Deleted from the sheet
   d. On another tab

4. What do the arrows on a cutting plane line indicate?

   a. Direction in which section is viewed
   b. Placement of center lines

   c. The top of the sheet
   d. None of the above

5. What is the name of the **HATCH** setting that controls the density of the hatch pattern?

   a. Volume
   b. Scale
   c. Hue
   d. All of the above

Matching

Column A

a. Half

b. Removed

c. One-fourth

d. Revolved

e. Offset

Column B

1. Section view that requires a label

2. Percentage of an object removed to create a half section

3. A cutting plane line is not necessary on this type of section

4. Percentage of an object removed to create a full section

5. Section that cuts through features that do not lie on a straight line

UNIT PROJECTS

Project 8-1: Tool Holder Sectioning

In this project, you will open the **Tool Holder** drawing you created in Unit 4 and convert its front view to a full section.

Studying Figure 8-33 will help you visualize the tool holder after it has been cut along the cutting plane line. Figure 8-34 shows the location of the cutting plane line in the top view.

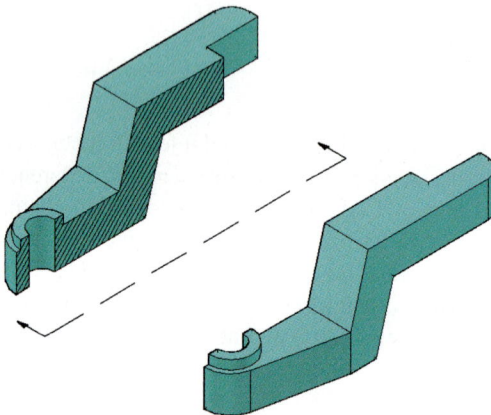

Figure 8-33   3D sketch of the tool holder shown as a full section

Directions

1. Create two new layers, one for the cutting plane line and another for the hatch pattern.
2. Set the linetype for the hatch pattern layer to **Continuous** and the lineweight to **Default**. Use the **ANSI31** hatch pattern to draw the section lines.
3. Set the linetype for the cutting plane layer to **DashedX2** and set the lineweight to **.60mm**.

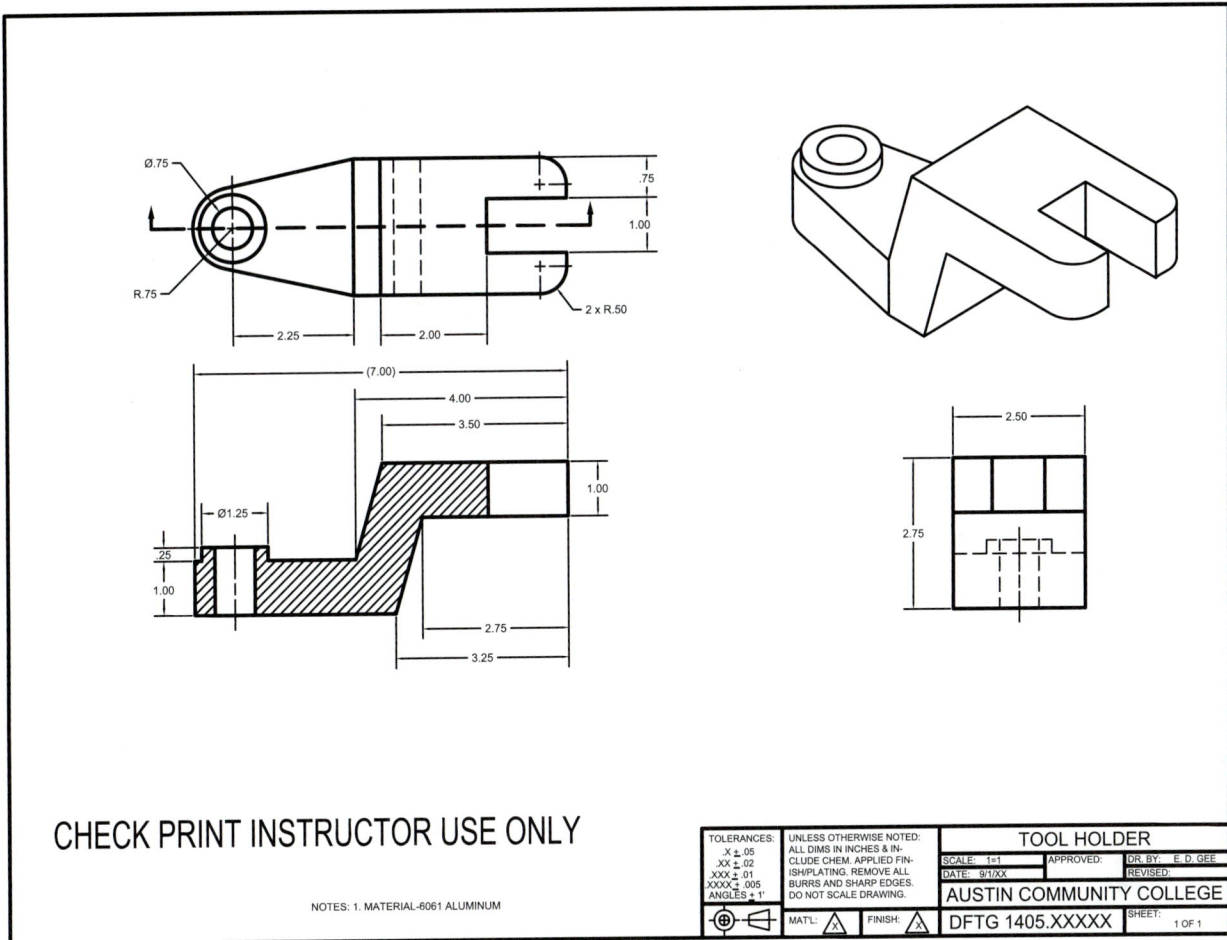

CHECK PRINT INSTRUCTOR USE ONLY

NOTES: 1. MATERIAL-6061 ALUMINUM

TOLERANCES:	UNLESS OTHERWISE NOTED:	TOOL HOLDER		
.X ±.05 .XX ±.02 .XXX ±.01 .XXXX ±.005 ANGLES ± 1'	ALL DIMS IN INCHES & IN-CLUDE CHEM. APPLIED FIN-ISH/PLATING. REMOVE ALL BURRS AND SHARP EDGES. DO NOT SCALE DRAWING.	SCALE: 1=1 DATE: 9/1/XX	APPROVED:	DR. BY: E. D. GEE REVISED:
		AUSTIN COMMUNITY COLLEGE		
	MAT'L:	FINISH:	DFTG 1405.XXXXX	SHEET: 1 OF 1

Figure 8-34   Completed tool holder drawing with full section

Project 8-2: Flange Bearing Sectioning

Directions

1. Select **FILE** and **OPEN**.
2. Select the **Student Resource CD**.
3. Select the **Prototype Drawings** folder.
4. Select the **MechC1-1** drawing (open as "read-only").
5. **SAVE AS** to your **Home** directory and rename the new drawing **FLANGE BEARING**.
6. Refer to the designer's sketch in Figure 8-35 and draw the front and top views of the flange bearing but construct the front view as a full section. If you need help visualizing the sectioned view, refer to Figure 8-36. Dimension the views, including any notes that are needed. The interpretation and construction of the countersunk and counterbored holes are presented on the following pages.

Print the drawing following your instructor's directions when you are finished.

Set the following **Dimension Styles** variables:

Text height:	.125
Arrow size:	.125
Center marks:	Line
Extend beyond dim lines:	.125

Precision:	Varies—match precision of dimensions on sketch
Zero suppression:	Leading
Offset from origin:	.062

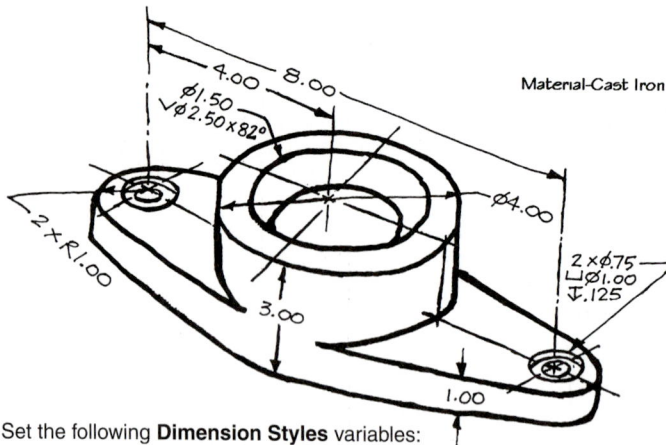

Set the following **Dimension Styles** variables:

Figure 8-35   Designer's sketch of the flange bearing

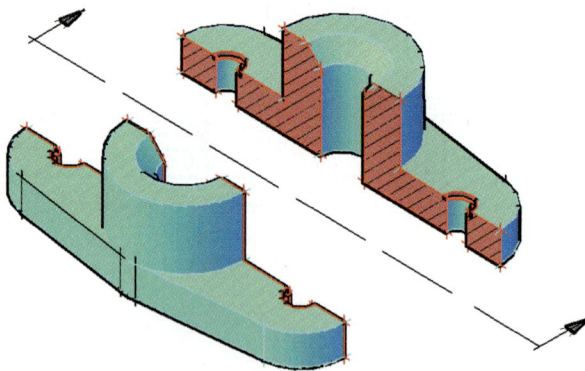

Figure 8-36   3D flange bearing

Constructing a Counterbored Hole

The following steps detail the process of drawing a counterbored hole, but before you can draw a counterbored hole, you must first interpret the symbology used in notating the hole. An interpretation of the note attached to the leader pointing to the counterbored hole in Figure 8-37 follows:

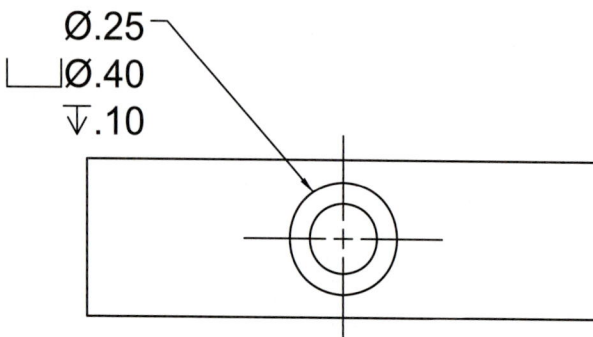

Figure 8-37   Counterbored hole specification

> **Note:**
> When constructing the counterbored hole for the flange bearing, substitute the dimensions noted on the designer's sketch in Figure 8-35 for those used in these examples.

Line 1: A **.25″** diameter hole passes all the way through the part.

Lines 2 and 3: A **.40″** diameter, flat-bottomed hole (a counterbore) is to be bored into the part to a **depth** of **.10″**. The symbols for notating *counterbore* and *depth* in technical drawings are ⌴ and ↧ respectively.

Step 1. Begin the construction by drawing two concentric circles in the top view, **.25** diameter and **.40** diameter, respectively.

Step 2. Project lines from the quadrants of these circles to the front view to construct the through hole and **.10** deep counterbore as shown in Figure 8-38.

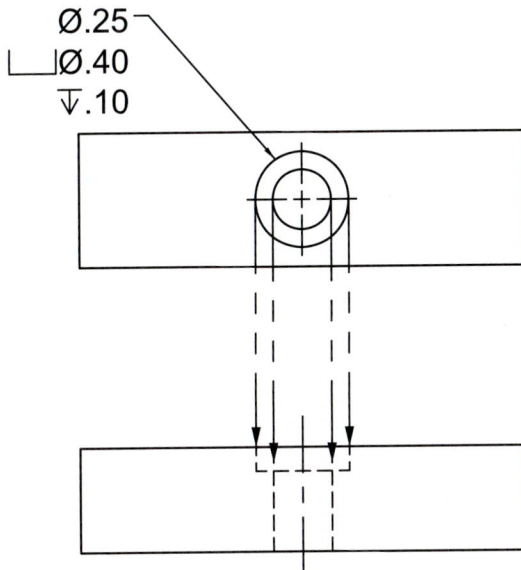

Ø.25
⌴Ø.40
↧.10

Figure 8-38 Multiview drawing of counterbored hole

Adding Annotation to a Counterbored Hole

The following steps are required to add notation to a counterbored hole. The desired notation is shown in Figure 8-39.

Step 1. Select the **Diameter Dimension** icon from the **Dimension** toolbar (see Figure 8-40) and place a diameter dimension on the larger circle.

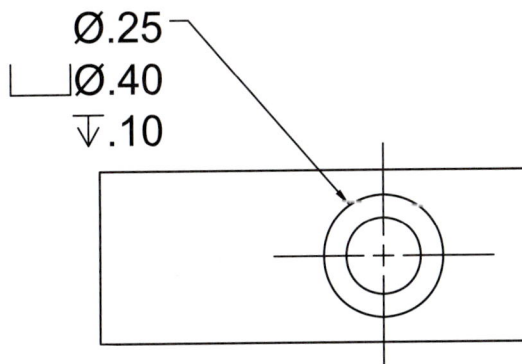

Ø.25
⌴Ø.40
↧.10

Diameter Dimension

Figure 8-39   Counterbored hole note

Figure 8-40   Diameter Dimension icon

Step 2. Select the diameter dimension and select the **Text Editor** (or type **ED** and press **<Enter>**) to create the three lines of text shown in Figure 8-41.

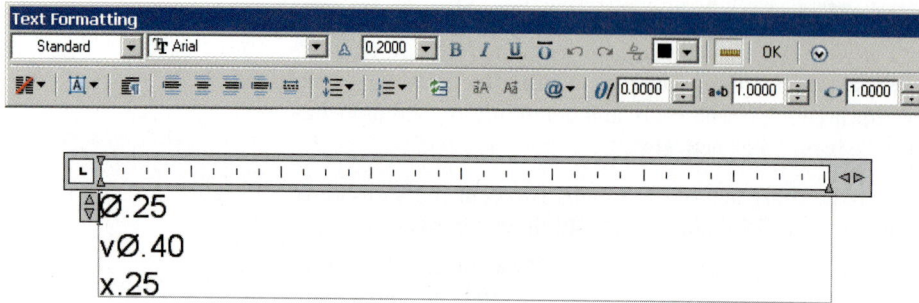

Figure 8-41   Multiline text for a counterbored hole

TIP   When editing text with AutoCAD's text editor, typing **%%C** will result in a diameter symbol and typing **%%d** will result in a degree symbol.

Step 3.   Highlight the **v** and change the font to **GDT**. The **v** will be replaced by the **counterbore** symbol (⌴). Repeat this step for the **x** and it will be replaced with a **depth** symbol (↧). See Figure 8-42.

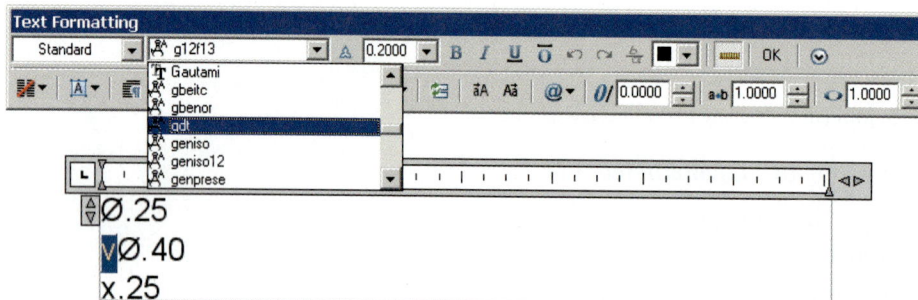

Figure 8-42   Changing Arial text characters to the GDT font

Constructing a Countersunk Hole

The following steps detail the process of drawing a countersunk hole. Before you can draw a countersunk hole, you must first interpret the symbology used in notating the hole.

An interpretation of the note attached to the leader pointing to the countersunk hole in Figure 8-43 follows:

Line 1: A **1.00″** diameter hole passes all the way through the part.

Line 2: A **1.50″** diameter countersunk hole is to be drilled into the part. The angle between the sloping sides of the countersink is **82°**. The symbol for notating a countersink on a technical drawing is ∨.

Step 1.   Begin the construction by drawing two concentric circles in the front view, **1.00″** diameter and **1.50″** diameter, respectively. Project construction lines from the quadrants of these circles to the top view as shown in Figure 8-44.

Ø1.00
Ø1.50 X 82°

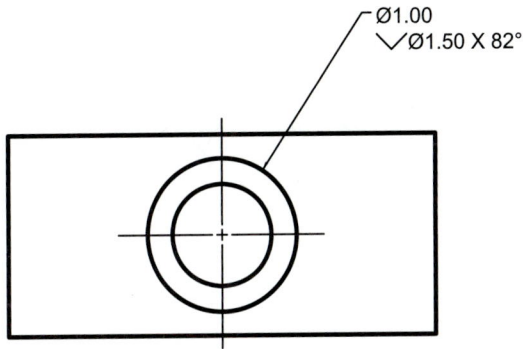

Figure 8-43   Countersunk hole specification

Note:
When constructing the counter-sunk hole for the flange bearing, substitute the dimensions noted on the designer's sketch in Figure 8-35 for those used in these examples.

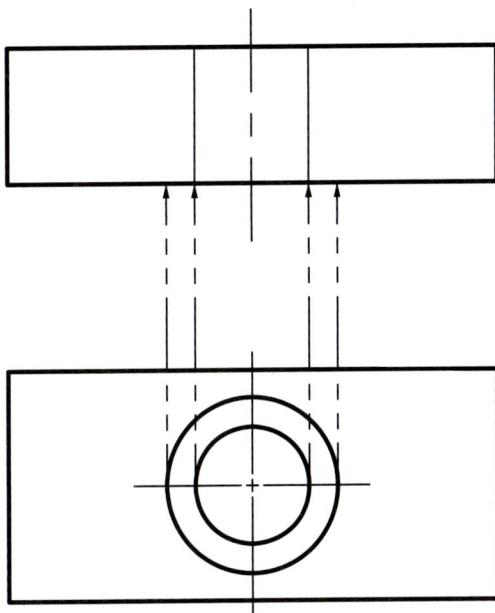

Figure 8-44   Constructing a countersunk hole

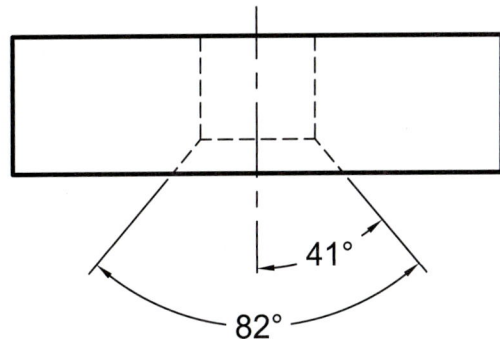

41°

82°

Figure 8-45   Drawing the 82° angle

Step 2.   From the lines projected from the circle quadrants in the front view, construct the **82°** angled sides of the countersunk hole and the sides of the through hole in the top view (see Figure 8-45). For help constructing the angles, see Figures 8-46 through 8-48.

Use the protractor in Figure 8-46 to determine countersink angles. Once you have determined the beginning points of the countersunk diameter, draw lines by using either Polar coordinates or **Polar Tracking** with the required angles added in, as illustrated in Figures 8-47 and 8-48. Do not be concerned with the length of the lines; concentrate on the angle. If the lines are too short, use **Extend** to extend them to the vertical (or horizontal) lines. If they overlap, use **Trim** to trim back to the vertical (or horizontal) lines.

The completed front and top views of the countersunk hole are shown in Figure 8-49.

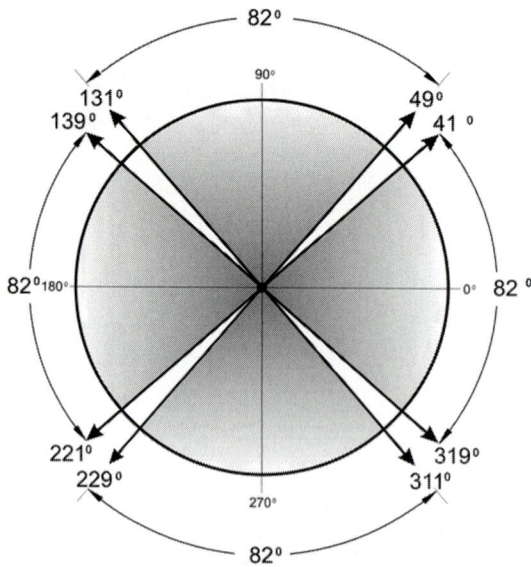

Figure 8-46   Countersunk hole protractor

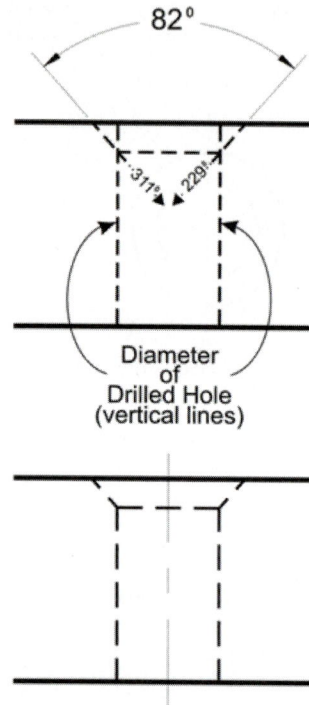

Figure 8-47   Constructing a vertical countersunk hole

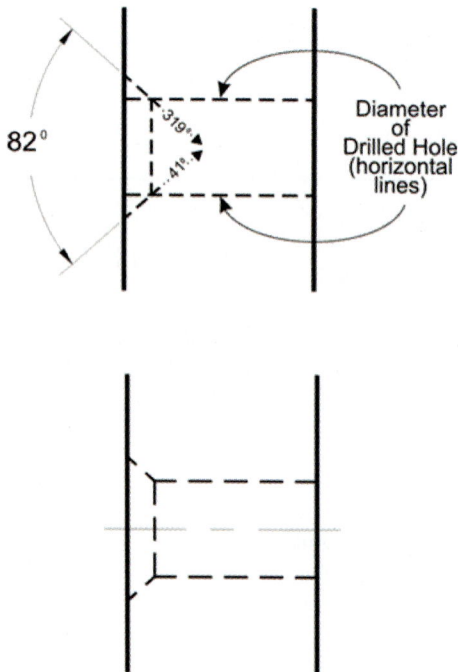

Diameter
of
Drilled Hole
(vertical lines)

Diameter
of
Drilled Hole
(horizontal
lines)

Figure 8-48   Constructing a horizontal countersunk hole

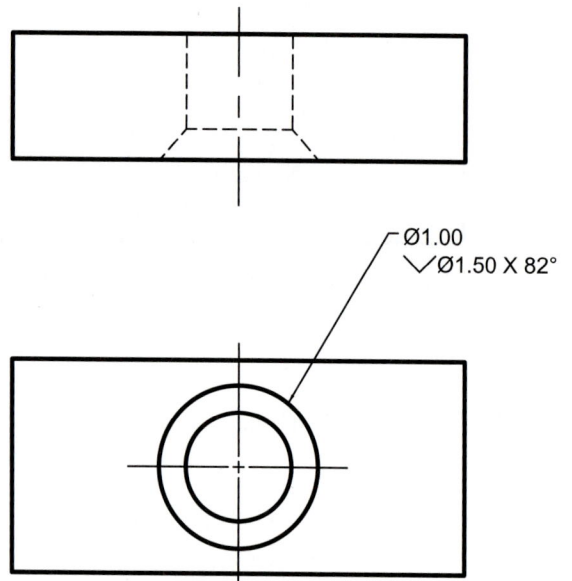

Figure 8-49   Completed multiviews of a countersunk hole

Ø1.00
Ø1.50 X 82°

Adding Annotation to a Countersunk Hole

The steps required to add notation to a countersunk hole are as follows. The desired notation is shown in Figure 8-50.

Step 1.   Select the **Diameter Dimension** icon from the **Dimension** toolbar (see Figure 8-51) and place a diameter dimension on the larger circle.

Ø1.00
Ø1.50 X 82°

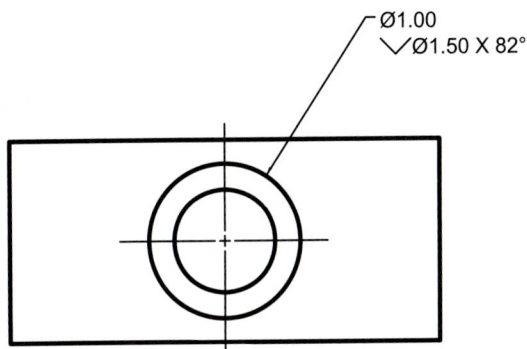

Figure 8-50 Countersunk hole specification

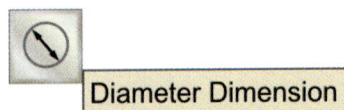

Diameter Dimension

Figure 8-51 Diameter Dimension icon

Step 2. Select the diameter dimension and select the **Text Editor** (or type **ED** and press **<Enter>**) to create the two lines of text shown in Figure 8-52.

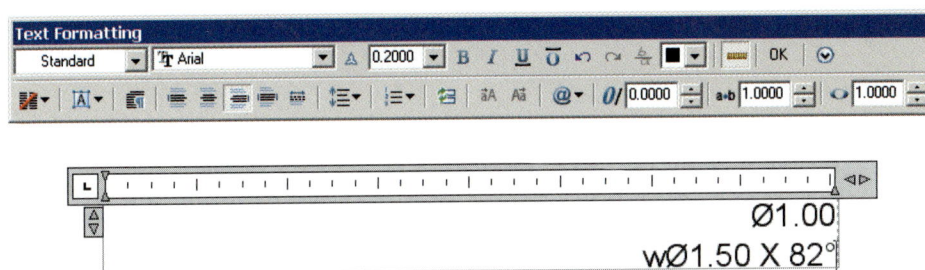

Ø1.00
wØ1.50 X 82°

Figure 8-52 Multiline text for a countersunk hole

TIP Typing **%%D** will result in a degree symbol.

Step 3. Highlight the **w** and change the font to **GDT**. The **w** will be replaced by the countersink symbol (∨). See Figure 8-53.

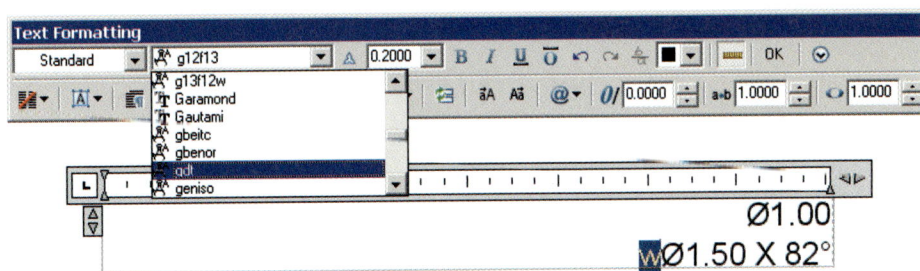

Ø1.00
Ø1.50 X 82°

Figure 8-53 Changing Arial text characters to the GDT font

Auxiliary Views

9

Unit Objectives

- Define what auxiliary views are and how they are used in technical drawings.
- Explain the glass box theory of visualizing an auxiliary view.
- Use AutoCAD to create a primary auxiliary view for an inclined surface.

INTRODUCTION

In some instances, such as when an object has features on an inclined plane, the regular multi-views may not describe these features in their true size or shape. In Figure 9-1, the holes and slot located on the inclined plane labeled *Plane A* are not shown in true shape in either the top or right-side view because the plane is foreshortened in both views. This situation may present problems when attempting to dimension these features.

In such cases, a drafter may decide to draw an auxiliary view of the inclined plane. The auxiliary view is drawn as if the viewer's line of sight were perpendicular to the inclined plane. The features of the inclined plane will appear true size and shape in the auxiliary view.

If the drafter is working from a 3D CAD model of the object, preparing an auxiliary view is a relatively easy process of rotating the model until the inclined plane is parallel to the plane of projection.

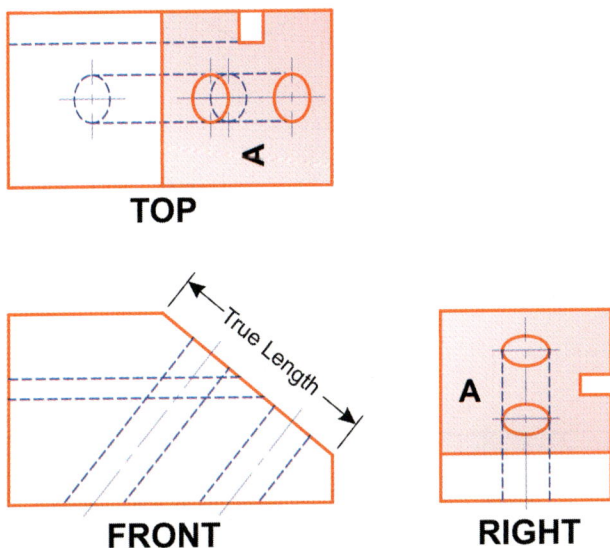

TOP

FRONT

RIGHT

Figure 9-1 Multiview drawing of an object including inclined Plane A

If the drafter is working with 2D geometry, the process of creating an auxiliary view is more complicated. This unit discusses the techniques used to add an auxiliary view to a 2D multiview drawing.

> **TIP** The American Society of Mechanical Engineers standard governing the creation and labeling of auxiliary views is *ASME Y14.3M-1994 Multiview and Sectional View Drawings*.

VISUALIZING AN AUXILIARY VIEW

primary auxiliary view: A view that is adjacent to, and aligned with, a principal view of the object showing the true shape of features that are not parallel to any of the principal projection planes (front, top, side, etc.).

In Figure 9-2, the object shown in Figure 9-1 has been placed inside a glass box. The box has a projection plane labeled *primary auxiliary view*. This projection plane is parallel to the object's inclined plane.

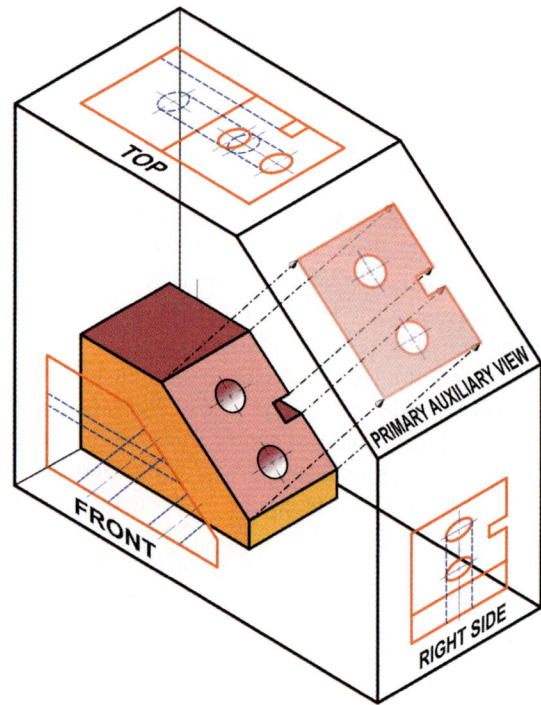

Figure 9-2 Glass Box with an auxiliary view of Plane A

Viewed through this projection plane, the inclined surface is true size and shape. This is because its features are projected perpendicular to the projection plane.

In Figure 9-3, the projection planes of the glass box are unfolded. In the resulting views, you can see the position of the auxiliary view relative to the other views.

> **TIP** The auxiliary view is in-line with the inclined plane.

secondary auxiliary view: A view that is adjacent to, and aligned with, a primary auxiliary view.

The view in Figure 9-3 is called a primary auxiliary view because it is adjacent to, and aligned with, a principal view of the object.

A *secondary auxiliary view* would be adjacent to, and aligned with, a primary auxiliary view.

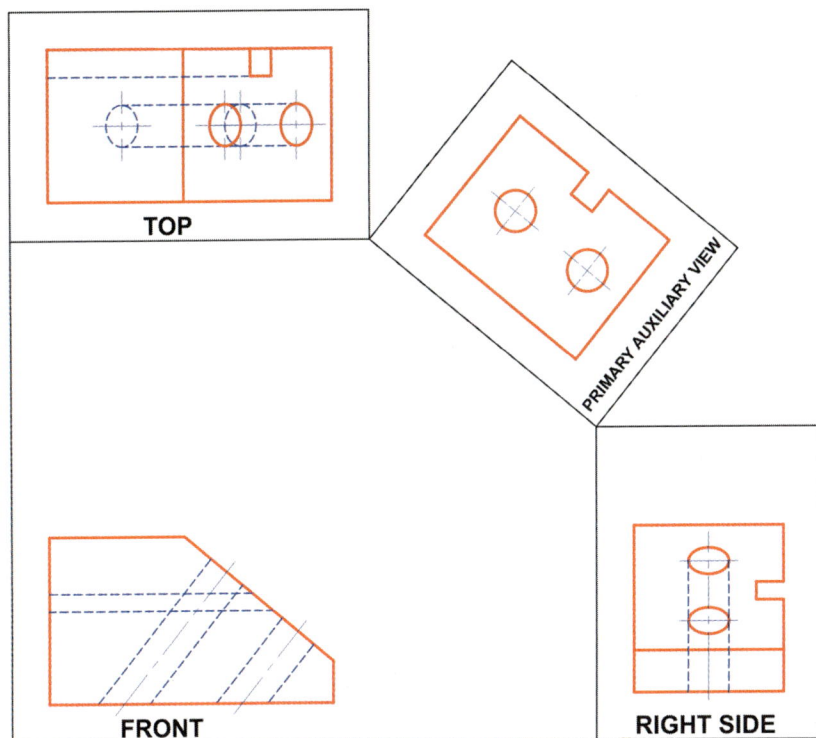

Figure 9-3   Glass box unfolded to show the primary auxiliary view

PROJECT 9-1: CONSTRUCTING A PRIMARY AUXILIARY VIEW—DESCRIPTIVE GEOMETRY METHOD

In this project you will use AutoCAD, and the principles of descriptive geometry, to create a primary auxiliary view for the object shown in Figure 9-3.

Directions

1. Select **FILE** and **OPEN**.
2. Select the **Student Resource CD**.
3. Select the **Prototype Drawings** folder.
4. Select the **Auxiliary View Prototype** drawing (open as "read-only").
5. **SAVE AS** to your **Home** directory and rename the new drawing **AUXILIARY VIEW PROJECT**.
6. Follow Steps 1 through 33 to create a primary auxiliary view of the inclined plane.

Step 1. Create two new layers named Reference and Projections. Assign the linetype for the Reference layer to **Phantom**. On the Reference layer, you will draw two reference lines from which distances are measured. You will use the Projections layer for laying out construction lines. The prototype drawing already contains Layers 0, Center, Hidden, and Visible.

Step 2. At the command line, type **SNAP** and press **<Enter>**. When prompted with Specify snap spacing or [ON/OFF/Aspect/Style/Type] <0.5000>:, type **R** for **ROTATE** and press **<Enter>**.

Step 3. When prompted with Specify a base point:, select the top end of the line in the front view labeled *Plane A* (see Figure 9-4).

Step 4. When prompted with Specify rotation angle:, select the bottom end of the line labeled Plane A. The **GRID** and crosshair will rotate and be perpendicular to *Plane A.* The grid and cursor should be turned as in Figure 9-4.

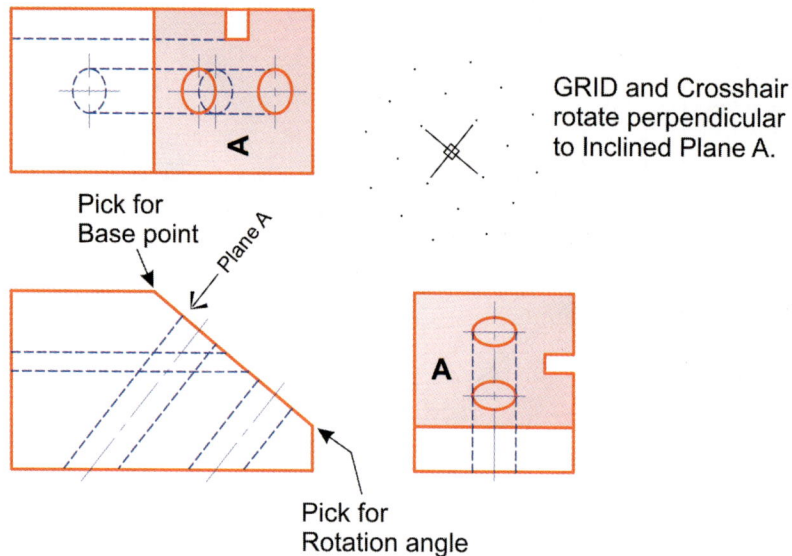

GRID and Crosshair
rotate perpendicular
to Inclined Plane A.

Pick for
Base point

Plane A

Pick for
Rotation angle

A

Figure 9-4   Rotating the grid

Step 5. Set the Projections layer current, turn **Ortho On**, and draw two **12″** lines, extending from each endpoint of inclined *Plane A* as shown in Figure 9-5.

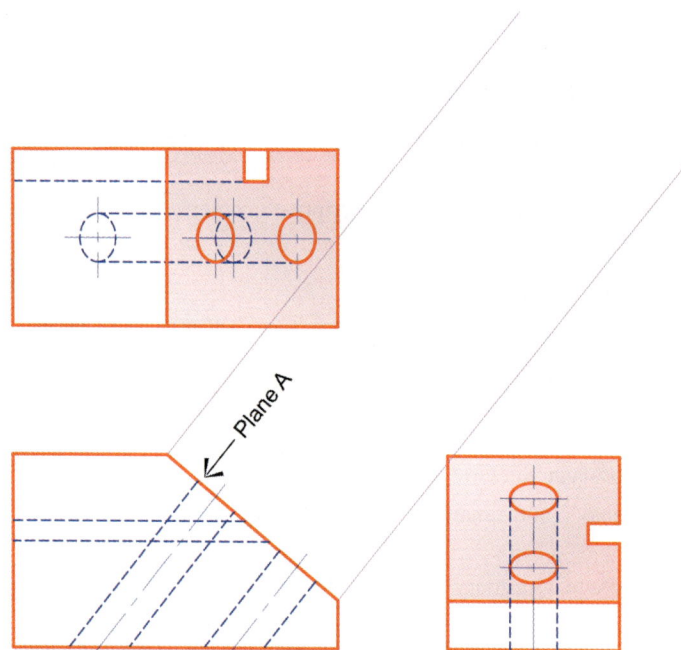

Plane A

Figure 9-5   Extending lines from Plane A

Step 6. Set the Reference layer current. Draw a line parallel to inclined *Plane A* as shown in Figure 9-6. An easy way to do this is by drawing a line from *midpoint* to *midpoint* of the two construction lines drawn in **Step 5**. This line will be referred to as *Reference line 1*.

Step 7. Turn **Ortho Off**, and draw a second line on the left edge of the right-side view as shown in Figure 9-6. This line will be referred to as *Reference line 2*.

Step 8. Use the **DISTANCE** command to measure the distances from *Reference line 2* to the points labeled *a* and *b* on the right-side view in Figure 9-7 by selecting the **DISTANCE** button from the **Inquiry** toolbar, or type in **DI**, at the command prompt. When prompted

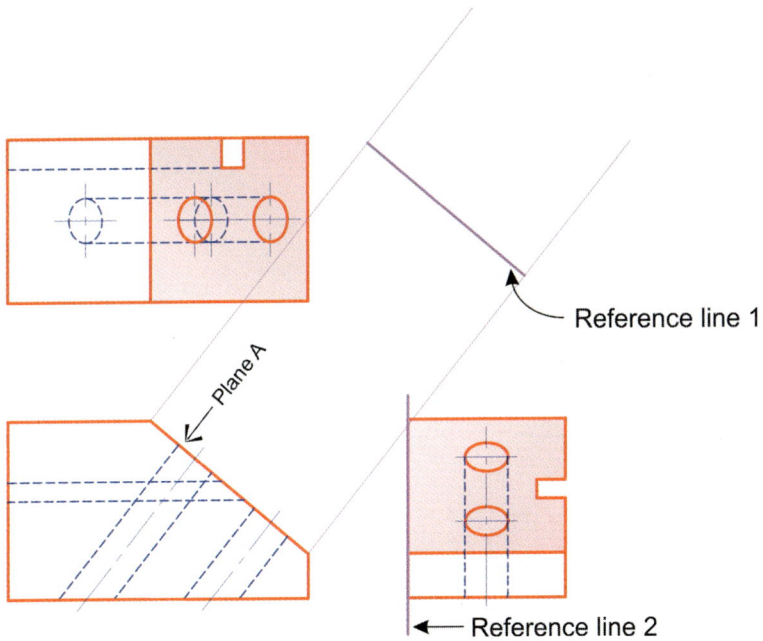

Figure 9-6   Adding Reference line 1 and Reference line 2

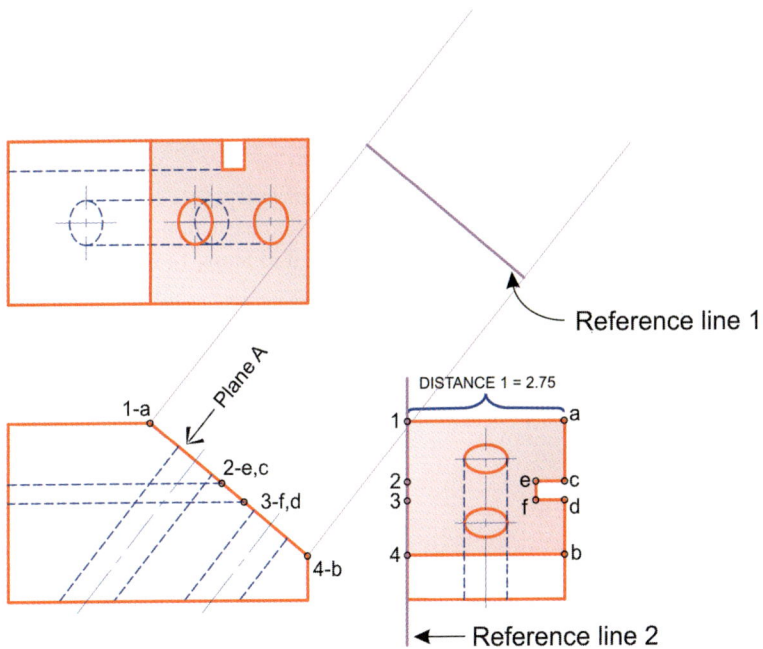

Figure 9-7   Measuring distance between Point 1 and Point a.

with '_dist Specify first point:, select *Point 1* on the right-side view where the object intersects *Reference line 2*.

Step 9. When prompted with Specify second point:, select *Point a* on the right-side view. The **DISTANCE** command determines that the distance between these points, measured along the *X*-axis, equals **2.75**.

Step 10. Use the **DISTANCE** command to measure the distance between *Point 4* and *Point b* in the right-side view as you did in Step 9.

Step 11. To add lines *1-a* and *4-b* to the auxiliary view as shown in Figure 9-8, set the Visible layer current. Transfer the distance found between *Point 1* and *Point a* in Step 8 by drawing a line **2.75″** in length from the top point of *Reference line 1* along the angled projection line constructed in Step 6.

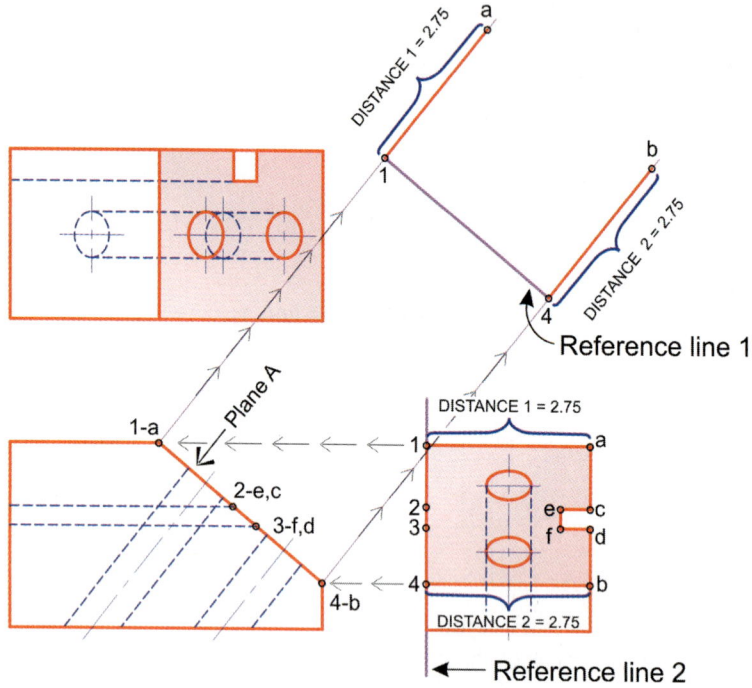

Figure 9-8   Transferring distance 1-2 and 4-b

Step 12. Draw another line **2.75″** in length from the bottom point of *Reference line 1* along the other angled projection line to represent the distance between *Point 4* and *Point b*. See Figure 9-8.

Step 13. To locate the top and bottom edges of the slot in the auxiliary view (see Figure 9-9), set the Projections layer current, turn **Ortho On**, and project construction lines from the points in the front view labeled *2-e,c* and *3-f,d,* as described in Steps 14 and 15.

Step 14. Draw a construction line **12″** long from the point labeled *2-e,c* in the front view that is perpendicular to *Reference line 1*.

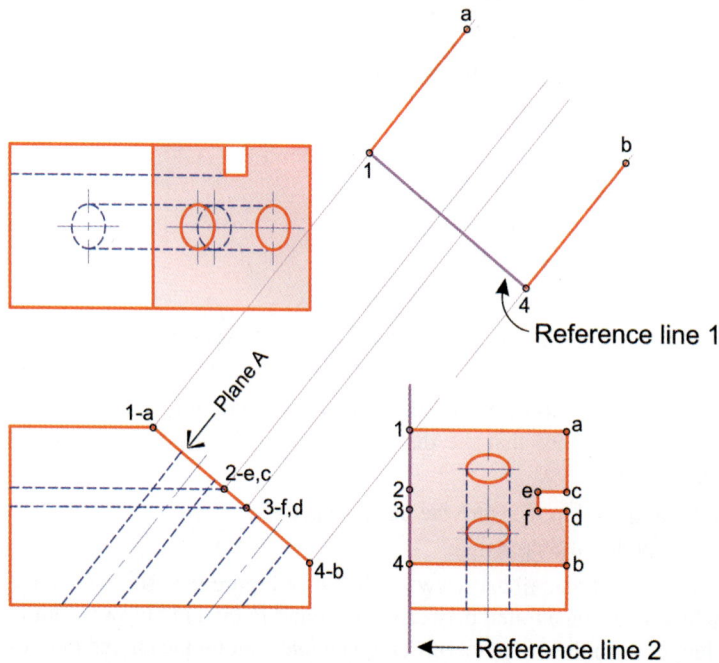

Figure 9-9   Extending lines from Point 2-e,c and Point 3-f,d

Step 15. Draw a construction line **12″** long from the point labeled *3-f,d* in the front view that is perpendicular to *Reference line 1*.

Step 16. Use the **POINT** command and the **Snap From** object snap setting to locate the points labeled *c* and *d* on the angled construction lines in Figure 9-10. Set the Reference layer current, and follow Steps 17 through 24.

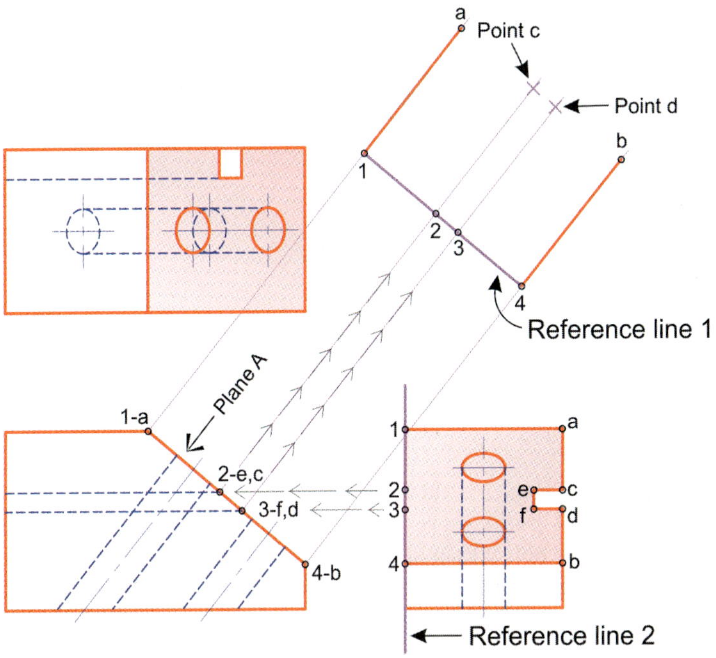

Figure 9-10   Locating Point c and Point d in auxiliary view

Step 17. Select **Point Style** from the **Format** pull-down menu. When the **Point Style** dialog box opens, select the **X** point style (fourth from the right on the top row), select the **Set Size in Absolute Units** button, set the **Point Size** to **.125**, and click **OK**.

Step 18. Turn **On** the following **Osnaps**: *Endpoint, Midpoint, Center, Node,* and *Intersection*. **Dynamic Input** should also be **On**. Refer to Figure 9-10 and note that the distance in the right-side view from *Point 2* to *Point c* is the same distance as measured earlier between *Point 1* and *Point a* (**2.75″**). This is true of the distance between *Point 3* and *Point d* as well.

Step 19. Open the **Object Snap** toolbar and select the **POINT** command from the **Draw** toolbar. When prompted with Specify a point:, select the **Snap From** icon (see Figure 4-101) on the **Object Snap** toolbar and select the point where the angled line drawn from *Point 2-e,c* in the front view intersects *Reference line 1*. When prompted with from Base point, move the cursor to the top end of the angled projection line until the **Endpoint** object snap lights up (do not select the point), type **2.75**, and press **<Enter>**.

Step 20. Repeat the **POINT** command as in Step 19, except this time, when prompted with Specify a point:, select the **Snap From** icon on the **Object Snap** toolbar and select the point where the angled line drawn from *Point 3-f,d* in the front view intersects *Reference line 1*. When prompted with from Base point, move the cursor to the top end of the angled projection line until the **Endpoint** object snap lights up (do not select the point), type **2.75**, and press **<Enter>**.

Step 21. Using the **POINT** command and the **Snap From** object snap setting, locate the points labeled *e* and *f* on the angled construction lines as shown in Figure 9-11.

Step 22. Use the **DISTANCE** command to find the distance between *Point 2* and *Point e* in the right-side view.

Step 23. Select the **POINT** command from the **Draw** toolbar. When prompted with Specify a point:, select the **Snap From** icon on the **Object Snap** toolbar and select the point

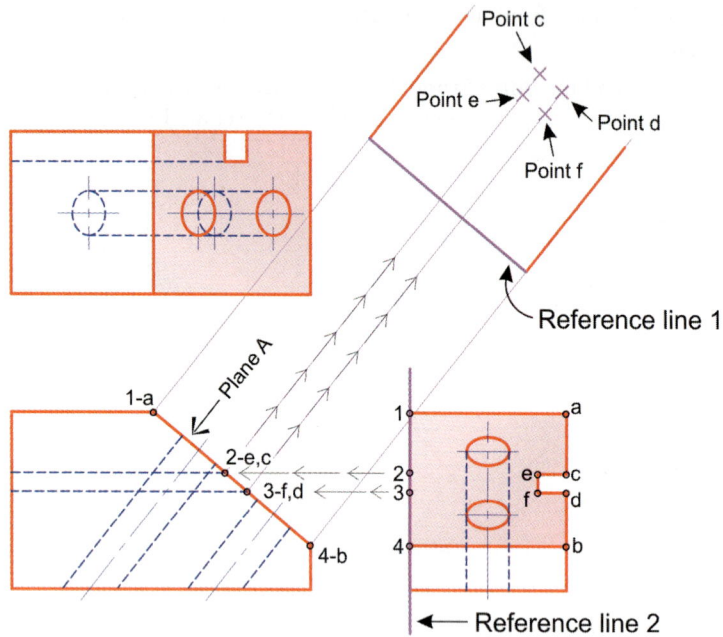

Figure 9-11   Locating Point e and Point f in auxiliary view

where the angled line drawn from *Point 2-e,c* in the front view intersects *Reference line 1.* When prompted with from Base point, move the cursor to the top end of the angled projection line until the **Endpoint** object snap lights up (do not select the point), type the distance between *Point 2* and *Point e* in the right-side view, and press **<Enter>**.

Step 24.   Repeat the **POINT** command as in Step 23, except this time, when prompted with Specify a point:, select the **Snap From** icon on the **Object Snap** toolbar and select the point where the angled line drawn from *Point 3-f,d* in the front view intersects *Reference line 1.* When prompted with from Base point, move the cursor to the top end of the angled projection line until the **Endpoint** object snap lights up (do not select the point), type the distance between *Point 2* and *Point e* in the right-side view, and press **<Enter>**.

Step 25.   Set the Visible layer current and use the **LINE** command to connect *Points a,c,e,f,d,* and *b* in the auxiliary view as shown in Figure 9-12. Change **Reference line 1** to the Visible layer. This completes the true shape profile of inclined *Plane A.*

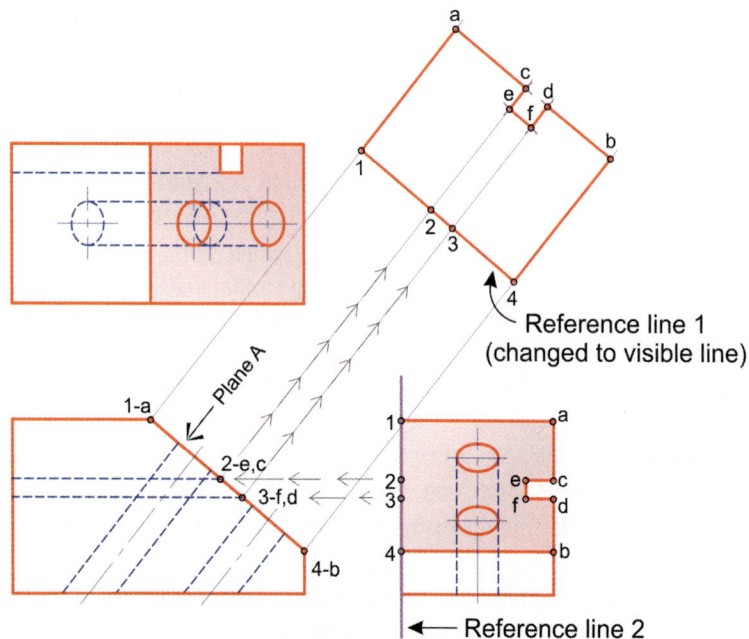

Figure 9-12   Connecting Points a,c,e,f,d and b in auxiliary view

Step 26. To locate the centers of holes marked *g* and *h* on the right-side view in Figure 9-13 to their positions in the auxiliary view, set the Projections layer current.

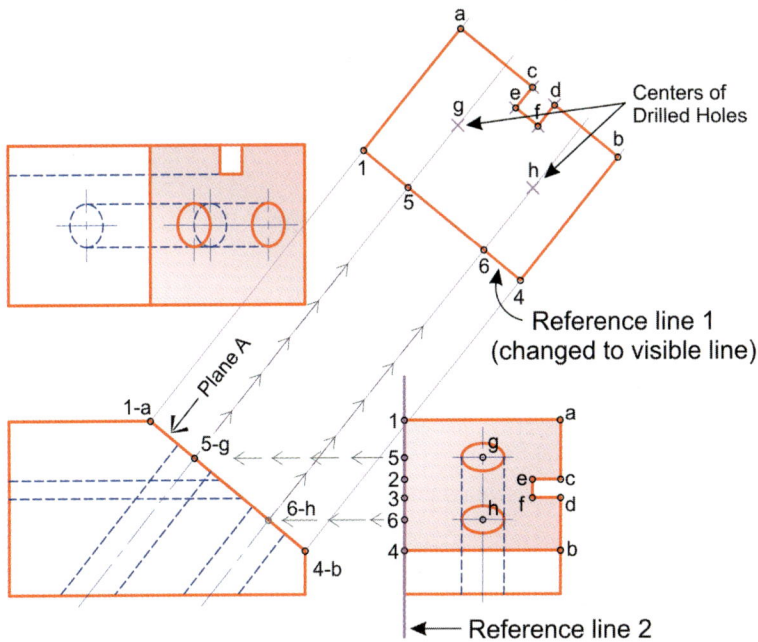

Figure 9-13   Locating Points g and h in auxiliary view

Step 27. Draw a construction line **12″** long from the point labeled *5-g* in the front view that is perpendicular to *Reference line 1* as shown in Figure 9-13.

Step 28. Repeat Step 27, but this time, draw a construction line **12″** long from the point labeled *6-h,* in the front view that is perpendicular to *Reference line 1* as shown in Figure 9-13.

Step 29. Use the **DISTANCE** command to find the horizontal distance between *Reference line 2* and *Point g* in the right-side view.

Step 30. Select the **POINT** command from the **Draw** toolbar. When prompted with Specify a point:, select the **Snap From** icon on the **Object Snap** toolbar and select the point where the angled line drawn from *Point 5-g* in the front view intersects *Reference line 1*. When prompted with from Base point, move the cursor to the top end of the angled projection line until the **Endpoint** object snap lights up (do not select the point), type the distance between *Reference line 2* and *Point g* in the right-side view, and press <**Enter**>.

Step 31. Repeat the **POINT** command as in Step 30, except this time, when prompted with Specify a point:, select the **Snap From** icon on the **Object Snap** toolbar. Select the point where the angled line drawn from *Point 6-h* in the front view intersects *Reference line 1*. When prompted with from Base point, move the cursor to the top end of the angled projection line until the **Endpoint** object snap lights up (do not select the point), type the horizontal distance between *Reference line 2* and *Point h* in the right-side view (this distance is the same as between *Reference line 2* and *Point g*), and press <**Enter**>.

Step 32. Set the Visible layer current and draw two circles **.75″** in diameter at the points identified in Steps 26–31 for *g* and *h* as shown in Figure 9-14.

Step 33. Turn off the Reference and Projections layers. Add center lines to the holes in the auxiliary view. Complete the view by adding center lines between the front view and the holes in the auxiliary view as shown in Figure 9-15.

Technically, the view shown in Figure 9-15 would be considered a ***partial auxiliary view*** because the other planes of the object are not shown in this view. If the drafter wished, broken lines could be added to the view to show more of the object as in Figure 9-16.

partial auxiliary view: An auxiliary view that is simplified by omitting planes and other features that are not shown true shape in the view.

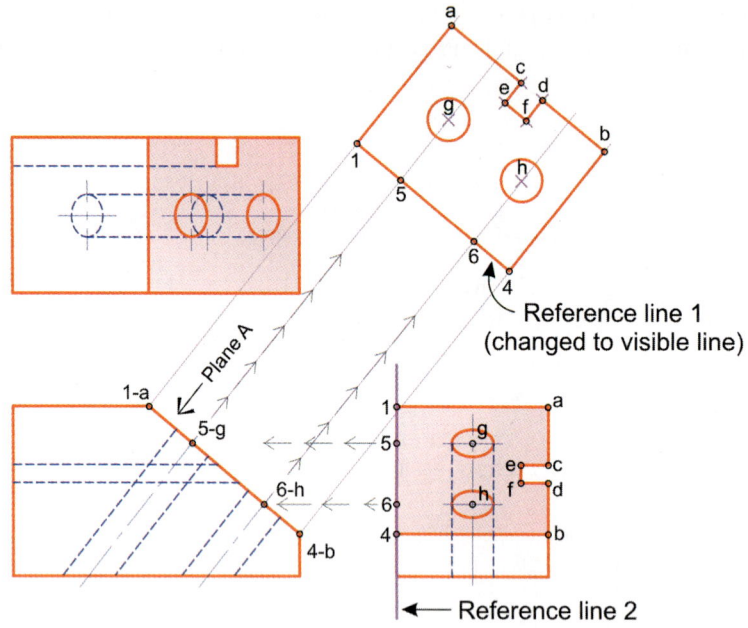

Figure 9-14 Drawing circles at Points g and h in auxiliary view

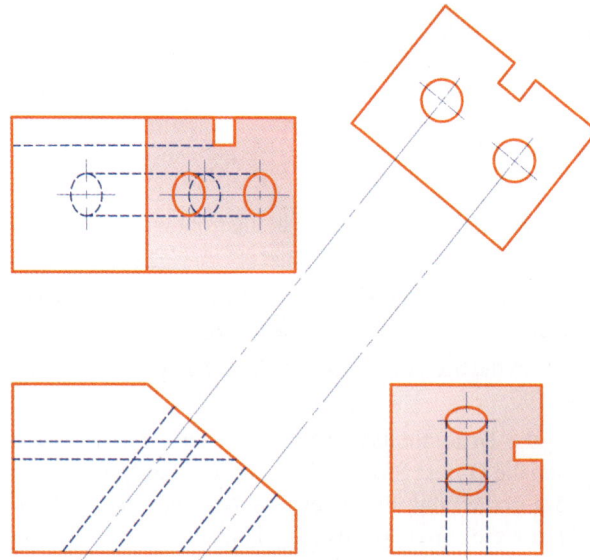

Reference line 1 (changed to visible line)

Reference line 2

Figure 9-15 The completed auxiliary view of Plane A

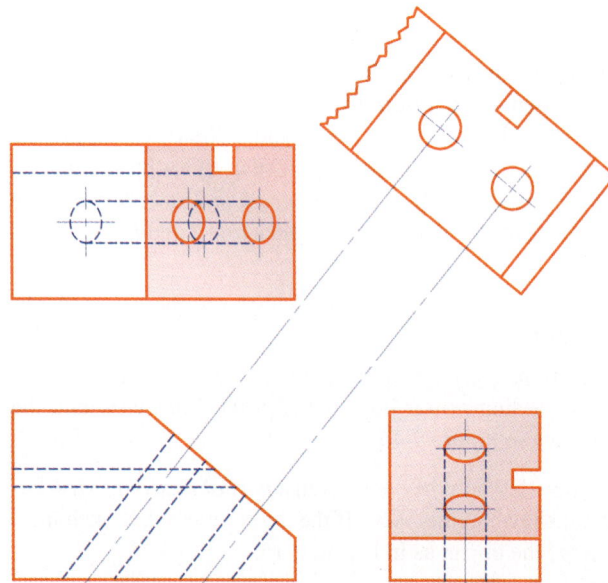

Figure 9-16 Partial auxiliary view showing broken lines

CONSTRUCTING AN AUXILIARY VIEW WITH THE OFFSET COMMAND—

A simpler, and quicker, way to construct an auxiliary view takes advantage of AutoCAD's **OFFSET** and **TRIM** commands. Also, instead of the **DISTANCE** command, the **LINEAR DIMENSION** command is used to measure distances in the right-side view.

The tutorial accompanying Project 9-2 shows how to quickly create an auxiliary view using these CAD Commands.

PROJECT 9-2: CONSTRUCTING A PRIMARY AUXILIARY VIEW—QUICK CAD METHOD

In this project, you will use the quick CAD method to create the primary auxiliary view for the object shown in Figure 9-3.

Directions

1. Select **FILE** and **OPEN**.
2. Select the **Student Resource CD**.
3. Select the **Prototype Drawings** folder.
4. Select the **Auxiliary View Prototype** drawing (open as "read-only").
5. **SAVE AS** to your **Home** directory and rename the new drawing **QUICK AUXILIARY VIEW PROJECT**.
6. Follow Steps 1 through 10 to create a primary auxiliary view of the inclined plane.

Step 1. Create a new layer named Projections and set the Visible layer current. Select the **OFFSET** command and when prompted to Specify offset distance, type **6.5** and press <Enter>. Select the line marked *Plane A* in the front view as the object to offset and press <Enter>. Pick a point above the right-side view when prompted to specify point on side to offset. The resulting line represents the left edge of the auxiliary view. See Figure 9-17.

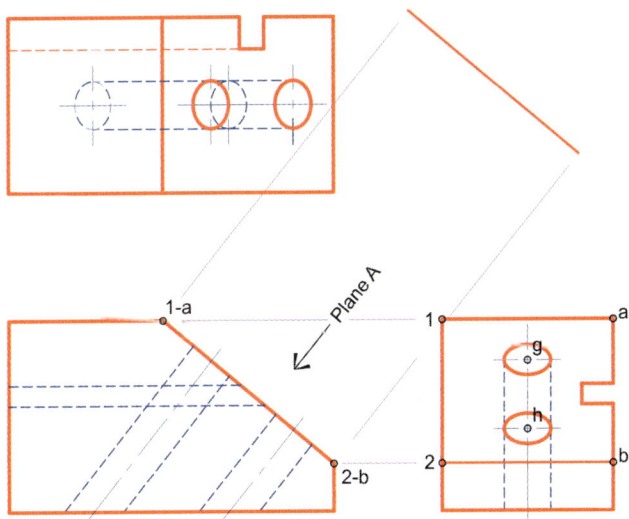

Figure 9-17   Offsetting line 1-a, 2-b

Step 2. The width of the object in the right side view is **2.75″**. Select the **OFFSET** command and type **2.75** as the offset distance. Select line created in Step 1 and offset it to the right side. The resulting line represents the right edge of the auxiliary view as shown in Figure 9-18.

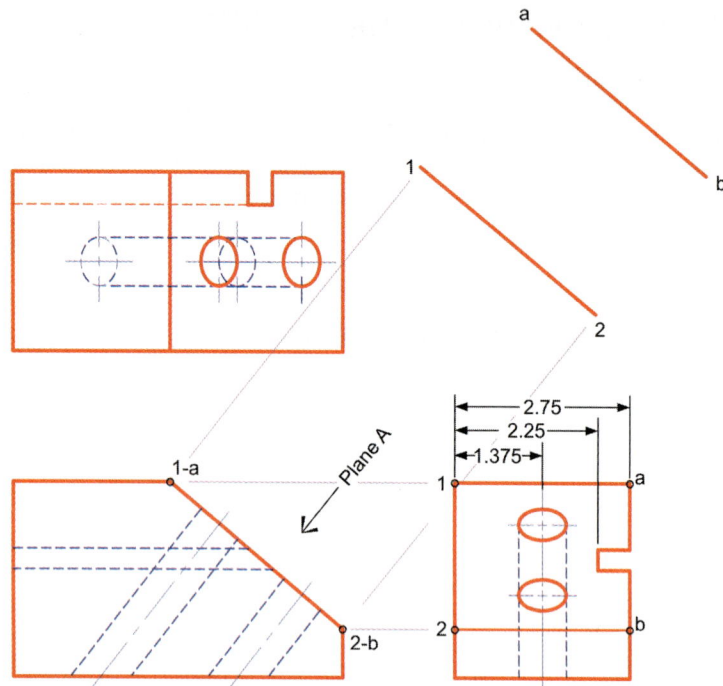

Figure 9-18   Offsetting line 1,2 to create line a,b

Step 3. Draw a line connecting *Point 1* and *Point a* in the auxiliary view as shown in Figure 9-19. Draw another line connecting *Point 2* and *Point b*.

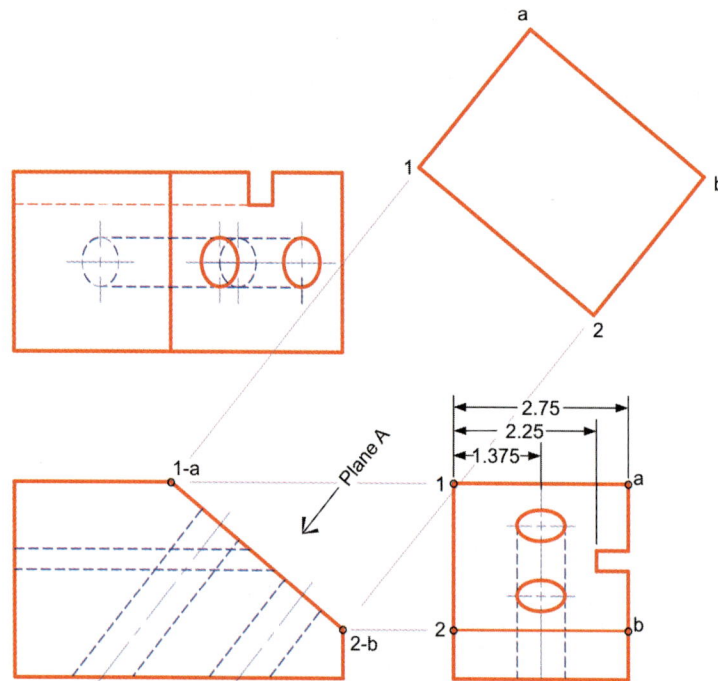

Figure 9-19   Connecting Point 1 to a, and Point 2 to b

Step 4. Select the **OFFSET** command and pick *Line 1-2* in the auxiliary view as the line to offset, enter **1.375** as the offset distance (the distance shown in the right-side view from the part's left edge to the center of the holes). Pick to the right of *Line 1-2* when prompted to specify point on side to offset. See Figure 9-20.

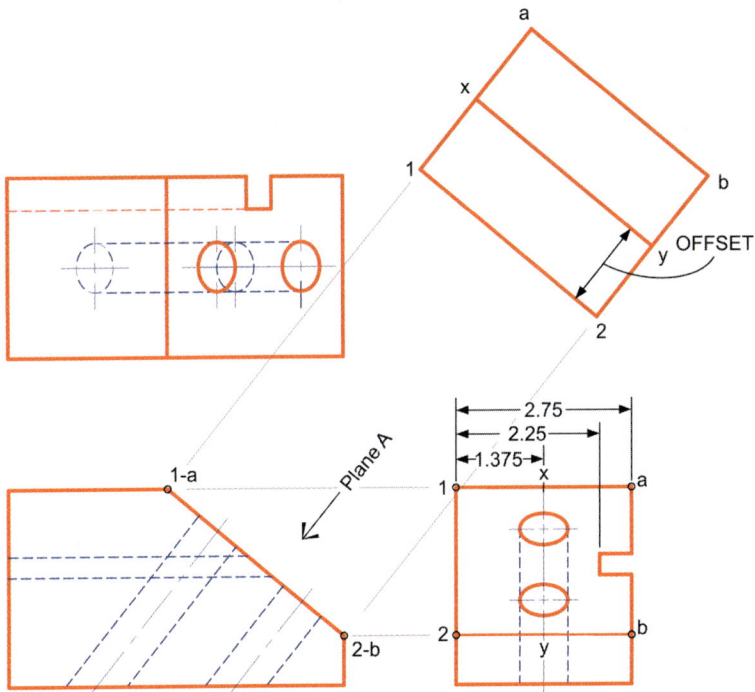

Figure 9-20   Offsetting line 1,2 to create line x,y

Step 5. Set the Projections layer current and draw a line from *Point 3-g* on *Plane A* perpendicular to *Line x-y* in the auxiliary view as shown in Figure 9-21. Repeat this step, drawing a
line from *Point 4-h* perpendicular to *Line x-y*.

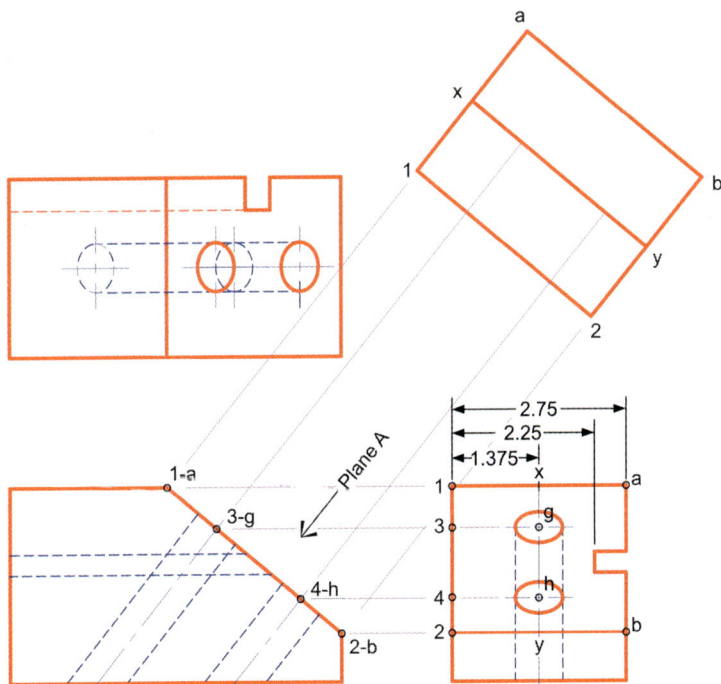

Figure 9-21   Projecting Points 3-g and 4-h to line x,y

Step 6. The points where *Line x-y* and the projection lines from *Points 3-g* and *4-h* intersect are
the centers for the two drilled holes. Set the Visible layer current and draw a **.75″** diameter circle at each intersection as shown in Figure 9-22.

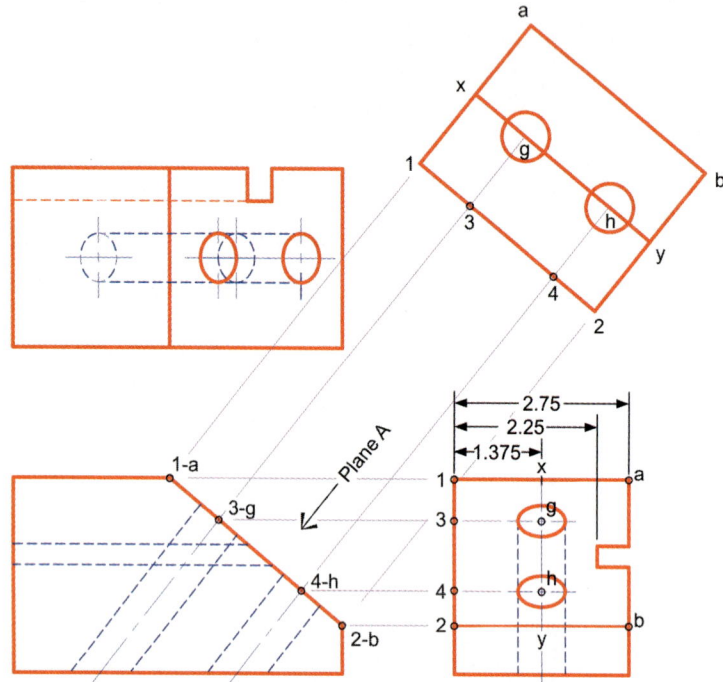

Figure 9-22   Adding circles at Points g and h

Step 7. Select the **OFFSET** command and type **2.25** for the distance to offset. Select *Line 1-2* and offset it to the right of the line as shown in Figure 9-23.

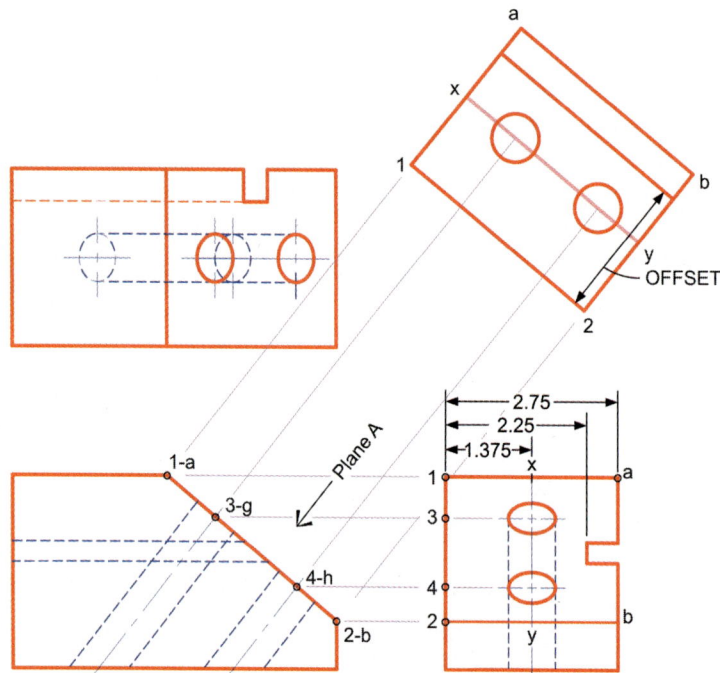

Figure 9-23   Offsetting line 1,2 2.25 inches

Step 8. Set the Projections layer current and project two construction lines from the intersections of *Points 5-e,c* and *6-f,d* perpendicular to *Line a-b* in the auxiliary view as shown in Figure 9-24.

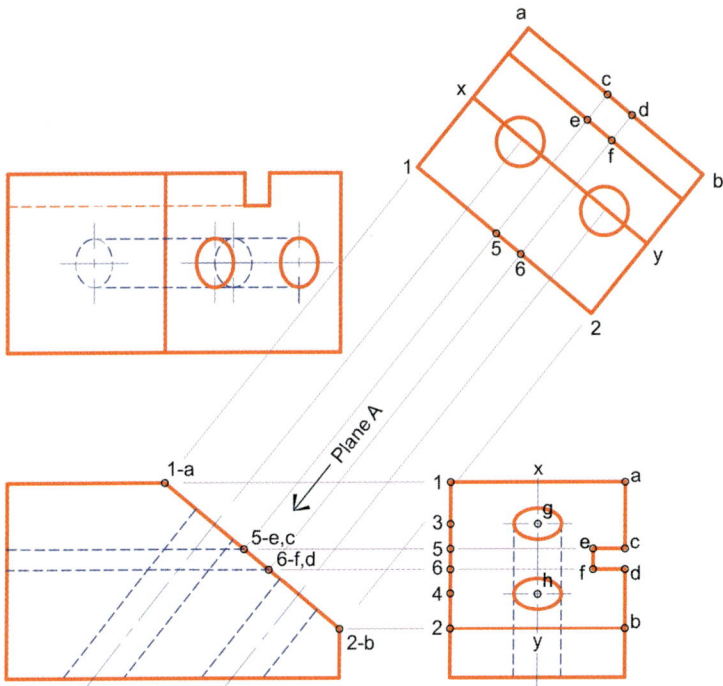

Figure 9-24   Projecting Points 5-e,c and 6-f,d to locate Points e and f

Step 9. Use the **TRIM** command to trim the lines representing the slot in the auxiliary view as shown in Figure 9-25. Set the Visible layer current and draw *Lines c-e* and *d-f*. Erase *Line x-y*.

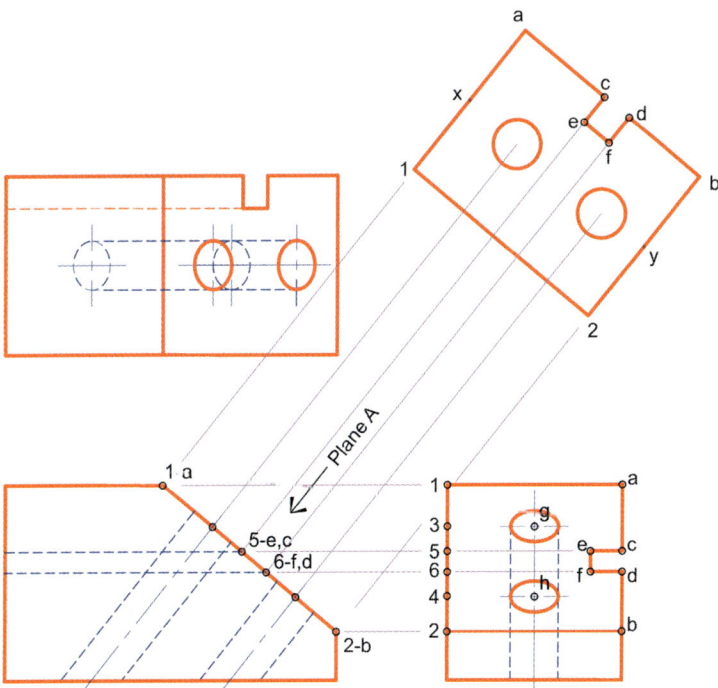

Figure 9-25   Connecting lines c,e,f, and d and trimming line between Points c and d

Step 10. Turn off the Projections layer and complete the drawing by adding center lines to the holes and between the front view and the holes in the auxiliary view as shown in Figure 9-26. This step completes the construction of the auxiliary view.

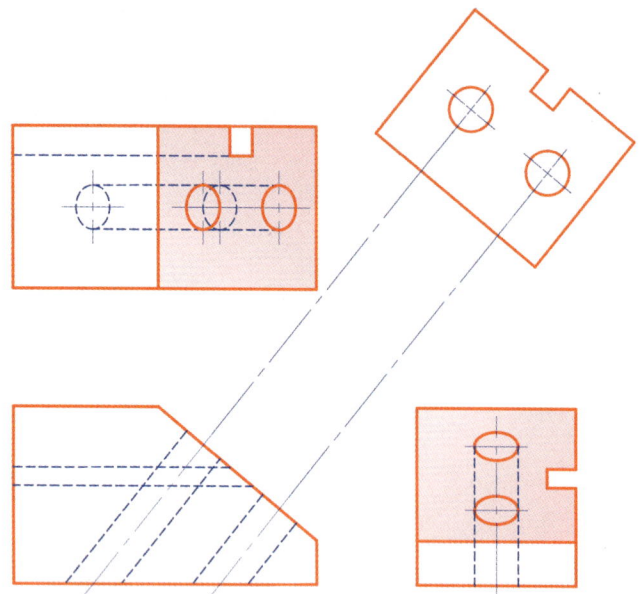

Figure 9-26   The completed auxiliary view of Plane A

SUMMARY

The techniques presented in this unit have their origins in Gaspard Monge's *Geometrie Descriptive*, written in the eighteenth century. As mentioned in Unit 1, many of Monge's ideas later became the foundation of modern technical drawing.

However, with the advent of CAD software capable of 3D modeling, auxiliary views would more likely be created by rotating a 3D model of the object until the inclined plane is perpendicular to the projection plane.

UNIT TEST QUESTIONS

Multiple Choice

1. In an auxiliary view, the plane of projection is _____ relative to the inclined plane.

    a. Oblique
    b. Parallel
    c. Skewed
    d. Perpendicular

2. A primary auxiliary view is drawn adjacent to, and aligned with, a(n) _____ view.

    a. Principal
    b. End
    c. Secondary auxiliary
    d. Perpendicular

3. An auxiliary view may be needed if the features of an object do not appear true shape in

    a. An oblique view
    b. An isometric view

    c. Any of the regular views
    d. All of the above

4. A partial auxiliary view omits

    a. Phantom lines
    b. Cutting plane lines
    c. Foreshortened features of the object
    d. None of the above.

5. What linetype is assigned to the lines that are drawn between the principal view and an auxiliary view?

    a. Center
    b. Phantom
    c. Hidden
    d. Visible

OPTIONAL UNIT PROJECT

Project 9-3: Auxiliary View

Use the quick CAD method to create a partial, primary auxiliary view for the object in the prototype drawing specified in the directions below. Project the view from the inclined plane in the front view of the object.

Directions

Student Files

1. Select **FILE** and **OPEN**.
2. Select the **Student Resource CD**.
3. Select the **Prototype Drawings** folder.
4. Select the **Auxiliary View Prototype 2** drawing (open as "read-only").
5. **SAVE AS** to your **Home** directory and rename the new drawing **AUXILIARY VIEW PROJECT 3**.

References

ASME Y14.3M-1994 Multiview and Sectional View Drawings

Blocks

10

Unit **Objectives**

- Describe what blocks are and how they are used in technical drawings created with AutoCAD.
- Create, insert, and edit blocks with AutoCAD software.
- Create a block library of architectural symbols and use them to produce a floor plan.

INTRODUCTION

A **block** is an AutoCAD term that refers to a pre-drawn object that is stored in the AutoCAD drawing file which can be placed or inserted into the drawing whenever it is needed. For example, an architectural firm may create a **block library** of door and window symbols. Later, when a door or window symbol is needed in a floor plan, the firm's drafters can select the symbol from the block library and insert it into the drawing rather than drawing each door and window from scratch. This can result in saving drafters a huge amount of time and lower the cost of producing a drawing. Another advantage of using blocks is that when inserting a block, the drafter can change its scale, proportion, and rotation without redrawing the object.

Sometimes, architectural and engineering firms purchase pre-made block libraries from vendors, or in some cases, download them directly from vendor's web-sites.

block: AutoCAD term that refers to a predefined object, or symbol, that is stored in the AutoCAD drawing file which can be inserted into the drawing whenever it is needed.

block library: A block library is a group of block definitions stored in a drawing file. For example, an architectural firm might create a block library of symbols like doors and windows that are frequently used on floor plans. See *block*.

CONSIDERATIONS FOR CREATING BLOCKS

An important consideration when you create a block is the layer that the entities of the block are drawn on prior to creating the block. This is important because objects that are on Layer **0** when the block is created will assume the properties (color, linetype, etc.) of the layer that is current when the block is inserted into the drawing. Conversely, if the entities of a block are on a layer other than **0** when the block is created, the block will retain the characteristics (color, linetype, etc.) of the original layer, even if the block is inserted on a layer with different layer properties.

Another important consideration in block creation is the definition of the block's **base point**. The base point is the point on the block that aligns to the location of the **insertion point** that AutoCAD prompts you to define when the block is inserted into a drawing. For example, if you were making a block of the door shown in Figure 10-1, the corner of the door where the hinge would meet the door's opening would be a logical base point. With this corner defined as the door's base point, the hinged corner of the door can be placed precisely into the floor plan by snapping to the corner of the door opening when you are prompted to define the block's insertion point.

Figure 10-2 shows the drawing for a ceiling fan. When making a block of the fan, a logical base point would be its center. So when prompted to define the base point of the fan, you would select the center of the circle around which the blades are arrayed. Later, when the fan is inserted into the drawing, the fan's center point will align to the pick point, or coordinate, that you define when prompted by AutoCAD to define the block's insertion point.

base point: The point on a block that aligns to the insertion point in the field of the drawing defined when placing the block.

insertion point: The point defined in the field of the drawing that a block's base point will align to upon insertion.

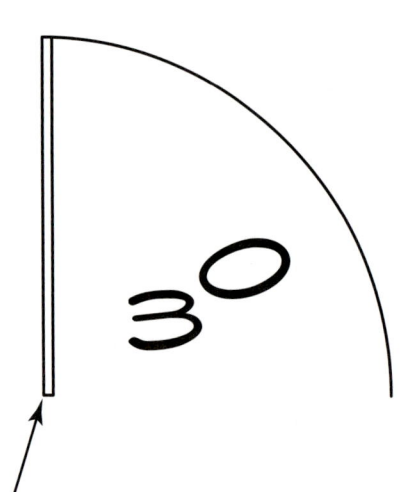

Select this corner to be the base point for the
block of the door.

Figure 10-1   Defining the base point of a
block of a door

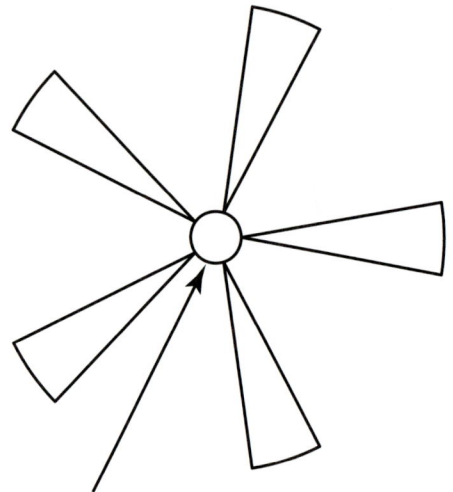

Select the center of this circle as the
base point for the ceiling fan block.

Figure 10-2   Defining the base point of a
block of a ceiling fan

CREATING BLOCKS

Step 1. Draw the object to be blocked.

> **TIP** The layer on which the entities of a block are drawn is very important. For example, objects that are on Layer 0 when the block is created will assume the color, linetype, and lineweight of the layer that is current when the block is inserted. Blocks created on a layer other than Layer 0 will retain the characteristics of that layer, even when inserted on a different layer.

Make Block

Figure 10-3   Block icon

Step 2. Type **BLOCK** or select the **MAKE BLOCK** icon from the **Draw** toolbar (see Figure 10-3). AutoCAD will display the **Block Definition** dialog box shown in Figure 10-4.

Step 3. Enter the block name in the **Name:** text box.

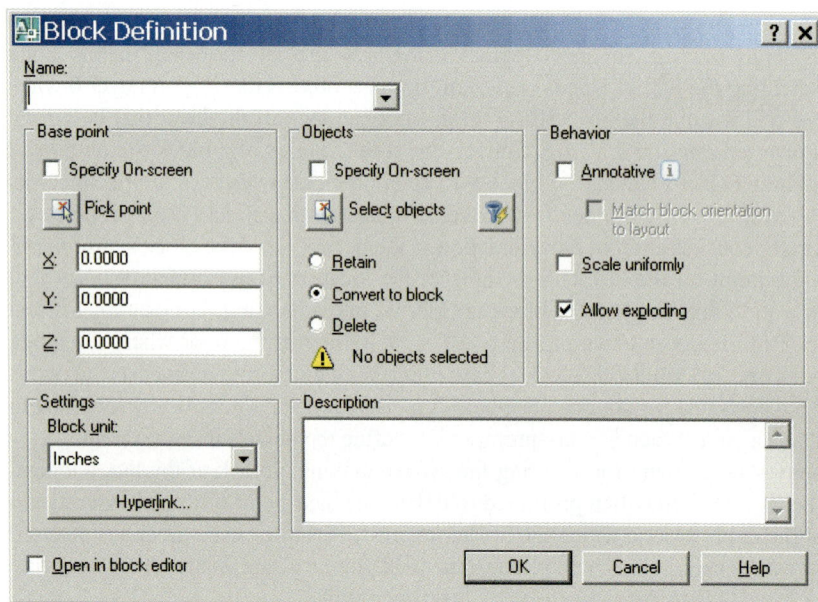

Figure 10-4   Block
Definition dialog box

Step 4. Specify the **Base point** of the block by picking on the **Pick point** button and selecting the desired base point on the object. Remember to select a point that would make a "logical" base point.

Step 5. Pick the **Select objects** button and select all of the objects that comprise the block. Press **<Enter>** when you are finished selecting objects.

Step 6. Checking the **Allow exploding** box allows the finished block to be exploded later if necessary.

Step 7. Click **OK** to exit the **Block Definition** dialog box.

INSERTING BLOCKS INTO A DRAWING

Figure 10-5 Insert icon

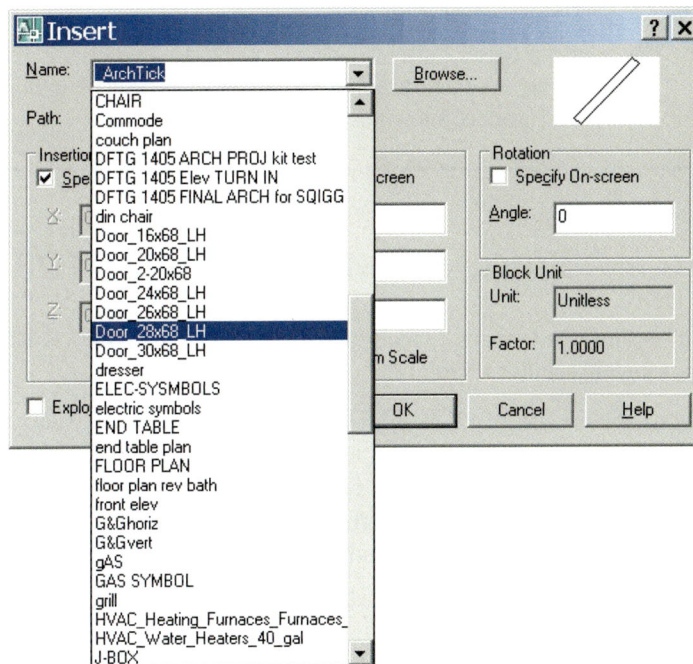

You place a block in a drawing by initiating the **INSERT BLOCK** command from the **Draw** tool bar (see Figure 10-5) or by typing **INSERT** at the command line. The steps involved in inserting a block into a drawing are detailed as follows:

Step 1. Select the **INSERT BLOCK** icon from the **Draw** toolbar or type **INSERT** and press **<Enter>** (see Figure 10-5).

Step 2. After the **Insert** dialog box opens (Figure 10-6), select the **Browse** button and choose the block you wish to insert from the drop-down list.

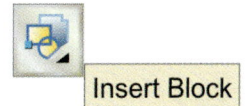

Figure 10-6 Insert dialog box

> **TIP**
>
> AutoCAD will allow you to insert an entire existing AutoCAD drawing into your current drawing. To do this, select the **Browse...** button from the **Insert** dialog box, and browse to the AutoCAD drawing file you wish to insert, select its file name, and click **OK**. The **0,0** point on the inserted drawing will default as its base point. The inserted drawing will behave like a block (it will be inserted as one entity) and will bring its layers, linetypes, and blocks into the current drawing.

Step 3. If the block needs to be rotated or scaled on placement, check the **Rotation Specify On-screen** box and/or the **Scale Specify On-screen** box in the **Insert** dialog box shown in Figure 10-7.

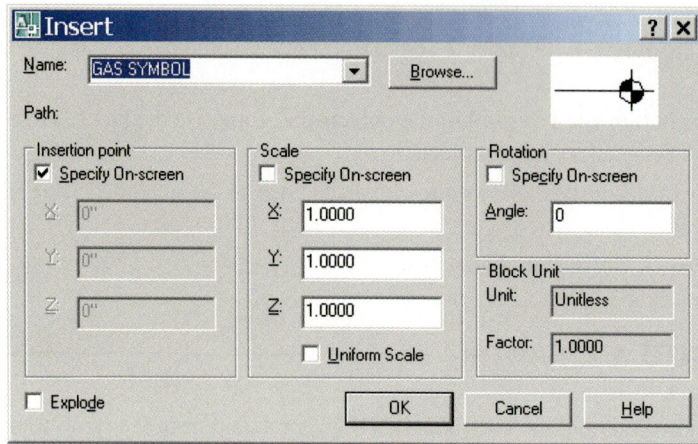

Figure 10-7   Insert
dialog box

Step 4. Click **OK** and select the insertion point on the drawing where you want the block's base point to be placed. You will also be prompted to define the block's scale and rotation if either of these boxes were checked in Step 3.

> TIP   In most cases, exact placement of the block is important—at the endpoint of a line, for example. For this reason, use an appropriate **OSNAP** setting to facilitate accurate block placement whenever possible.

EDITING BLOCKS WITH THE **BLOCK EDITOR** COMMAND

The easiest way to make changes to a block is with the **BLOCK EDITOR** command. The steps for using this command follow.

Step 1. Select the **BLOCK EDITOR** command from the **Tools** pull-down menu by double-clicking on a block that has already been inserted, or by typing **BEDIT** and pressing **<Enter>**.

Step 2. When the **Edit Block Definition** dialog box opens (see Figure 10-8[a]), choose the block name you wish to edit from the list on the left side of the box and click **OK**.

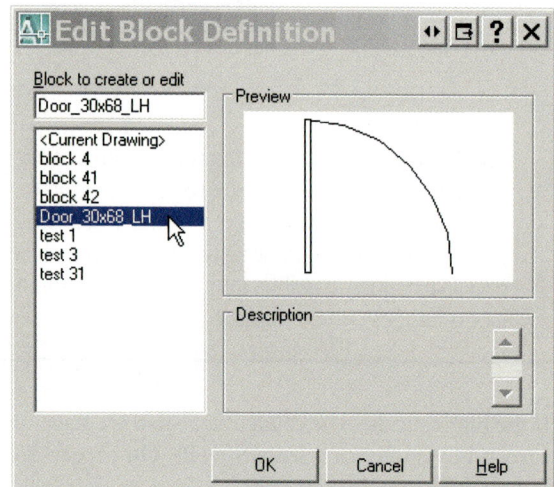

Figure 10-8(a)   Edit Block Definition dialog box

Step 3. A screen will open that displays the block you wish to edit on a yellow background (see Figure 10-8[b]). Using the tools from the **Draw** and **Modify** toolbars, make the desired changes to the block.

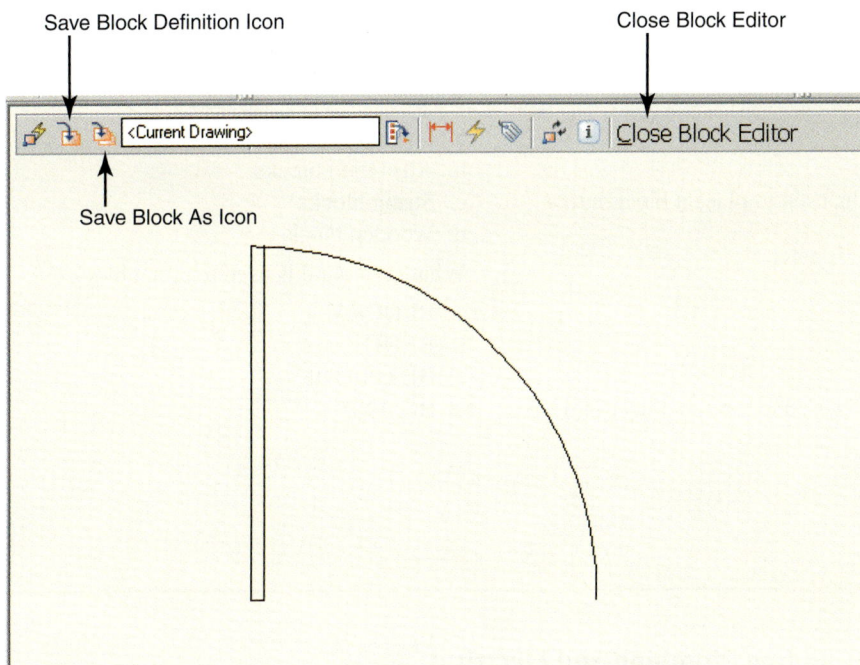

Save Block Definition Icon   Close Block Editor

Save Block As Icon

Figure 10-8(b)   Block Editor screen

Step 4. When you are finished editing the block, choose the **Save Block Definition** icon from the toolbar at the top of the yellow screen and then choose **Close Block Editor** from the same toolbar.

> **TIP**
>
> Blocks with the same name as the redefined block, that have been inserted into the drawing prior to editing, will also update to reflect the redefined properties. To avoid this, choose the **Save Block As** icon from the toolbar and give the edited block a different name.

> **TIP**
>
> In versions of AutoCAD prior to Release 2006, blocks are edited in one of two ways. The first is by inserting and exploding the block and making changes to it. It should be noted that block entities that were on Layer 0 when the block was created will revert back to Layer 0 after the block is exploded. Block entities that were on layers other than Layer 0 when the block was created will revert back to their original layers after the block is exploded.
>
> The second method employed to edit blocks is to use the **REFEDIT** command (type **REFEDIT** and press <**Enter**>). This technique is often covered in advanced CAD courses.

SUMMARY

The creation and use of blocks is another method employed by AutoCAD users to work more efficiently. This unit presented the basics of creation, usage, and editing of blocks. Other, more powerful, ways to use blocks, including the creation of *attributed blocks* (an attribute is a label or tag that attaches data to a block), and *dynamic blocks* (blocks that can be edited in place rather than redefining them), are usually covered in more advanced AutoCAD classes.

Later in this text, you will learn how to use an AutoCAD feature called *Design Center* to insert blocks from other drawings into the current drawing.

UNIT TEST QUESTIONS

Multiple Choice

1. What is a *block*?

   a. One set of a cube
   b. Pre-drawn object
   c. Group of drawing files
   d. None of the above

2. What AutoCAD command is used to place a block into a drawing?

   a. INSERT
   b. ADD
   c. COPY & PASTE
   d. All the above

3. What AutoCAD command is used to turn an object into a block?

   a. GROUP
   b. BLOCK

   c. Both A and B
   d. INSERT BLOCK

4. A label or tag that attaches data to a block is called what?

   a. Dynamic blocks
   b. Attributed blocks
   c. String blocks
   d. Wooden blocks

5. What command is used to edit a block?

   a. BLOCKX
   b. BEDIT
   c. BEXPLODE
   d. BCLOSE

UNIT PROJECT

Project 10-1: Adding Plumbing and Electrical Blocks to the Guest Cottage

In this project, you will add plumbing and electrical symbols to the floor plan of the guest cottage you created in Unit 4. This project will require you to create a block for each plumbing and electrical symbol. These blocks will be inserted into the floor plan in the locations shown in the designer's sketch in Figure 10-9.

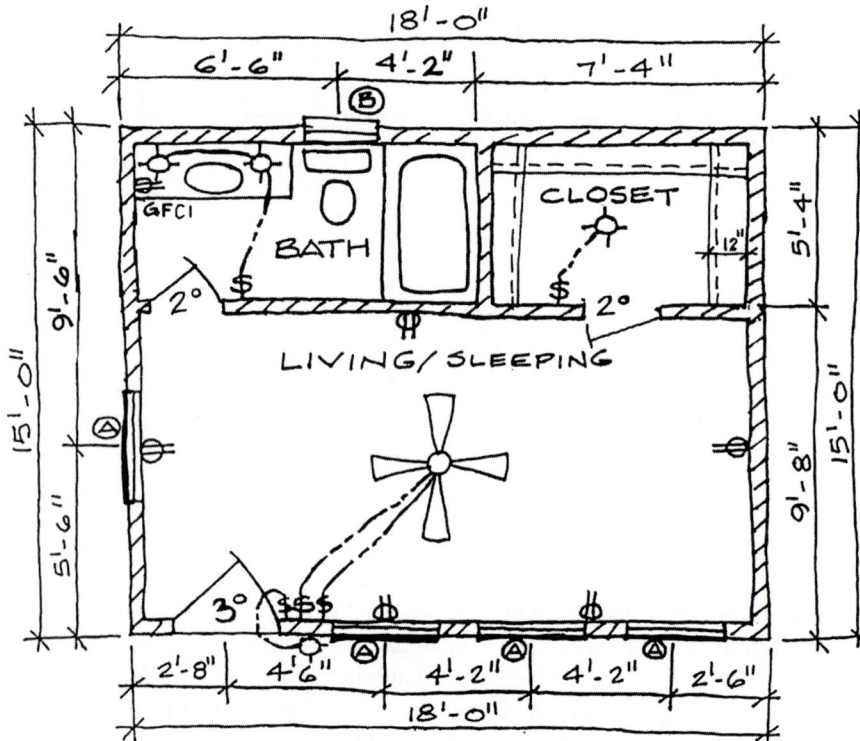

Figure 10-9   Designer's sketch of the guest cottage showing plumbing, electrical and wiring symbols

Directions

Student Files

1. Open the **Guest Cottage** drawing file you created in Unit 4.
2. Create the following layers: Plumbing, Electric, and Wiring. Assign a color to each layer and set the Wiring layer's linetype to **Phantom**.
3. Follow the steps on the next pages to create and place blocks for the plumbing and electrical symbols and to add switch lines between the switches, light fixtures, and ceiling fan.

Creating Blocks for the Guest Cottage

Follow the directions in Steps 1 and 2 to make blocks for the plumbing and electrical symbols. You will make five blocks in all. Insert the blocks into the **Guest Cottage** drawing in the locations shown in the designer's sketch in Figure 10-9.

TIP Draw and create the blocks on Layer 0. Insert the plumbing symbols on the Plumbing layer and the electrical symbols on the Electric layer.

Drawing the Plumbing Symbols

Step 1. To draw the tub, draw a **2′6″ × 5′0″** rectangle and Offset it **4″** to the inside. Use the **FILLET** command to fillet the corners of the inside rectangle. Fillet the top corners **R6″** and the bottom corners **R2″** as shown in Figure 10-10.

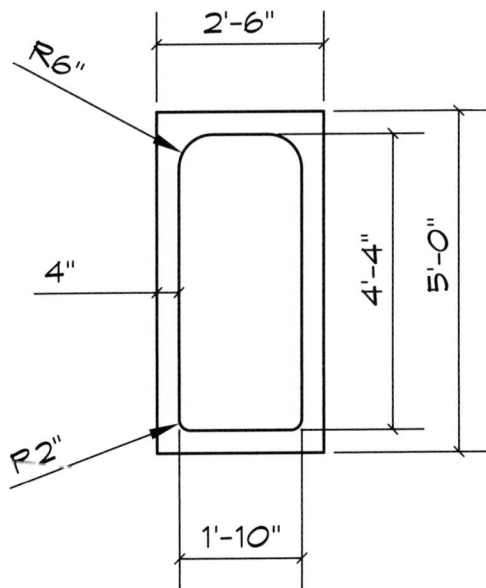

Figure 10-10   Tub dimensions

Step 2. Next, make a block for the tub, selecting the tub's upper right corner as the base point. Set the Plumbing layer current and insert the block of the tub into the corner of the bathroom as shown in Figure 10-9.

Step 3. To draw the lavatory, draw a **4′-6″ × 1′-10″** rectangle for the sink top. Next, select the **ELLIPSE** command, and when prompted for **specify axis endpoint**, pick a point and draw a horizontal line **1′-8″** long, then, at the length of the other axis prompt, type **7″** (1′-2″ divided by 2) to create the bowl. Move the center of the bowl to the location on the rectangle shown in Figure 10-11.

Step 4. Next, make a block for the lavatory, selecting the lavatory's upper left corner as the base point. Insert the block of the lavatory into the corner of the bathroom as shown in Figure 10-9.

Step 5. To draw the commode, draw a **1′8″ × 6″** rectangle for the toilet tank. Next, draw a vertical line from the midpoint of the top edge of the rectangle to a point **1′-8″** below the first point. Select the **ELLIPSE** command and draw a horizontal line **1′-4″** long; then, at the length of the other axis prompt, type **10″** (1′-8″ divided by 2) to create the toilet bowl. Move the ellipse by its center point to the endpoint of the **1′-8″** line drawn earlier and erase the line. See Figure 10-12.

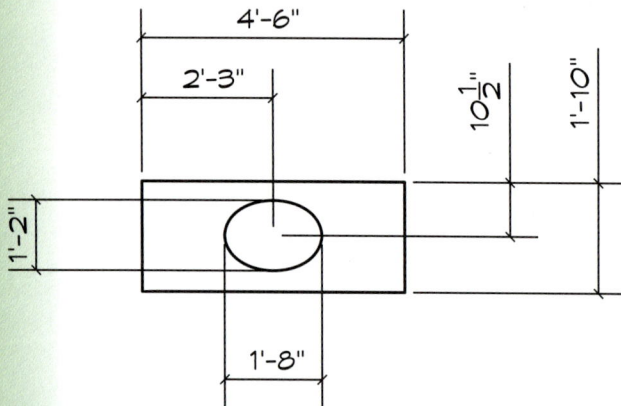

Figure 10-11   Lavatory dimensions

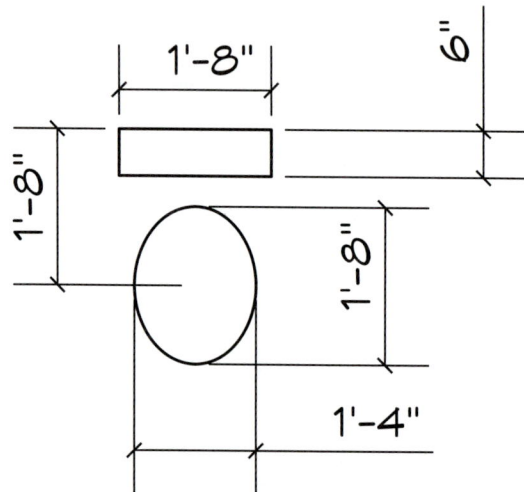

Figure 10-12   Commode dimensions

Step 6. Next, make a block for the commode, selecting the midpoint of the commode tank's top edge as the base point. Insert the block of the commode into the **Guest Cottage** floor plan by centering it between the right edge of the lavatory and the left edge of the tub as shown in Figure 10-9. Move the commode **2″** away from the inside edge of the wall.

Drawing the Electrical Symbols

Step 1. To create the ceiling fan block, draw the fan blade and circle shown in Figure 10-13. Construct the fan inside a circle **3′** diameter.

Step 2. Perform a polar **ARRAY**, selecting the fan blade as the object to array and the center of the **4″** circle as the center point of the array. Enter **5** as the total number of items and **360** as the angle to fill. Check the box for **Rotate items as copied**. The finished fan should look like the one shown in Figure 10-14. Make a block of the fan, choosing the center of the array as the base point. Set the Electric layer current and insert the fan in the location shown in Figure 10-9.

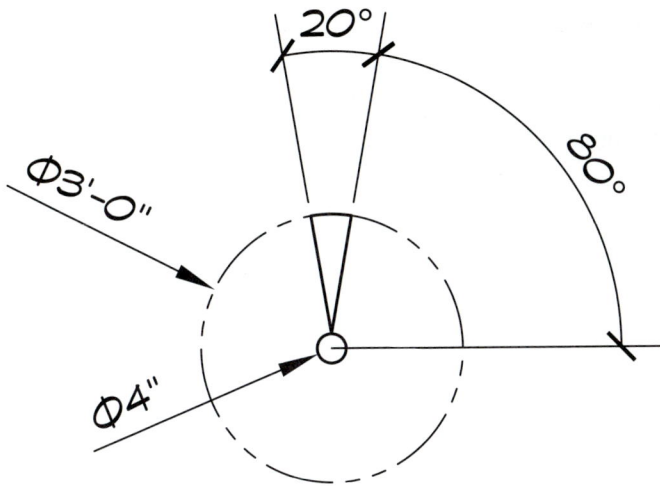

Figure 10-13   Fan blade construction

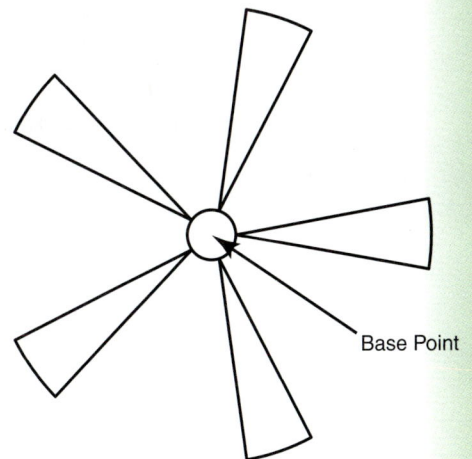

Figure 10-14   Fan blade array

Step 3. To draw the light fixture, draw a **6″** diameter circle and then draw a **1-1/2″** line from each of the circle's quadrants as shown in Figure 10-15. Make a block of the light fixture, choosing the center of the circle as the base point. Insert the light fixture block into the plan as shown on the sketch of the floor plan in Figure 10-9.

Step 4. To add switches and outlets, you can use the blocks for the switches and outlets that are included in the prototype drawing for the **Guest Cottage** drawing. Use the **INSERT** command to place the electrical outlets (see Figure 10-16) and the switches (see Figure 10-17) from the list of blocks. Refer to the floor plan sketch in Figure 10-9 for placement of these symbols.

Set the Labels layer current and label the outlet in the bathroom GFCI using **3″** text as shown in Figure 10-9. GFCI indicates that the type of outlet is a *Ground Fault Circuit Interrupt*. GFCI outlets are required near plumbing fixtures where an electric appliance, like a hair dryer, might come in contact with water and pose a risk of electric shock.

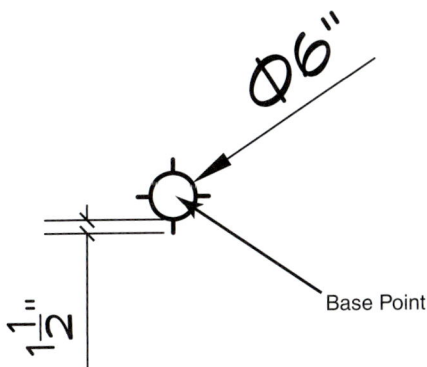

Figure 10-15   Light fixture dimensions

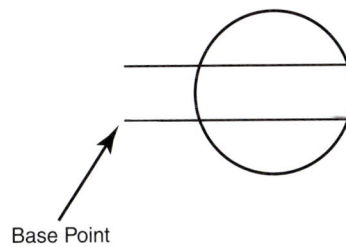

Figure 10-16   Electrical outlet symbol

Figure 10-17   Wall switch symbol

Adding the Switch Lines

Step 1. Set the Wiring layer current and use the **SPLINE** command to create the switch lines as shown in Figure 10-18.

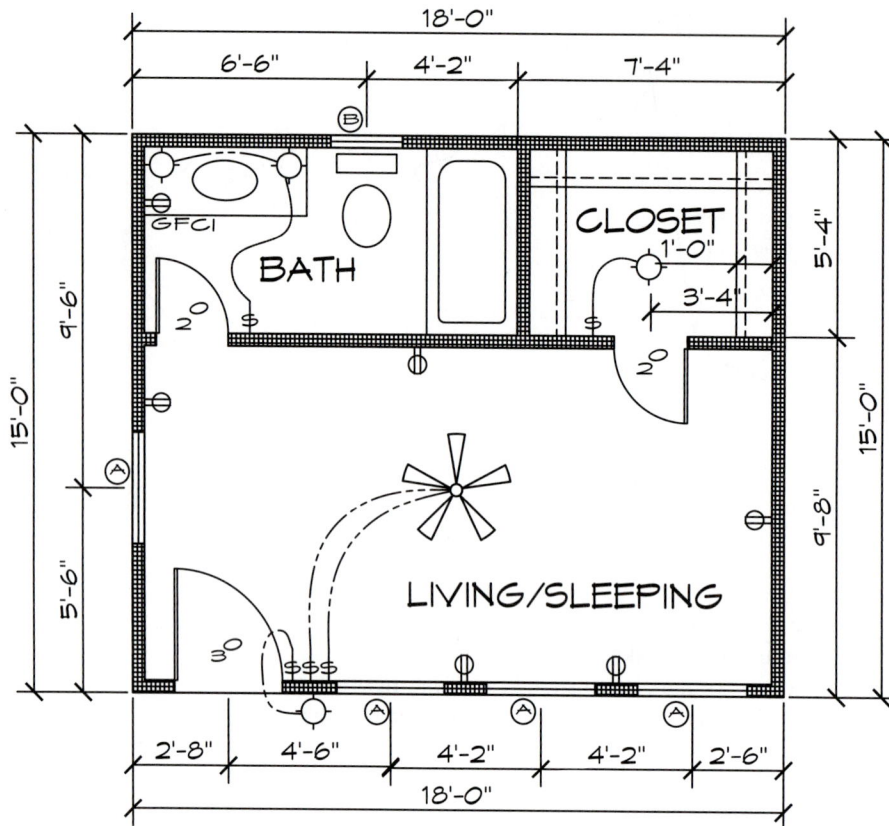

Figure 10-18   Switch lines added to lights and ceiling fan

Step 2. You may need to set the **LTSCALE** (Linetype Scale) to a larger value in order for dashes to appear in the phantom lines. **LTSCALE** controls the size of dashes in noncontinuous lines such as hidden, center, and phantom lines. To change the **LTSCALE**, type **LTSCALE** and press <Enter>, enter a new value (the default value is 1), and press <Enter>. To change the **LTSCALE** of only one line, select the line, right-click, and select **Properties** from the menu; this will open the **Properties** dialog box. In the field next to **Linetype Scale**, set the value to a larger number.

This step completes the **Guest Cottage** project. Follow your instructor's directions to print the drawing. Be sure to save the drawing file when you close AutoCAD.

OPTIONAL UNIT PROJECT

Project 10-2: FM Tuner

Draw the schematic of the FM Tuner shown in Figure 10-19. The schematic symbols needed in this drawing have already been drawn, but you will need to create a block for each symbol.

Directions

1. Select **FILE** and **OPEN**.
2. Select the **Student Resource CD**.
3. Select the **Prototype Drawings** folder.
4. Select the **Electronic Schematic Template** drawing (open as "read-only").

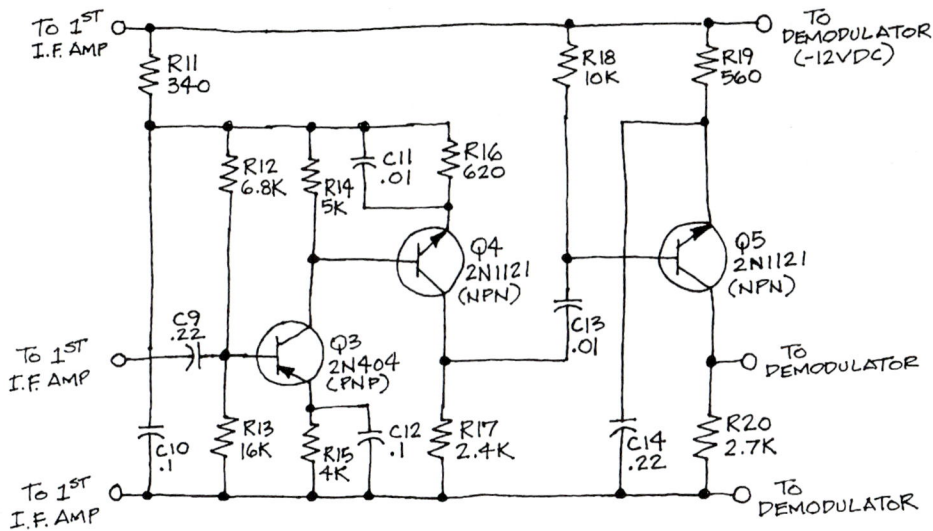

Figure 10-19   Designer's sketch of the FM tuner schematic

5. **SAVE AS** to your **Home** directory renaming the new drawing as **FM TUNER**.
6. Create two new layers named Circuit and Symbols, and following the next steps, draw and label the schematic diagram shown in Figure 10-19.
7. Place Notes 1 and 2 on the Text layer in the area to the left side of the title block using .125″ text height.

Creating Blocks for the FM Tuner

Make a block for each of the symbols shown on the prototype drawing. There will be six blocks in all. Use the block names shown in Figure 10-20 and select the point displayed in red that accompanies each drawing as the block's base point.

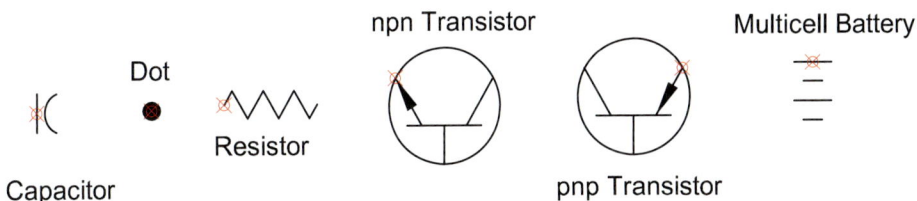

Figure 10-20   Schematic symbols

> **TIP**
> You will need to have the **Node** object snap setting selected in order to snap to the point but do not include the point when selecting the objects to be included in the block.

When you have finished making the blocks, turn the Nodes layer off and erase the symbols remaining on the screen.

Drawing the Schematic Diagram

Step 1. Set the Circuit layer current and use **OFFSET** or rectangular **ARRAY** to draw a grid like the one shown in Figure 10-21

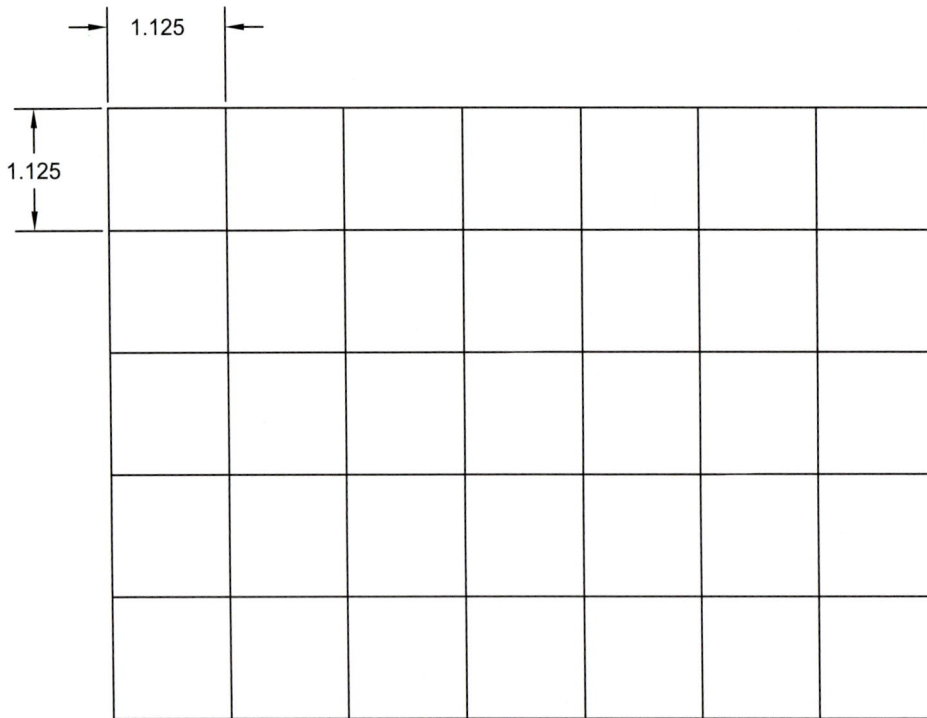

Figure 10-21   Construction of a schematic grid

using **1.125″** spacing between lines. Begin the lower left corner of the grid at absolute coordinate **1.25,2.5**.

Step 2. Make the Symbols layer current and insert the blocks on the grid lines as shown in Figure 10-22. Visually center the symbols between the lines. It may be necessary to move some lines at this point to reflect the sketch of the circuit accurately.

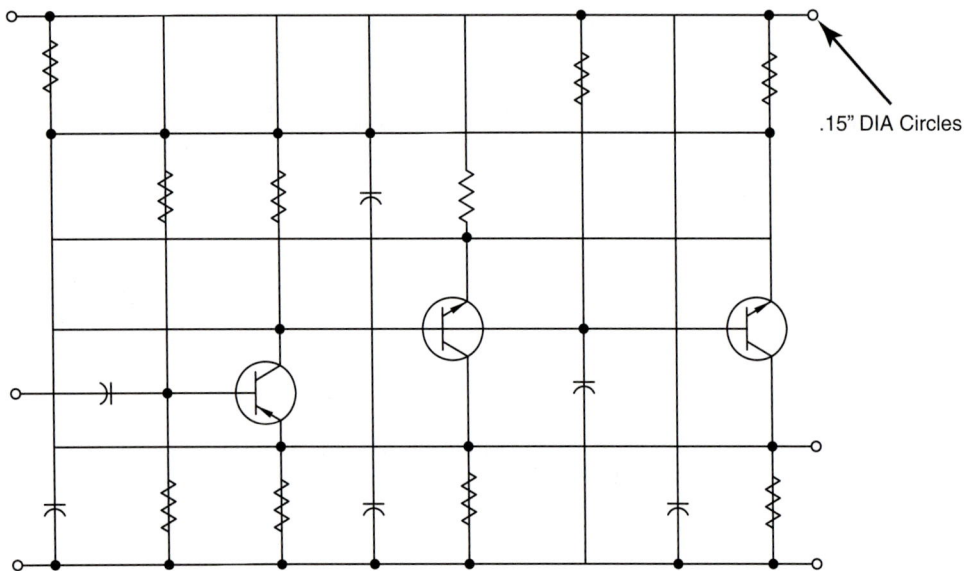

Figure 10-22   Block placement

Step 3. Trim the lines that run through the blocks of the electronic symbols. Next, trim or delete other layout lines as needed to create the circuit configuration shown in Figure 10-23.

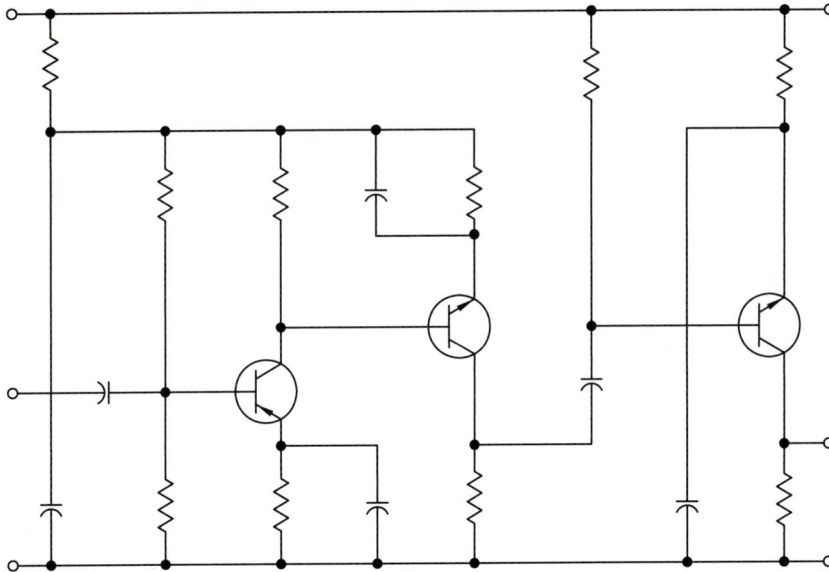

Figure 10-23   Trimming circuit lines

Step 4. Refer to the sketch of the schematic shown in Figure 10-19 and add **.125″** labels on the Text layer as shown in Figure 10-24.

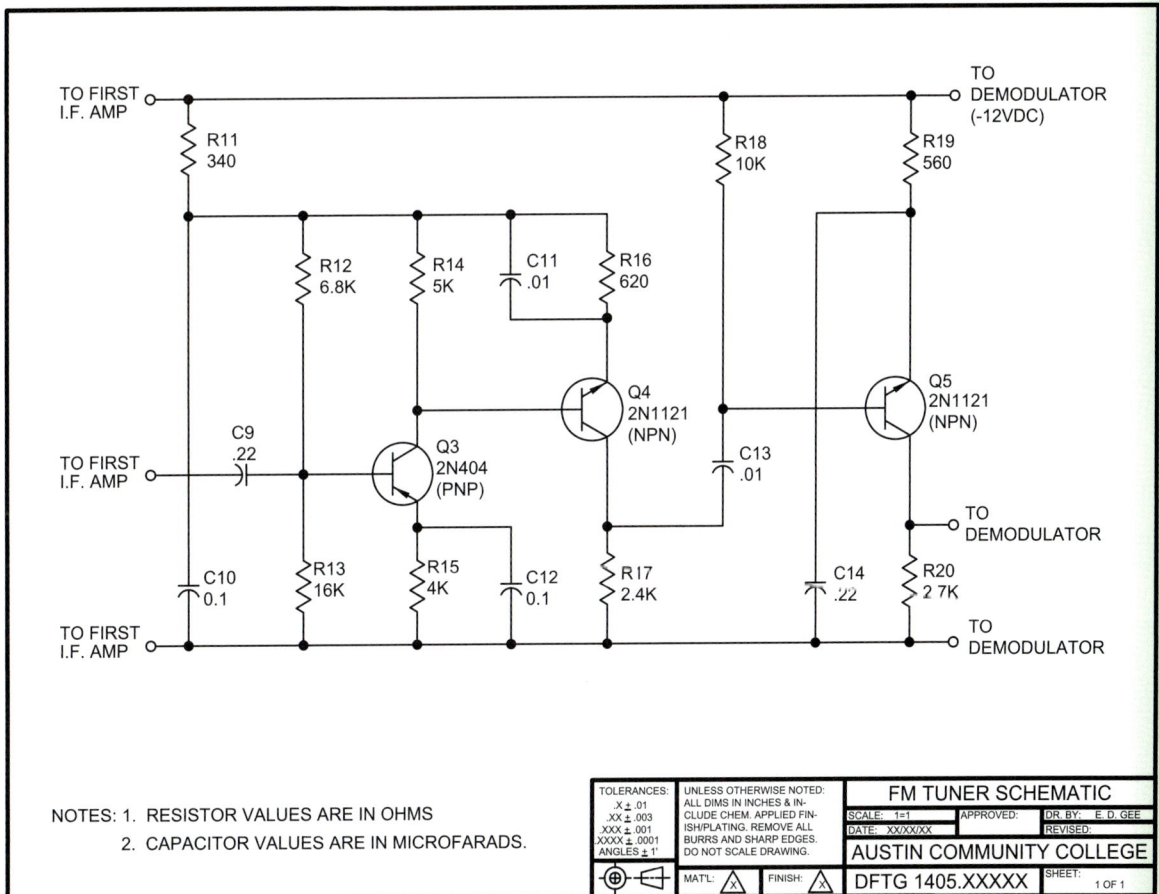

Figure 10-24   Finished FM tuner schematic diagram

TIP

Text height should be **.125″** for both labels and notes.

This completes the FM Tuner project. Follow your instructor's directions to print the drawing. Be sure to save the drawing file when you close AutoCAD.

Mechanical Working Drawings

11

Unit Objectives

- Describe what mechanical working drawings are and how they are produced.
- Use AutoCAD to create an exploded, isometric assembly view of a mechanism including balloons, part numbers, and a parts list.
- Use AutoCAD to create detail drawings of mechanical parts, including all of the necessary multiviews, dimensions, and notations required to manufacture each part.
- Represent and specify fasteners and other hardware in a mechanical working drawing.

INTRODUCTION

In the mechanical engineering field, drafters are often required to create complex sets of **mechanical working drawings** for entire mechanical assemblies. To better understand the creation of these drawings, it is helpful to step back and take a wider view of the mechanical design process itself.

    For most mechanical designers, the first phase of a design project involves clearly defining the design problem and specifying the criteria that the finished design must meet to be considered a success. As stated earlier in Unit 1, designers often refer to this phase in the design process as *problem identification*. For example, before beginning a design for a machine part, a mechanical designer must have a clear understanding of the following: the function the part serves; the ability of the part to work in conjunction with other parts; an idea of the shape, size, strength, and material of the finished part; any safety and reliability concerns the part may present; and an estimated budget for the project.

    After the design problem is clearly defined, the *preliminary design* phase begins. This stage is also referred to by designers as the *ideation*, or *brainstorming*, phase of the process. During this phase, multiple solutions to the design problem are generated. The documentation for these preliminary designs may be in the form of freehand sketches or 2D and 3D CAD models.

    In the next phase, the preliminary designs are analyzed and evaluated by the design team to decide which one best meets the design criteria defined in the first phase. During this process, the design may be further refined, and the best features of some of the rejected designs may be incorporated into the final design solution. The analysis of the designs may involve computer modeling or the preparation of actual prototypes of the part that are subjected to performance testing. It is not unusual for a design to go through many revisions, or *iterations*, during this phase.

    After the team decides on the best solution, the designer begins preparing design inputs that more clearly define the details of the project. Design inputs may include freehand sketches or CAD models that provide dimensional information and detailed notes about the project.

    When the design inputs are finished, they are given to the drafter(s) responsible for preparing the working drawings for the project. In this phase, drafters usually work closely with designers, checkers, engineers, and other drafters to create the set of drawings. Drafters must follow any applicable drawing standards (such as ASME or ISO) during this phase.

    As the drafter finishes each sheet in the set of drawings, the sheet is checked carefully for mistakes by designers or checkers. If mistakes are found, or if the design needs to be revised, the

mechanical working drawings: Drawings used in the fabrication and assembly of machine parts.

drafter will make the necessary corrections or revisions to the drawings. This process is repeated until the drawings are complete. Revisions are such an integral part of the design process that they are usually noted in a Revision History Block located on the upper right corner of the sheet.

The finished set of working drawings represents the master plan for the project. All of the information required to manufacture and assemble the project should be included in these drawings.

PREPARING MECHANICAL WORKING DRAWINGS

assembly drawing: A drawing that illustrates how the separate parts of an assembly are related to each other, for example, how mating parts fit together.

detail drawings: Drawings that provide the information required to manufacture or purchase each part in an assembly including the necessary views, dimensions, notations, and specifications.

Working drawings usually include assembly and detail drawings. *Assembly drawings* show how the separate parts of the assembly are related to each other, for example, how mating parts fit together. *Detail drawings* provide all of the information required to manufacture or purchase each part in the assembly including the necessary views, dimensions, notations, and specifications.

Assembly Drawings

The assembly drawing usually acts as the "cover" sheet in a set of working drawings and is numbered as the first sheet in the set of plans. For example, if a set of working drawings contains a total of 20 sheets, the drafter will label **SHEET 1 OF 20** in the assembly drawing's title block. The details for each part in the assembly are drawn on subsequent sheets, and these sheets are numbered sequentially, for example, **SHEET 2 OF 20** and **SHEET 3 OF 20** and so on through **SHEET 20 OF 20**. **Note:** The layout of title blocks for the first sheet and continuation sheets, including the information that goes into the fields of the title block, is covered in the *ASME Y14.1-2005* Standard.

Note:
The ASME standards governing the size and format of drawing sheets, including borders, title blocks, and revision history blocks, are ASME Y14.1-2005 and ASME Y14.1M-2005 (metric).

Assembly drawings are often drawn pictorially with the parts pulled apart, or "exploded," to show how the device is assembled. Figure 11-1 shows an example of an exploded isometric drawing that is used to show how the parts in the assembly fit together.

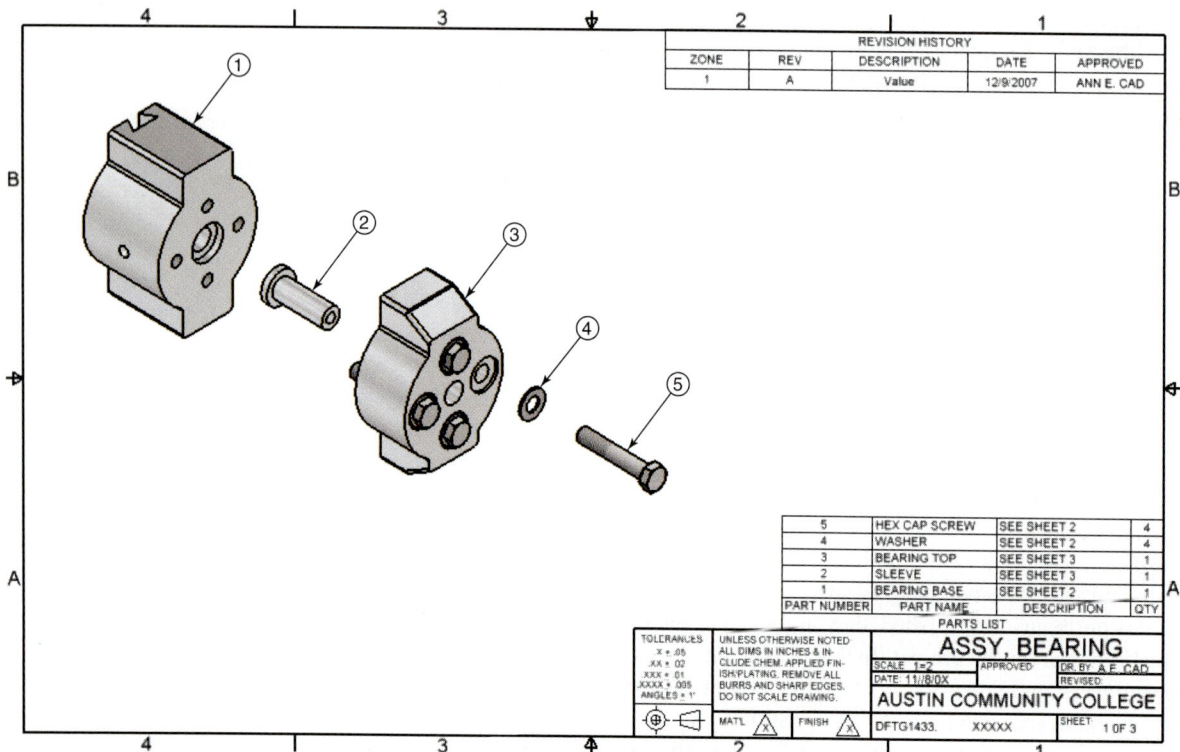

Figure 11-1   Exploded assembly drawing

The assembly drawing often includes a **parts list** (sometimes referred to as a *Bill of Material or BOM*) itemizing all of the parts in the assembly. In Figure 11-1, the Parts List is located above the title block, and a Revision History Block is located in the upper right corner of this drawing.

Creating an Exploded Assembly Drawing

Step 1. Study the individual parts of the assembly and determine how they fit together. Make an isometric planning sketch of the assembly similar to the one in Figure 11-2. Sketch the parts of the assembly as if they were pulled apart along isometric axes.

Figure 11-2   Planning sketch of exploded assembly

Assign part numbers to the separate parts and enclose each number in a circle with a leader attached. Part numbers are also referred to as *find* numbers. In an assembly drawing, this combination of a part number, circle, and leader is called a *balloon*.

Refer to the part numbers assigned to the sketch and plan a parts list. The categories the parts list must include are the part number, description, and the quantity required. The parts list may also include categories for material, Commercial and Governmental Entity Code (CAGE Code), and other information about the assembly.

Step 2. Begin a new AutoCAD drawing using an appropriate sheet size, border, and title block. Refer to the planning sketch made in Step 1 and construct an isometric view for each part in the assembly. Whenever possible, orient the isometric view to show as much detail about the assembly as possible. Align the parts as they would appear if they had been pulled apart from their normal assembled arrangement as shown in Figure 11-3.

Step 3. Add balloons and part numbers to each part.

TIP
Duplicate parts in an assembly receive the same part number.

Add phantom lines along isometric axes to show the relationship between mating parts. See Figure 11-4.

parts list: A table placed on a technical drawing that itemizes all of the parts in an assembly (sometimes referred to as a *Bill of Material or BOM*). The Parts List may include columns for part number, part name, description, quantity, and material.

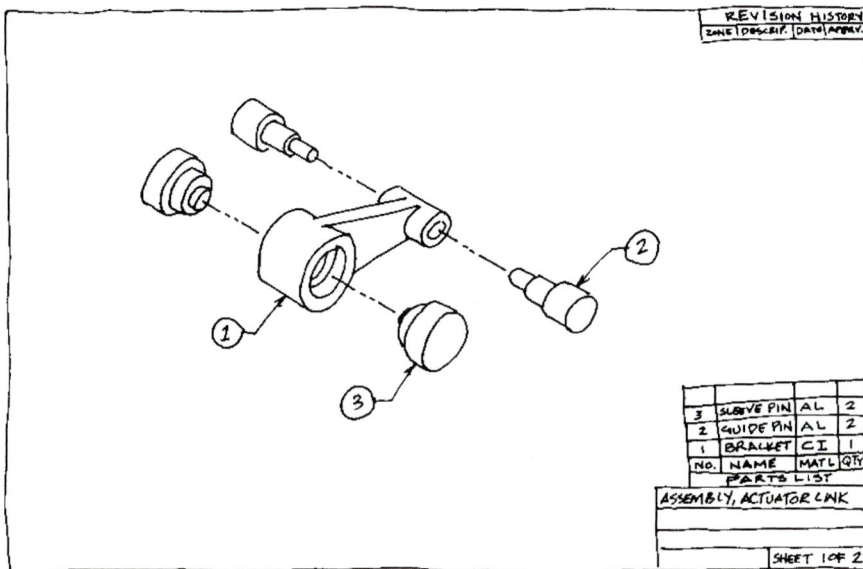

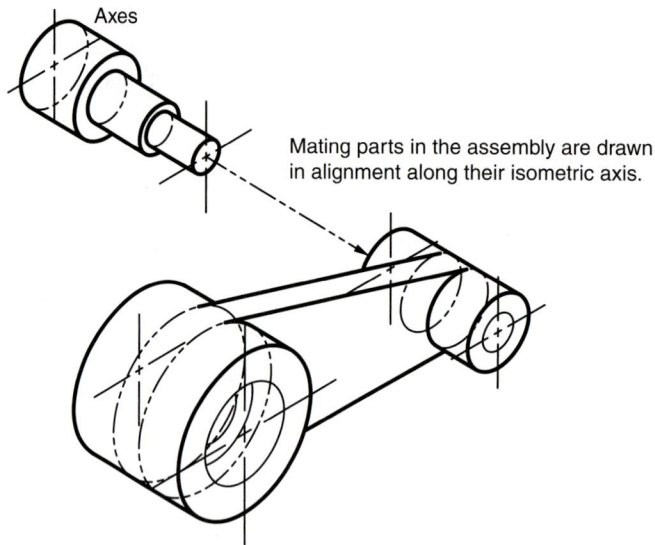

Figure 11-3 Aligning mating parts along their isometric axes

Axes

Mating parts in the assembly are drawn in alignment along their isometric axis.

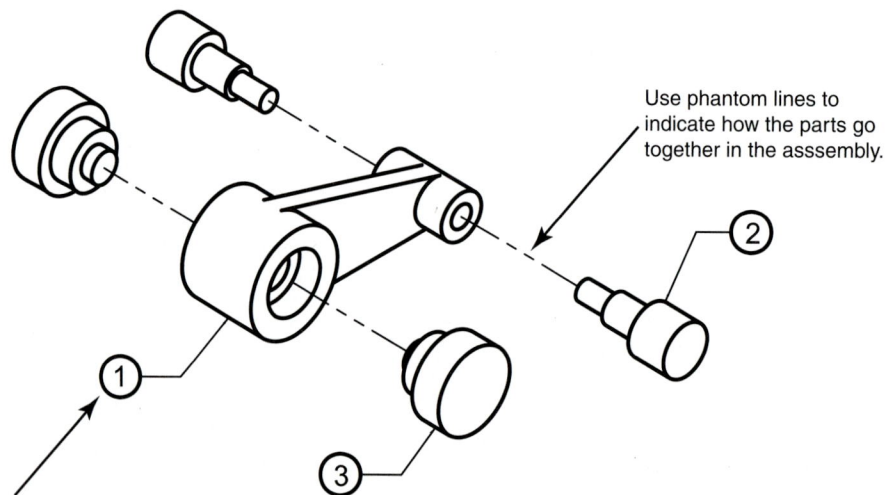

Use phantom lines to indicate how the parts go together in the asssembly.

Balloons are circles that enclose the part number. The balloons in this example were drawn .50″ in diameter and .25″ text height was used for the part numbers.

Figure 11-4 Adding balloons, part numbers, and phantom lines

Step 4. Add a Parts List to the drawing. An example of a parts list is shown in Figure 11-5. Parts lists are often placed above the title block in assembly drawings but can be placed in other areas of the drawing.

Step 5. Complete the information in the title block including the name, scale, and sheet number of the assembly. Compare the planning sketch shown in Figure 11-2 to the completed CAD drawing shown in Figure 11-6.

Note:
The ASME standard governing location and format of a parts list is ASME Y14.34M-1996. This standard also governs other types of lists used in engineering drawings.

Detail Drawings

An example of a detail drawing is shown in Figure 11-7. To create this drawing, the drafter works from design inputs provided by an engineer or designer. Often, the design input is in the form of a rough sketch; however, in the modern practice of technical drawing, the drafter may receive this input in the form of a three-dimensional CAD drawing file created by the designer.

5	HEX CAP SCREW	SEE SHEET 2	4
4	WASHER	SEE SHEET 2	4
3	BEARING TOP	SEE SHEET 3	1
2	SLEEVE	SEE SHEET 3	1
1	BEARING BASE	SEE SHEET 2	1
PART NUMBER	PART NAME	DESCRIPTION	QTY
PARTS LIST			

Figure 11-5   Example of a Parts List

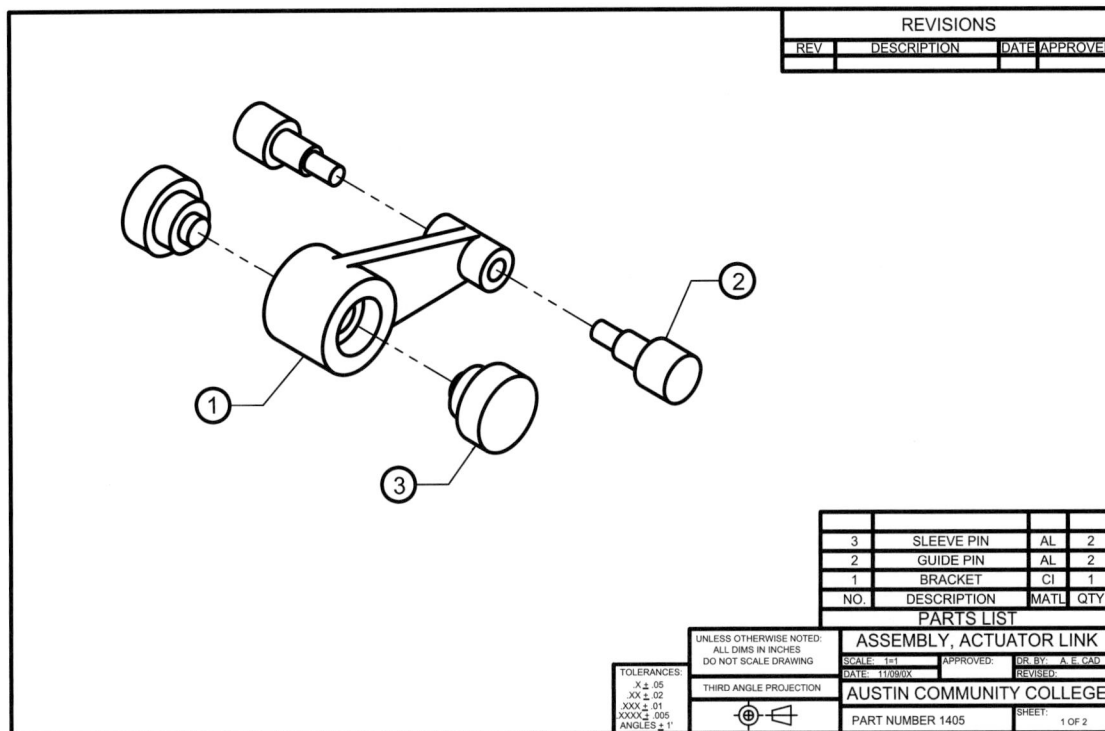

	REVISIONS		
REV	DESCRIPTION	DATE	APPROVED

3	SLEEVE PIN	AL	2
2	GUIDE PIN	AL	2
1	BRACKET	CI	1
NO.	DESCRIPTION	MATL	QTY
PARTS LIST			

UNLESS OTHERWISE NOTED:
ALL DIMS IN INCHES
DO NOT SCALE DRAWING

ASSEMBLY, ACTUATOR LINK

SCALE: 1=1   APPROVED: DR. BY: A. E. CAD
DATE: 11/09/0X   REVISED:

THIRD ANGLE PROJECTION

AUSTIN COMMUNITY COLLEGE

PART NUMBER 1405   SHEET: 1 OF 2

TOLERANCES:
.X ± .05
.XX ± .02
.XXX ± .01
.XXXX ± .005
ANGLES ± 1'

Figure 11-6   Completed exploded assembly

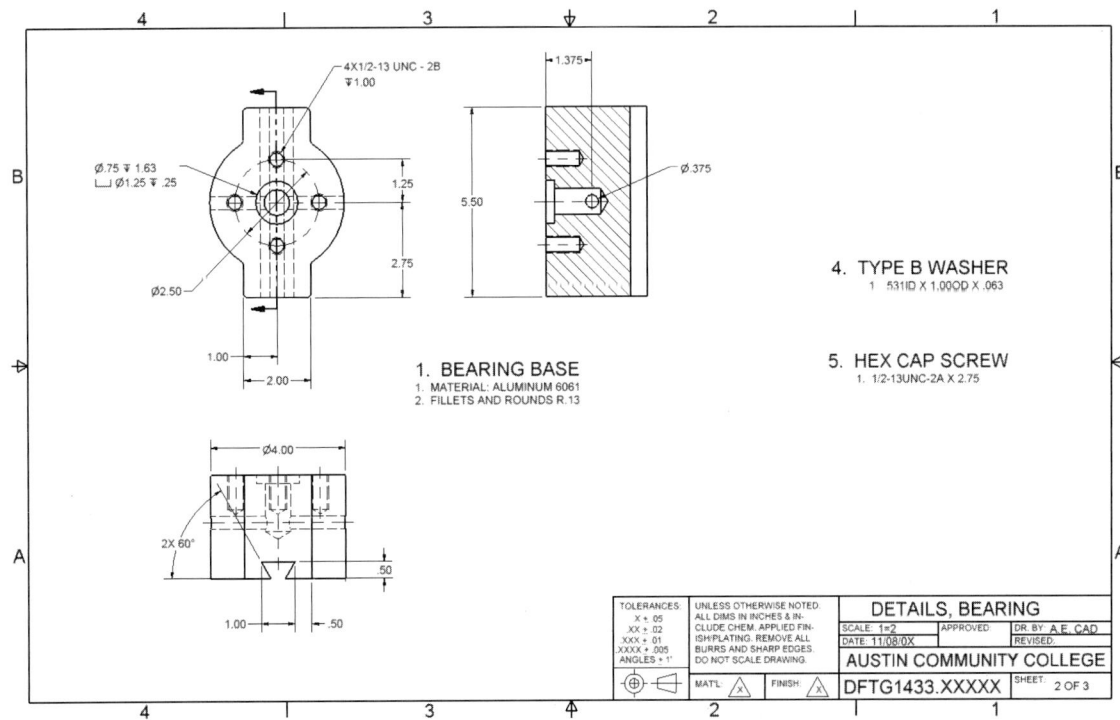

4X1/2-13 UNC - 2B
▼1.00

Ø.75 ▼1.63
⌴ Ø1.25 ▼.25

1.25
2.75
1.00
2.00
Ø2.50

1.375
Ø.375
5.50

1. BEARING BASE
1. MATERIAL: ALUMINUM 6061
2. FILLETS AND ROUNDS R.13

4. TYPE B WASHER
1   .531ID X 1.00OD X .063

5. HEX CAP SCREW
1. 1/2-13UNC-2A X 2.75

Ø4.00
2X 60°
.50
1.00   .50

TOLERANCES:
X ± .05
.XX ± .02
.XXX ± .01
.XXXX ± .005
ANGLES ± 1'

UNLESS OTHERWISE NOTED:
ALL DIMS IN INCHES & IN-
CLUDE CHEM. APPLIED FIN-
ISH/PLATING. REMOVE ALL
BURRS AND SHARP EDGES.
DO NOT SCALE DRAWING.

DETAILS, BEARING

SCALE: 1=2   APPROVED: DR. BY: A.E. CAD
DATE: 11/08/0X   REVISED:

AUSTIN COMMUNITY COLLEGE

MATL: △   FINISH: △   DFTG1433.XXXXX   SHEET: 2 OF 3

Figure 11-7   Example of a detail drawing

When the individual part is "detailed out," all of the views, dimensions, and notes necessary to manufacture the part are included. Because the drafter's expertise is in creating technical drawings, decisions regarding the necessary views and correct dimensioning of the object are often left up to the drafter's judgment. The drafter may also be responsible for determining the sheet size and making sure that the drawing complies with any drafting or dimensioning standards that might be appropriate (such as *ASME Y14.5M-1994*).

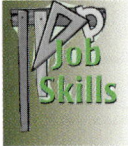

When creating a detail drawing of an object, the drafter must be careful not to change the design intent of the designer by incorrectly noting dimension values, changing the precision of dimensions, or referencing different datum features than the ones on the designer's sketch.

Creating a Detail Drawing

Step 1. Make a planning sketch of the multiviews of the parts to be detailed similar to the one shown in Figure 11-8. This sketch may also include dimensions and notations.

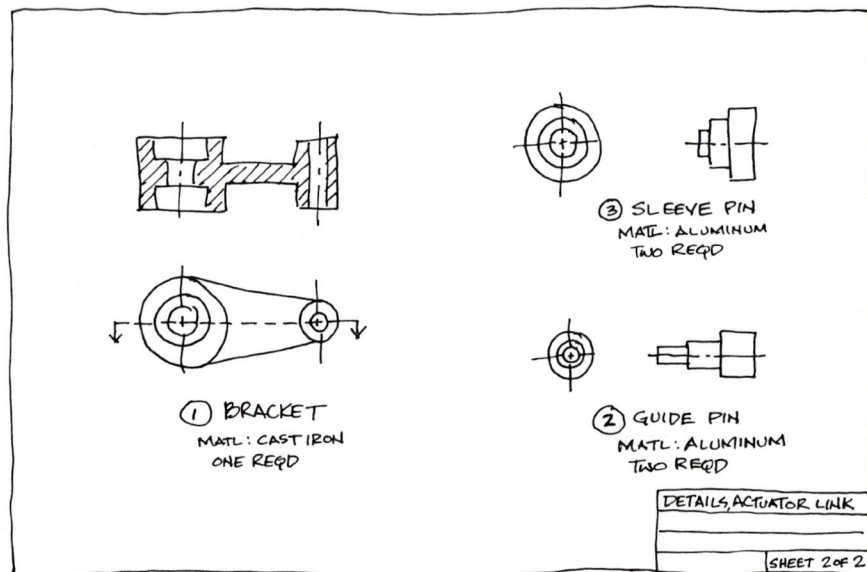

Figure 11-8 Planning sketch for a Detail Sheet

If multiple parts are drawn on the same sheet, space the views of separate parts so there is no confusion concerning which view goes with which part.

Using the same part names and numbers defined in the assembly drawing, label each part. The part's material, quantity (the number required for the assembly), and any other notes required for manufacture should also be included in this drawing.

Step 2. Begin a new AutoCAD drawing using the same sheet size, border, and title block used in the assembly drawing. Refer to the planning sketch and draw the necessary views for each part. Consider drawing section views of parts with complex interior details.

Include all the dimensions necessary to manufacture the part. When dimensioning, follow guidelines established by the *ASME Y14.5-1994M* Standard (refer to Unit 5).

Add part numbers and part names. These are often placed using .25″ text height for emphasis.

Include any notes required for manufacture and the total number of each part required for the assembly. These notes are often placed beneath the Part Name and Number using the same text height as the dimension text.

Step 3. Complete the information in the title block including the name, scale, and sheet number of the detail. Compare the planning sketch shown in Figure 11-8 to the completed CAD drawing shown in Figure 11-9.

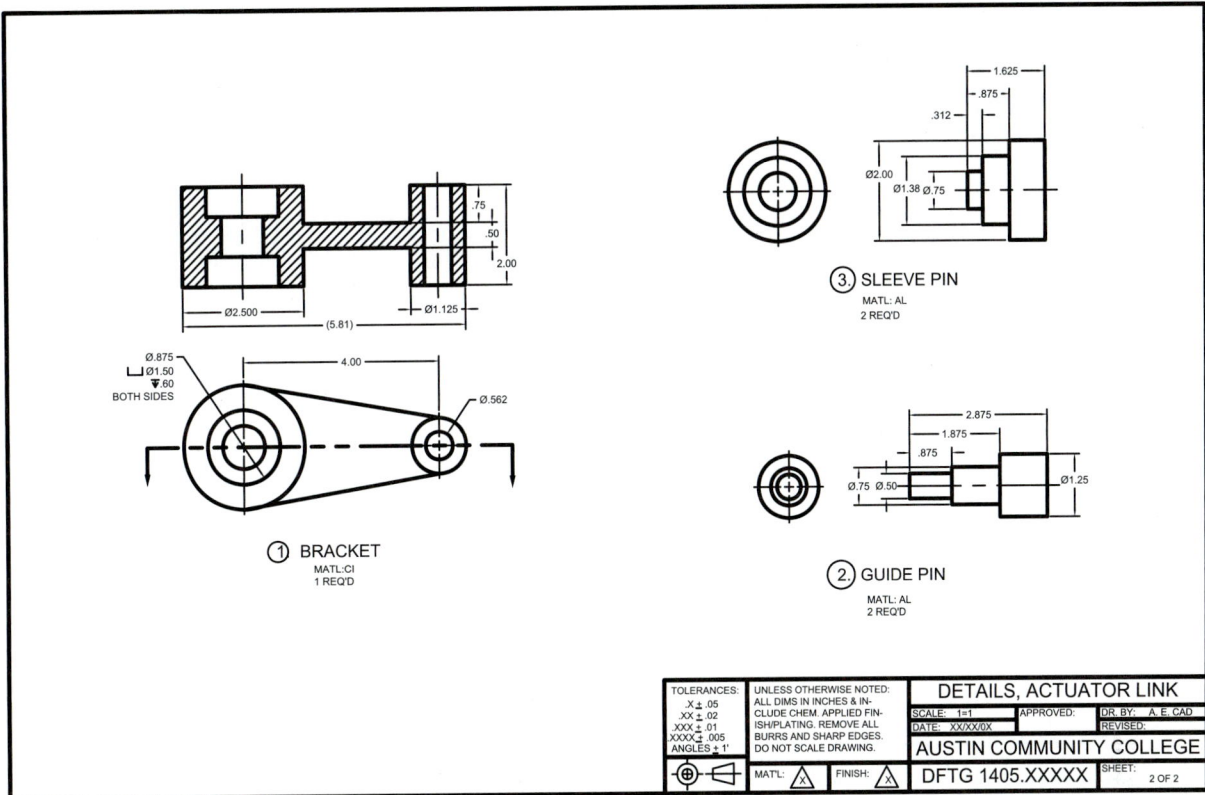

Figure 11-9   Completed Detail Drawing

Summary

To be successful in the mechanical design field, drafters and CAD operators must be able to create working drawings consisting of both assembly and detail drawings. Assembly drawings show how the parts in the assembly fit together and should include part numbers, balloons, and a parts list. Detail drawings should include all the views, dimensions, and notations required to manufacture each part of the assembly. Drafters and designers should also be familiar with specifying and representing threaded holes, shafts, and fasteners. It is important that the drafter represent the assembly exactly as the designer intended to ensure that the project is fabricated correctly.

Mechanical drafters must also be familiar with the numerous standards that control the format, dimensioning, and tolerancing of engineering drawings.

Unit Test Questions

Short Answer

1. Define what is meant by the term *assembly drawing*.
2. Define what is meant by the term *detail drawing*.
3. Explain the sheet numbering system used in mechanical working drawings.
4. Name four columns that might be included on a parts list.
5. Explain how planning sketches are used in the creation of mechanical working drawings.

Unit Project

Project 11-1: Toe Stop Assembly

Create mechanical working drawings for the toe stop assembly detailed in Figures 11-10, 11-11, and 11-12. The assembly should be drawn as an exploded isometric view including balloons, part numbers, and a Parts List. On the detail sheets, fully dimension each part and provide specifications for the standard hardware (Hex Screw, Hex Nut, and Washer).

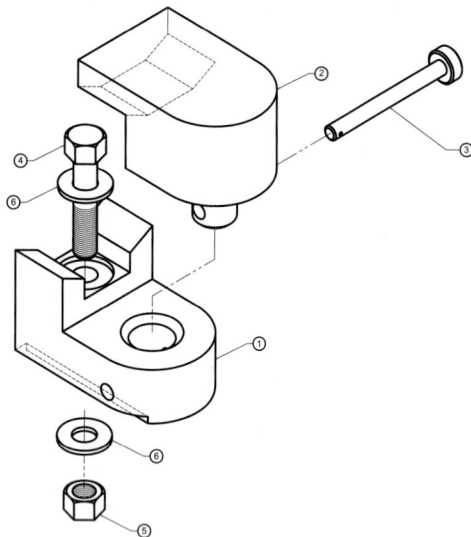

Figure 11-10   Toe stop assembly

6	TYPE A PLAIN WASHER	SEE SHEET 3	2
5	HEX NUT	SEE SHEET 3	1
4	HEX CAP SCREW	SEE SHEET 3	1
3	CLEAT PIN	MILD STEEL	1
2	CLEAT	MILD STEEL	1
1	TOE STOP BASE	MILD STEEL	1
PART NO	DESCRIPTION	MATERIAL	QTY
PARTS LIST			

Figure 11-11   Toe Stop Assembly Parts List

Directions

1. Select **FILE** and **OPEN**.
2. Select the **Student Resource CD**.
3. Open the **Prototype Drawings** folder.
4. Open the **Mechanical Working Drawing Prototype** drawing (it will open as "read-only").
5. Use **SAVE AS** to save the drawing to your **Home** directory and rename the new drawing **TOE STOP ASSEMBLY**.
6. Draw and dimension the necessary views of the Toe Stop Base, Cleat, and Cleat Pin.
7. Create a full section view of the Toe Stop Base.
8. Draw an exploded isometric assembly showing all of the parts, add balloons and part numbers, and complete the parts list. Insert the pre-existing blocks of the Hex Cap Screw, Hex Nut, and Washer contained in the prototype drawing.

Creating the Assembly and Detail Drawings for the Toe Stop Project

Planning the Sheets.   This project will require three D-size (34″ × 22″) sheets. Sheet 1 will be an exploded assembly (including a parts list, balloons, and part numbers), and Sheets 2 and 3 will contain the details (multiviews, dimensions, and notations) for each part.

① TOE STOP BASE

MATERIAL – MILD STEEL
ONE REQUIRED

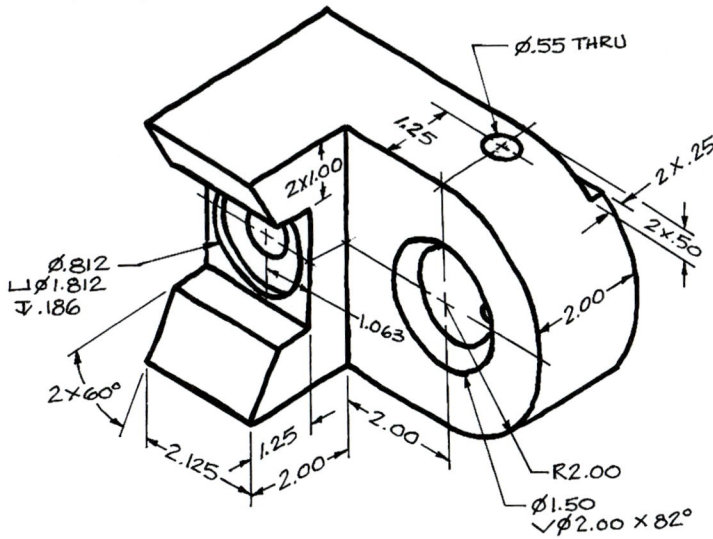

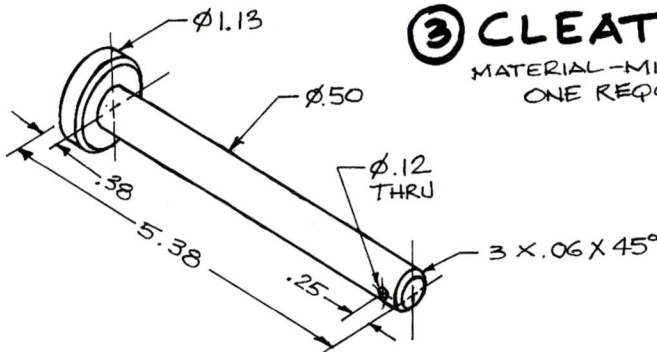

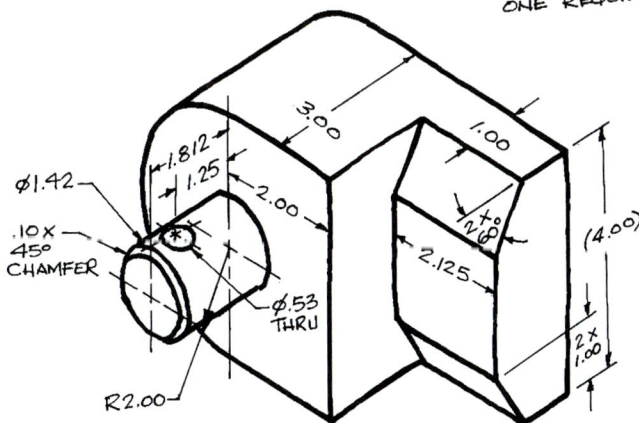

Ø.55 THRU

1.25

2X.25

2X.50

2X1.00

2.00

Ø.812
⌴Ø1.812
⌵.186

1.063

2X60°

2.125   1.25   2.00

2.00

R2.00

Ø1.50
⌵Ø2.00 X 82°

③ CLEAT PIN

MATERIAL – MILD STEEL
ONE REQUIRED

Ø1.13

Ø.50

Ø.12
THRU

.38

5.38

.25

3 X .06 X 45°

② CLEAT

MATERIAL – MILD STEEL
ONE REQUIRED

3.00

1.00

1.812

1.25

2.00

2X60°

(4.00)

Ø1.42

.10 X
45°
CHAMFER

2.125

Ø.53
THRU

2X
1.00

R2.00

Figure 11-12   Designer's sketch of toe stop assembly parts

The exploded assembly will be the first sheet in the set and will be numbered Sheet 1 of 3. A planning sketch for this sheet might resemble the one shown in Figure 11-13. In this sketch, the parts are sketched isometrically and positioned as they would look if the assembly had been pulled apart. Balloons, part numbers, and a parts list are also included on this sheet.

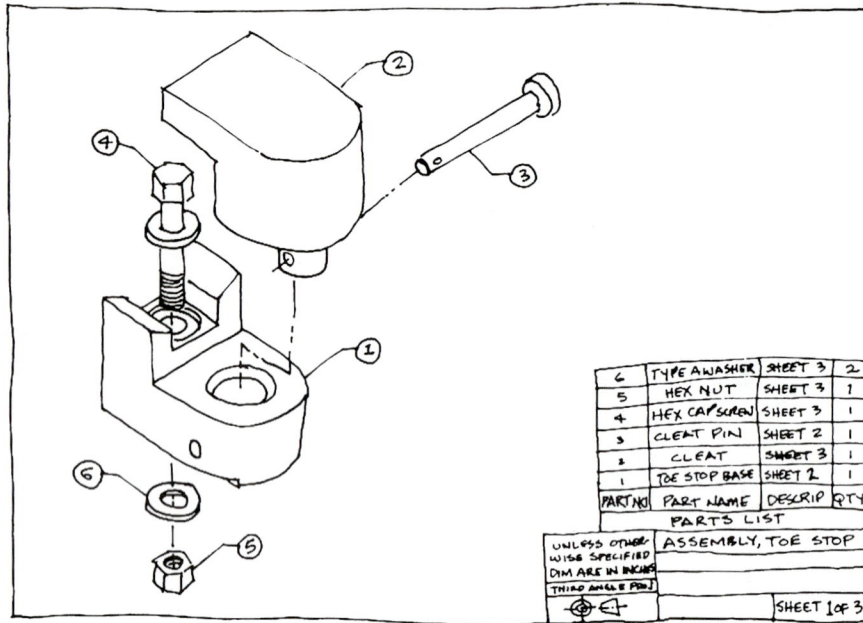

Figure 11-13 Sheet 1 planning sketch

> **TIP**
>
> The CAD drawing of the exploded assembly should be drawn *after the detail sheets are drawn* because you will need to transfer measurements from the views of the detail drawings in order to construct isometric drawings of features such as the inclined planes and the countersunk hole.

Begin the project by drawing the first sheet of details. Because the sheet containing the assembly drawing is numbered Sheet of 1 of 3, this sheet will be numbered Sheet 2 of 3.

This sheet will include the details of the Toe Stop Base and the Cleat Pin. The Toe Stop Base is considered the principal part in this assembly because all of the other parts are aligned to, or mate with, it. Because it is the principal part, it will be labeled Part 1.

Figure 11-14 shows a planning sketch made for Sheet 2. Begin this sheet by determining which view of the Toe Stop Base will be drawn as the front, or principal, view and then determine the other multiviews necessary to describe the part. For the Toe Stop Base, a full section should be included to show the interior detail of the object. Add the necessary dimensions and notations and follow the *ASME Y14.5M-1994* dimensioning guidelines described in Unit 5. Next, draw the necessary views of the Cleat Pin and add the required dimensions and notations to these views.

The next sheet drawn in this set will be the second sheet of details which will be numbered Sheet 3 of 3. On this sheet, you will draw the details of the Cleat and include the specifications for the Hex Cap Screw, Hex Nut, and Washer. A planning sketch for this sheet is shown in Figure 11-15. Begin this sheet by determining which view of the Cleat will be drawn as the front, or principal, view and then determine the other required views. Add the dimensions and notations required to manufacture the part.

In Figure 11-15, you will note that no detail drawings are included for the Hex Cap Screw, Hex Nut, and Washer. This is because

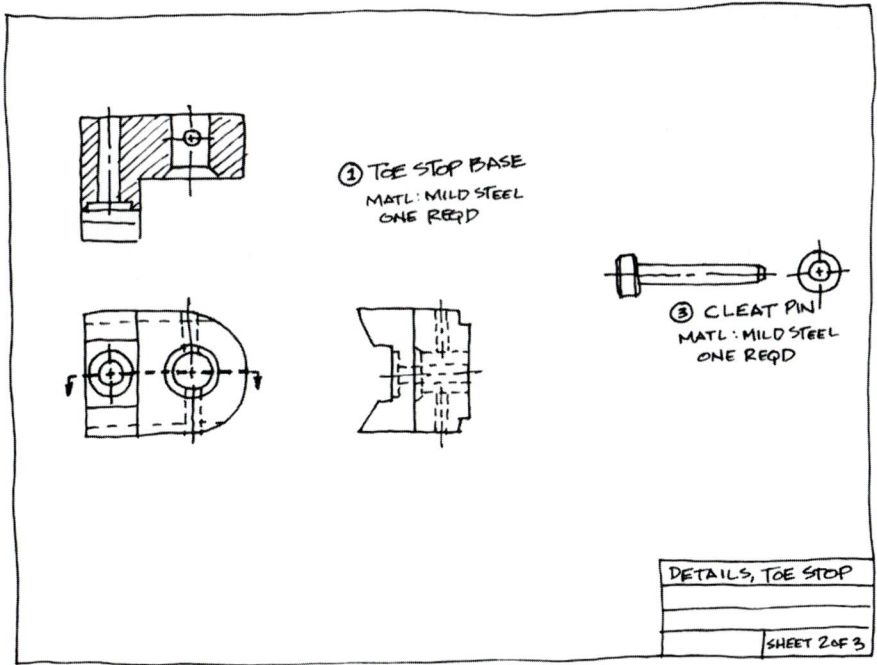

Figure 11-14   Sheet 2 of the planning sketch

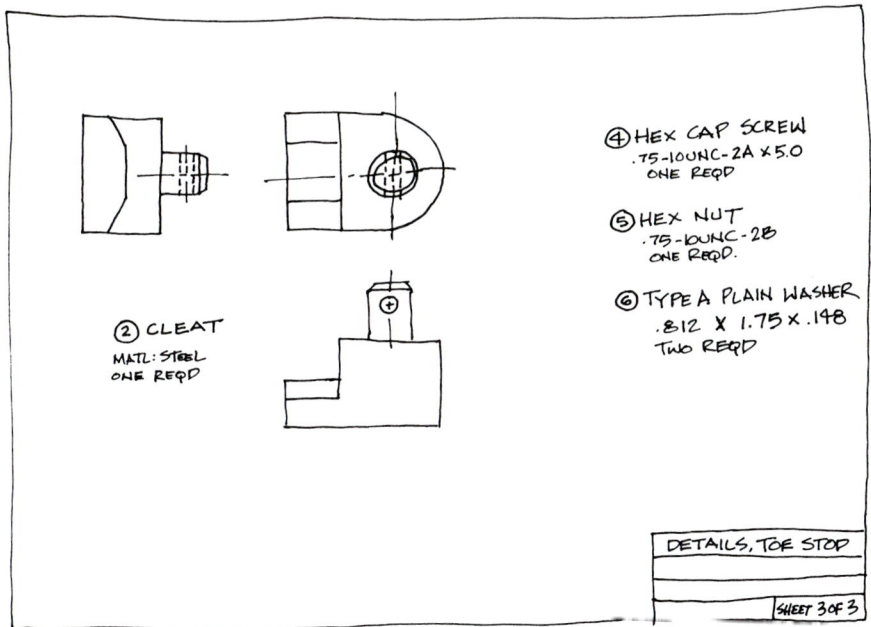

Figure 11-15   Sheet 3 of the planning sketch

these parts are standard hardware which can purchased from an outside vendor. However, it is necessary to note on the drawing the part name, part number, specification, and quantity for each part so that the correct hardware can be ordered.

> **TIP**
>
> The specification for the washer indicates the washer's insider diameter (.812), outside diameter (1.75), and thickness (.148). The thread specifications provided for the Hex Cap Screw and Hex Nut are explained in the next section.

Threads and Fasteners in Mechanical Working Drawings

Mechanical drawings often include threaded holes, shafts, and fasteners (hex nuts, bolts, screws, etc.). The threads on these features are specified with a thread note detailing the size and type of thread.

The Toe Stop Assembly project calls for a Hex Cap Screw and a mating Hex Nut (see Figure 11-12). The thread specification for the screw is .75-10UNC-2A. This thread note is interpreted in the following way: The major diameter of the thread is **.75**, the number of threads per inch is **10**, the thread series is **Unified National Coarse (UNC)**, the thread class is **2**, and the **A** indicates it is an external thread. The thread specification for the mating hex nut is exactly the same, except the **A** is replaced with a **B** to indicate an internal thread. Figure 11-16 shows how to interpret the thread specification for the Hex Cap Screw.

Note:
The ASME standards governing the specification and dimensioning screw threads are ASME Y14.6 and ASME Y14.6M (metric supplement).

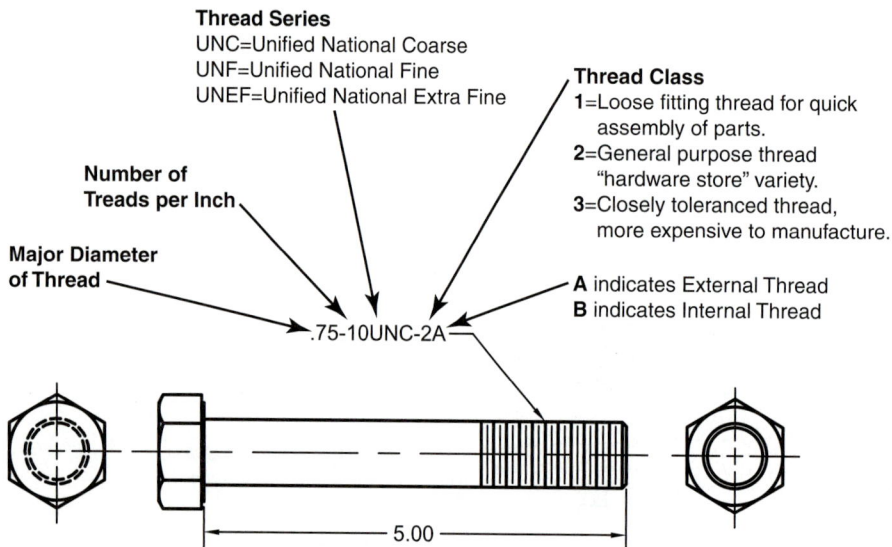

Thread Series
UNC=Unified National Coarse
UNF=Unified National Fine
UNEF=Unified National Extra Fine

Thread Class
1=Loose fitting thread for quick assembly of parts.
2=General purpose thread "hardware store" variety.
3=Closely toleranced thread, more expensive to manufacture.

Number of Treads per Inch

Major Diameter of Thread

.75-10UNC-2A

A indicates External Thread
B indicates Internal Thread

— 5.00 —

Figure 11-16   Interpreting a thread note for a hex cap screw

Representing External Screw Threads on Mechanical Drawings.

There are three methods of representing external screw threads on a drawing: *schematic, detailed,* and *simplified.* The schematic method shown in Figure 11-17 is widely used because it can be drawn quickly using AutoCAD tools such as **OFFSET** and **ARRAY**. The detailed method shown in Figure 11-18 is used less frequently because it is more time consuming to construct. The simplified method shown in Figure 11-19 is also commonly used to represent screw threads on mechanical drawings.

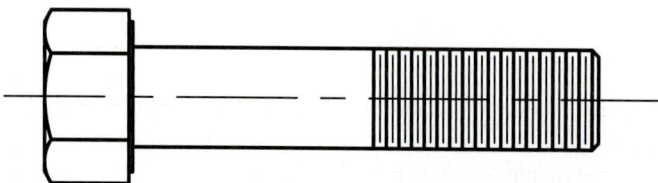

Figure 11-17   Schematic method of representing external screw threads

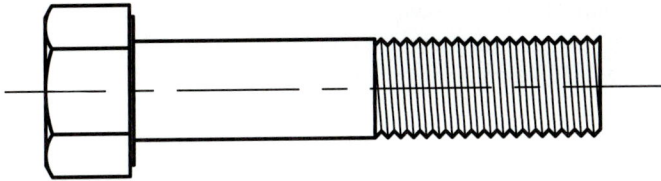

Figure 11-18   Detailed method of representing external screw threads

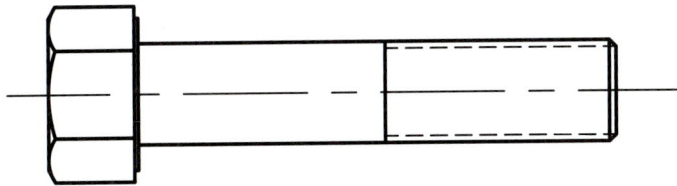

Figure 11-19   Simplified method of representing external screw threads

Representing Internal Screw Threads on Mechanical Drawings. The end view of a threaded hole is shown in Figure 11-20(a). In this view, the major diameter of the thread is represented by a circle drawn with a hidden line. The circle drawn with a visible line represents the minor diameter of the thread. The threaded hole is represented in the side view by four hidden lines as shown in Figure 11-20(b). These hidden lines are projected from the quadrants of the circles in the front view.

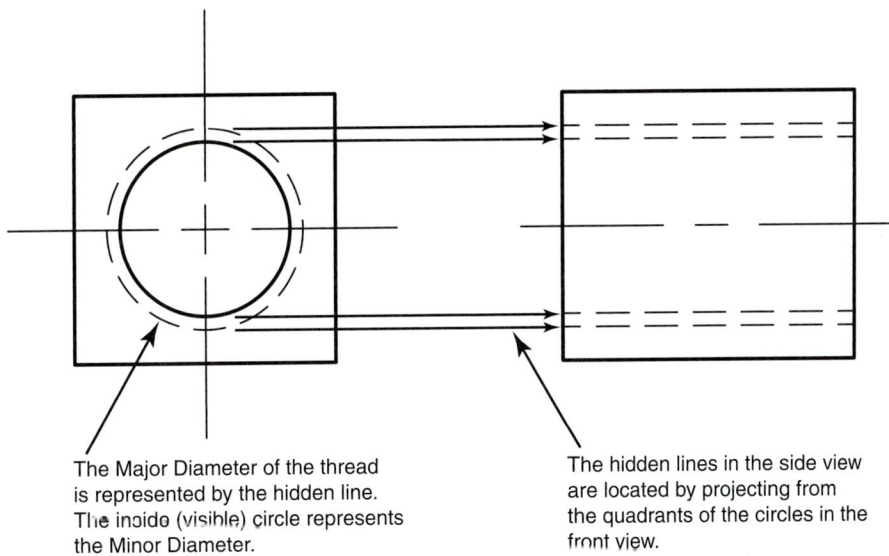

The Major Diameter of the thread is represented by the hidden line. The inside (visible) circle represents the Minor Diameter.

The hidden lines in the side view are located by projecting from the quadrants of the circles in the front view.

Figure 11-20(a)   Front view of a threaded hole   **(b)** Side view of a threaded hole

Drawing an Isometric Counterbored Hole

In creating the exploded assembly drawing for the Toe Stop Base, you will need to construct an isometric view of a countersunk hole and a counterbored hole. Figures 11-21 through 11-25 illustrate the steps in constructing an isometric, counterbored hole.

> **TIP** When constructing the isometric counterbored hole for the Toe Stop Base, substitute the dimensions noted on the designer's sketch in Figure 11-12 for those used in these examples.

The specifications for this counterbored hole are shown in Figure 11-21. In this example, a **1.00″** diameter hole goes all the way through the object. Centered on this hole is a **1.50″** diameter counterbore which has a depth of **.25″**.

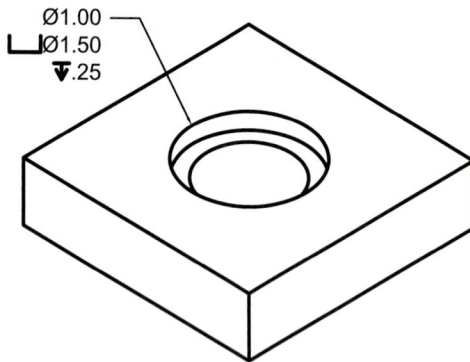

Ø1.00
⌴Ø1.50
▼.25

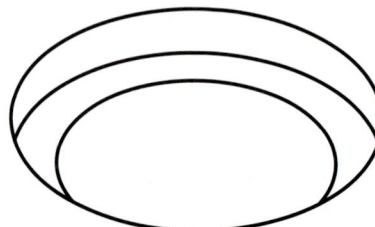

Figure 11-21   Isometric view of a counterbored hole

Step 1. Draw a **1.50″** diameter isometric ellipse and copy it directly below the first ellipse at a distance of **.25″** as shown in Figure 11-22.

Step 2. Trim the bottom ellipse to the edge of the top ellipse as shown in Figure 11-23.

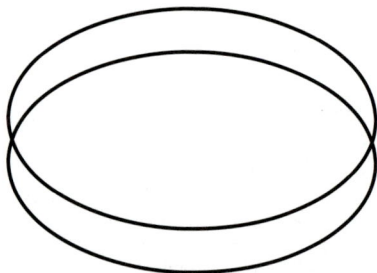

Figure 11-22   Isometric ellipses   **Figure 11-23**   Ellipses after trimming

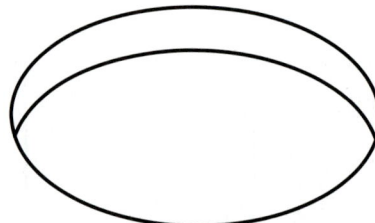

Step 3. Draw a **1.00″** diameter ellipse at the center of the lower ellipse as shown in Figure 11-24.

Step 4. Trim the part of this ellipse that extends below the first ellipse as shown in Figure 11-25. This completes the construction of the counterbored hole.

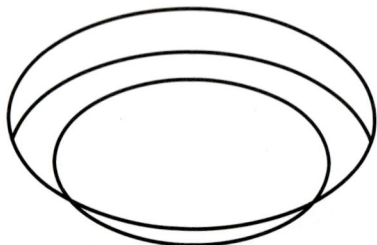

Figure 11-24   Adding the 1.00″ ellipse   **Figure 11-25**   Finished isometric counterbored hole

Drawing an Isometric Countersunk Hole

Figures 11-26 through 11-29 illustrate the steps in constructing an isometric countersunk hole.

The specifications for this countersunk hole are shown in Figure 11-26. In this example, a **1.00″** diameter hole passes all the way through the object. Centered on this hole is a **1.50″** diameter countersunk hole. The angle between the sides of the countersunk hole is **82°**.

Step 1. Draw two concentric isometric ellipses, one with a diameter of **1.00″**, and the second with a diameter of **1.50″** as shown in Figure 11-27.

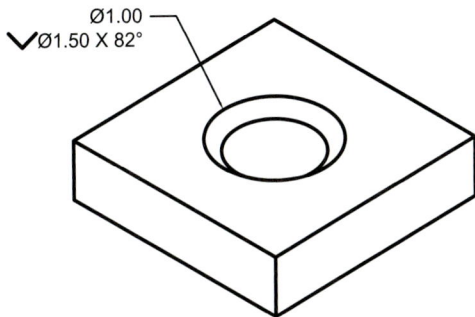

Ø1.00
VØ1.50 X 82°

Figure 11-26   Isometric countersunk hole

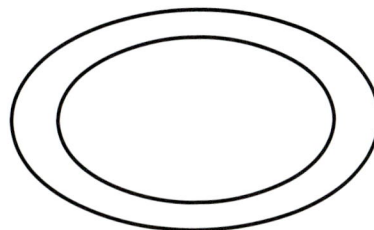

Figure 11-27   Isometric ellipses

Step 2. Determine the value for **D** by measuring from the multiview of the countersunk hole's side view as shown in Figure 11-28(a) and move the **1.00″** diameter ellipse straight down at distance **D** as shown in Figure 11-28(b).

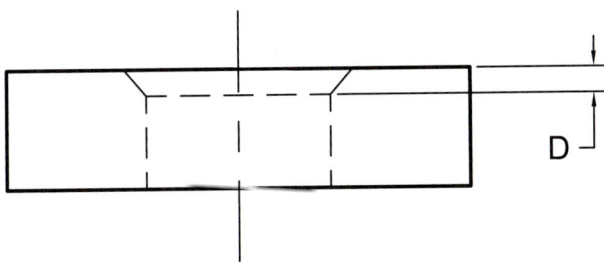

D

D

Figure 11-28(a)   Finding the depth (D) of the countersunk hole

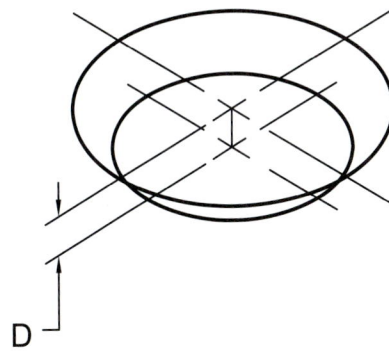

Figure 11-28(b)   Move small ellipse down depth (D)

Step 3. Trim the smaller ellipse where it extends beyond the larger ellipse to complete the construction of the countersunk hole (see Figure 11-29).

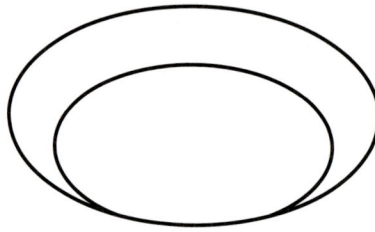

Figure 11-29 Completed countersunk hole

Constructing the Slots in the Isometric View of the Toe Stop Base

On each side of the bottom of the Toe Stop Base is a slot that is **.50″** wide by **.25″** tall. The construction for the isometric drawing of these slots is shown in Figure 11-30(a). Figure 11-30(b) shows the Toe Stop Base after the construction lines for the slots have been trimmed.

Note:
Hidden lines showing the profile of the slot in the front side of the Toe Stop Base have been shown for clarity.

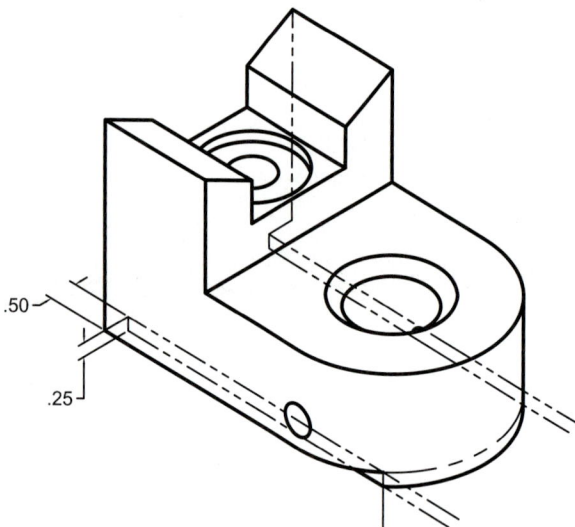

Figure 11-30(a) Constructing slots on Toe Stop Base

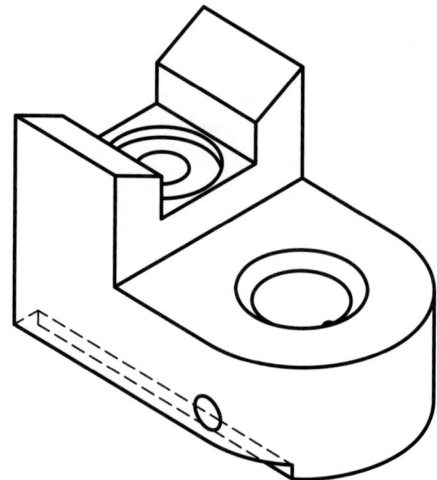

Figure 11-30(b) Completed isometric drawing of Toe Stop Base

Plotting the Sheets

Follow your instructor's directions to plot the sheets. If your plotter or printer allows, each sheet can be plotted on a 17″ × 11″ sheet using the monochrome setting at a scale of **1=2**.

TIP Follow the steps presented in Unit 4 to create a Page Setup for plotting each sheet of this project.

OPTIONAL UNIT PROJECT

Project 11-2: Test Fixture Assembly

Create mechanical working drawings for the Test Fixture Assembly
detailed in Figures 11-31, 11-32, and 11-33(a)–(c). The assembly

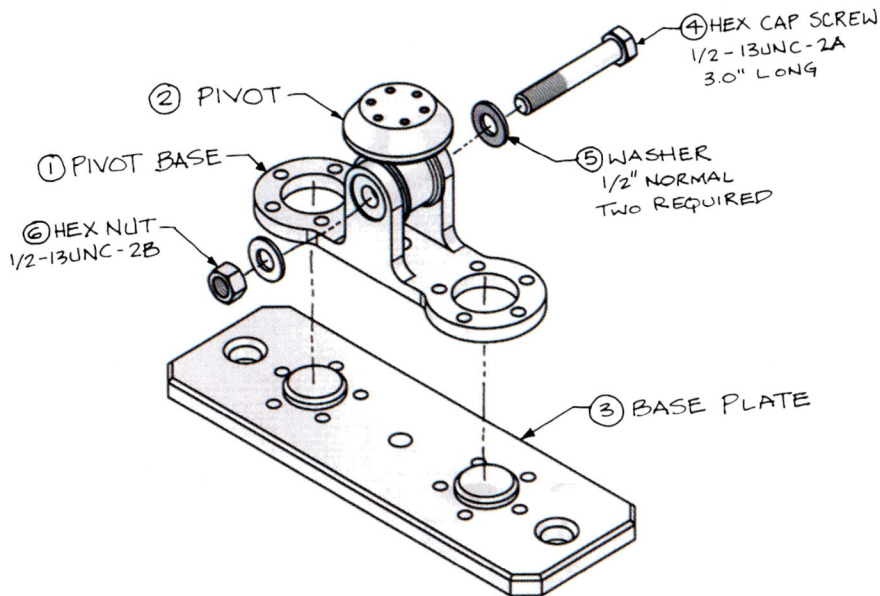

Figure 11-31   Test fixture
assembly

6	Hex Nut	SEE SHEET 3	1
5	Washer	SEE SHEET 3	2
4	Hex Cap Screw	SEE SHEET 3	1
3	Base Plate	SEE SHEET 3	1
2	Pivot	SEE SHEET 2	1
1	Pivot Base	SEE SHEET 2	1
PART NO	PART NAME	DESCRIPTION	QTY

PARTS LIST

Figure 11-32   Test fixture assembly parts list

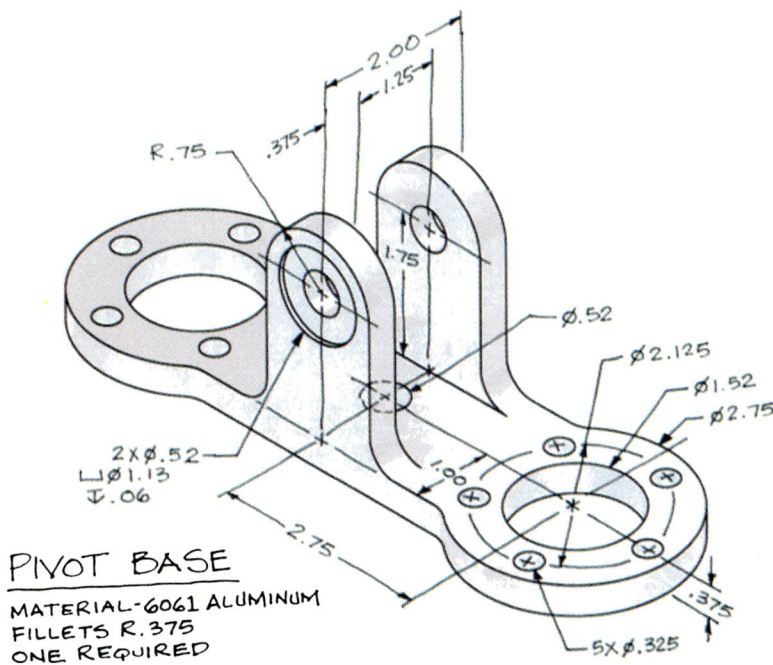

PIVOT BASE

MATERIAL-6061 ALUMINUM
FILLETS R.375
ONE REQUIRED

Figure 11-33(a)   Designer's sketch
of the pivot base

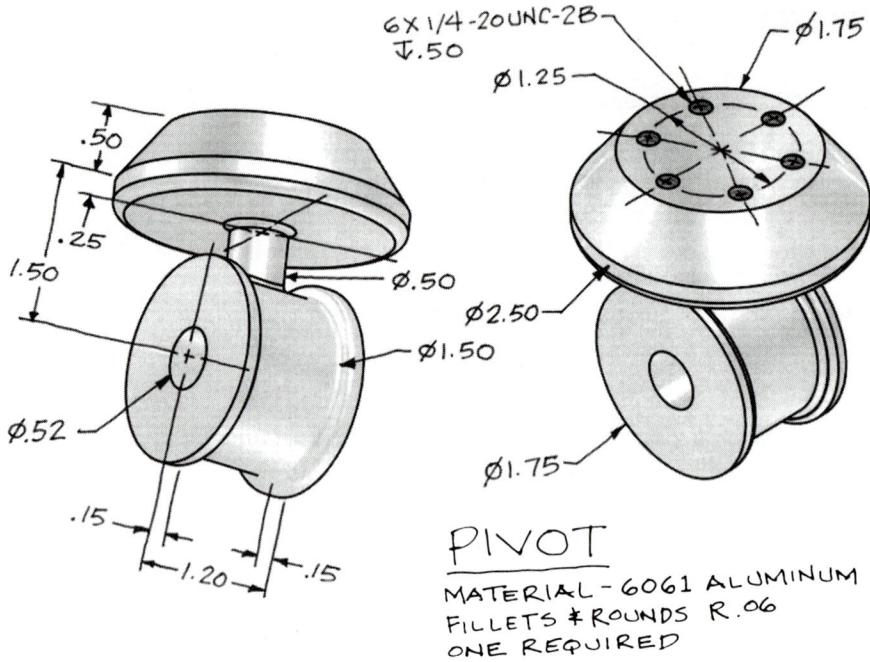

Figure 11-33(b)   Designer's sketch of the pivot

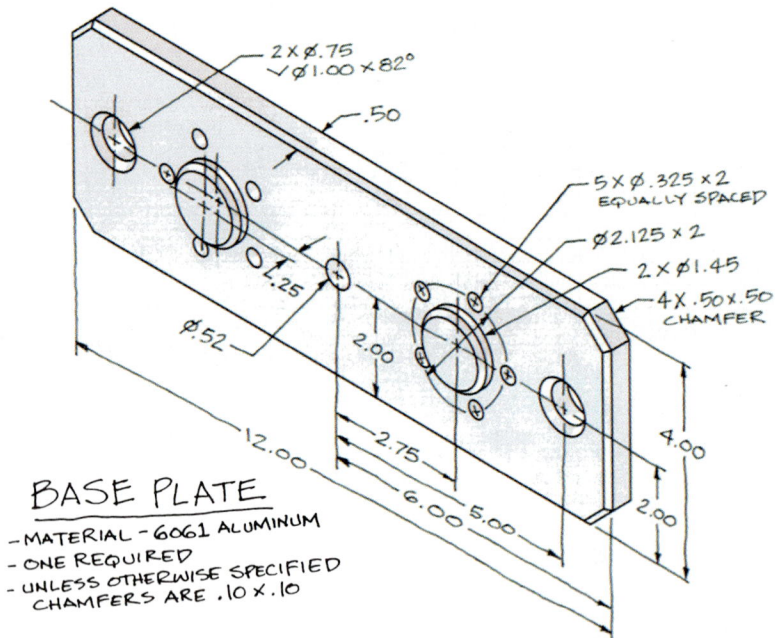

Figure 11-33(c)   Designer's sketch of the base plate and hardware

should be drawn as an exploded isometric view including balloons, part numbers, and a parts list (see Figure 11-32). On the detail sheets, fully dimension each part and provide specifications for the standard hardware (Hex Screw, Hex Nut, and Washer).

Directions

1. Select **FILE** and **OPEN**.
2. Select the **Student Resource CD**.
3. Open the **Prototype Drawings** folder.
4. Open the **Mechanical Working Drawing Prototype-Test Fixture** drawing (it will open as "read-only").

5. Use **SAVE AS** to save the drawing to your **Home** directory and rename the new drawing. **TEST FIXTURE ASSEMBLY**.

6. Draw and dimension the necessary views of the Pivot, Pivot Base, and Base Plate. Create a full section view of the Base Plate.

7. Draw an exploded isometric assembly showing all of the parts, add balloons and part numbers, and complete the parts list. Insert the blocks of the isometric views of the Hex Cap Screw, Hex Nut, and Washer contained in the prototype drawing.

Plotting the Sheets

Follow your instructor's directions to plot the sheets. If your plotter or printer allows, each sheet can be plotted on a 17″ × 11″ sheet using the monochrome setting at a scale of **1=2**.

> **TIP** Follow the steps presented in Unit 4 to create a Page Setup for plotting each sheet of this project.

Architectural Working Drawings

Unit Objectives

- Describe architectural working drawings and their importance to the field of architecture.
- Describe how floor plans and elevation drawings are planned and prepared.
- Use AutoCAD to create a floor plan for a small house.
- Use AutoCAD's **DesignCenter** to place blocks of electrical and plumbing symbols into the floor plan.
- Use AutoCAD to create elevation drawings for a small house.

INTRODUCTION

In many architectural offices, drafters work with architects and designers to prepare the drawings used in the construction of residential and commercial buildings. These drawings, which may include floor plans, elevations, foundations, wall sections, and roof framing plans, are called *construction documents (CDs)*. Often, the separate sheets for a full set of plans are created on different CAD layers within the same CAD drawing file so that the drafter can selectively view and print the layers as needed.

FLOOR PLANS

Floor plans, like the one shown in Figure 12-1, provide home builders and contractors with the necessary information to lay out the building, including the locations of features such as walls, doors, electrical components (switches, lamps, etc.) and plumbing fixtures (tubs, commodes, sinks, etc.). Floor plans should include all of the dimensions and notations required by the workers on the jobsite. Doors and windows are dimensioned to their centers and continuous (also known as *chain*) dimensioning is typically employed on floor plans. Dimensions are labeled above the dimension line, and tick marks replace arrowheads.

Architectural firms create, or purchase, block libraries of doors, windows, electrical, plumbing, and other symbols frequently used on floor plans.

> **construction documents (CDs):** Drawings used in the construction of residential and commercial buildings. These drawings may include floor plans, elevations, foundations, wall sections, and roof framing plans.

> **floor plans:** Drawings that provide home builders and contractors with the necessary information to lay out the building, including the locations of features such as walls, doors, electrical components (switches, lamps, etc.) and plumbing fixtures (tubs, commodes, sinks, etc.). Floor plans usually include all of the dimensions and notations required by the workers on the jobsite. Doors and windows are dimensioned to their centers and continuous (also known as *chain*) dimensioning is typically employed on floor plans.

The efficient use of blocks and layering techniques by drafters can increase productivity and lower the cost of creating a set of plans.

Job Skills

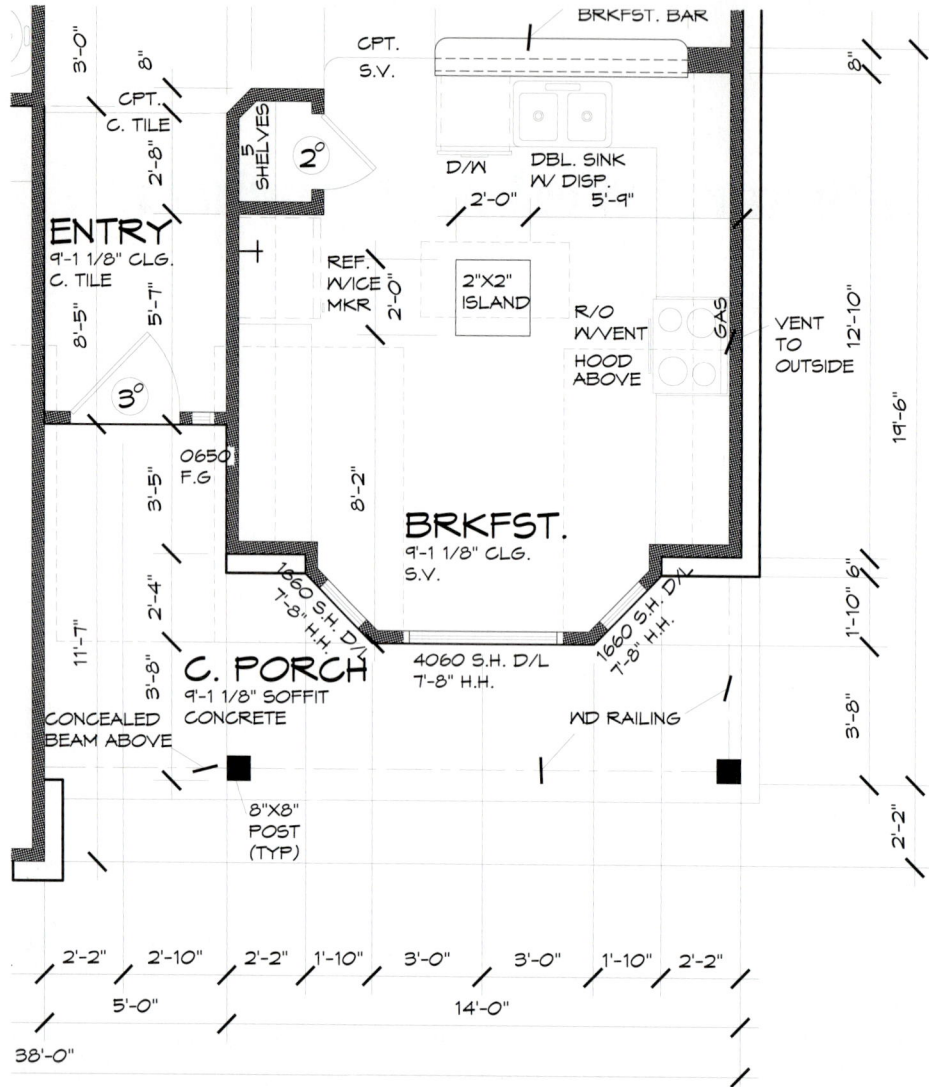

Figure 12-1   Example detail from a floor plan

electrical plans: Drawings that provide electrical contractors information about the type, location, and installation of electrical components (switches, lamps, ceiling fans, electrical outlets, cable TV jacks, etc.) used in the project. All of the information needed by the electrical contractor to wire the building should be provided by this plan.

elevation drawings: Drawings that provide information about the exterior details of a building. This information may include roof pitch, exterior materials and finishes, overall heights of features, and window and door styles. All of the dimensions and notations required by workers on the jobsite should be included on this sheet.

ELECTRICAL PLANS

An **electrical plan** provides electrical contractors information about the type, location, and installation of electrical components (switches, lamps, ceiling fans, electrical outlets, cable TV jacks, etc.) used in the project. All of the information needed by the electrical contractor to wire the building should be provided by this plan.

Drafters can create a block library of electrical symbols to speed the process by which electrical plans are created. The electrical components and wiring are usually drawn on a separate layer that is superimposed on the floor plan layer(s). Figure 12-2 shows a detail from an electrical plan. A legend is included on the electrical plan to help workers identify all of the components on the floor plan. Figure 12-3 shows an example of an electrical legend.

ELEVATIONS

Elevation drawings provide information about the exterior details of a building. This information may include roof pitch, exterior materials and finishes, overall heights of features, and window and door styles as shown in Figure 12-4. All of the dimensions and notations required by workers on the jobsite should be included on this sheet.

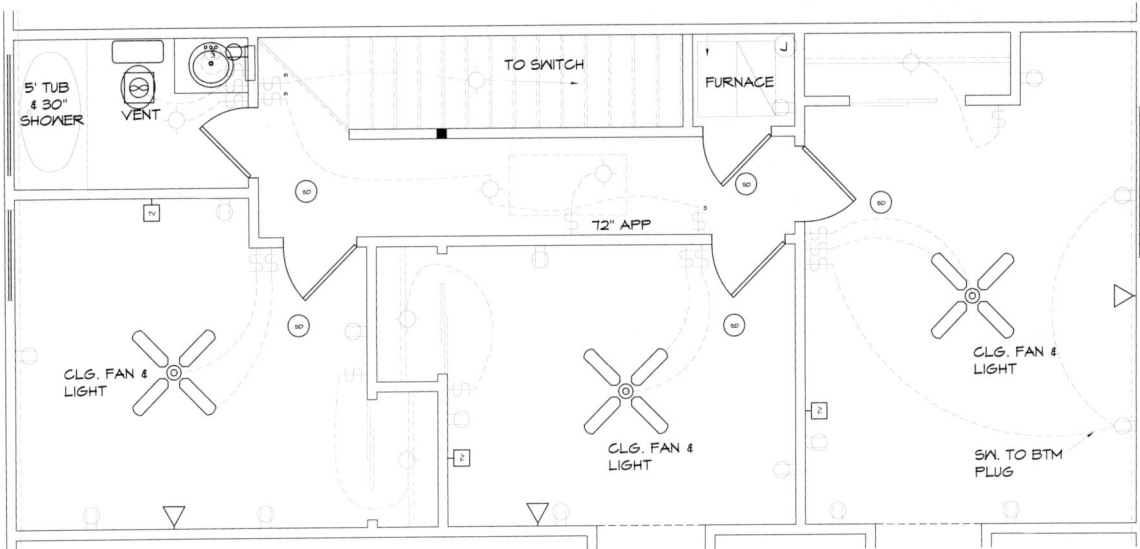

Figure 12-2   Electrical plan detail

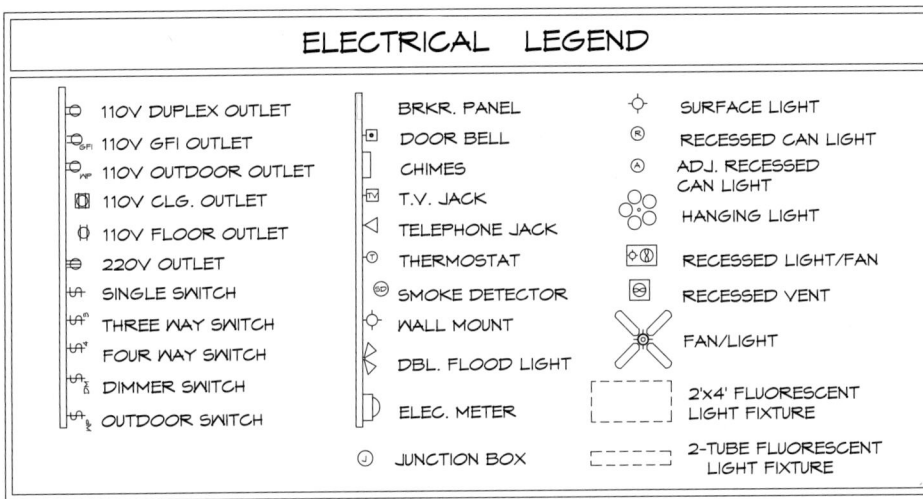

ELECTRICAL   LEGEND

110V DUPLEX OUTLET	BRKR. PANEL	SURFACE LIGHT
110V GFI OUTLET	DOOR BELL	RECESSED CAN LIGHT
110V OUTDOOR OUTLET	CHIMES	ADJ. RECESSED CAN LIGHT
110V CLG. OUTLET	T.V. JACK	HANGING LIGHT
110V FLOOR OUTLET	TELEPHONE JACK	
220V OUTLET	THERMOSTAT	RECESSED LIGHT/FAN
SINGLE SWITCH	SMOKE DETECTOR	RECESSED VENT
THREE WAY SWITCH	WALL MOUNT	
FOUR WAY SWITCH	DBL. FLOOD LIGHT	FAN/LIGHT
DIMMER SWITCH	ELEC. METER	2'X4' FLUORESCENT LIGHT FIXTURE
OUTDOOR SWITCH	JUNCTION BOX	2-TUBE FLUORESCENT LIGHT FIXTURE

Figure 12-3   Electrical legend

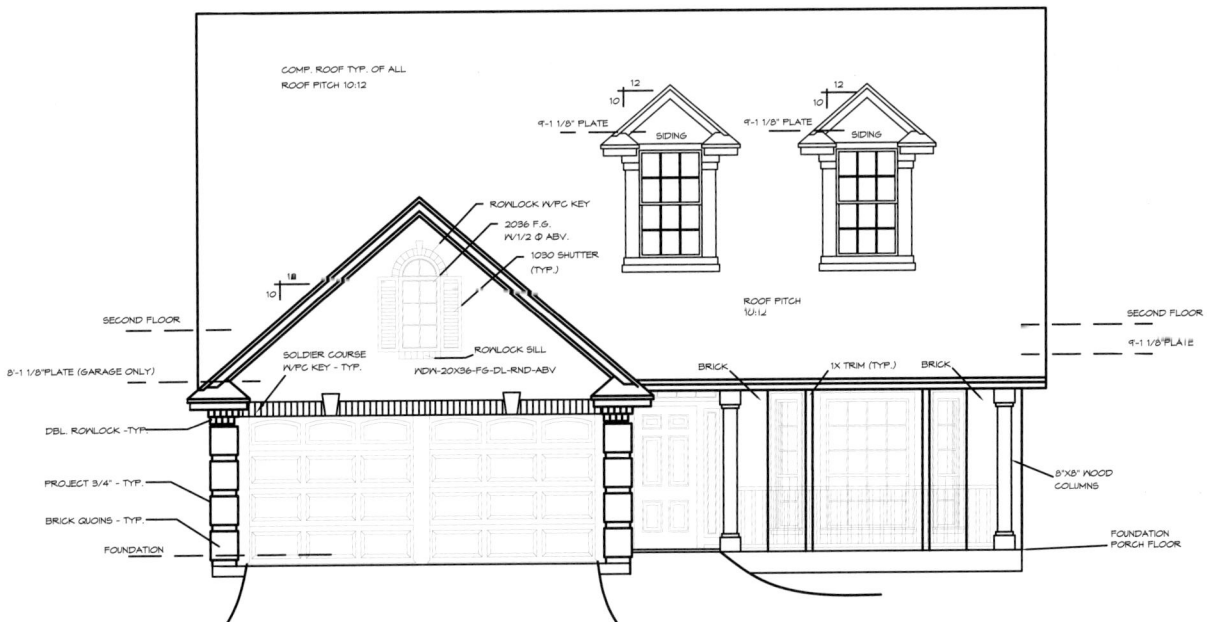

Figure 12-4   Front elevation of a house

Creating Elevations Using Multiview Drawing Techniques

In Figure 12-5, lines and arrows have been drawn between the views to show how the location and size of features on the house's exterior can be projected from one view to another using multiview drawing techniques. In fact, sometimes it is not possible to complete the construction of one elevation view without constructing a neighboring elevation and projecting information from the new view back to the original view. For example, in Figure 12-5, it would not be possible to draw the top line of the roof plane in either of the side elevations without first drawing the front elevation and projecting the roof peak to the side elevations.

Note:
Although **Figure 12-5** shows the top view of the house, this view would not be included on the elevations sheet because the top view reflects the roof plan of the building which is typically drawn on a separate sheet.

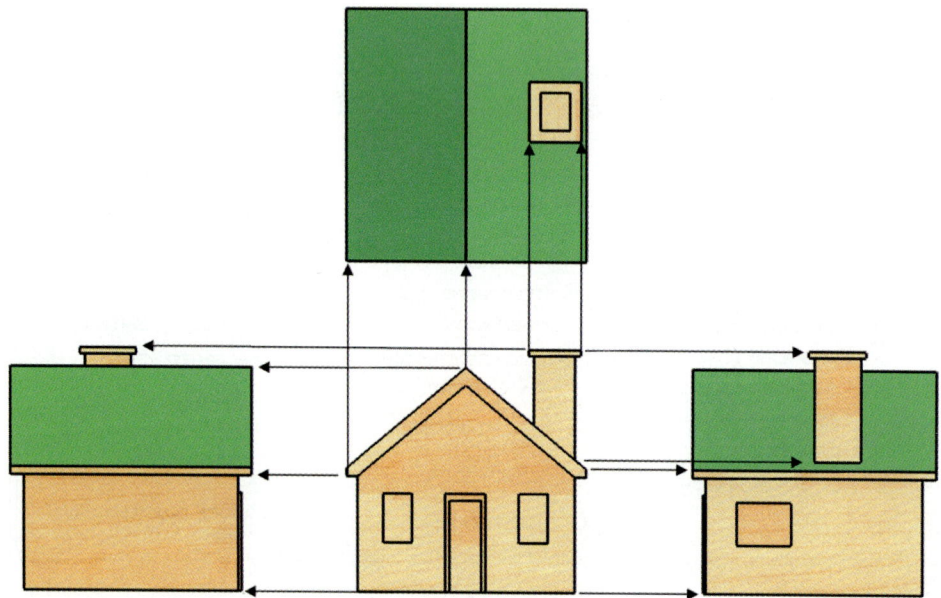

Figure 12-5   Projecting points and planes between views of a house using multiview drawing techniques

Architectural Wall Sections

In architectural drawings, *wall sections* are often included to specify the composition of a wall as shown in Figure 12-6. Drafters often refer to an exterior wall section when determining roof angles, overhangs of rafters, and heights of walls and ceilings in the elevation view.

Sections are also drawn on *foundation* plans to show details of the composition of foundation beams, slabs, and footings.

Roof Profiles on Architectural Elevations

The angle of a roof is called its ***roof pitch***. Pitch is specified as a ratio of the vertical *rise* of the roof (measured in inches) to the horizontal *run* of the roof (measured in inches). Using this notation, a roof with a "four-twelve" pitch (labeled as **4/12** on the drawing) would rise 4″ for every 12″ of horizontal run. A roof with a **12/12** pitch would rise 12″ for every 12″ of run. A **12/12** pitch would result in a roof angle of 45°.

roof pitch: The angle of a roof. Pitch is expressed as a ratio of the vertical *rise* of the roof (measured in inches) to the horizontal *run* of the roof (measured in inches). Using this notation, a roof with a "four-twelve" pitch (labeled as **4/12** on the drawing) would rise 4″ for every 12″ of horizontal run. Roof pitch is noted on elevation drawings in a set of construction documents.

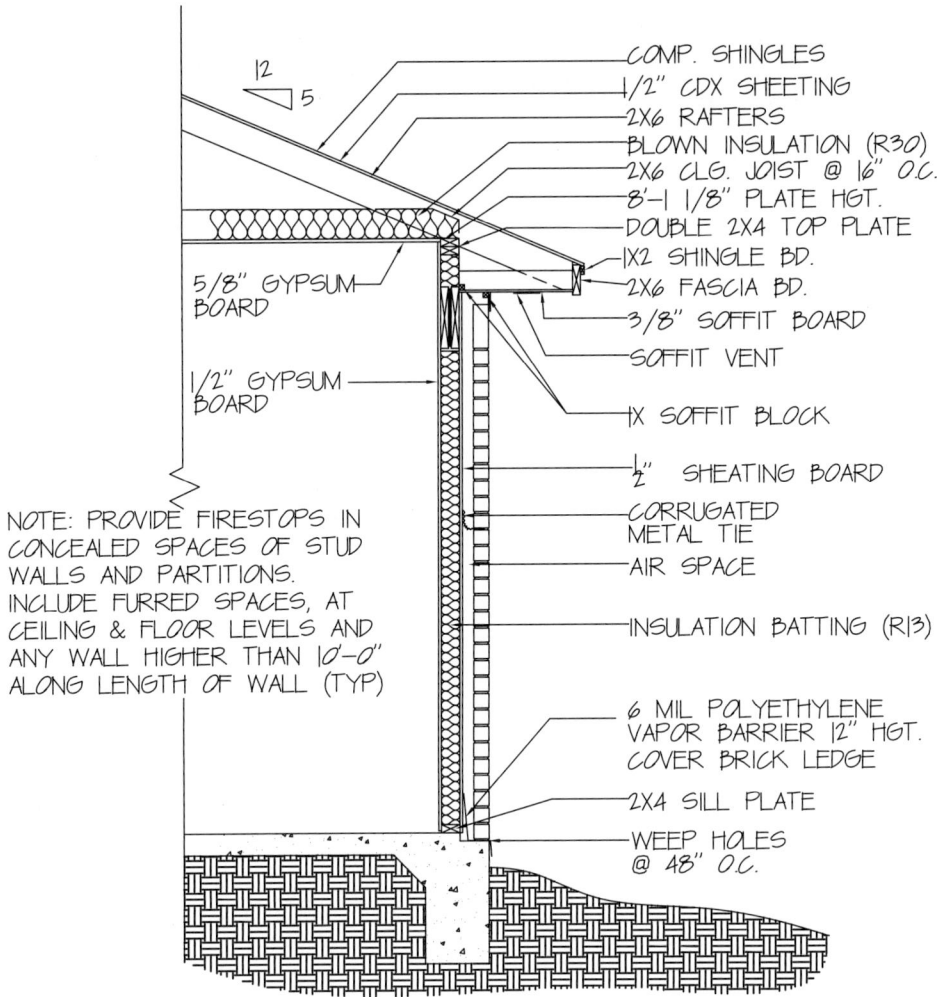

COMP. SHINGLES
1/2" CDX SHEETING
2X6 RAFTERS
BLOWN INSULATION (R30)
2X6 CLG. JOIST @ 16" O.C.
8'-1 1/8" PLATE HGT.
DOUBLE 2X4 TOP PLATE
1X2 SHINGLE BD.
2X6 FASCIA BD.
3/8" SOFFIT BOARD
SOFFIT VENT
1X SOFFIT BLOCK
½" SHEATING BOARD
CORRUGATED METAL TIE
AIR SPACE
INSULATION BATTING (R13)
6 MIL POLYETHYLENE VAPOR BARRIER 12" HGT. COVER BRICK LEDGE
2X4 SILL PLATE
WEEP HOLES @ 48" O.C.

5/8" GYPSUM BOARD

1/2" GYPSUM BOARD

NOTE: PROVIDE FIRESTOPS IN CONCEALED SPACES OF STUD WALLS AND PARTITIONS. INCLUDE FURRED SPACES, AT CEILING & FLOOR LEVELS AND ANY WALL HIGHER THAN 10'-0" ALONG LENGTH OF WALL (TYP)

ONE STORY DETAIL
3/8" = 1'-0"

Figure 12-6   Architectural wall section

A roof pitch symbol is created by drawing a horizontal line that is crossed near one end by a vertical line like the ones shown along the roof profiles in Figure 12-7. The rise is labeled next to the vertical line and the run (usually 12″) is noted above the horizontal line.

In the roof profile shown in Figure 12-7, for every 12″ the roof runs along its horizontal axis, it rises 10″. A drafter would label this pitch specification as **10/12** on the drawing.

Using the Floor Plan to Locate Features on Elevations

When creating elevation drawings, the physical location of features on the floor plan, such as doors and windows, can be used to locate these same features in the elevation drawing.

Figure 12-8 shows the floor plan of the Guest Cottage you drew earlier in this course. Also shown in this figure are the front, back, and side elevations of the Guest Cottage. To speed the creation of the elevation views, information about the size and location of the windows, outside walls, and the front door was projected from the floor plan to the elevation views.

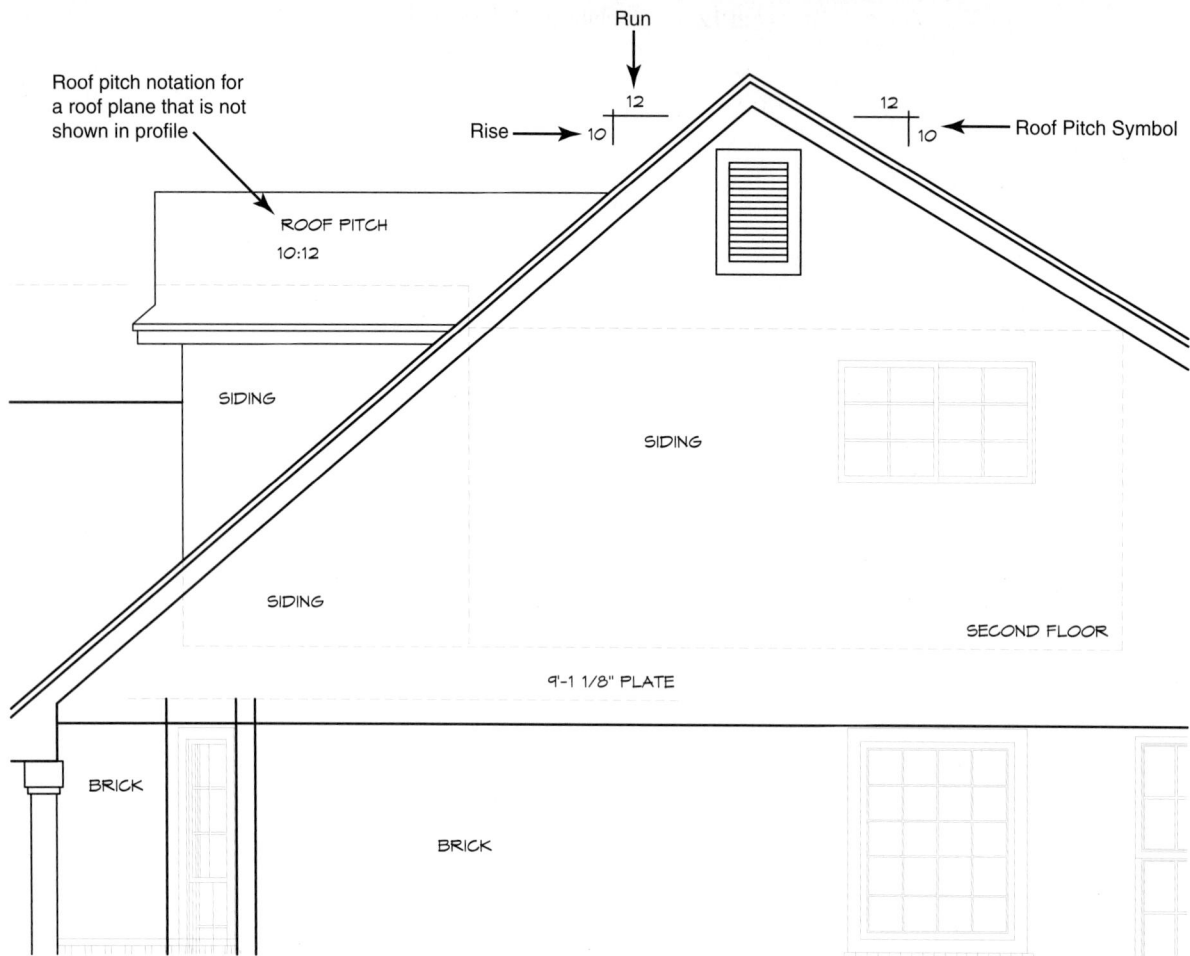

Figure 12-7 Notating roof pitch in an elevation view

Run

Rise

12
10

12
10

Roof pitch notation for a roof plane that is not shown in profile

ROOF PITCH
10:12

Roof Pitch Symbol

SIDING

SIDING

SIDING

SECOND FLOOR

9'-1 1/8" PLATE

BRICK

BRICK

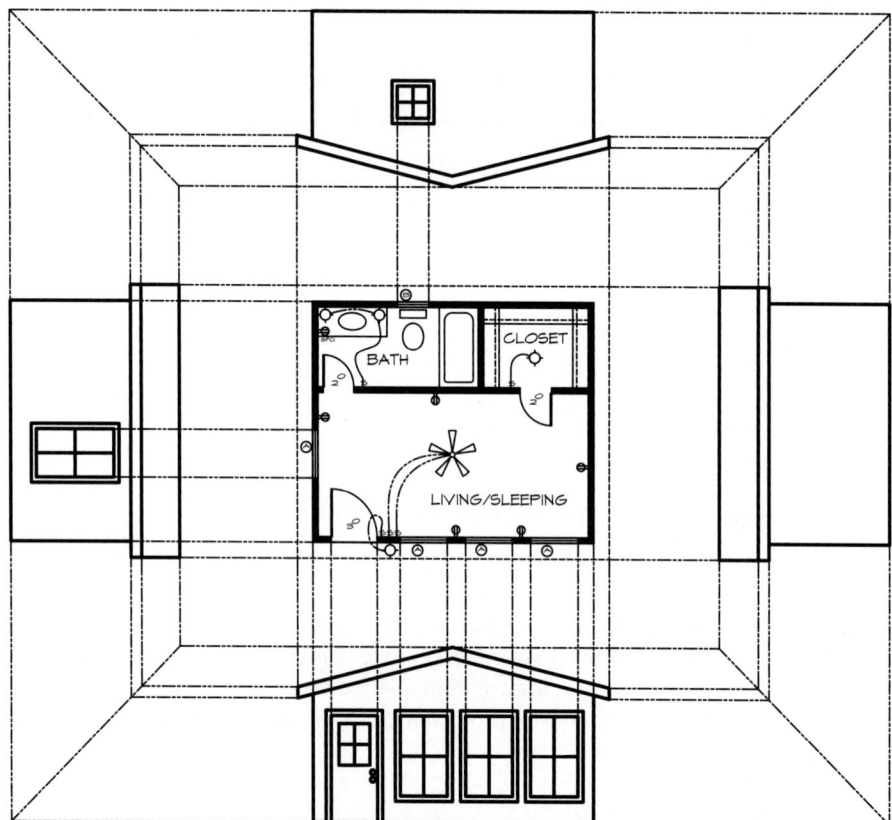

Figure 12-8 Projecting elevation features of the guest cottage from the floor plan

BATH

CLOSET

LIVING/SLEEPING

Locate the 45° miter lines in the corners of Figure 12-8 and note how information is projected among the front, back, and side views through these miter lines.

The multiview projection technique shown in Figure 12-8 is commonly used by architectural drafters to create elevation views.

SUMMARY

In the early stages of an architectural project, designers and clients work together to produce a design that meets the client's needs *and* budget. During the design stage, the designer may communicate with the client through sketches, rendered CAD models, or scale models built of cardboard and foam-core.

When the client is satisfied with the initial design, drafters work under the supervision of the designer to create a set of construction documents containing all of the information necessary to build the project. Construction documents (CDs) are the centerpiece of every construction project, and almost all who have a role in the construction of the project rely on architectural working drawings to accomplish their jobs, from lenders who review CDs to determine the levels of funding for the project to contractors who use CDs during the bidding and construction phases of the project.

Drafters must understand how to apply CAD techniques, such as the use of block libraries, in order to produce CDs quickly without sacrificing detail and accuracy.

UNIT TEST QUESTIONS

Short Answer

1. What are construction documents?

2. Name four block libraries that might be used by architectural firms to produce drawings.

3. What is meant by the term *pitch* when used in the context of a roof?

4. Describe what is meant by a **3/12** notation on a roof plan.

5. Name three types of architectural drawings that may be drawn as section views.

UNIT PROJECTS

Project 12-1: Cabin

In this project you will create the floor plan and elevations for a small cabin. The finished sheets will resemble the ones shown in Figure 12-9 and Figure 12-10.

Directions

1. Select **FILE** and **OPEN**.
2. Select the **Student Resource CD**.
3. Select the **Prototype Drawings** folder.
4. Select the **Arch Cabin Prototype** drawing (open as "read-only").
5. **SAVE AS** to your **Home** directory and rename the drawing **CABIN PROJECT**.
6. Create the following layers: Floor Plan, Doors, Electric Plan, Switch Lines, Plumbing Plan, Kitchen, Labels, Dimensions, Wall Hatch, Elevations, and Notes. Assign a color to each layer and set the linetype for the Switch Lines layer to Phantom.
7. Following the directions on the next pages, create the floor plan and the front and right elevations of the cabin.

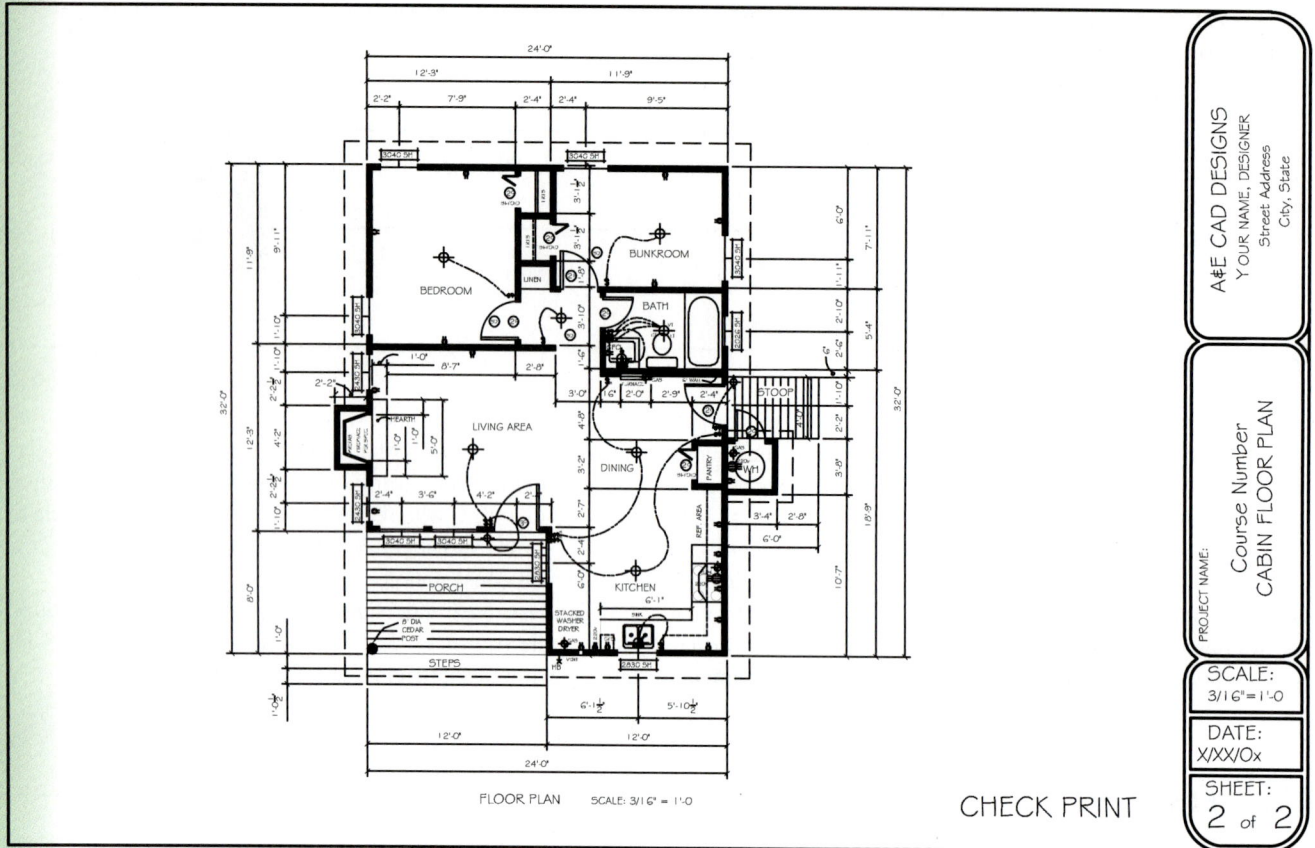

Figure 12-9   Cabin floor plan

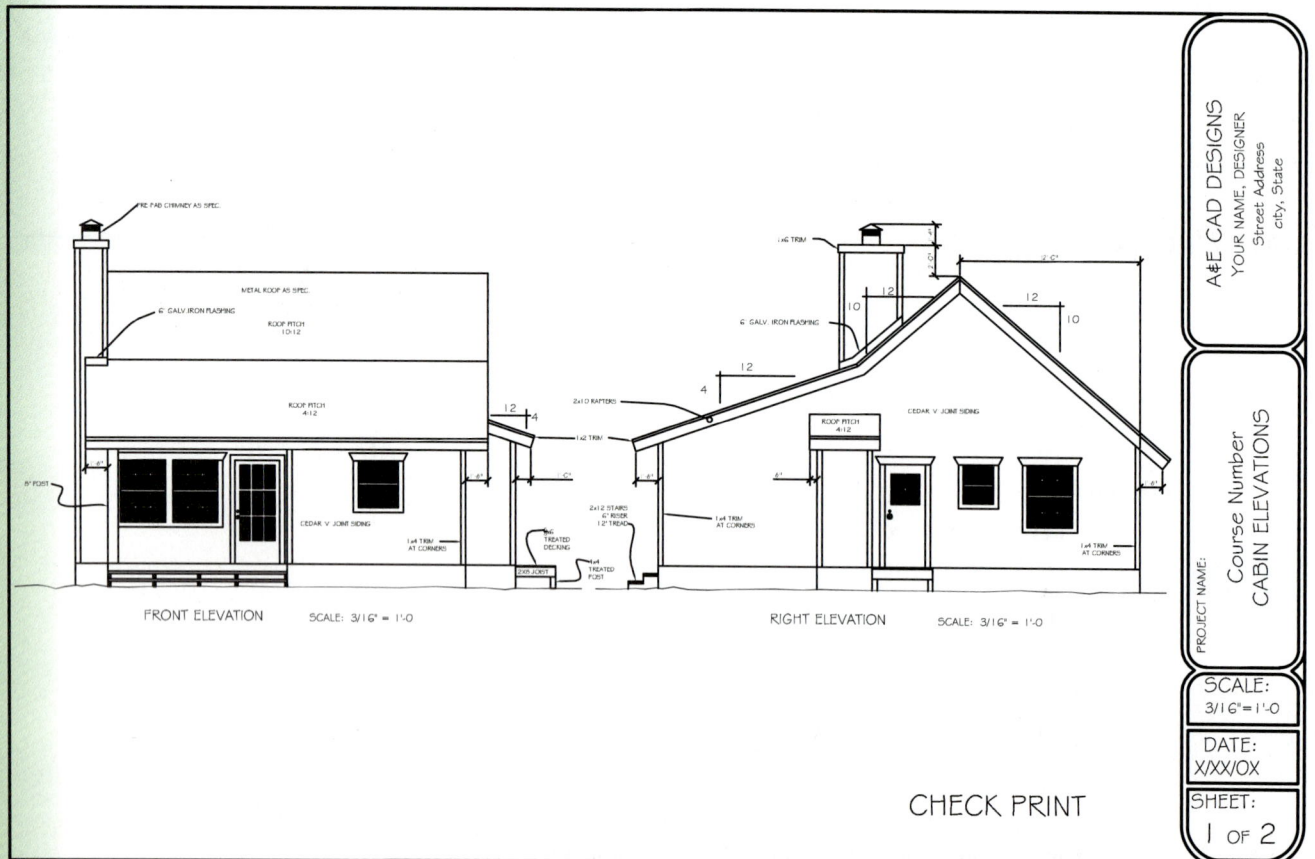

Figure 12-10   Cabin elevations

Drawing the Floor Plan

Step 1. To draw the cabin's perimeter walls, set Floor Plan as the current layer and draw the perimeter of the cabin using the dimensions shown in Figure 12-11. Use the **OFFSET** command to draw the walls **4"** thick.

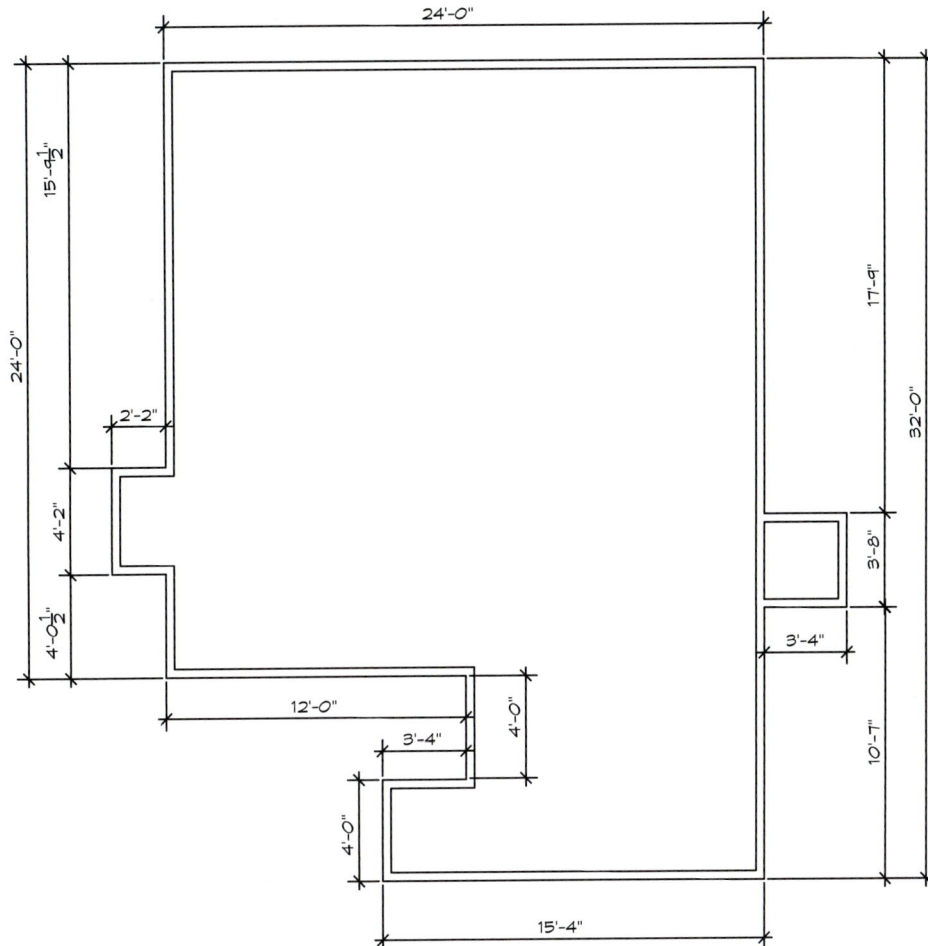

Figure 12-11   Perimeter walls

> **TIP** Use the **POLYLINE** command to draw the perimeter as one entity. This will facilitate using the **OFFSET** command to create the wall thickness.

When Architectural units are in effect, distances will default to inches unless you enter a foot mark ('). For example, for a line 24' 6" in length, enter 24'-6 (you do not need to type the inch mark after 6 because AutoCAD will default to inches).

Enter dimensions with fractions by typing a dash between the inch value and the fractional value, for example, 15'9-1/2.

Note:
Do not add dimensions to the floor plan until you are instructed to do so in Step 12.

Step 2. To draw the interior walls, use the dimensions shown in Figure 12-12. You can locate these walls by offsetting their edges from the edges of known walls. Draw all interior walls 4" wide except for the 6" wide wall noted in the figure.

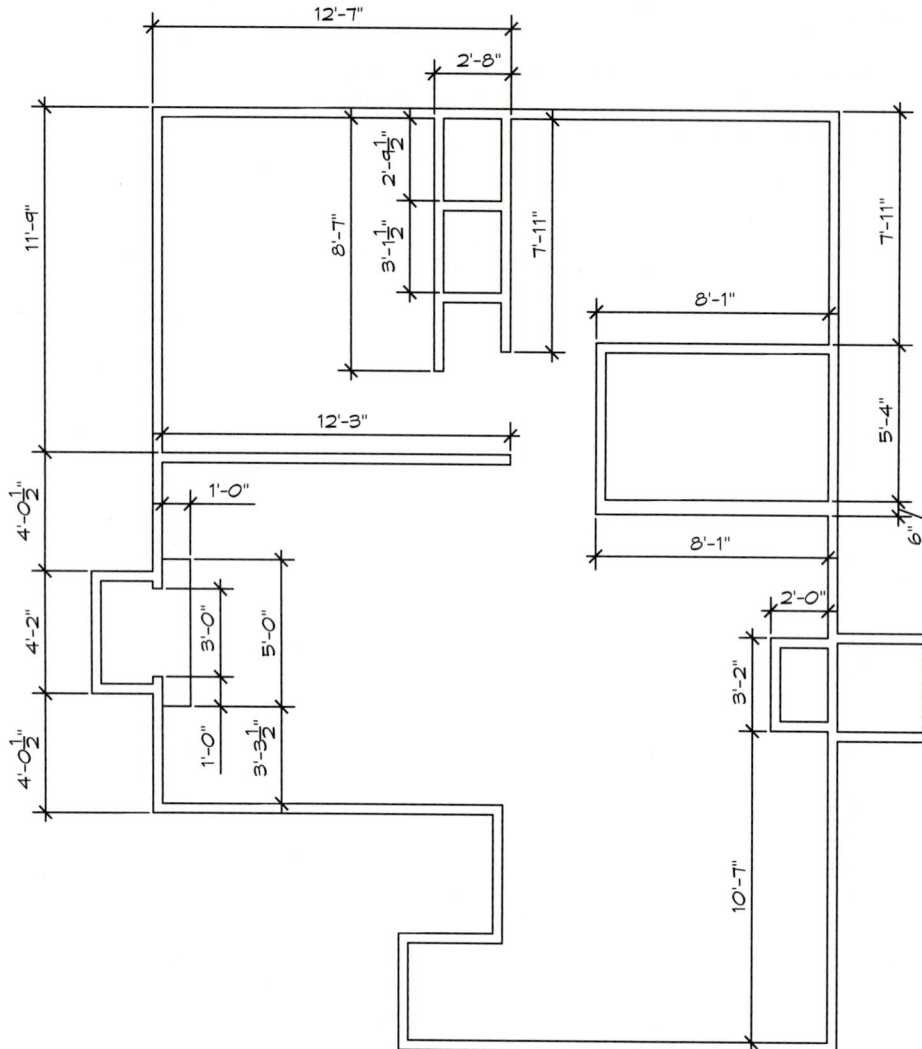

Figure 12-12 Interior walls

Step 3. To create a block library, open the drawing file named **Cabin Symbols** that is located on the **Student Resource CD** inside the **Prototype Drawings** folder and use **SAVE AS** to save the file to your **Home** folder. Make blocks for each of the symbols shown in Figure 12-13 (refer to Unit 10 on block creation and editing if necessary).

Make the blocks on Layer 0 and assign base points that you think will facilitate placing the blocks into the drawing.

Accessing the Cabin's Blocks Through Design Center

DesignCenter is an AutoCAD feature that allows drafters to insert the blocks, layers, linetypes, dimension styles, and text styles located in one drawing file into a different drawing file. By following Steps A through D below, you can use **DesignCenter** to insert the blocks created in Step 3 (which were saved in the **Cabin Symbols** drawing file) into the **Cabin Project** drawing file.

Step A. Pick the **DesignCenter** icon located on the **Standard** toolbar (see Figure 12-14).

Step B. When the **DesignCenter** window opens, browse through the file tree in the pane on the left side of the window and find the

Directions:

For each of the architectural symbols shown below, create a block. Select insertion points that will facilitate placement of the block.

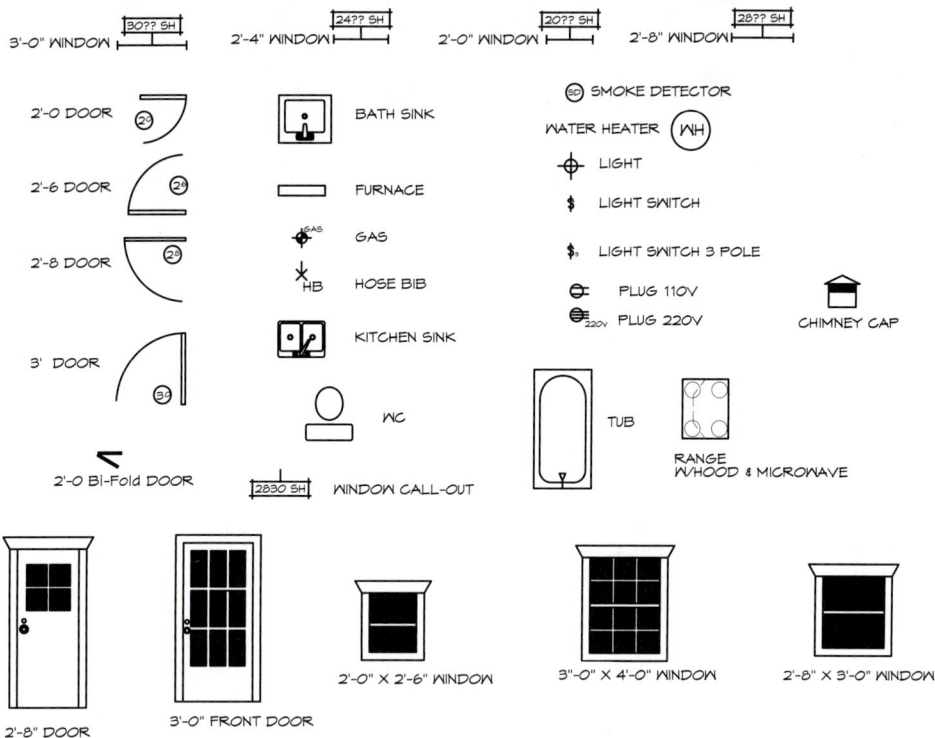

Figure 12-13   Block library of architectural symbols

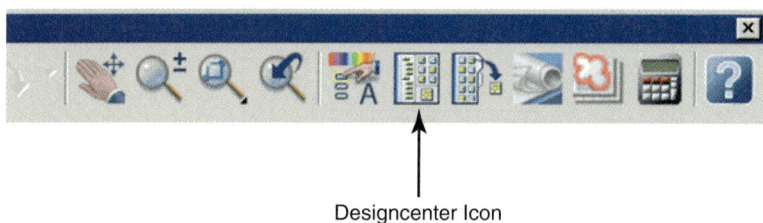

Designcenter Icon

Figure 12-14   Locating the DesignCenter icon on the Standard toolbar

Cabin Symbols drawing located in your **Home** folder (see Figure 12-15).

Step C. Double-click on the **Cabin Symbols** drawing file and double-click on **Blocks** from the tree below the **Cabin Symbols** file name. The **Cabin Symbols** blocks will become visible in the content area in the pane to the right of the file tree pane.

Step D. Insert the blocks in the **Content Area** pane by "dragging and dropping" them into the drawing by holding down the pick button of your mouse, or by selecting a block, right-clicking, choosing **Insert Block**, and inserting the block in the usual way.

Step 4. To locate the centers of the windows, offset the perimeter wall lines using the dimensions shown in Figure 12-16.

Use **DesignCenter** to access the block library created for the **Cabin Symbols** drawing and insert the window blocks into the walls of the floor plan. Trim the walls to the edges of the windows as shown in Figure 12-16.

Step 5. To place the doors, set the Doors layer current, and offset the wall lines to locate the centers and edges of doors—for

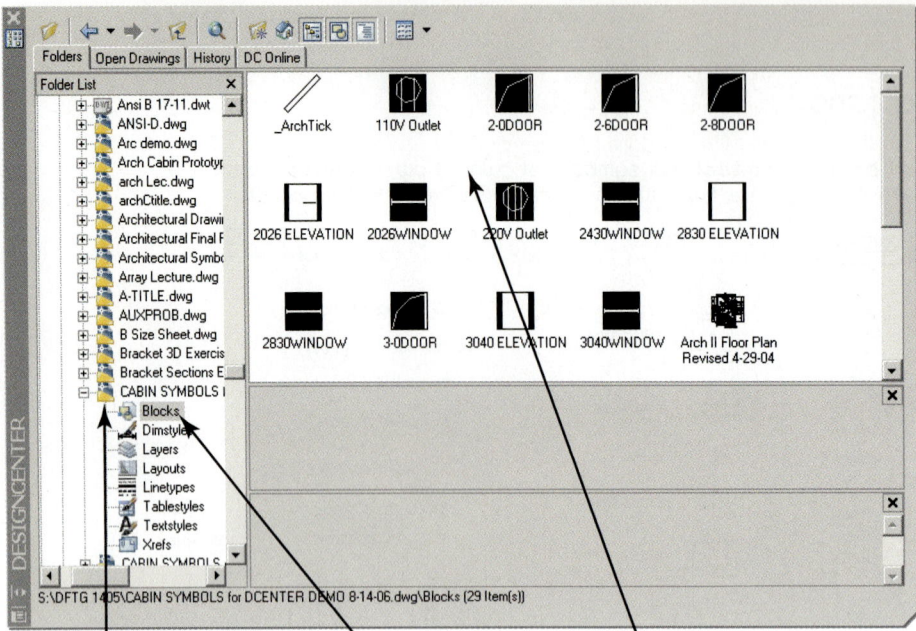

Select the Cabin Symbols Drawing File from the File Tree

Double click on Blocks

Drag and Drop Block from Content Area into the Drawing

Figure 12-15 DesignCenter window

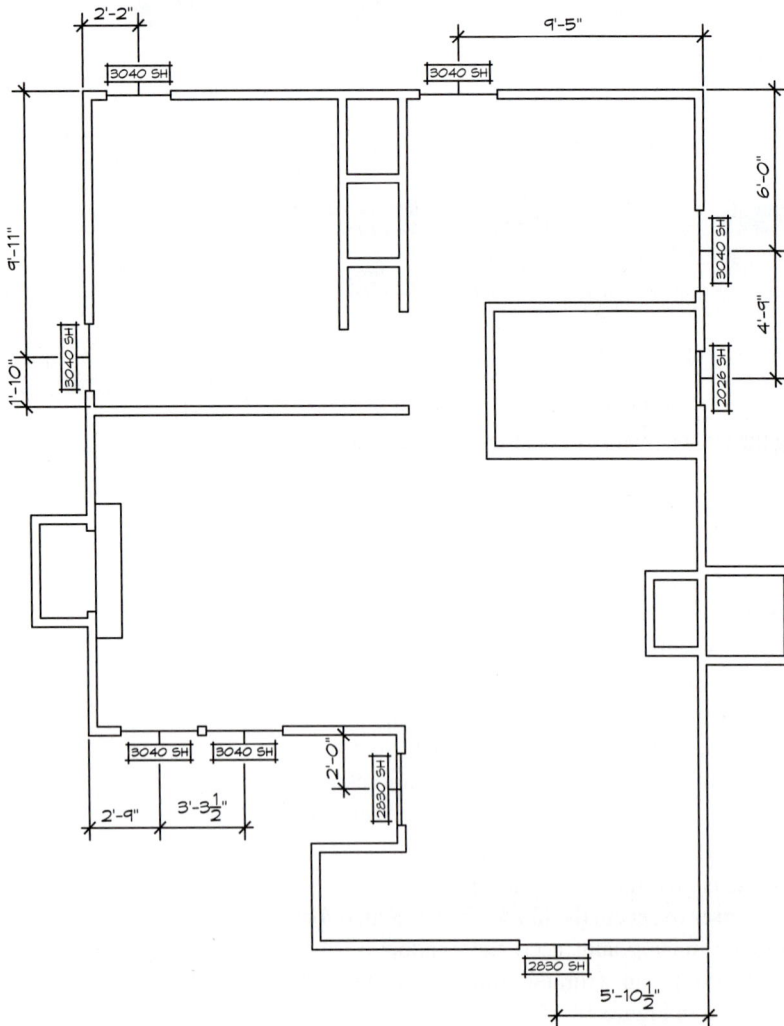

Figure 12-16 Window placement

example, the opening for a door marked 2^6 will be 2'-6" wide. Allow a minimum of **4"** on door returns. Insert the desired door block from the **DesignCenter** location and trim to the edges of the door block as shown in Figure 12-17. The construction details for the bi-fold door are shown in Figure 12-18.

Step 6. To draw the fireplace, refer to the details shown in Figure 12-19 to construct the fireplace. Add the fireplace to the

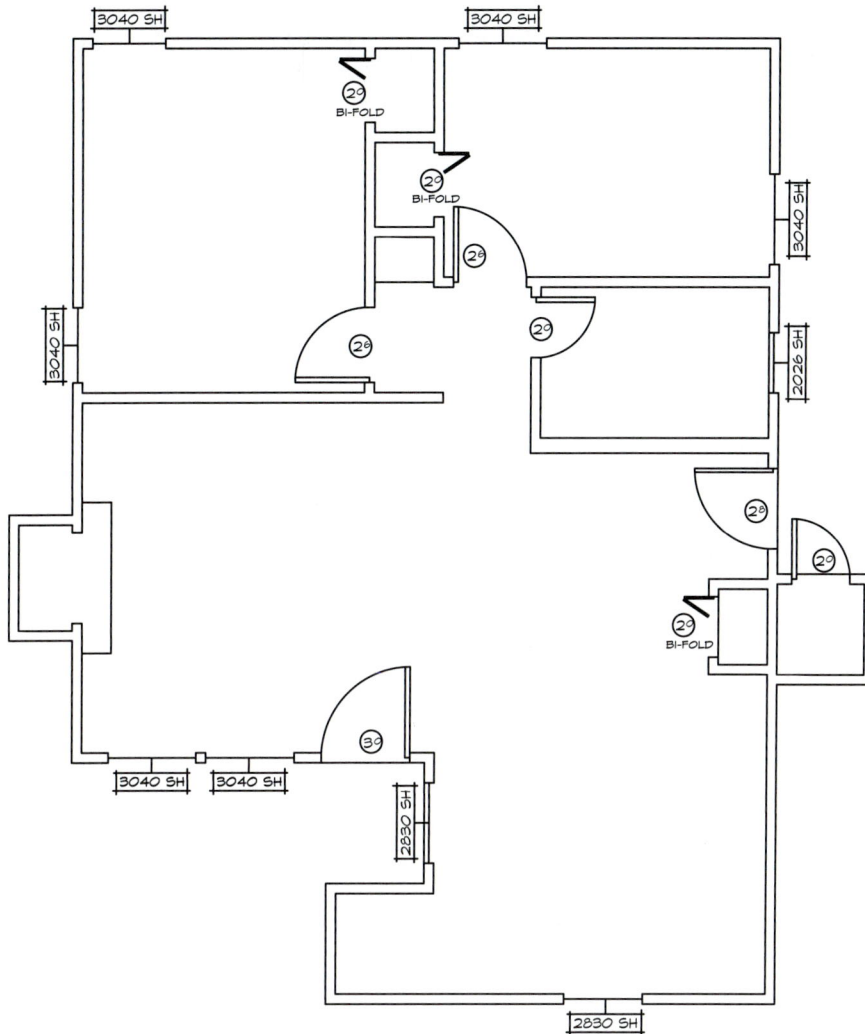

Figure 12-17   Door placement

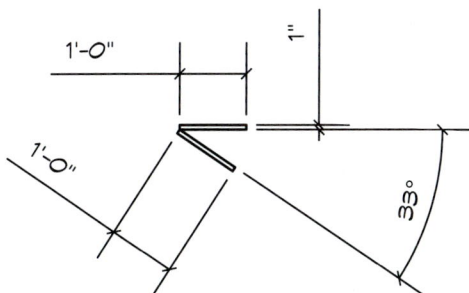

Figure 12-18   Bi-fold door

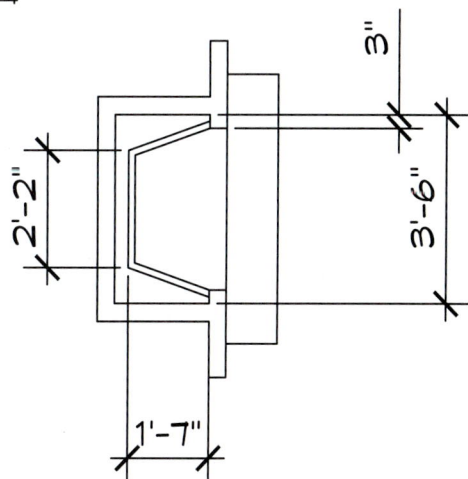

Figure 12-19   Fireplace details

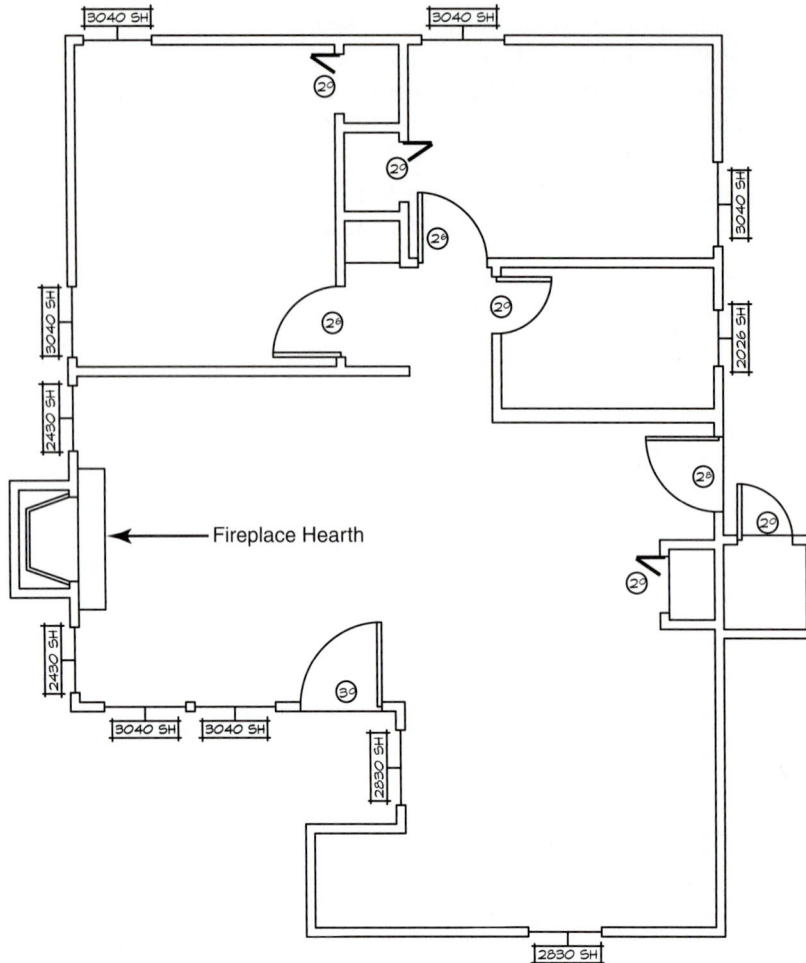

Figure 12-20   Adding the fireplace and hearth

drawing as shown in Figure 12-20. The hearth is drawn **1′** wide by **5′** long and should be centered on the fireplace.

Step 7. To draw the kitchen and bath, set the Kitchen layer current and add the kitchen cabinets and kitchen fixtures as shown in Figure 12-21. The lower kitchen cabinets should be **24″** wide. The upper cabinets represented with dashed lines should be **12″** wide. Open **DesignCenter** and insert the blocks of the kitchen fixtures by selecting them from the **Cabin Symbols** drawing file.

Next, set the Plumbing Plan layer current and insert the bathroom fixtures, furnace, and water heater as shown in Figure 12-21. Open **DesignCenter** and insert the blocks of the bathroom fixtures, furnace, and water heater from the **Cabin Symbols** drawing file. Add the dryer vent and hose bib as shown in Figure 12-21.

Step 8. To draw the electrical plan, set the Electric Plan layer current. Open **DesignCenter** and insert the blocks of the electric symbols from the **Cabin Symbols** drawing file into the floor plan as shown in Figure 12-22.

Then, set Switch Lines as the current layer and, using the **SPLINE** command, draw the *switch legs* from the switches to the lamps as shown in Figure 12-22. In an electrical plan, switch legs represent the electrical circuit that connects the switches to the lamps (or other electric fixtures) and are drawn as phantom lines.

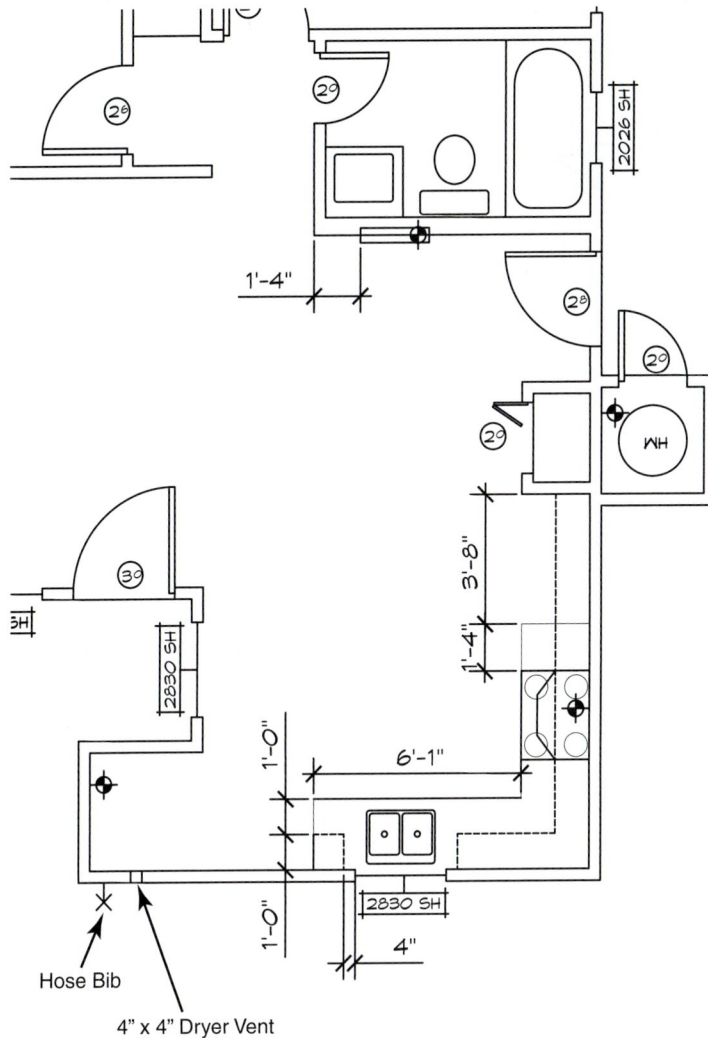

Figure 12-21 Placement of kitchen cabinets and plumbing fixtures

> **TIP** You will probably need to set the **LTSCALE** of the drawing to a larger value in order for dashes to appear in the phantom lines.

Step 9. To add labels, set the Labels layer current and add the labels shown in Figure 12-23. Change the **Standard** text style's font to **Stylus BT**, and use the following text heights: for room names, use **6″** text height; for detail notes or "call-outs," use **4″** text height; and for very small text use **3″** text height. In the title block, use **8″** text for the **Drawing Name** and **6″** text for the **Scale**.

Step 10. To add the porch, stoop, and hatching to the walls, set the Floor Plan layer current and draw the porch and stoop as shown in Figure 12-24. Draw the stoop **6′** long and **4′** wide. Use **6″** wide boards for the porch and stoop flooring. The treads of the porch steps should be drawn **12″** wide.

Add an **8″** diameter cedar post to the front left corner of the porch roof. This post supports the roof above the front porch.

Next, set Wall Hatch as the current layer and hatch the walls and the cedar post with the **Net** pattern. Set the hatch pattern to a scale of **10**.

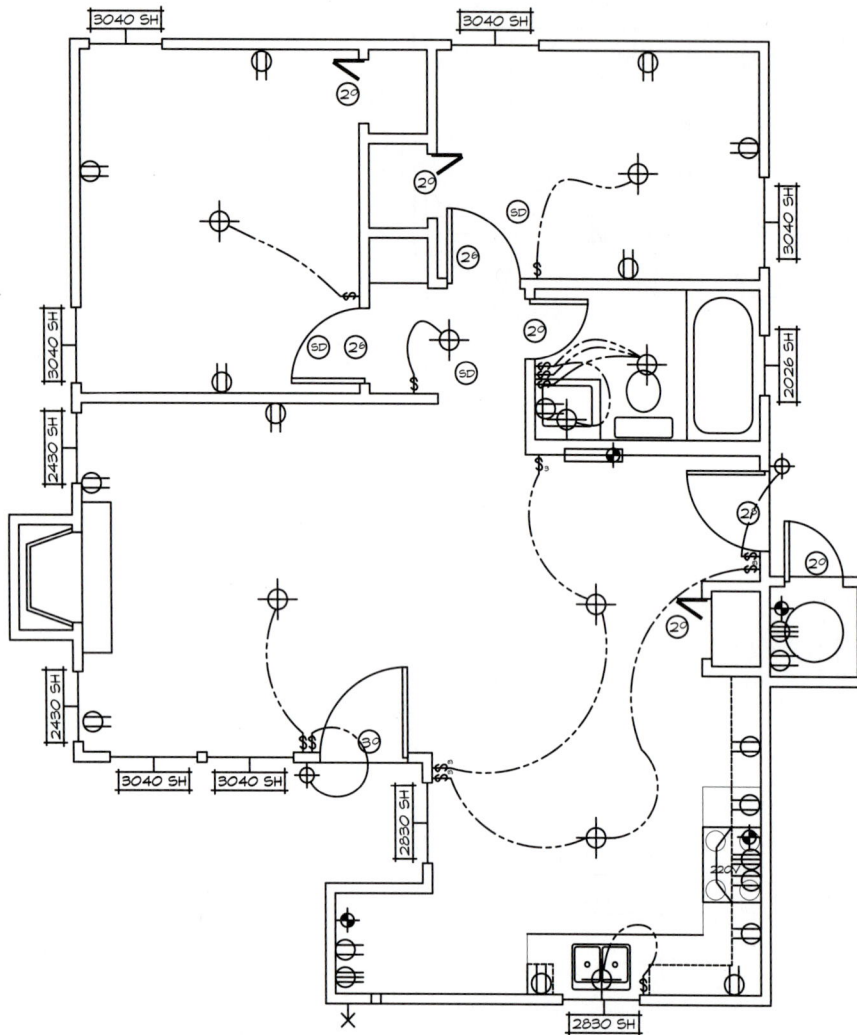

Figure 12-22   Electrical plan

Step 11. After consulting with the project architect, a client may decide to change the layout of the floor plan. Making the changes shown in Figure 12-25 at this point in the design process may increase the fees the architectural firm will charge the client for this project. Changes may also affect the final construction cost of the cabin.

The architect then has to issue a formal *Architectural Change Order* to the drafting department to request the changes. The change order will note the specific changes that need to be made to the floor plan. By following a formal process in making these changes, the architectural firm creates a record of what changes were made, when they were made, and by whom.

Edit your floor plan to reflect all of the changes to the living area and kitchen shown in Figure 12-25 and reapply the hatch pattern to the walls when the changes are complete.

Step 12. Set the Dimensions layer current and add dimensions as shown in Figures 12-26(a)–(b). Before adding dimensions, apply the dimension settings shown in Figure 12-27(a)–(e) to the tabs of the **Dimension Style Manager** dialog box.

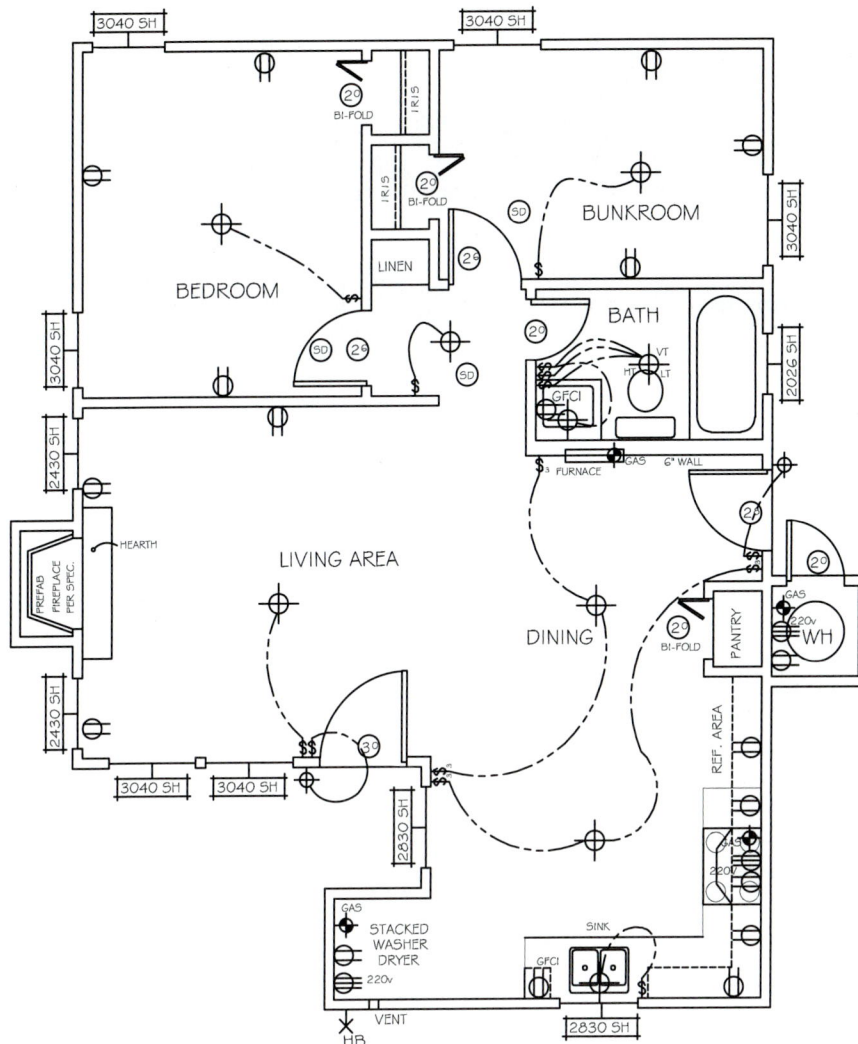

Figure 12-23   Adding labels to the floor plan

The dashed lines around the perimeter walls of the floor plan in Figures 12-26 (a)–(b) represent the edges of the roof overhang. Drafters can determine the placement of these lines by referring to the elevation sketches of the cabin and noting the rafter overhang distances. The roof overhang around the main part of the cabin is **18″**, but the roof overhang around the water heater closet is **6″** on the sides and **12″** along the outside edge.

Dimension Style Settings for the Cabin Project

The dimension style settings for the cabin project are shown in Figures 12-27(a)–(e).

The values shown in these tabs may seem large when compared to the settings for a mechanical drawing—for example, setting text height to 5″—but when the project is printed to a scale of 3/16″ = 1′-0″, the settings will be proportional to the size of the printed sheets.

When Step 12 is completed the floor plan of the cabin is finished.

Creating the Elevations of the Cabin Project

Using the same prototype drawing used to create the floor plan of the cabin, draw the front and right elevations as shown in Figure 12-28. Remember that architectural elevations are constructed using multiview

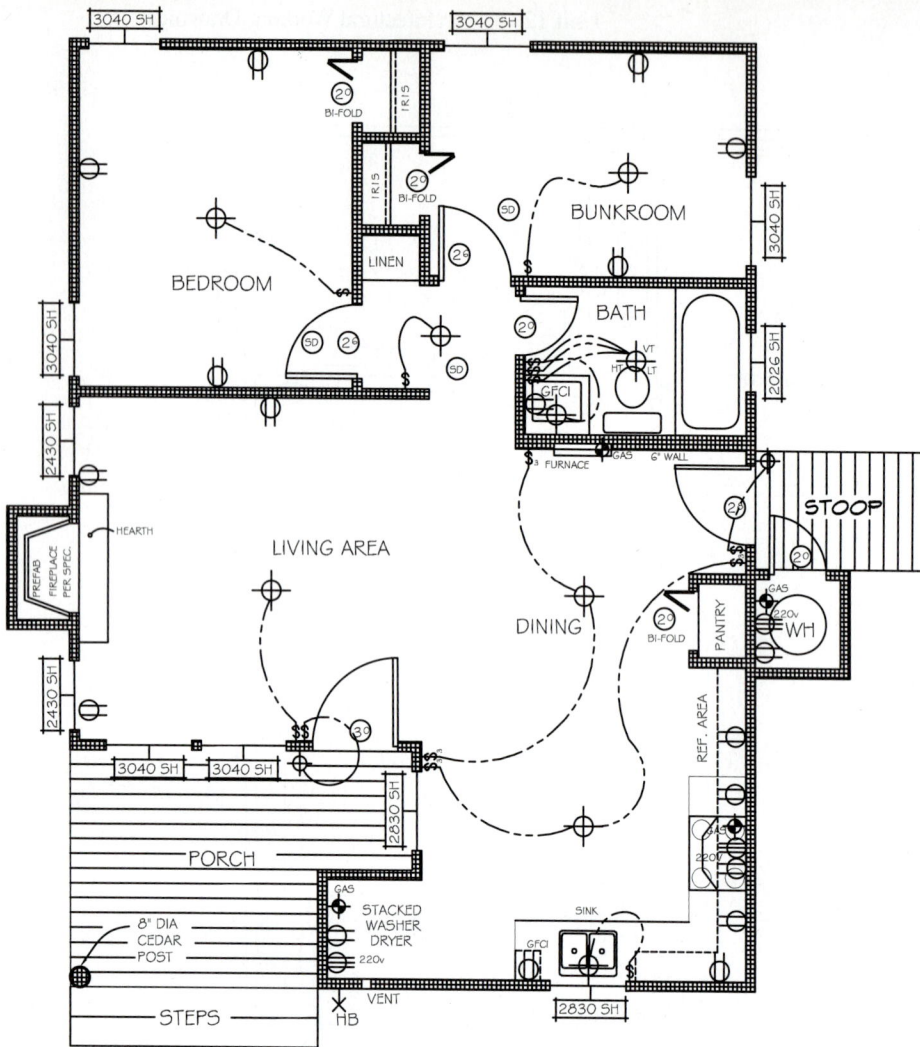

Figure 12-24 Floor plan with porch, stoop, and hatched walls

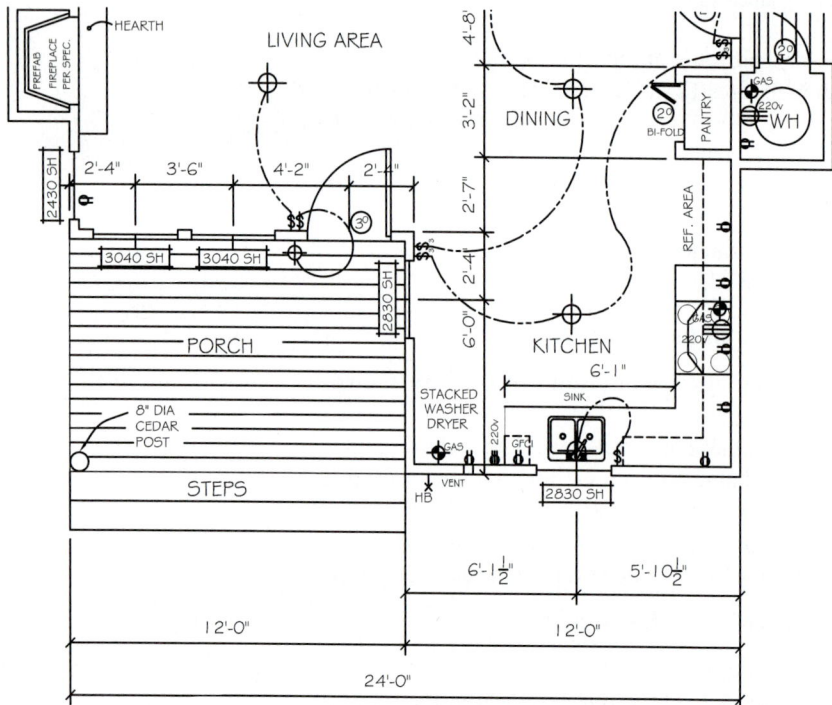

Figure 12-25 Cabin floor plan after an Architectural Change Order

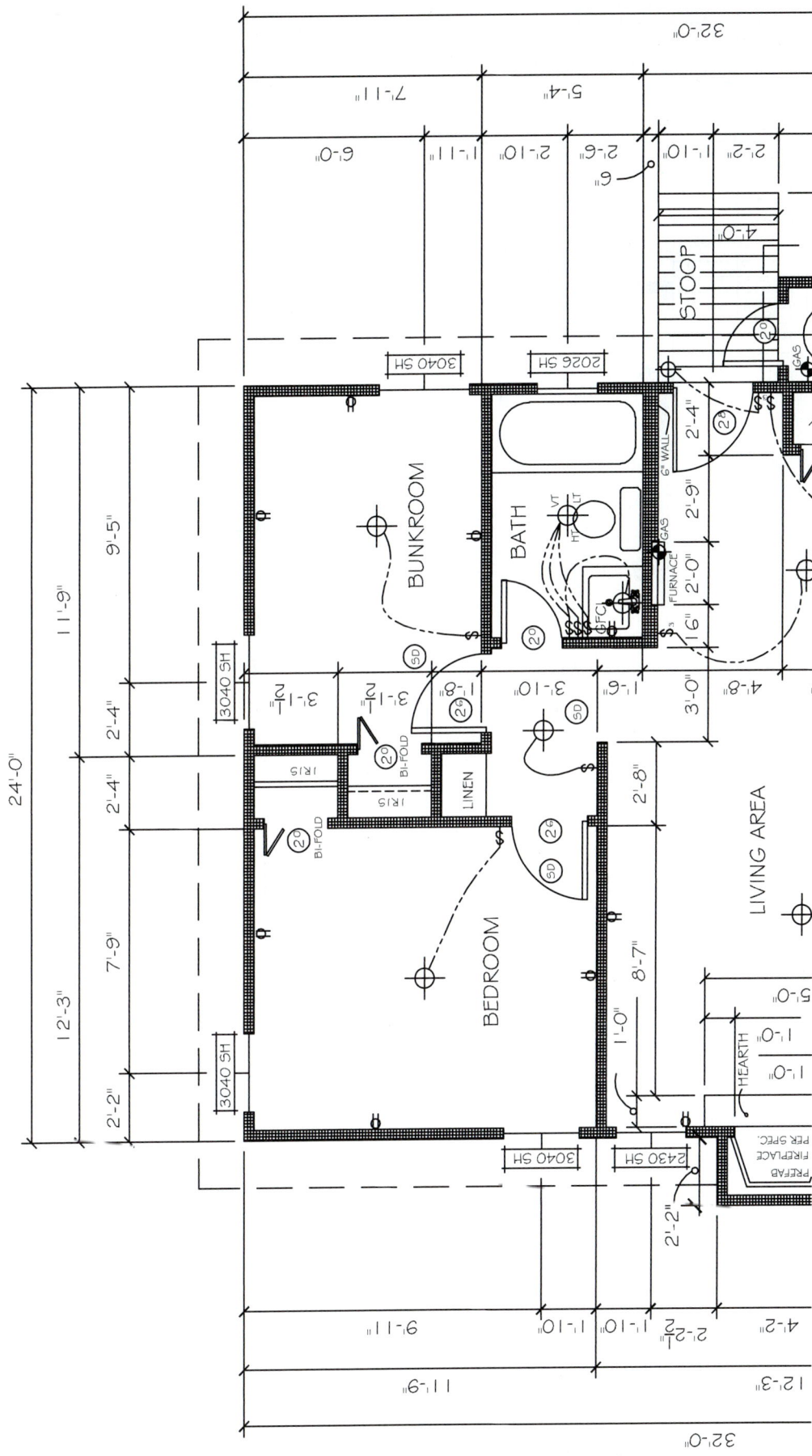

Figure 12-26(a)   Detail of floor plan dimensions

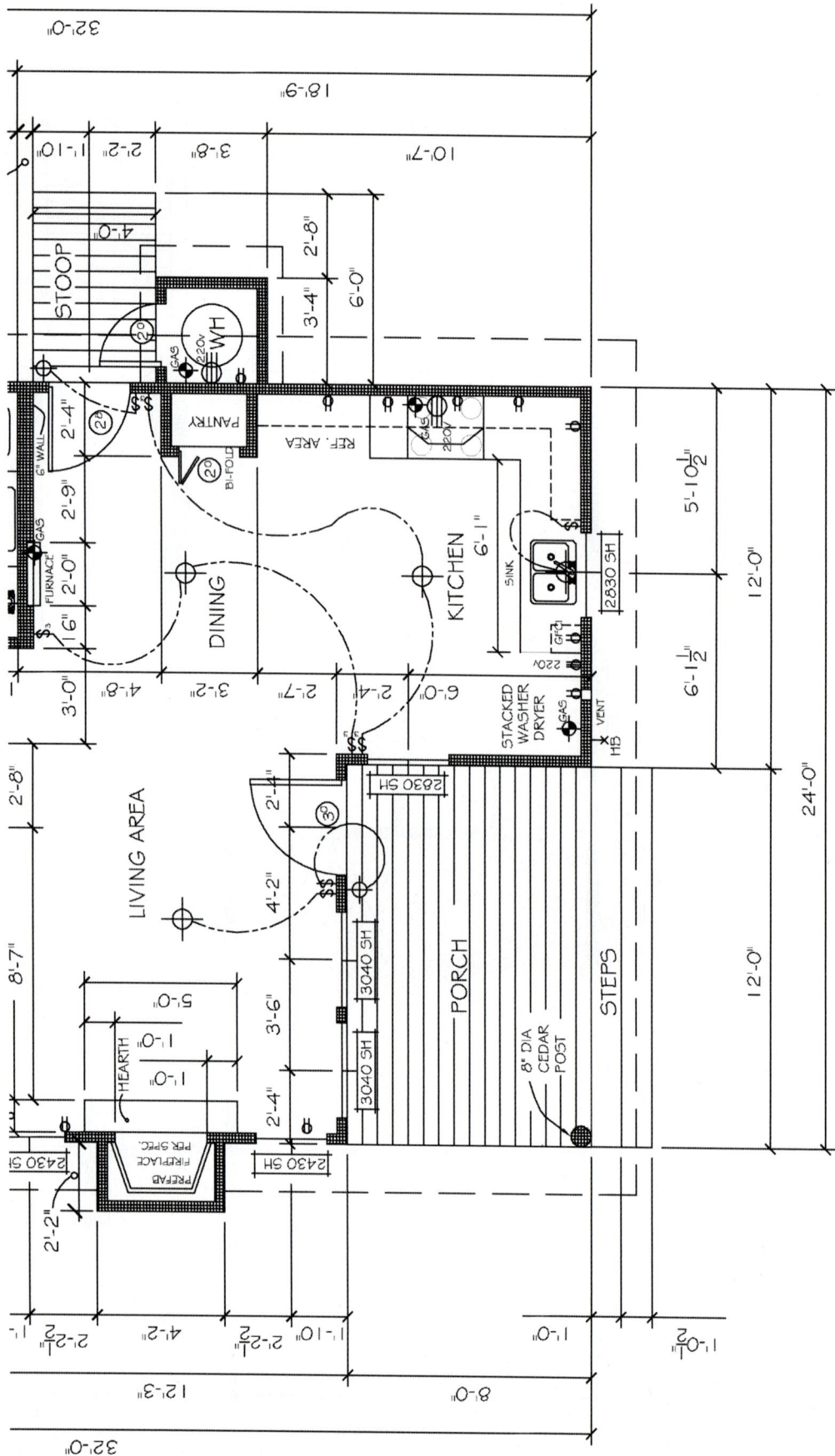

Figure 12-26(b) Dimensioning the floor plan

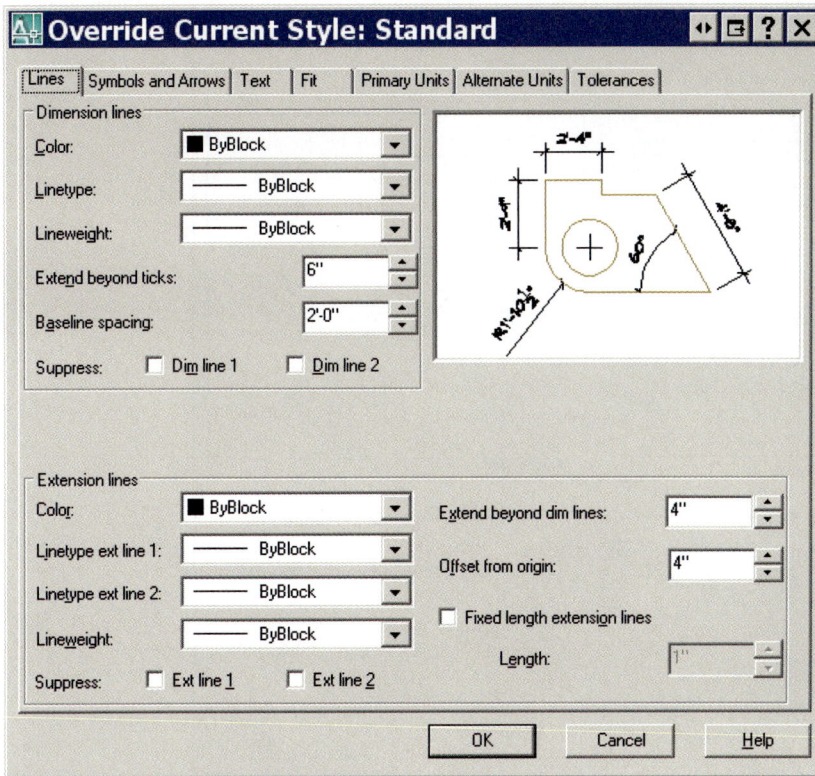

Figure 12-27(a)   Cabin project dimension style setting

drawing techniques and that information will be projected from one view to the next as the views are constructed.

The steps to create the elevations are presented on the following pages.

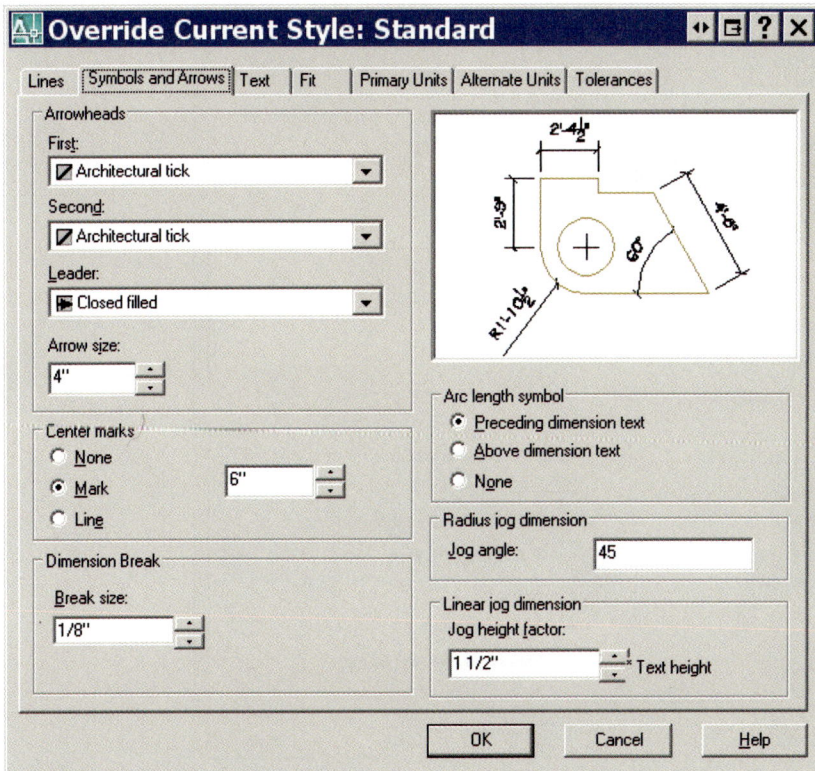

Figure 12-27(b)   Cabin project dimension style setting

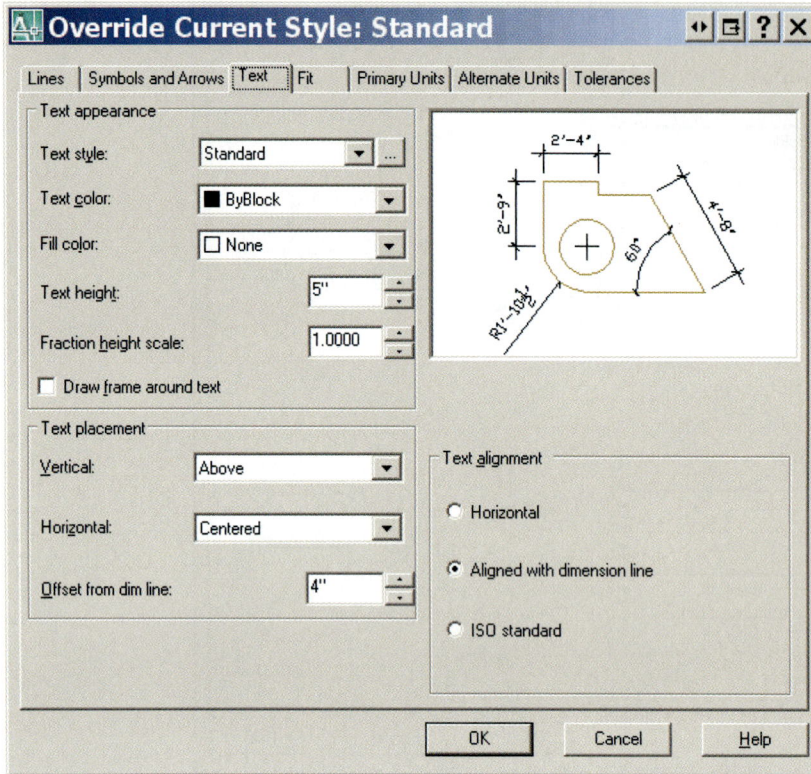

Figure 12-27(c)   Cabin project dimension style setting

Relating the Cabin Elevations to Features on the Floor Plan.   When creating elevation drawings of the cabin, the dimensions shown on the floor plan are used to locate exterior features such as doors, windows, porches, and the chimney.

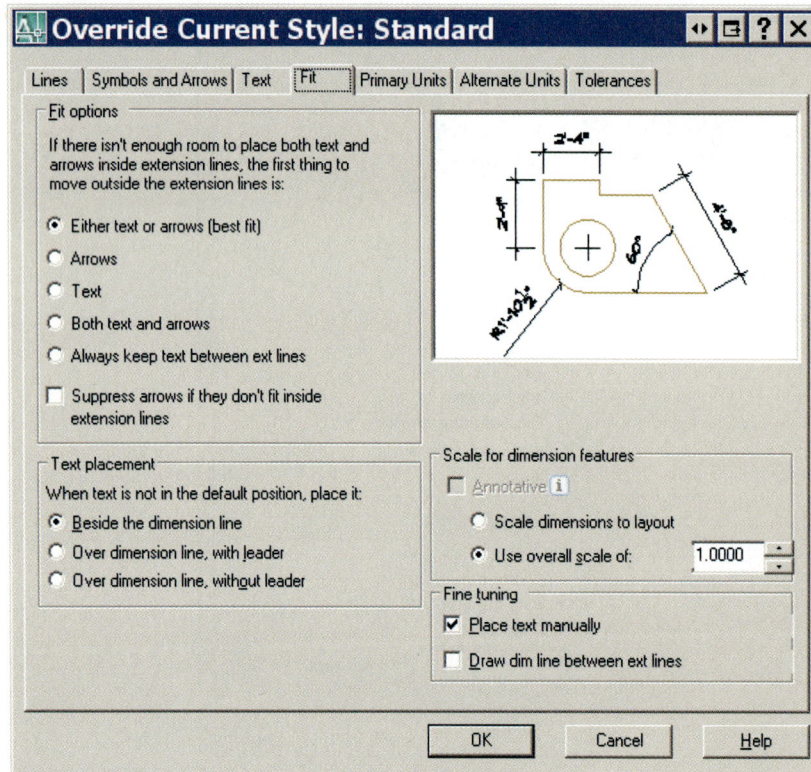

Figure 12-27(d)   Cabin project dimension style setting

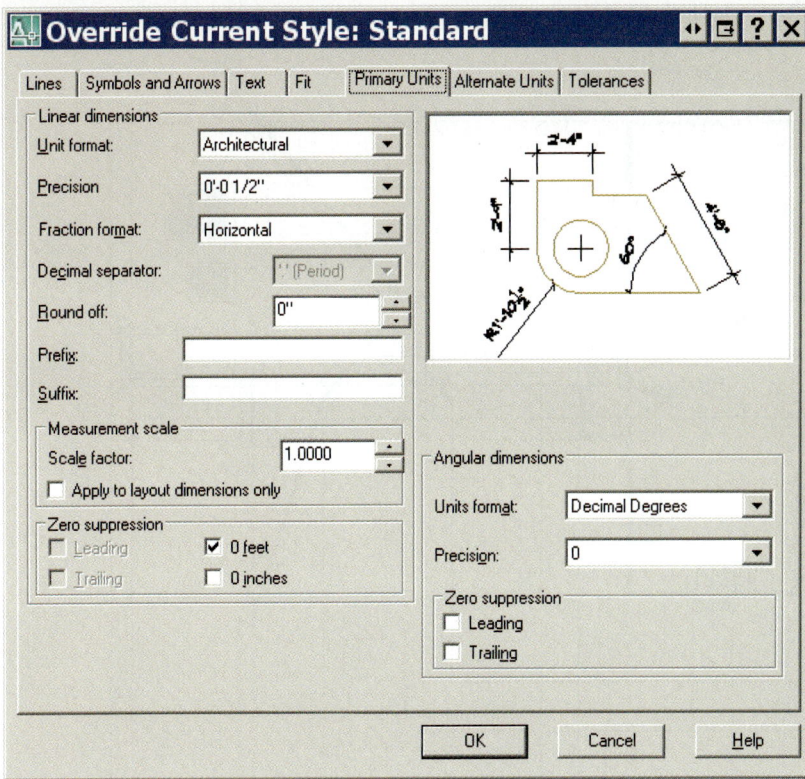

Figure 12-27(e) Cabin project dimension style setting

The dashed lines in Figure 12-29 illustrate how multiview drawing techniques can be applied to an architectural drawing to project the size and location of features in the floor plan to the front and side elevations. In Figure 12-29, locate the 45° miter line in the lower right corner and notice how information is projected between the front and side views through the miter line.

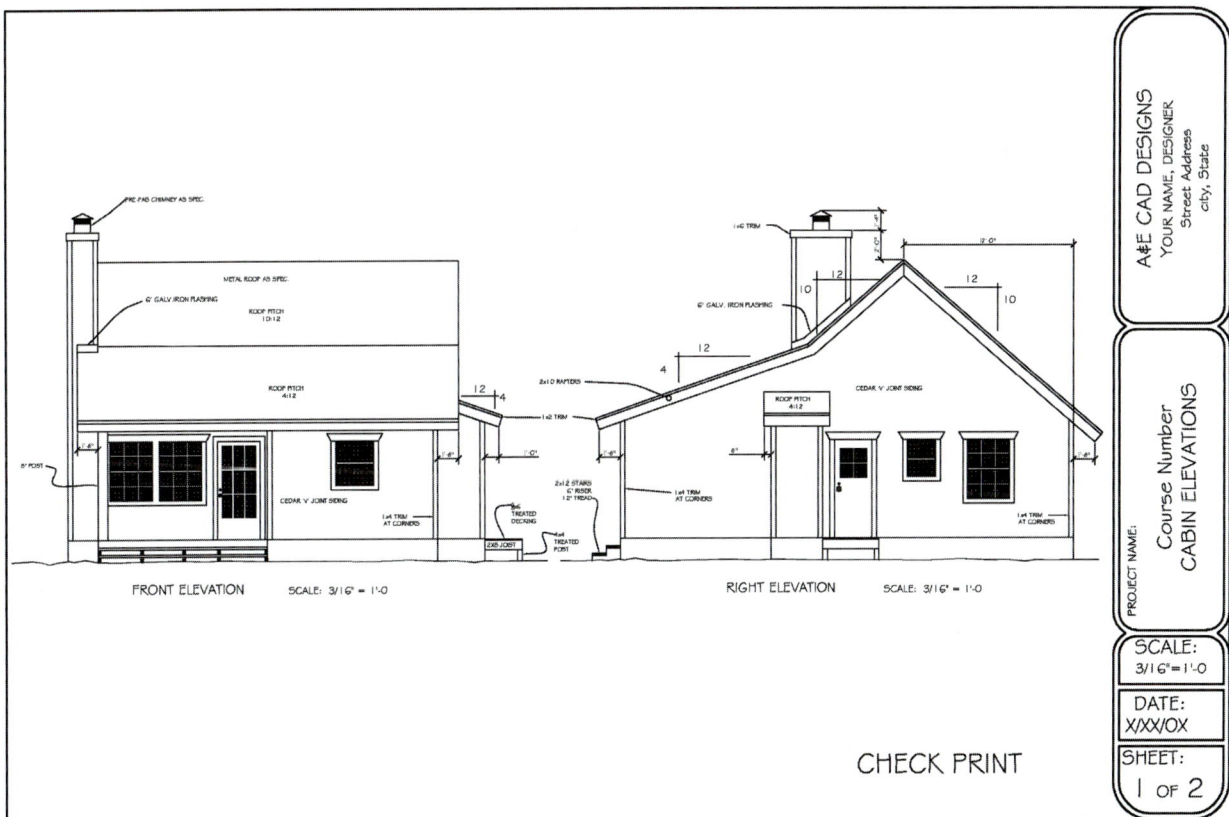

Figure 12-28 Front and right elevations

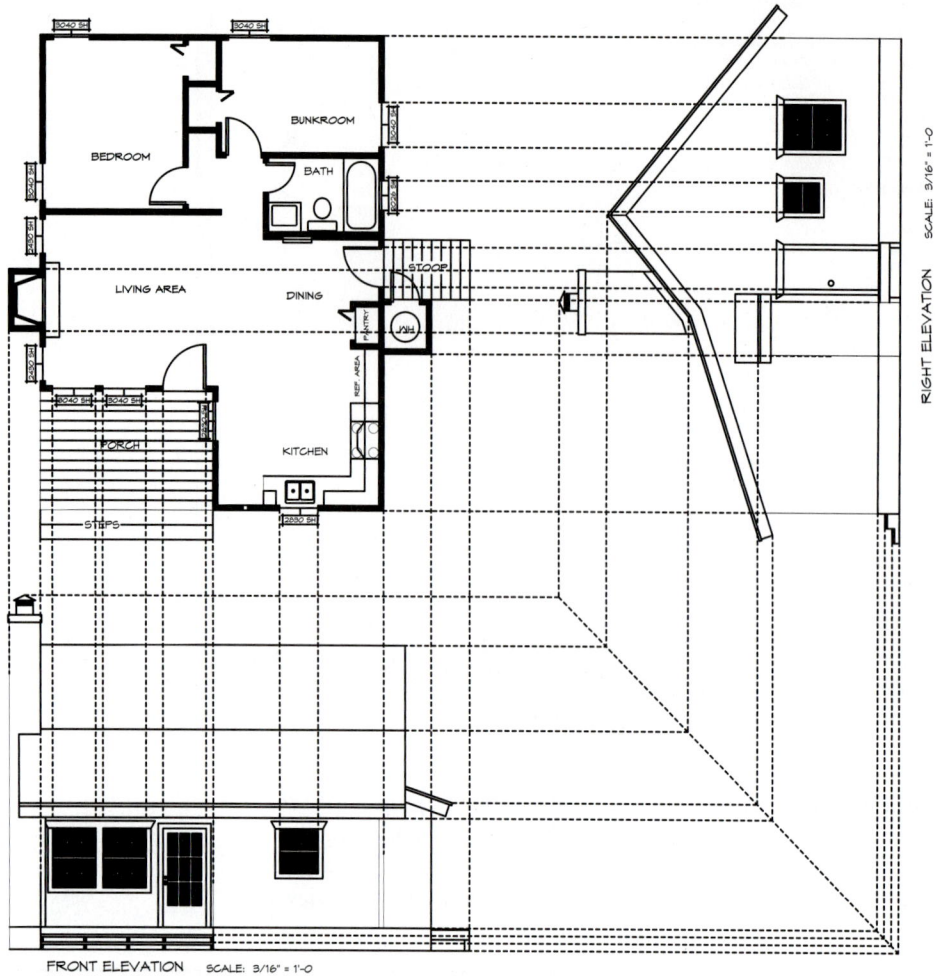

Figure 12-29   Relating the cabin's elevations to the floor plan

Wall Framing 101.   An understanding of the basics of framing is very helpful to drafters when constructing elevation drawings. The method used to frame the wall will determine the heights of ceilings, windows, and other exterior features.

The example in Figure 12-30 shows the front and end views of a length of wall framing. In this example, the studs are placed 16″ on

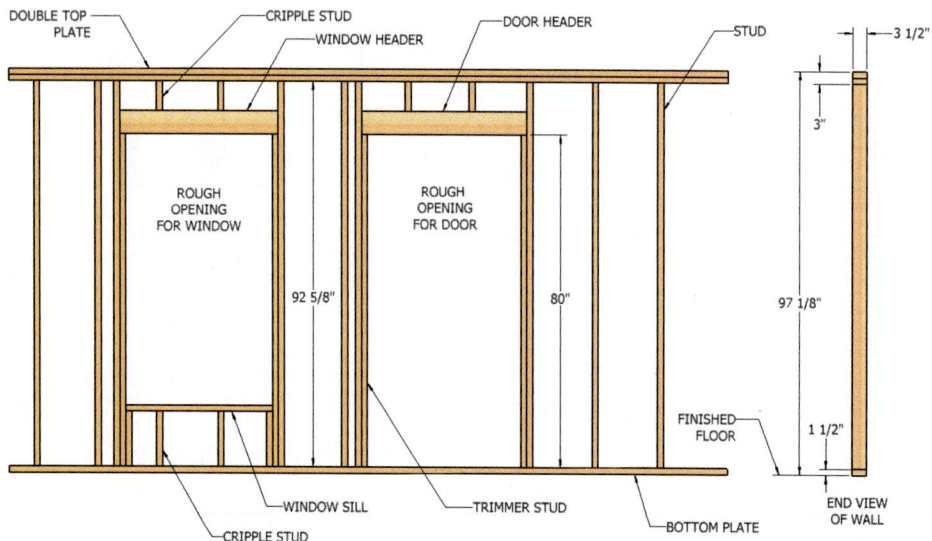

Figure 12-30   Wall framing example

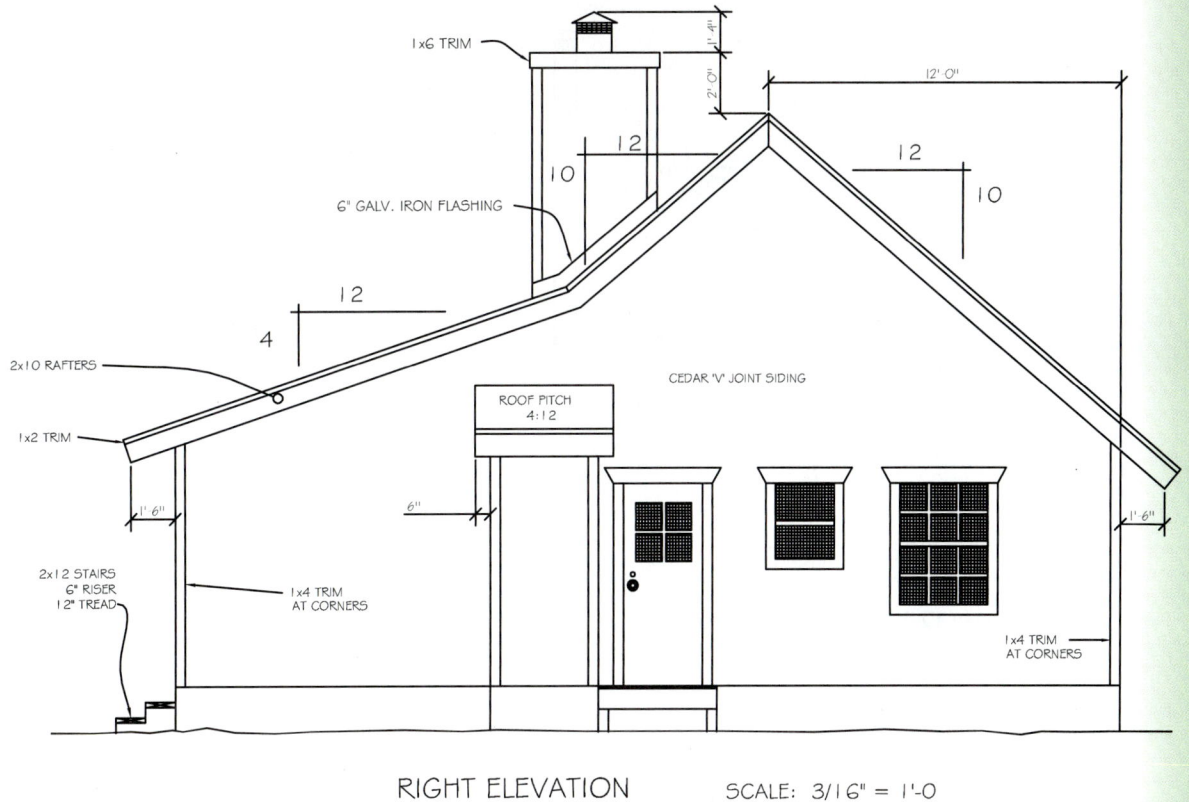

RIGHT ELEVATION   SCALE: 3/16" = 1'-0

Figure 12-31   Right-side elevation

center. Figure 12-30 also shows the framing for the rough openings for a window and a door. The rough opening is sized by the framers to accommodate the size of the window or door specified on the plan. Generally, rough openings will be about 2 1/2″ wider and taller than the window or door specified on the floor plan.

Constructing the Right-Side Elevation.   Follow Steps 13 through 23 to draw the elevations of the cabin. The first view to be drawn is the right-side elevation shown in Figure 12-31. This view was chosen because it shows the profile of the roof's pitch. When this view is complete, construction lines will be projected from its features to assist in the construction of the front view of the cabin.

Step 13.   To draw the right elevation, set the Elevations layer current and begin the drawing of the right elevation by constructing the cabin's exterior wall as shown Figure 12-32.

  The line labeled *Finish Grade* in Figure 12-32 is the point where the foundation of the cabin meets the ground line.

Step 14.   To draw the rafters, follow the directions shown in Figure 12-33 to define the **10/12** rafter pitch that begins at the upper left corner of the exterior wall constructed in Step 13.

  Next, offset the rafter pitch line **10″** and **8″** as shown in Figure 12-34. The first offset line represents the rafter's width of 10″. The second offset line is used to represent a 2″ wide piece of trim attached along the rafter's top edge.

Step 15.   The rafter overhangs the outside edge of the wall **1′-6″** as shown in Figure 12-35. To create this overhang, offset the outside wall line **1′-6″** inches to the right and extend the bottom edge of the rafter pitch line to the offset line. From the end of the rafter pitch line, draw a line perpendicular to the

$3\frac{1}{2}"$

$1'-6"$

$3"$

$8'-1 \ 1/2"$

$1\frac{1}{2}"$

FINISH FLOOR

FINISH GRADE

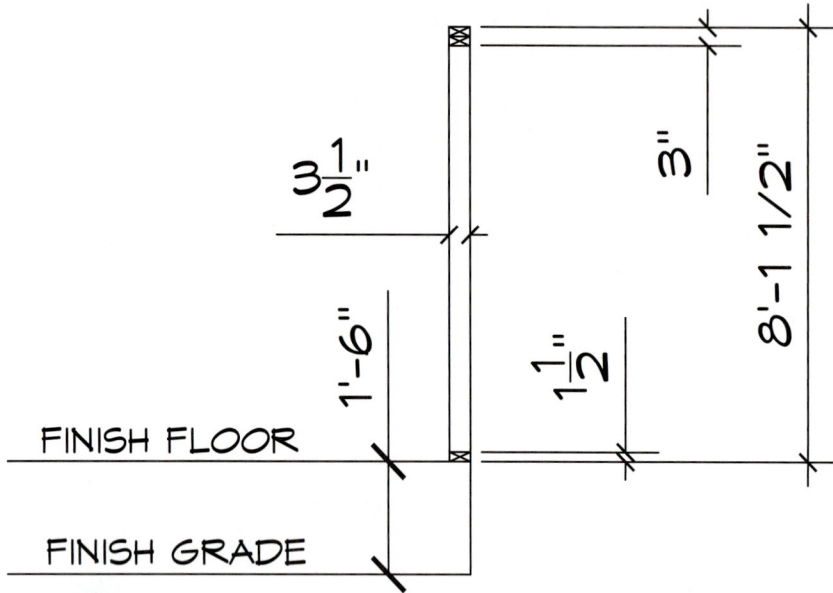

Figure 12-32   Exterior wall dimensions

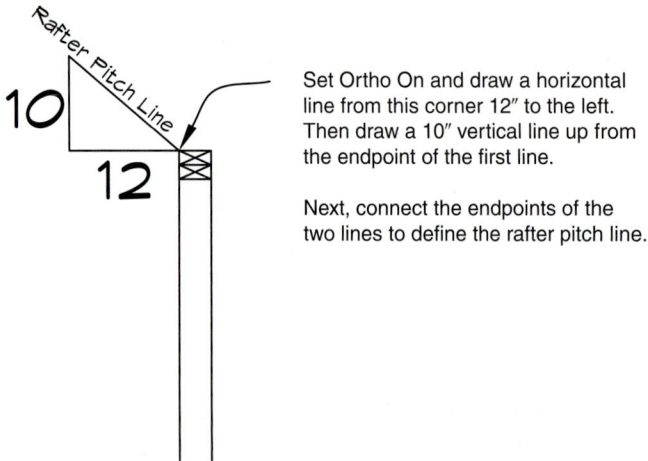

Rafter Pitch Line

10

12

Set Ortho On and draw a horizontal line from this corner 12″ to the left. Then draw a 10″ vertical line up from the endpoint of the first line.

Next, connect the endpoints of the two lines to define the rafter pitch line.

Figure 12-33   Drawing a 10/12 rafter pitch

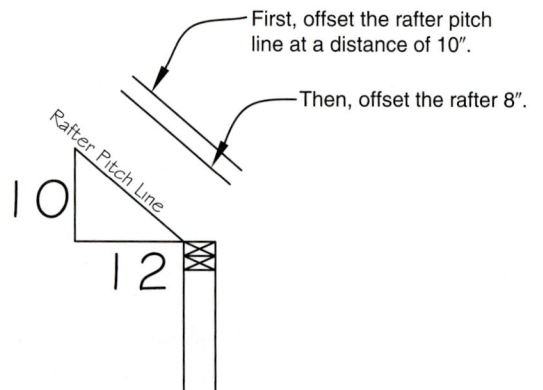

First, offset the rafter pitch line at a distance of 10″.

Then, offset the rafter 8″.

Rafter Pitch Line

10

12

Figure 12-34   Offsetting the rafter pitch Line

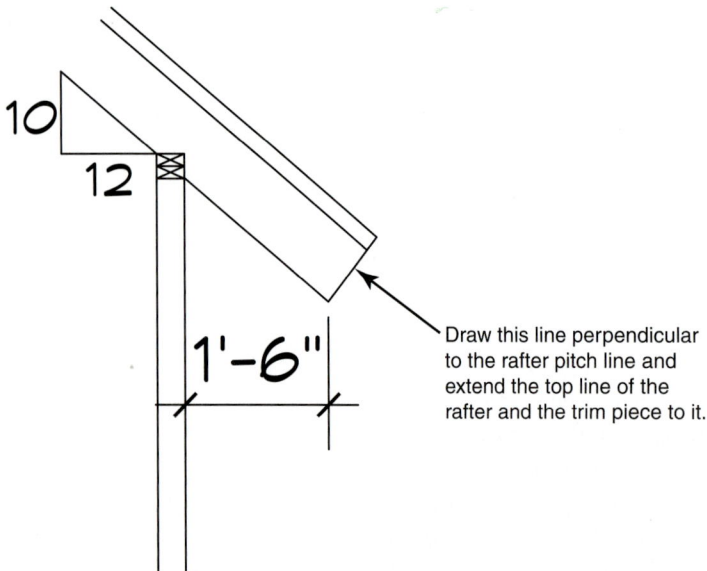

10

12

1'-6"

Draw this line perpendicular to the rafter pitch line and extend the top line of the rafter and the trim piece to it.

Figure 12-35   Detail of the rafter overhang

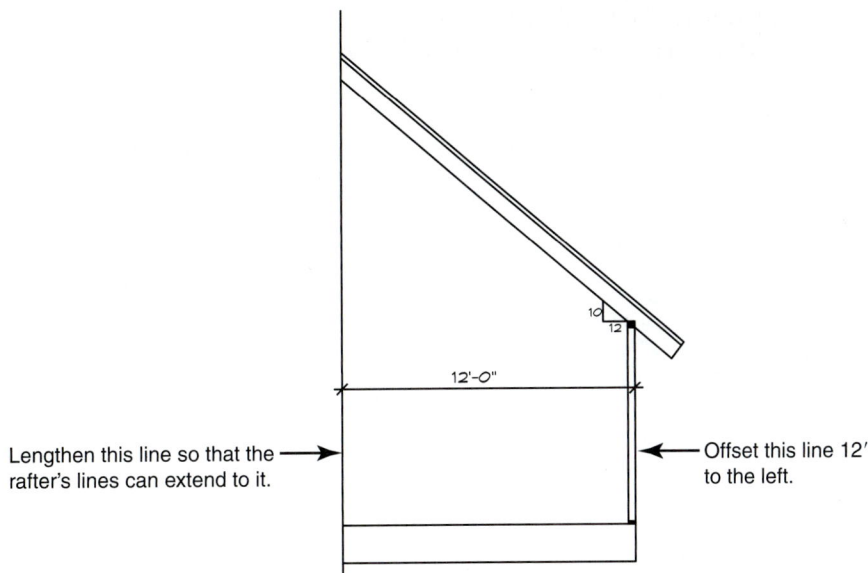

Figure 12-36   Extending the rafter

10/12 pitch and extend the lines that were offset in Step 14 to the perpendicular line to define the rafter's end.

Step 16. To extend the rafter to the roof peak, offset the outside right edge of the exterior wall **12′** to the left. Turn **Ortho On,** and use *grips edit* to lengthen the line so that the lines representing the rafter and the finished floor and grade can be extended to it.

Extend the rafter and the finished floor and grade lines to the offset line as shown in Figure 12-36.

Step 17. Mirror the rafter, foundation, and right outside wall as shown in Figure 12-37. Select the offset line created in Step 15 as the mirror line.

Step 18. To construct the left side wall and roof profile, copy the side wall created with the **MIRROR** command in Step 17, **8′** toward the left (**Ortho** should be **On**) as shown in Figure 12-28. Then, construct the rafter as you did in Steps 14 and 15 but

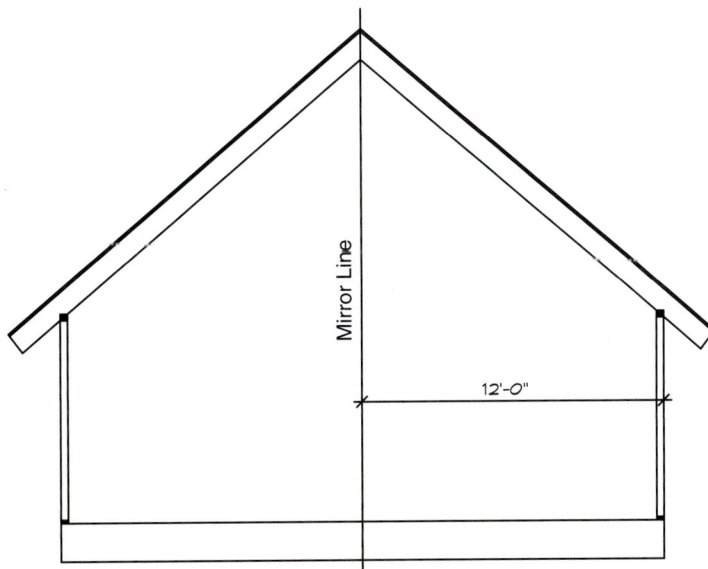

Figure 12-37   Mirroring the roof line

this time with a **4/12** pitch. Extend the rafter lines until they intersect with the **10/12** pitch roof section as shown in Figure 12-38.

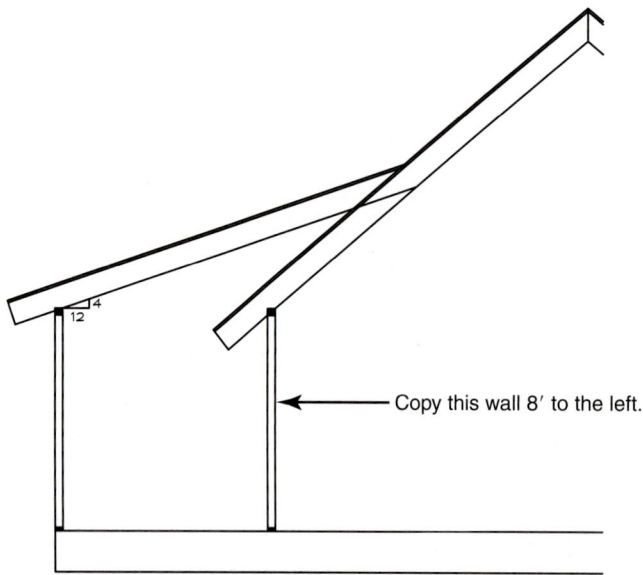

Figure 12-38   Constructing the left side wall and roof

Step 19. Use **TRIM, EXTEND,** and **ERASE** to complete the profile of the right elevation as shown in Figure 12-39.

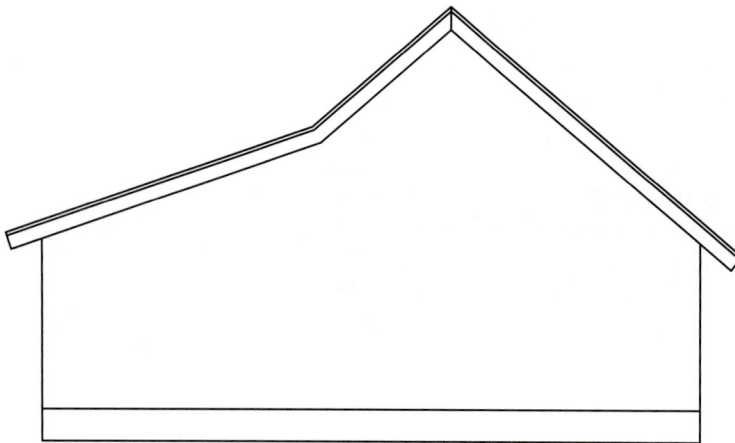

Figure 12-39   Right elevation after removal of construction lines

Step 20. Window dimensions for the **3⁰4⁰** window are shown in Figure 12-40. The dimensions for the wood trim around the windows will be the same for other windows and exterior doors.

To add windows, doors, and trim, open **Design Center** and insert blocks of the exterior door and windows from the **Cabin Symbols** drawing into the elevation view.

Use a **6″** wide trim piece along the top edges of doors and windows and draw the angled cuts at **15°** from vertical.

Next, using dimensions from the floor plan, add the chimney and locate the edges of the water heater closet in the right elevation (refer to Figure 12-29).

Note:

You will not be able to add the roof to the water heater closet at this time because you will need information projected from the front elevation.

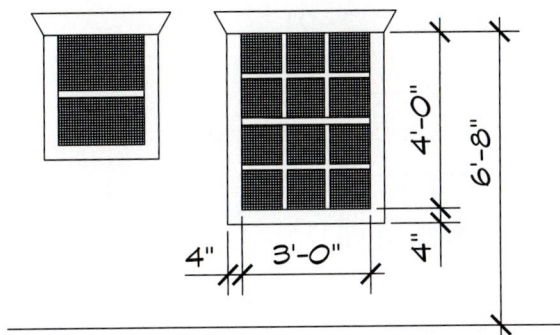

Figure 12-40   Window Dimensions

Step 21. Information from features in the right elevation, such as the roof height and location and height of the porch steps and chimney, can be projected to the front elevation. The dashed lines in Figure 12-41 show where geometry is projected between the views.

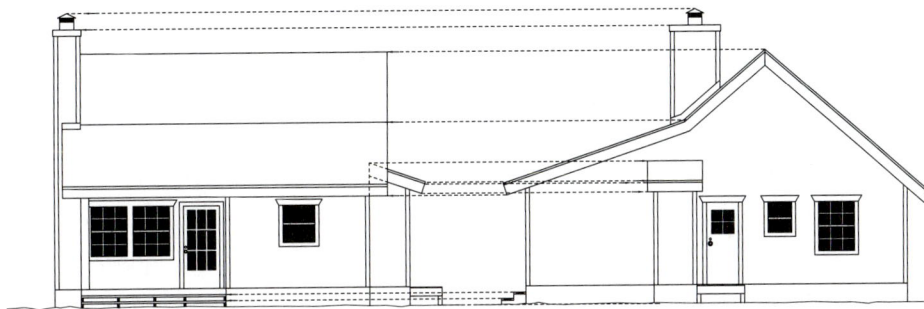

Figure 12-41   Projecting information between the right elevation and the front elevation

Construct the side wall and the **4/12** roof pitch for the water heater closet in the front elevation using the same techniques used during the construction of the right elevation. Notice how the water closet's roof extends to the cabin's right side wall in the front elevation. It will be necessary to project the water heater closet's roof geometry from the front elevation back to the right elevation in order to complete the construction of the water heater closet in the right elevation.

Step 22. Figure 12-42 shows the correct spacing for the front door and windows. The dimensions reflect the distances between centers. These dimensions are for reference only and should not be shown on the elevation view. To complete the front elevation, open **DesignCenter** and insert blocks for the front door and windows from the **Cabin Symbols** drawing into the elevation view.

Step 23. Add notes to the elevations as shown in Figures 12-31 and Figure 12-43. Where leaders are needed, use the **Spline** option of the **MULTILEADER** command found on the **Multileader** toolbar. To set **MULTILEADER** to the **Spline** option, choose the **MULTILEADER STYLE** icon from the **Multileader** toolbar and when the **Multileader Style Manager** dialog box opens, pick the **Modify** button. From the **Leader Format** tab, change the **Type** from **Straight** to **Spline**, click **OK**, and **Close**. Next, select the **MULTILEADER** command from the **Multileader** toolbar and place the leaders as needed.

2'-4"   3'-6"   4'-2"   2'-0"

1'-6"

8" Post

6'-8"

3'-0"

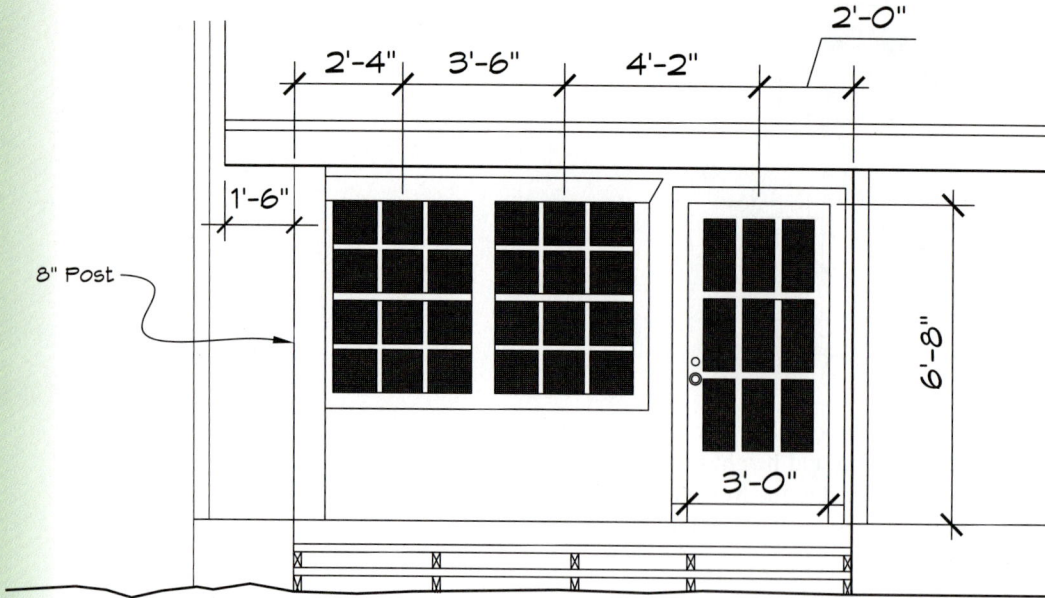

Figure 12-42 Front door and window placement

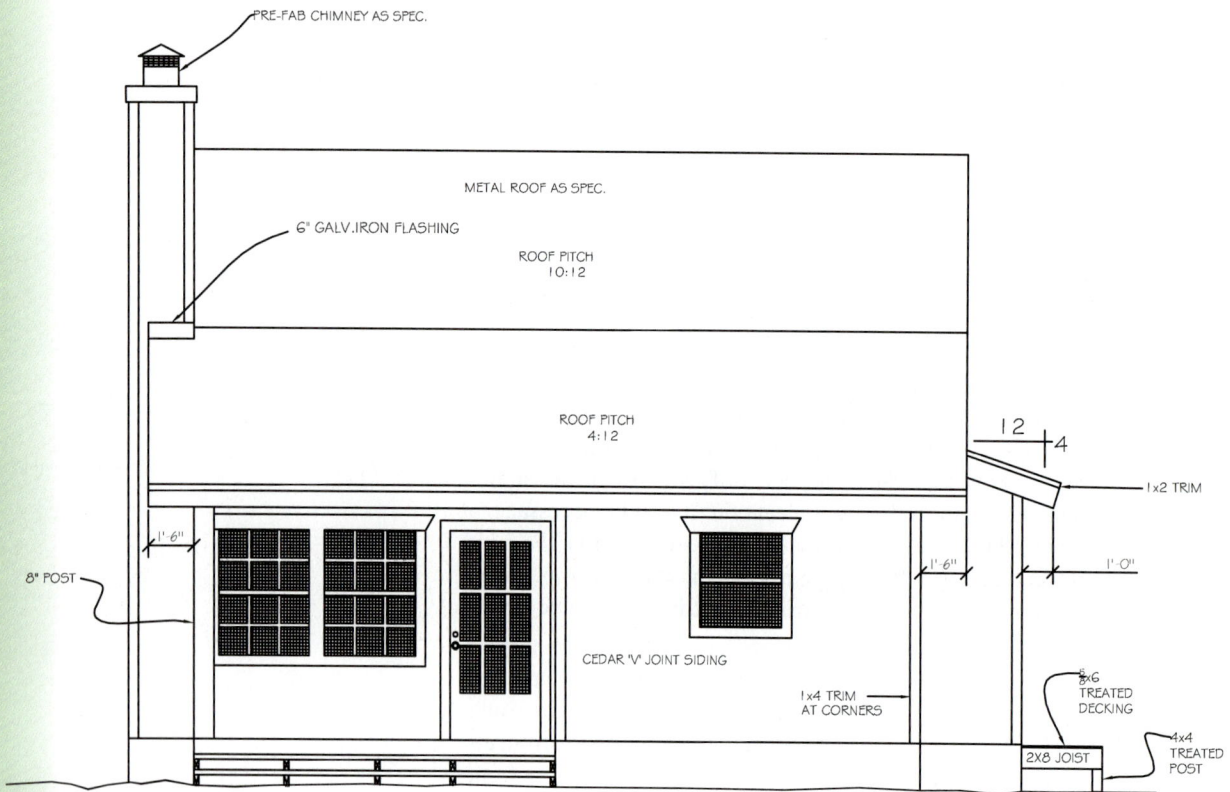

PRE-FAB CHIMNEY AS SPEC.

METAL ROOF AS SPEC.

6" GALV. IRON FLASHING

ROOF PITCH
10:12

ROOF PITCH
4:12

12   4

1x2 TRIM

8" POST

1'-6"

1'-6"   1'-0"

CEDAR 'V' JOINT SIDING

1x4 TRIM
AT CORNERS

8x6
TREATED
DECKING

2X8 JOIST

4x4
TREATED
POST

FRONT ELEVATION   SCALE: 3/16" = 1'-0

Figure 12-43 Front elevation notes

Step 24. Follow your instructor's directions to plot the sheets. If your plotter or printer allows, each sheet can be plotted on a 17 × 11 sheet using the monochrome setting at a scale of 3/16″ = 1′0″.

TIP

Follow the steps presented in Unit 4 to create a **Page Setup** for plotting each sheet of this project. The plotted sheets should resemble the examples shown in **Figure 12-9** and **Figure 12-10**.

13

3D Modeling Basics

Unit Objectives

- Define coordinates along *X*-, *Y*-, and *Z*-axes.
- Change the viewpoint to reveal the *Z*-axis of the User Coordinate System in an AutoCAD drawing.
- Use the REGION and EXTRUDE commands to convert two-dimensional entities into three-dimensional objects.
- Use the ROTATE 3D command to rotate objects as needed in the creation of 3D models.
- Employ the SUBTRACT and UNION commands to create complex three-dimensional objects with AutoCAD.
- Represent 3D models in wireframe or shaded form.

INTRODUCTION

For years, drafters and designers in the mechanical engineering field have used CAD programs such as Inventor, Solidworks, and ProE to produce two-dimensional (2D) working drawings from three-dimensional (3D) models. With the release of CAD software products such as Civil 3D and Revit, this technique is rapidly becoming the norm for civil and architectural drafters and designers as well. Working with 3D models gives designers a much more powerful way to conceptualize, edit, and analyze their designs.

Most 3D CAD software programs use a technique known as ***parametric modeling*** to create 3D geometry. In parametric models, the geometry of the model is driven by the dimensions associated with the geometry. This allows designers to modify the features of the model by simply editing the dimensions. When a parametric dimension is changed, the 3D model updates to reflect the new dimension value.

parametric modeling: A method of creating 3D CAD models in which the geometry of the model is driven by the dimensions associated with the geometry. This allows designers to modify the features of a model by simply editing its dimensions. When the parametric dimension is changed, the 3D model updates to reflect the new dimension value.

Finite Element Analysis (FEA): Software that is used in conjunction with CAD modeling software to perform advanced design analysis on 3D models. FEA allows engineers and designers to calculate such properties as an object's mass, center of gravity, strength, distribution of stresses, and bending moments.

Finite Element Analysis (FEA) software is often used in conjunction with CAD modeling software to perform advanced design analysis on 3D models. FEA allows engineers and designers to calculate such properties as an object's mass, center of gravity, strength, distribution of stresses, and bending moments.

2D VERSUS 3D

Two-dimensional objects drawn with AutoCAD are described by specifying *X* and *Y* coordinates. In a 2D drawing, the value of points located on the *Z*-axis is zero. Figure 13-1 shows a 2D drawing of an object created with AutoCAD. In the lower left corner of the screen, the **User Coordinate System (UCS)** icon is visible. As you learned in Unit 3, the UCS icon orients the drafter to the location of the **0,0** coordinate and the direction of the positive *X*- and *Y*-axes.

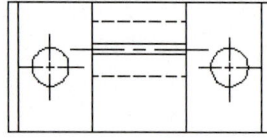

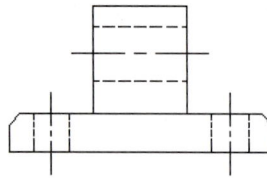

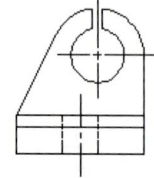

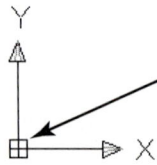

The **USC Icon** displays the location of the drawing's lower left limit and the direction of the (positive) **X** and **Y** coordinates. The viewer is looking straight down the **Z** axis of the icon at the graphics area.

Figure 13-1   2D views of an object

In 3D drawing, objects are defined with coordinates that have *X, Y, and Z* values. For example, the coordinate of the endpoint of a 3D line may be described as **6, 2, 4** with **6** representing the *X* value, **2** representing the *Y* value, and **4** representing the *Z* value. However, this line would not appear to be 3D if the graphics area is viewed from directly overhead as in **Figure 13-1**. In order to view the *Z*-axis, it would be necessary to change the point of view from which the 3D line were viewed.

CHANGING THE POINT OF VIEW OF AN AutoCAD DRAWING

Changing the point of view to show the *Z*-axis can be accomplished by opening the **View** toolbar and selecting the **SE Isometric** icon (see Figure 13-2). The AutoCAD screen will adjust to resemble the example shown in Figure 13-3. In this figure, the 2D objects appear as if they are viewed from the South East side of the drawing. In fact, the *SE* in **SE Isometric** refers to this South East point of view. Likewise, *SW* refers to South West, *NE* refers to North East, and *NW* refers to a viewpoint located North West of the viewed object.

Note:
Figure 13-1 shows the objects as they would appear if the **Top** icon were selected from the **View** toolbar.

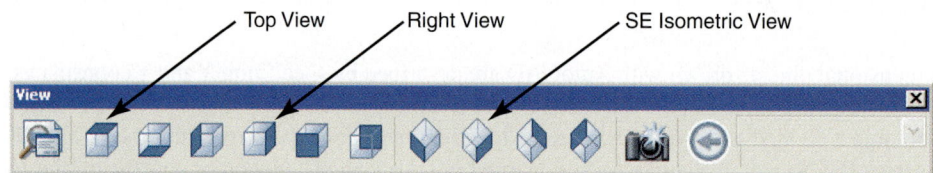

Top View   Right View   SE Isometric View

Figure 13-2   View toolbar

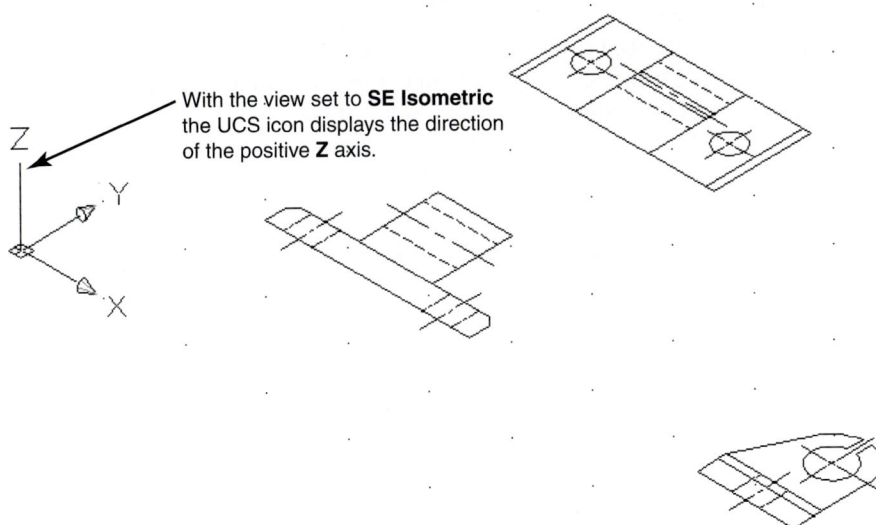

With the view set to **SE Isometric** the UCS icon displays the direction of the positive **Z** axis.

Figure 13-3   SE isometric view of 2D objects

SETTING THE AUTOCAD ENVIRONMENT FOR 3D MODELING

With the release AutoCAD 2007, the 3D environment was given a unique workspace named **3D Modeling** (see Figure 13-4).

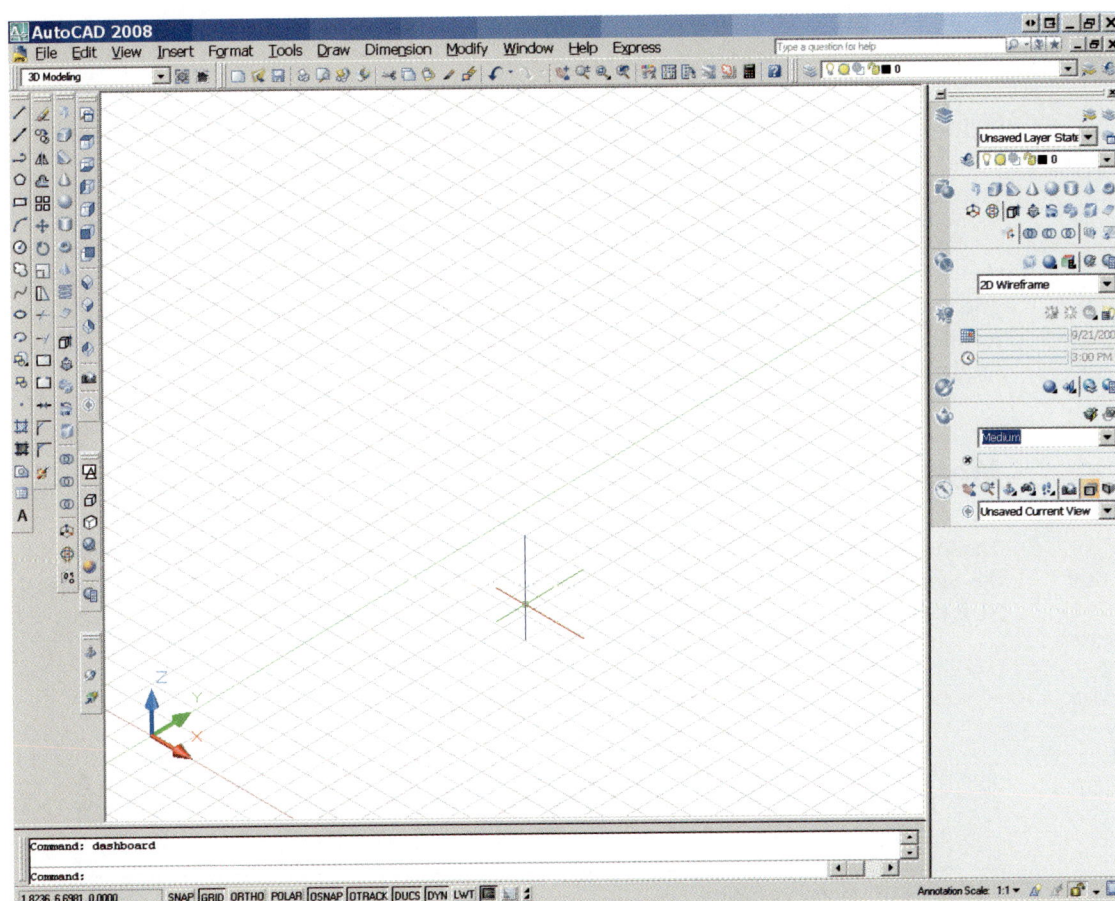

Figure 13-4   3D Modeling workspace environment

As you can see in Figure 13-4, the **3D Modeling** workspace looks different from AutoCAD's traditional 2D workspace (which is now referred to as **AutoCAD Classic**). The area shown along the right side of the graphics window in Figure 13-4 is called the **Dashboard**. The **Dashboard** is a tool palette that displays many of the commands that are used to create, edit, and view 3D models. Beginners may find the Dashboard palette confusing and may prefer instead to close the Dashboard and open the **Draw**, **Modify**, **Modeling**, **View**, **Visual Styles**, and **Orbit** toolbars shown along the left side of the graphics window in **Figure 13-4.**

To begin a new drawing with the **3D Modeling** workspace as the default environment, select **File**, then pick **New**, and select the **acad3D.dwt** template file.

Steps to Setting the 3D Modeling Environment

To change the environment of an existing drawing to reflect the **3D Modeling** workspace, follow these steps:

Step 1. Select **3D Modeling** from the **Workspaces** toolbar (see **Figure 13-4**).

Step 2. Open the **Visual Styles** toolbar shown in Figure 13-5 and select the **Conceptual Visual Style** icon.

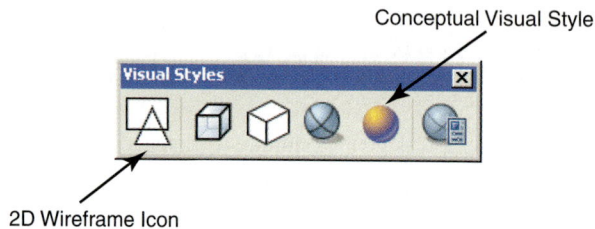

Figure 13-5   Visual Styles toolbar

Step 3. Open the **Drafting Settings** dialog box. Under the **Grid Behavior** settings, check the box next to **Display Grid Beyond Limits** and turn the **Grid On**.

Step 4. Next, open the **View** toolbar and select the **SE Isometric** icon.

When Steps 1 through 4 have been completed, the workspace should resemble the one shown in Figure 13-4. Note in this figure that the grid is no longer represented by dots but by grid lines running along the X- and Y-axes. Also note how the **UCS** icon is displayed as a 3D object with a color assigned to each of its axes (red for the X-axis, green for the Y-axis, and blue for the Z-axis).

MODELING TOOLBAR

The **Modeling** toolbar holds many of the commands necessary to create 3D models. In Figure 13-6, the icons for four very useful *modeling commands* are identified: **EXTRUDE**, **UNION**, **SUBTRACT**, and **3D ROTATE**.

modeling commands: Used to create 3D models in an AutoCAD drawing. These commands are located on the **Modeling** toolbar and include **UNION, SUBTRACT, 3D ROTATE**, and **EXTRUDE**.

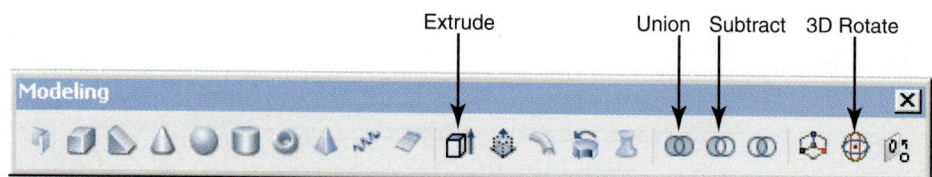

Figure 13-6   Modeling toolbar

Extruding 2D Entities to Create 3D Objects

In many cases, 2D objects can be converted to 3D objects by *extruding* them along their Z-axis using AutoCAD's **EXTRUDE** command (see Figure 13-7). The **EXTRUDE** icon is located on the **Modeling** toolbar. Objects that can be extruded include circles, rectangles, ellipses, polygons, closed polylines, and shapes that have been *regioned* with the **REGION** command. The **REGION** command is located on the **Draw** toolbar.

Using the EXTRUDE Command.   Select the **EXTRUDE** command from the **Modeling** toolbar and at the Select objects prompt, pick the 2D object(s) to be extruded. Press **<Enter>** after the object(s) to be extruded has been selected. At the Specify height of the extrusion (path) prompt, type in the extrusion height as measured along the Z-axis (the height may be given as either a positive or a negative value depending on which direction is desired for the extrusion) and press **<Enter>**. The 2D shape will extrude along its Z-axis, becoming a 3D object.

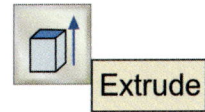

Figure 13-7   Extrude icon

Unioning 3D Objects

Two or more 3D objects can be combined to form one object using the **UNION** command located on the **Modeling** toolbar (see Figure 13-8).

Using the UNION Command.   Select the **UNION** command located on the **Modeling** toolbar. At the Select objects prompt, select the 3D objects you would like to combine. When all of the objects are selected, press **<Enter>.**

Figure 13-8   Union icon

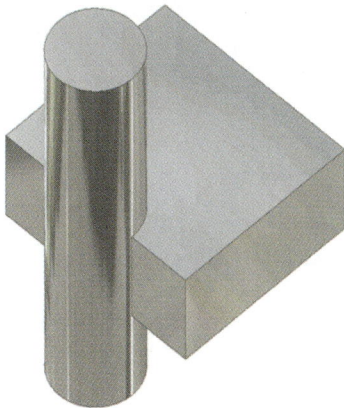

Figure 13-9   Object resulting from the union of a 3D cylinder and a 3D box

Note:
Unioning the objects combines their total volumes into one object, and they can no longer be edited as separate parts. An example of the union of a cylinder and a box is shown in Figure 13-9.

Subtracting 3D Objects

The volume of one 3D object can be removed from the volume of another 3D object by using the **SUBTRACT** command located on the **Modeling** toolbar (see Figure 13-10). For example, by subtracting a cylinder with a small diameter from a cylinder with a larger diameter, a hole is formed through the larger cylinder.

Using the SUBTRACT Command   Select the **SUBTRACT** command. At the Select solids or regions to subtract from prompt, pick the principal object and press **<Enter>**. At the Select solids or regions to subtract prompt, pick the object whose mass you would like to subtract from the mass of the principal object and press **<Enter>**. An example of the subtraction of a 3D cylinder from a 3D box is shown in Figures 13-11 and 13-12.

Figure 13-10   Subtract icon

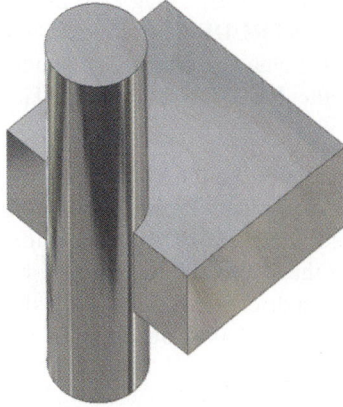

Figure 13-11   3D cylinder and 3D box before SUBTRACT command

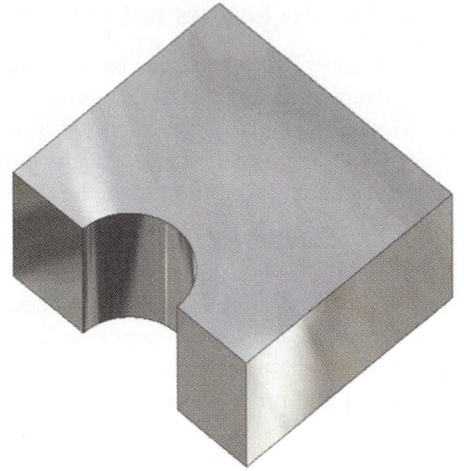

Figure 13-12   Object resulting from the subtraction of a 3D cylinder from a 3D box

Rotating 3D Objects

3D Rotate

Figure 13-13   Rotate 3D icon

The **3D ROTATE** command (see Figure 13-13) found on the **Modeling** toolbar is used to rotate 3D objects around the *X-*, *Y-*, or *Z-*axis. When using the **3D ROTATE** command, the drafter is prompted to Select objects, Specify base point, Pick a rotation axis, and Specify angle start point or type an angle. The selected object(s) rotates around the base point, and the axis of rotation passes through the base point. The angle of the rotation is specified in degrees.

Note:
The step-by-step instructions for using the 3D ROTATE command are presented in Unit Project 13-1.

VIEWING 3D OBJECTS

In addition to the viewpoints available on the **View** toolbar, the **FREE ORBIT** command is also helpful when viewing 3D objects. By using this command, a drafter can "orbit" around a 3D object and view it from any angle. The **FREE ORBIT** command is located on the **Orbit** toolbar shown in Figure 13-14

Using the FREE ORBIT Command

Free Orbit

Figure 13-14   Orbit toolbar

After selecting **FREE ORBIT** from the **Orbit** toolbar, hold down the left button of the mouse, and move the mouse up and down or side to side. The viewpoint will change dynamically as the mouse is moved. To end the command, right-click and select **Exit**. To reset the object to a **SE Isometric** view, choose the **SE Isometric** icon from the **View** toolbar.

REPRESENTING 3D OBJECTS AS SHADED OR WIREFRAME MODELS

The icons on the **Visual Styles** toolbar allow the drafter to represent a 3D object as either a *shaded, unshaded,* or *wireframe* form. See Figure 13-15.

Both the **Conceptual Visual Style** and **Realistic Style** icons create shaded 3D representations.

TIP   Shading works best on objects that have been assigned a color other than black or white. Selecting the **2D Wireframe**, **3D Wireframe**, or **3D Hidden Visual Style** icon will result in wireframe images of 3D objects.

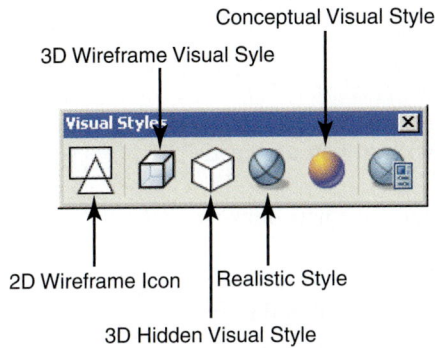

Conceptual Visual Style

3D Wireframe Visual Syle

Visual Styles

2D Wireframe Icon   Realistic Style

3D Hidden Visual Style

Figure 13-15   Visual Styles toolbar icons

SUMMARY

Compared to 2D CAD drafting, creating 3D models gives designers a more powerful and dynamic means to conceptualize, edit, and analyze their designs. The evolution of 3D CAD tools will allow many of the design functions currently performed by engineers to be performed by designers and designer/drafters. This will create career advancement opportunities for design drafters who adapt to these changes in CAD technology and master the design capabilities of these programs.

UNIT TEST QUESTIONS

Short Answer

1. Define the term *parametric modeling*.

2. What properties can be calculated using Finite Element Analysis software?

3. What is the name of the AutoCAD command that combines the volumes of two or more 3D objects into one object?

4. Name the AutoCAD toolbar on which the **SUBTRACT** command is located.

5. When using the **EXTRUDE** command, the height of the extrusion is defined along which axis (*X*, *Y*, or *Z*)?

UNIT PROJECTS

Project 13-1: 3D Tool Slide

Create a 3D model of the tool slide shown in the designer's sketch in Figure 13-16.

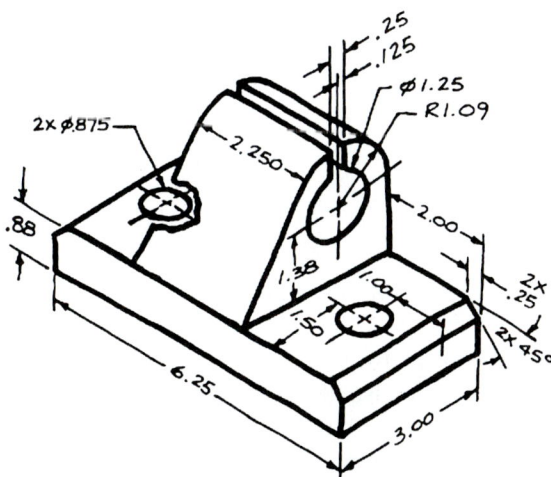

Figure 13-16   Designer's sketch of the tool slide

Directions

1. Select **FILE** and **OPEN**.
2. Select the **Student Resource CD** directory.
3. Select the **Prototype Drawings** folder.
4. Select the **Tool Slide 3D Exercise** drawing (open as "read-only").
5. **SAVE AS** to your **Home** directory and rename the new drawing **TOOL SLIDE 3D**.
6. The drawing should resemble the one shown in Figure 13-17. However, if it does not, follow the four steps outlined in the section "Setting the AutoCAD Environment for 3D Modeling" presented earlier in this unit.

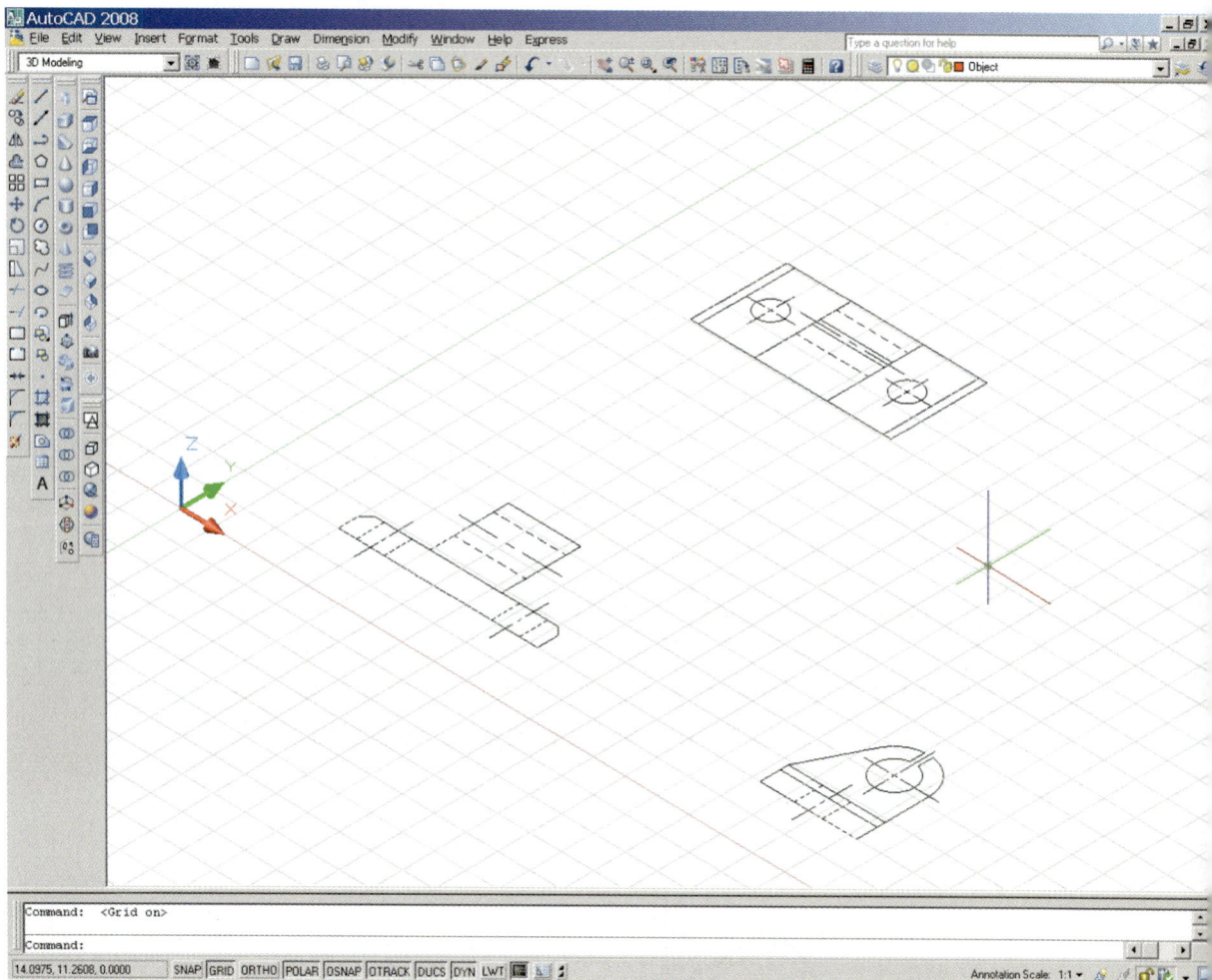

Figure 13-17   Tool Slide 2D drawing

Step 1. Make a copy of the front and side views of the tool slide. Edit the views to create the profiles shown in Figure 13-18.

Step 2. Select the **REGION** command (see Figure 13-19) from the **Draw** toolbar, and create a separate region for each of the two profiles created in Step 1. When prompted to select objects, select all of the entities comprising the profile to be regioned.

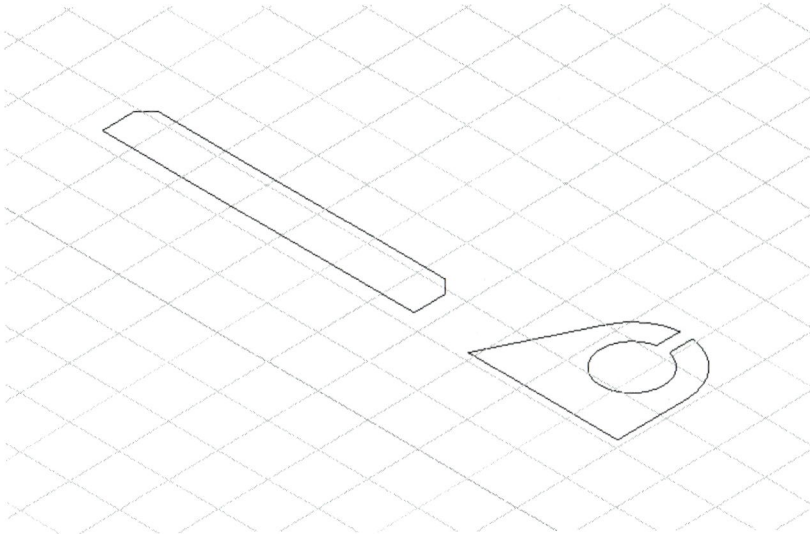

Figure 13-18   Profiles of the front and side view of the tool slide

Figure 13-19   REGION command icon

Step 3. Use the **EXTRUDE** command to extrude the regions created in Step 2 as shown in Figure 13-20. The height of the extrusion for the object on the left is **3.00″**, and the height of the extrusion of the object on the right is **2.250″**.

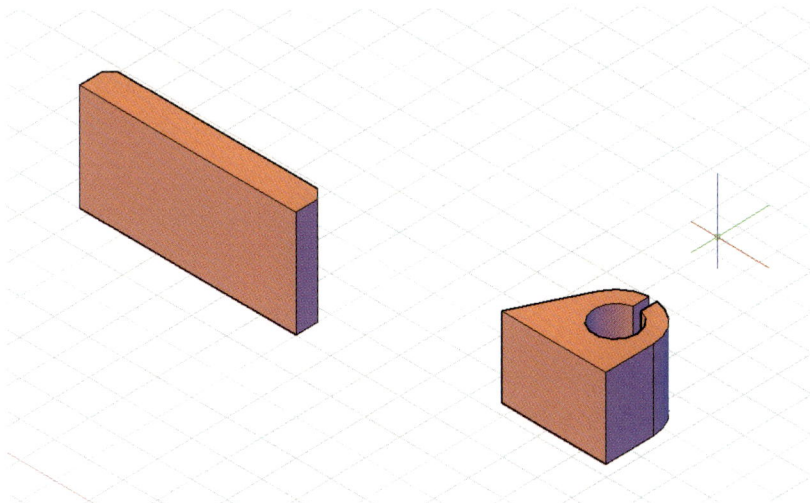

Figure 13-20   3D Models created by extruding

Step 4. To rotate the 3D objects shown created in Step 3, select the **3D ROTATE** icon from the **Modeling** toolbar (refer to Figure 13-6). When prompted to Select objects, select both of the 3D objects created in Step 3 and press **<Enter>**.

Step 5. When prompted to Specify base point, select the endpoint shown in Figure 13-21.

Step 6. When prompted to Pick a rotation angle. Pick the axis (ellipse) shown in Figure 13-22.

Step 7. At the Specify angle start point prompt, type **90** and press **<Enter>**. The object will rotate **90°** around the base point in the positive direction of the rotation axis.

Select these Objects to Rotate

Endpoint

Select this Endpoint
for the Base Point

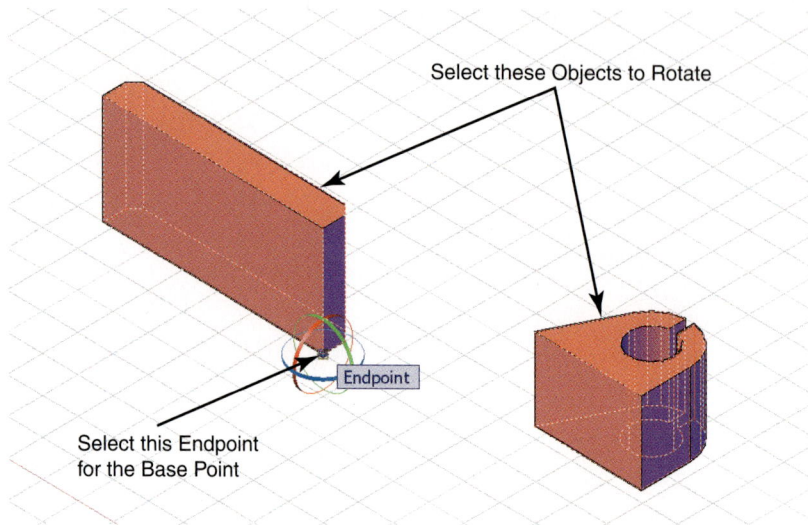

Figure 13-21   Specifying the base point for the 3D ROTATE command

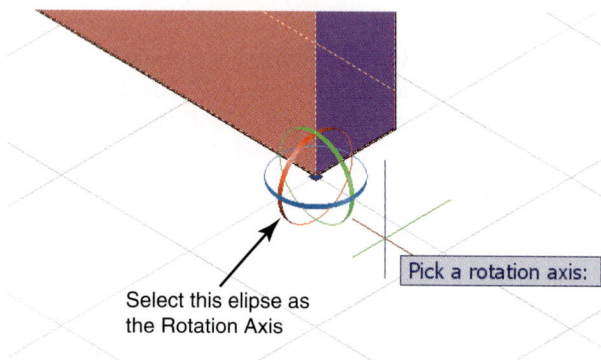

Figure 13-22   Selecting the rotation axis of the 3D ROTATE command

Select this elipse as
the Rotation Axis

Pick a rotation axis:

Step 8. Use the **3D ROTATE** command to rotate the object shown in Figure 13-23 again. This time, the rotation will be **90°** around the Z-axis.

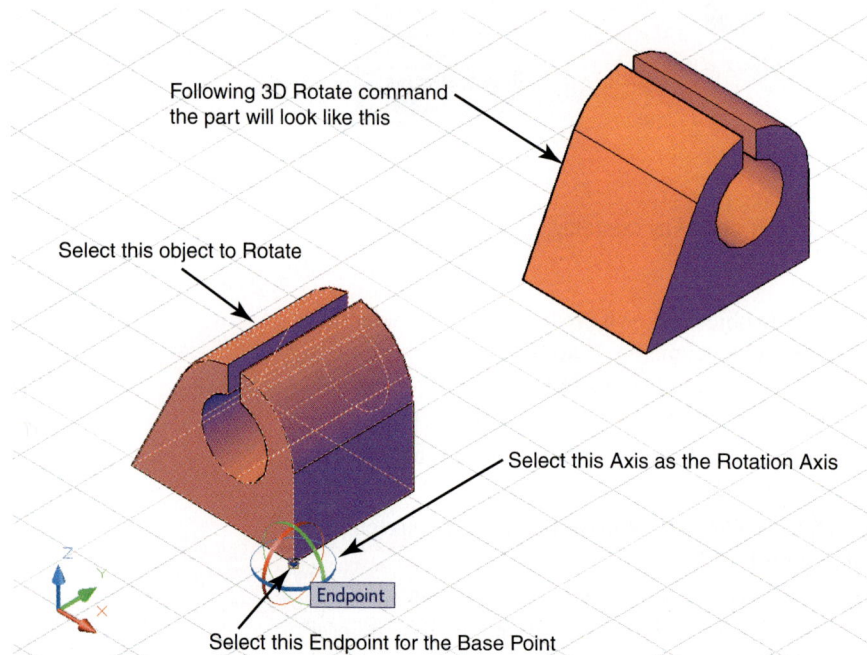

Following 3D Rotate command
the part will look like this

Select this object to Rotate

Select this Axis as the Rotation Axis

Endpoint

Select this Endpoint for the Base Point

Figure 13-23   Rotating the object 90° around the Z-axis

Step 9. Use the **MOVE** command to position the second object on top of the first object as shown in Figure 13-24.

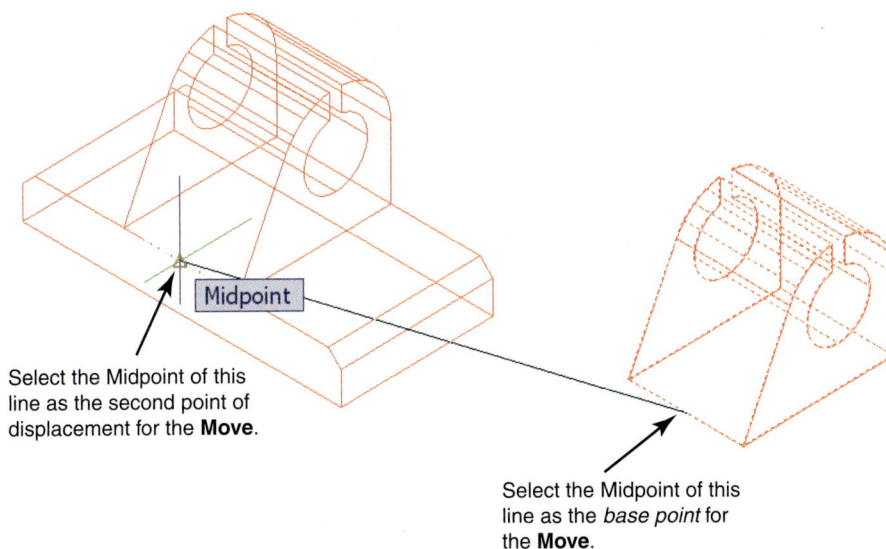

Select the Midpoint of this line as the second point of displacement for the **Move**.

Midpoint

Select the Midpoint of this line as the *base point* for the **Move**.

Figure 13-24   Positioning the objects with the MOVE command

Step 10. Use the **UNION** command to combine the two 3D objects into one 3D object.

Step 11. Copy the two circles from the top view of the **Tool Slide** drawing to the 3D model. Use the bottom left corner of the top view as the Base point and the front left corner of the 3D model as the Second point of displacement as shown in Figure 13-25.

Select the endpoint of this line for the *second point of displacement* for the **Copy** command.

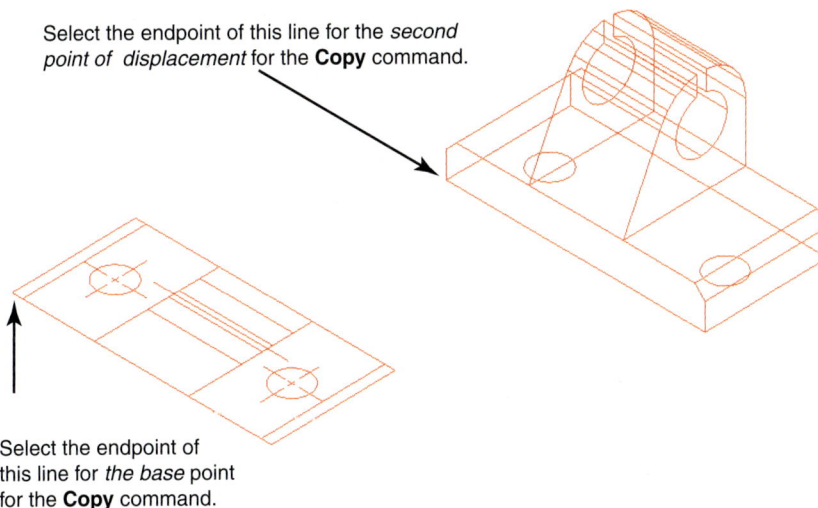

Select the endpoint of this line for *the base* point for the **Copy** command.

Figure 13-25   Copying the circles from the top view to the 3D model

Step 12. **EXTRUDE** the 2 circles copied in Step 11 to a height of **1.00"** to form two cylinders as shown in Figure 13-26.

Step 13. Use the **SUBTRACT** command to subtract the cylinders created in Step 12 from the 3D object to create the holes in the part. This step finishes the construction of the 3D model of the tool slide.

Figure 13-26   Model with extruded circles

Step 14. From the **Visual Styles** toolbar, select the **Conceptual Vi-sual Style** icon. The finished model should look like the one in Figure 13-27 after shading. Print the model per your instructor's directions.

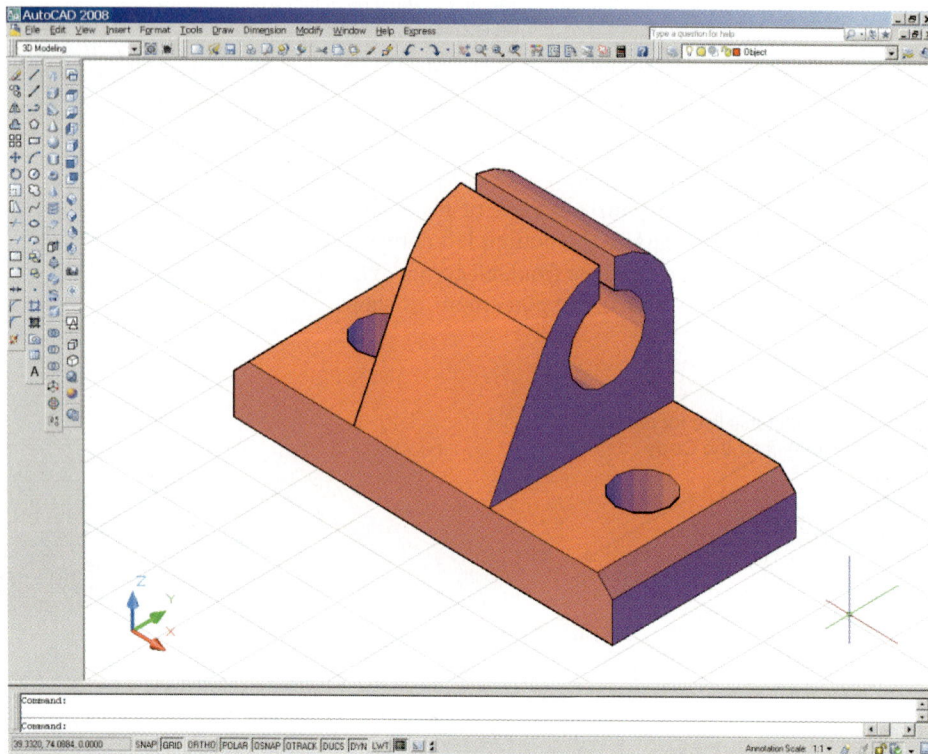

Figure 13-27   Completed 3D model of the tool slide

Project 13-2: 3D Bracket

Create a 3D model of the Bracket shown in Figure 13-28.

Directions

1. Select **FILE** and **OPEN.**
2. Select the **Student Resource CD**.
3. Select the **Prototype Drawings** folder.
4. Select the **Bracket 3D Exercise** drawing (open as "read-only").
5. **SAVE AS** to your **Home** directory and rename the new drawing **BRACKET 3D**.

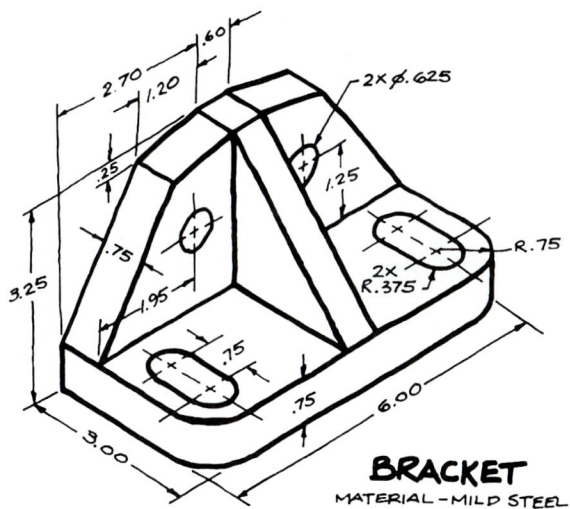

Figure 13-28   Designer's sketch of the bracket project

6. Use the 2D geometry in the prototype drawing to create 2D pro-
    files (set the **View** to *SE Isometric* if needed) and then use the
    REGION command to create a region for each profile. See
    Figure 13-29.

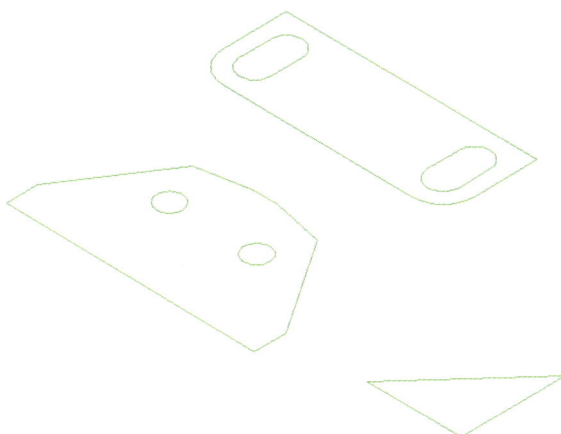

Figure 13-29   Creating profiles and regions

7. Use the **EXTRUDE** command to extrude the regions to the
    desired height. See Figure 13-30.

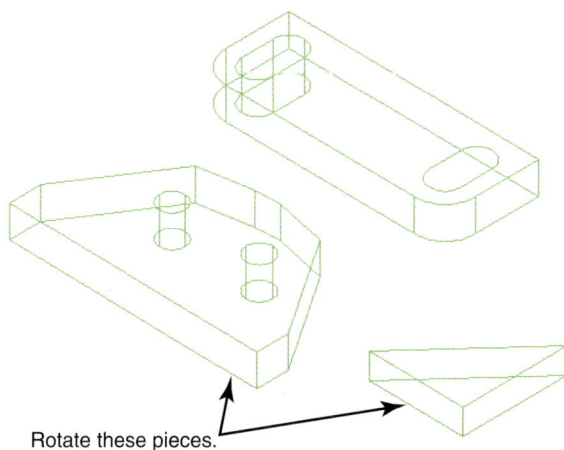

Rotate these pieces.

Figure 13-30   Extruded profiles

8. Use the **3D ROTATE, UNION, SUBTRACT**, and **MOVE** commands as needed to construct the 3D model. See Figure 13-31.
9. Print the model per your instructor's directions.

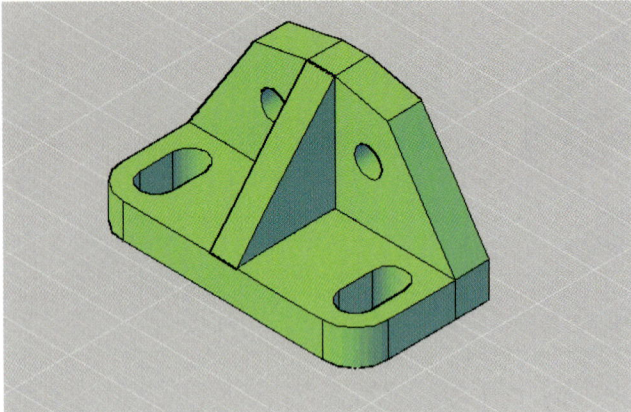

Figure 13-31   Completed 3D model of the bracket

Project 13-3: 3D Tool Holder

Create a 3D model of the tool holder shown in Figure 13-32.

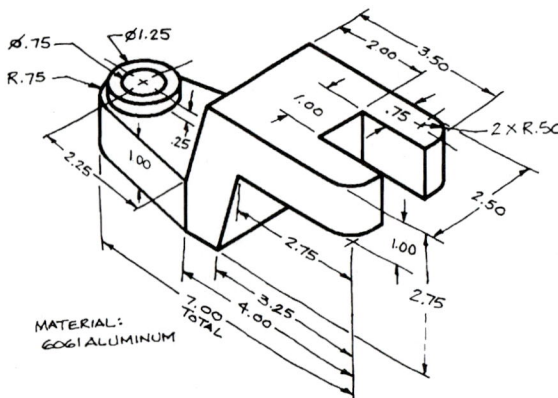

Figure 13-32   Designer's sketch of the tool holder project

Directions

1. Select **FILE** and **OPEN**.
2. Select the **Student Resource CD**.
3. Select the **Prototype Drawings** folder.
4. Select the **Tool Holder 3D Exercise** drawing (open as "read-only").
5. **SAVE AS** to your **Home** directory and rename the new drawing **TOOL HOLDER 3D**.
6. Use the 2D geometry in the prototype drawing to create 2D profiles (set the **View** to *SE Isometric* if needed) and then use the **REGION** command to create a region for each profile. See Figure 13-33.
7. Use the **EXTRUDE** command to extrude the regions to the desired height. See Figure 13-34.
8. Use the **3D ROTATE, UNION, SUBTRACT**, and **MOVE** commands as needed to construct the 3D model. See Figure 13-35.
9. Print the model per your instructor's directions.

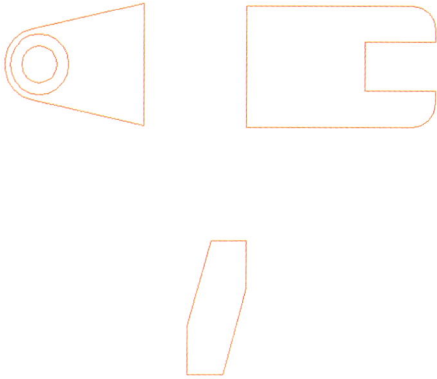

Figure 13-33   Creating profiles and regions

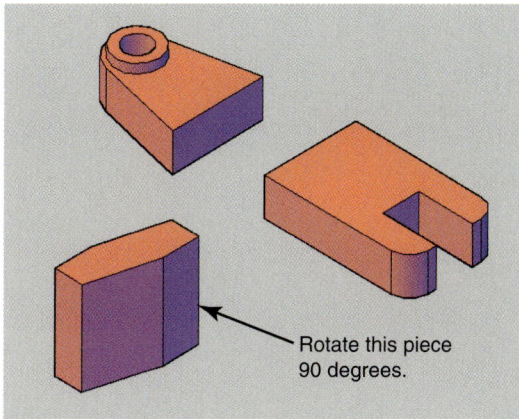

Rotate this piece 90 degrees.

Figure 13-34   Extruded profiles

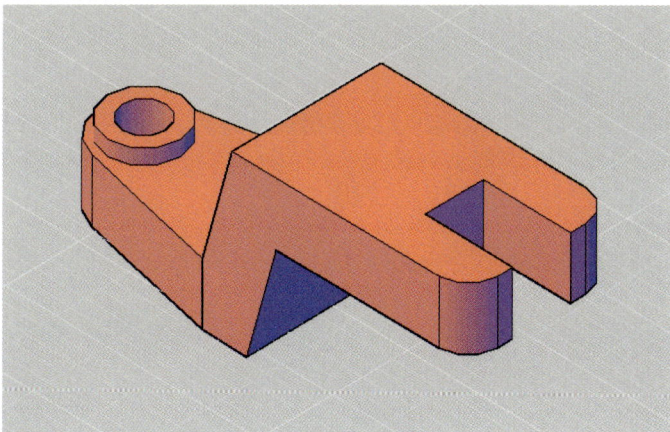

Figure 13-35   Finished 3D model

ANSI/ASME Standards

A

The American National Standards Institute (ANSI) oversees the creation of thousands of norms and guidelines that directly impact the manufacture and development of products and services in the United States. ANSI is also the United States' representative to ISO, the International Organization for Standards (see Appendix B).

A listing of all ANSI standards can be found at the ANSI website (www.ansi.org), but many of the the ANSI/ASME standards that are relevant to the creation of Engineering Drawings are as follows:

Y14.1-2005	*Decimal Inch Drawing Sheet Size and Format*
Y14.1M-2005	*Metric Drawing Sheet Size and Format*
Y14.2M-1992	*Line Conventions and Lettering*
Y14.3-2003	*Multiview and Sectional View Drawings*
Y14.34M-1996	*Parts List, Data List, Associated Lists*
Y14.4M-1989	*Pictorial Drawing*
Y14.5M-1994	*Dimensioning and Tolerancing*
Y14.6-2001	*Screw Thread Representation*

ISO Standards

B

The International Organization for Standardization (ISO) is a consortium of the national standards institutes from over 150 member countries. These institutes collaborate to establish standards for the development and manufacturing of products and services. The goal of ISO is not only to improve the quality of these products and processes but also to make them safer and more efficient. Another goal of international standards is to make trade between countries easier and fairer and to provide governments with a technical base for health, safety, and environmental legislation. ISO standards also help safeguard consumers and end users of products and services. The United States is represented on the ISO network by the American National Standards Institute (ANSI).

The first attempts at international standardization began in the early twenteeth century, but it was not until 1946 when delegates from 25 countries met and created an international organization. The purpose of this organization would be "to facilitate the international coordination and unification of industrial standards." The new organization, which was named ISO, officially began operations on February 23, 1947.

In English, the acronym ISO stands for the International Organization for Standardization; in French, ISO stands for *Organisation internationale de normalization.* Because of the many countries and languages represented on the ISO network, it was decided to use a word derived from the Greek *isos*, meaning "equal" to describe the organization. For this reason, regardless of the language, the organization is referred to as ISO.

A listing of most ISO standards can be found at the ISO website (http://www.iso.org), but a partial list of the ISO standards that relate specifically to the creation of technical drawings follows:

01.100.01	*Technical Drawings in General*
01.100.20	*Mechanical Engineering Drawings*
01.100.25	*Electrical and Electronics Engineering Drawings*
01.100.27	*Technical Drawings for Telecommunications/Information Technology*
01.100.30	*Construction Drawings Including Civil Engineering Drawings*
01.100.40	*Drawing Equipment*
01.100.99	*Other Standards Related to Technical Drawings*

United States National CAD Standard

The United States National CAD Standard® (NCS) coordinates CAD-related publications for the building design and construction industry in an attempt to make communication between design/construction teams and clients more consistent and direct.

Although adoption of the NCS by the building design and construction industry is voluntary, several government agencies have adopted the standard, and many public and private organizations are in the process of adopting it.

At its website (http://www.nationalcadstandard.org/), the NCS lists the following benefits as the advantages of adopting the NCS standard:

BENEFITS TO CLIENTS AND OWNERS

- Consistent organization of data for all projects, from all sources
- Greater clarity of communication of design intent to the client
- Streamlined electronic data management of Facility Management data
- Enhanced potential for automated document storage and retrieval
- Streamlined construction document checking process

BENEFITS TO DESIGN PROFESSIONALS

- Consistent data classification for all projects, regardless of the project type or client
- Seamless transfer of information among architects, engineers, and other design team members
- Predictable file translation results between formats; reduced preparation time for translation
- Reduced file formatting and setup time when adopted by software application vendors
- Reduced staff training time to teach "office standards"
- Streamlined checking process for errors and omissions
- New opportunities for expanded services and revenue beyond building design
- Added value to design services; firms can feature compliance with the NCS

BENEFITS TO CONTRACTORS AND SUBCONTRACTORS

- Consistent drawing sheet order and sheet organization; information appears in the same place in all drawing sets
- Consistent detail reference system

- Reduction of discrepancies, reducing the potential for errors, change orders, and construction delays
- Enhanced potential for automated payment process
- Consistent organization of data for all projects, from all sources

INDUSTRY-WIDE BENEFITS FOR NCS ADOPTION

- Reduced in-office training time with "collective professional memory" of a drawing standard
- Improved training at undergraduate and graduate levels
- Enhanced potential for automated training and distance learning
- Substantial reduction of barriers to seamless exchange of building construction data, leading to greater efficiency and decreased costs

Geometric Dimensioning and Tolerancing Basics

Applying plus/minus dimensioning to a drawing allows a mechanical designer to define the location and size of a part's features within a certain allowance, but applying the concepts of Geometric Dimensioning and Tolerancing (GDT) allows the designer to also define the *form* (flatness, straightness, circularity, and cylindricity), *orientation* (perpendicularity, angularity, and parallelism), or *position* of a part's features.

By applying GDT, the odds that a part will pass a quality control inspection are increased, and fewer rejected parts will result in lower production costs.

On a drawing, GDT symbols are shown inside a rectangular box called a *Feature Control Frame*. If the feature being defined by the GDT symbol is located relative to a datum feature (usually a surface or axis), the datum feature is defined with a *Datum Feature Symbol*.

Figure D-1 shows a view of an object with GDT symbols added. Study this figure and note how the objects identified as the Feature Control Frame and the Datum Feature Symbol are represented.

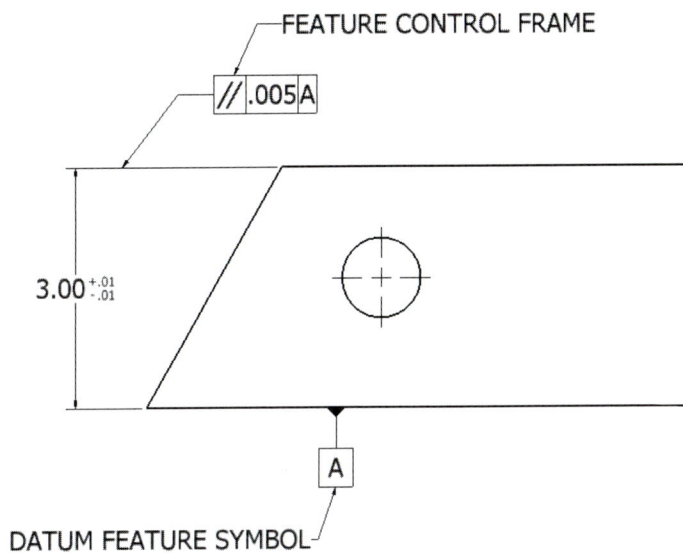

Figure D-1 A drawing with GDT symbols

INTERPRETING THE FEATURE CONTROL FRAME

The Feature Control Frame contains the GDT characteristic symbol (parallelism, perpendicularity, etc.), the tolerance, and, if the tolerance is referenced from a datum(s), the datum reference letter(s). Figure D-2 shows a Feature Control Frame with its components identified.

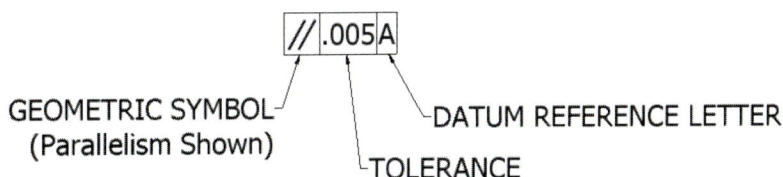

Figure D-2 Interpreting a Feature Control Frame

Geometric Characteristic Symbols

The GDT characteristic symbols that may be included in a feature control frame are shown in Figure D-3. The symbols used to represent these characteristics are defined in the American Society of Mechanical Engineers standard for dimensioning and tolerancing, *ASME Y14.5M-1994*. These symbols are included in the **TOLERANCING** command found on AutoCAD's **Dimension** toolbar (see Figures 5-42[a] and 5-42[b]).

Flatness-the amount that a surface can deviate from being perfectly flat.

Straightness-the amount a line on a surface can deviate from being perfectly straight.

Circularity-the amount a circle can deviate from being perfectly round.

Cylindricity-the amount a cylinder can deviate from being a perfect cylinder.

Profile of a Line-the amount a profile shape can deviate from the specified shape.

Profile of a Surface-the amount a surface can deviate from the specified shape.

Angularity-the amount an angled surface can deviate from a specified angle.

Perpendicularity-the amount that a surface can deviate from being perfectly perpendicular to a defined datum.

Parallelism-The amount a surface can deviate from parallel to a defined datum.

Position-The amount a feature can deviate from its true position as defined from a datum(s).

Concentricity-The amount the center axis of a revolved surface can deviate from being concentric with a defined datum.

Circular Runout-a measure of both the roundness and the location of circle.

Symmetry-the amount of deviation allowed from perfect symmetry.

Total Runout-the amount the surface of a cylinder can deviate as it is rotated around a center axis.

Figure D-3 Geometric characteristic symbols

Interpreting GDT on a Drawing

Figure D-4 shows a drawing of an object with a Feature Control Frame specifying the allowed deviation from perfect *flatness* for the bottom surface of the object. The distance between the two lines representing the .005 tolerance is exaggerated in this figure to better illustrate the concept.

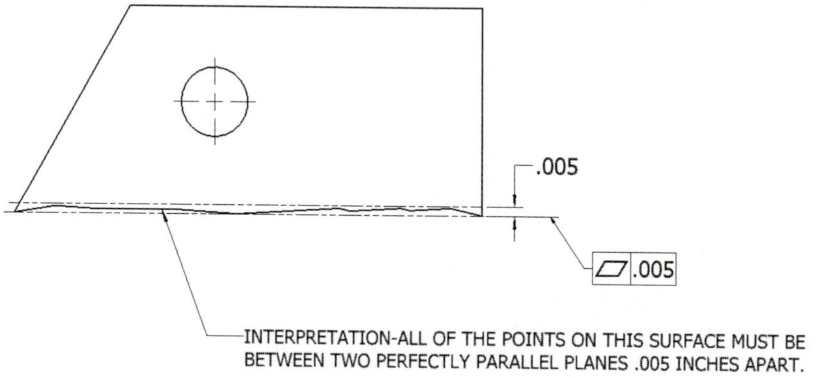
.005
.005
INTERPRETATION-ALL OF THE POINTS ON THIS SURFACE MUST BE BETWEEN TWO PERFECTLY PARALLEL PLANES .005 INCHES APART.

Figure D-4 Interpreting flatness on a drawing

Interpreting Parallelism

Figure D-5 shows a drawing of an object with a Feature Control Frame specifying the allowed deviation that the top surface can have from being perfectly parallel to the bottom surface—identified with a Datum Feature Symbol as *Datum A*. The distance between the two lines representing the .005 tolerance is exaggerated in this figure to better illustrate the concept.

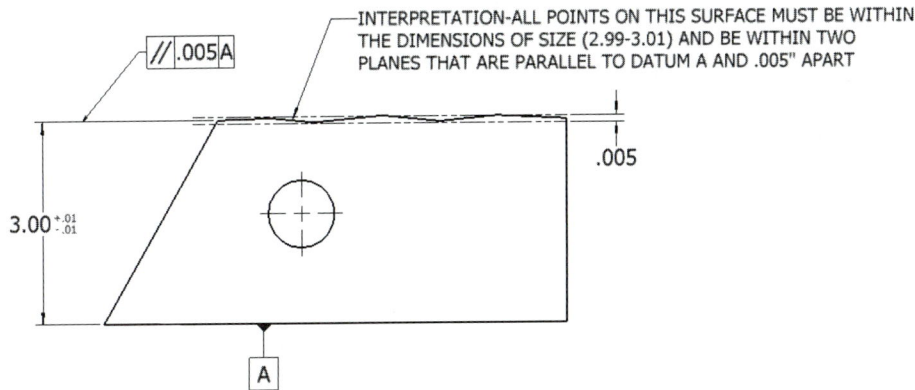

INTERPRETATION-ALL POINTS ON THIS SURFACE MUST BE WITHIN THE DIMENSIONS OF SIZE (2.99-3.01) AND BE WITHIN TWO PLANES THAT ARE PARALLEL TO DATUM A AND .005" APART

//.005A

.005

$3.00^{+.01}_{-.01}$

A

Figure D-5   Interpreting parallelism on a drawing

Adding GDT to an AutoCAD Drawing

Step 1. Select the **Tolerance** icon from AutoCAD's **Dimension** toolbar (see Figure D-6). This will open the **Geometric Tolerance** dialog box shown in Figure D-7.

⊕ .1

Tolerance

Figure D-6   Tolerance icon

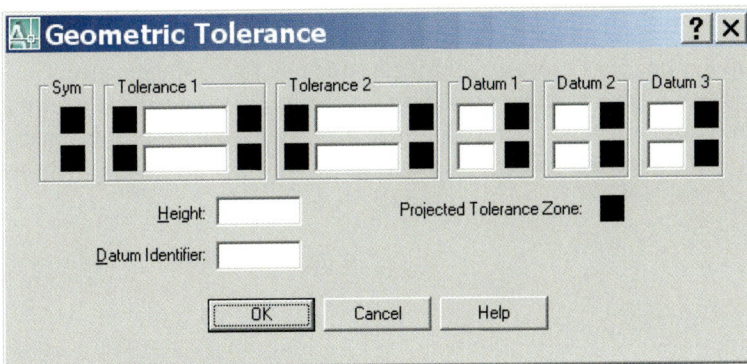

Geometric Tolerance

| Sym | Tolerance 1 | Tolerance 2 | Datum 1 | Datum 2 | Datum 3 |

Height:

Projected Tolerance Zone:

Datum Identifier:

OK   Cancel   Help

Figure D-7   Geometric Tolerance dialog box

Step 2. Select the first black box below the word **Sym** located in the **Geometric Tolerance** dialog box shown in Figure D-7. This will open the **Symbol** box shown in Figure D-8.

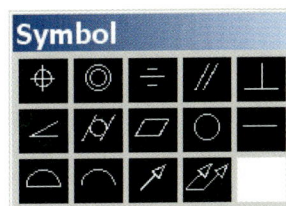

Symbol

⊕ ◎ = // ⊥

∠ ⌀ ▱ ○ —

⌒ ⌒ ⌀ ⌀

Figure D-8   Symbol box

Step 3. Select the geometric characteristic symbol from the symbols available in the **Symbol** box (see Figure D-8). After selecting a symbol, it will appear in the **Geometric Tolerance** dialog box in the window beneath **Sym** as shown in Figure D-9.

Step 4. Fill in the values for the desired Tolerance and Datum(s) as shown in Figure D-9 and click **OK**. Then, use the mouse to move the Feature Control Frame to its desired location in the drawing.

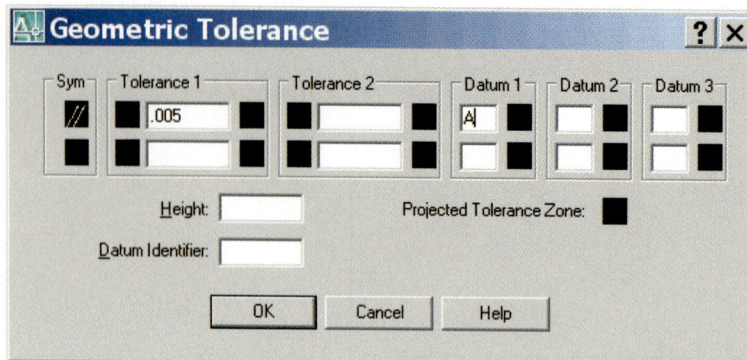

Figure D-9 Adding tolerance and datum information

ADDING A DATUM FEATURE SYMBOL

A Datum Feature Symbol can be added by following **Steps 1** and **2** in the preceding section, but instead of selecting the black box below the word **Sym**, type the desired datum identification character (an *A,* for example) into the **Datum Identifier** box and click **OK**. Then, use the mouse to move the Datum Feature Symbol to its desired location in the drawing.

LEARNING TO APPLY GDT TO DRAWINGS

Learning to apply GDT to drawings is a process requiring instruction, study, and practice. GDT techniques are often included in the curriculum of advanced mechanical engineering drawing courses. Many organizations also offer continuing education GDT workshops for professionals.

A good resource for GDT materials and certification can be found at the ASME website: www.asme.org.

AutoCAD 2009 Update

BEGINNING AN AUTOCAD 2009 DRAWING

Launch the AutoCAD 2009 program by *double clicking* on the **AutoCAD 2009** icon located on the desktop of your computer (see Figure E-1).

A new AutoCAD 2009 drawing session will open. The graphics window will appear similar to the one shown in Figure E-2.

AutoCAD 2009

Figure E-1 AutoCAD 2009 Icon

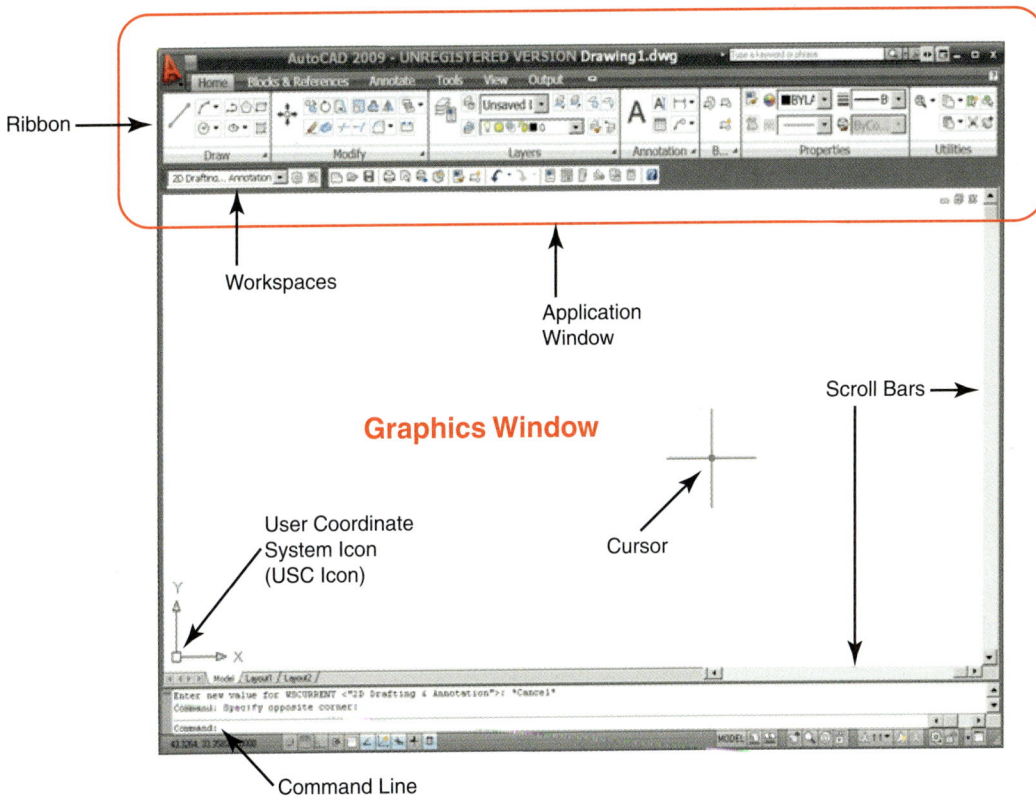

Figure E-2 AutoCAD 2009 Screen Layout

Study the AutoCAD Screen Layout shown in Figure E-2 and acquaint yourself with the terminology used to describe its features.

If you have used a previous version of AutoCAD, the biggest difference you will notice is that the pulldown menu has been replaced by an **Application Window** containing a **Ribbon**. Find the **Application Window** and **Ribbon** noted in Figure E-2.

SETTING THE DRAWING ENVIRONMENT TO AUTOCAD CLASSIC

Some users of AutoCAD 2009 may prefer to work in the more familiar *AutoCAD Classic* workspace environment which is nearly identical to the screen layout found in earlier releases of AutoCAD.

To set AutoCAD 2009 workspace to AutoCAD Classic, follow these steps:

- Click on the **Menu Browser Button** (the red *A* in the upper left corner of the AutoCAD screen) to display the pulldown menu (see Figure E-3).

- From the pulldown menu, select **Tools**, then **Workspaces**, and click on **AutoCAD Classic**.

Note:

The pulldown menu shown in the left column in Figure E-3 replaces the pulldown menu bar that was part of the screen layout of earlier versions of AutoCAD (see Figure 4-2). To make the pulldown menu visible in the graphics window of AutoCAD 2009, type **Menubar** on the command line and change the setting to 1.

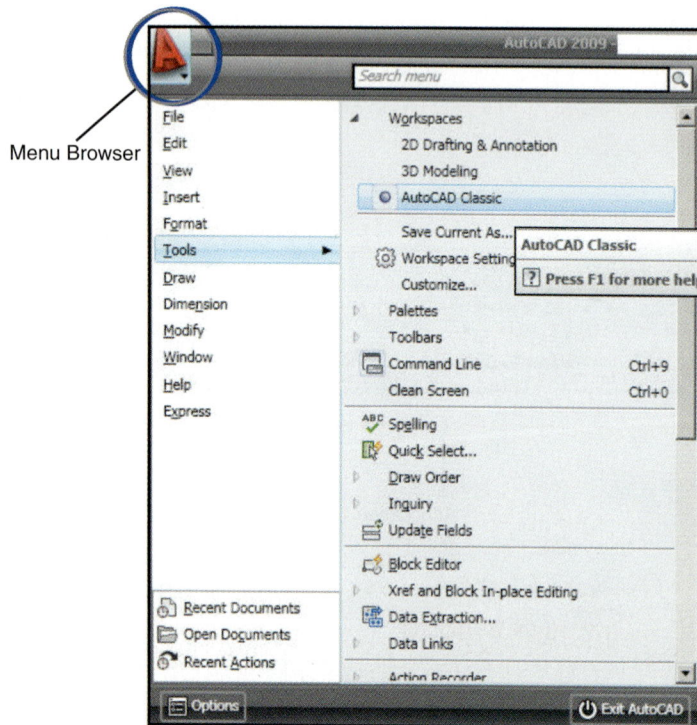

Menu Browser

Figure E-3 AutoCAD 2009 Menu Browser and Pulldown menu

With the environment set to AutoCAD Classic, your screen will resemble the one shown in Figure E-4. Study Figure E-4 and note the difference in the appearance of the **Explode** icon between AutoCAD 2009 and AutoCAD 2008.

In Figure E-4 note the location and appearance of the buttons on the **Application Status Bar**. In previous releases of AutoCAD, the buttons on this bar were labeled. In AutoCAD 2009, the labels have been replaced by icons. Figure E-5 shows a detailed view of the buttons on the AutoCAD 2009 Application Status Bar.

USING THE RIBBON

When the AutoCAD 2009 workspace is set to 2D Drafting and Annotation, or 3D Modeling, the ribbon will be visible in the application window. The ribbon maximizes the work area by grouping commands and operations into compact *toolbar panels*. Toolbar panels are clustered together according to function to create six *ribbon tabs*. These tabs are labeled **Home**, **Blocks & References**, **Annotate**, **Tools**, **View**, and **Output**. By selecting one of these tabs, the user can select from the commands and operations contained on the toolbar panels of that tab. For example,

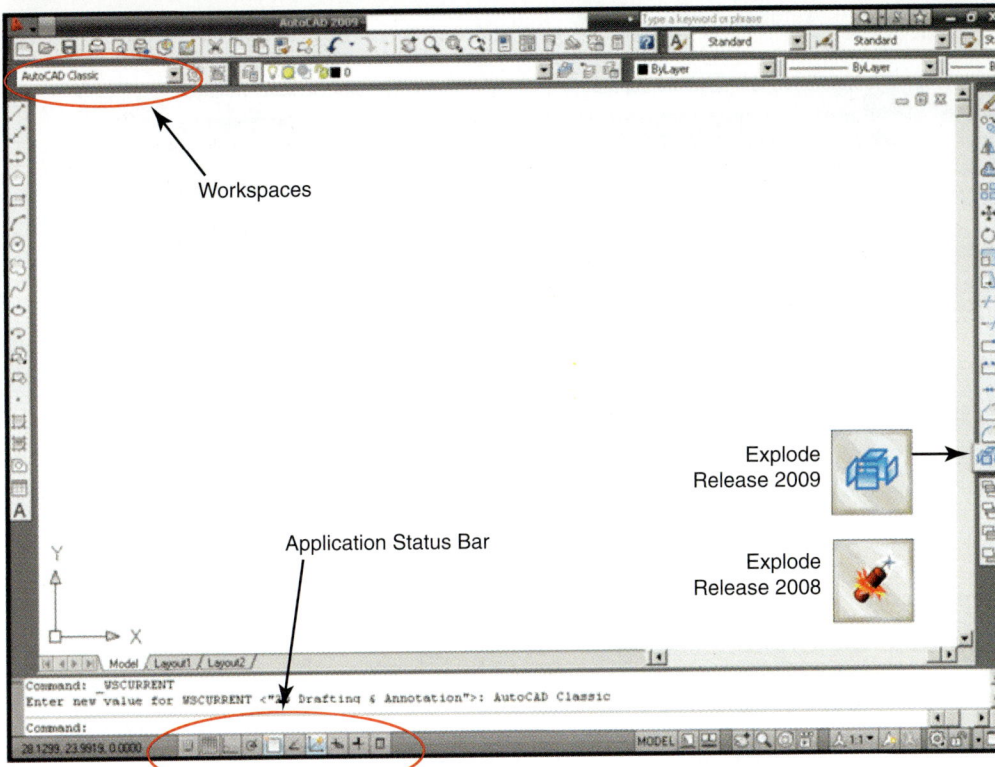

Figure E-4 AutoCAD 2009's Classic Workspace screen layout

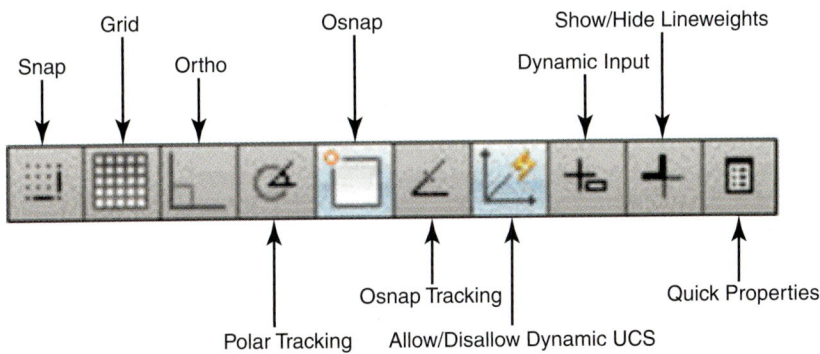

Figure E-5 Detail of AutoCAD 2009's Application Status Bar

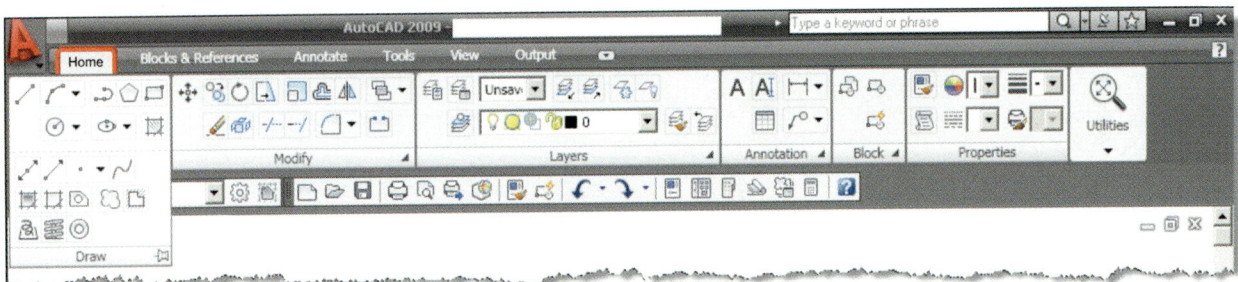

Figure E-6 The Home tab showing the expanded Draw toolbar

when the **Home** tab is selected, the **Draw**, **Modify**, **Layers**, **Annotation**, **Block**, **Properties**, and **Utilities** toolbar panels are visible as shown in Figure E-6.

By clicking on the label at the bottom of a toolbar panel, the panel expands to show other command icons contained in the panel. Figure E-6 shows the **Draw** toolbar panel expanded in this way.

Holding the cursor over the **CIRCLE** icon (do not click on the icon) on the **Draw** toolbar panel causes a tooltip to appear as shown in Figure E-7. If the cursor is held over the

Note:

The technique of holding the cursor over an icon can be used to access information about the commands located on other toolbar panels as well.

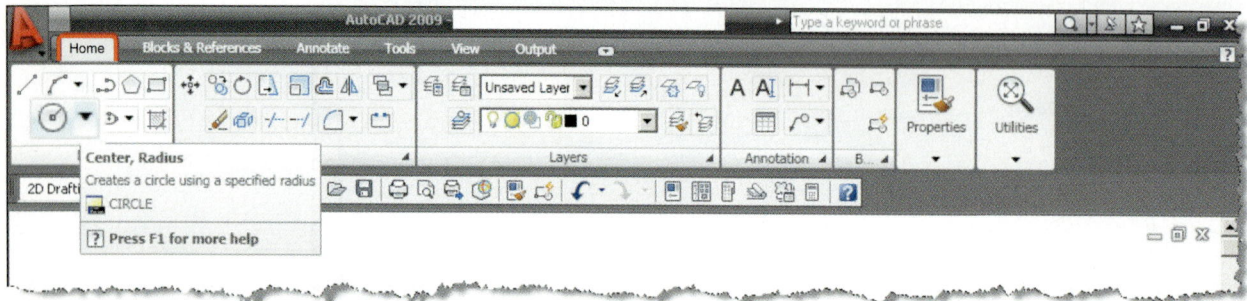

Figure E-7   CIRCLE command tooltip

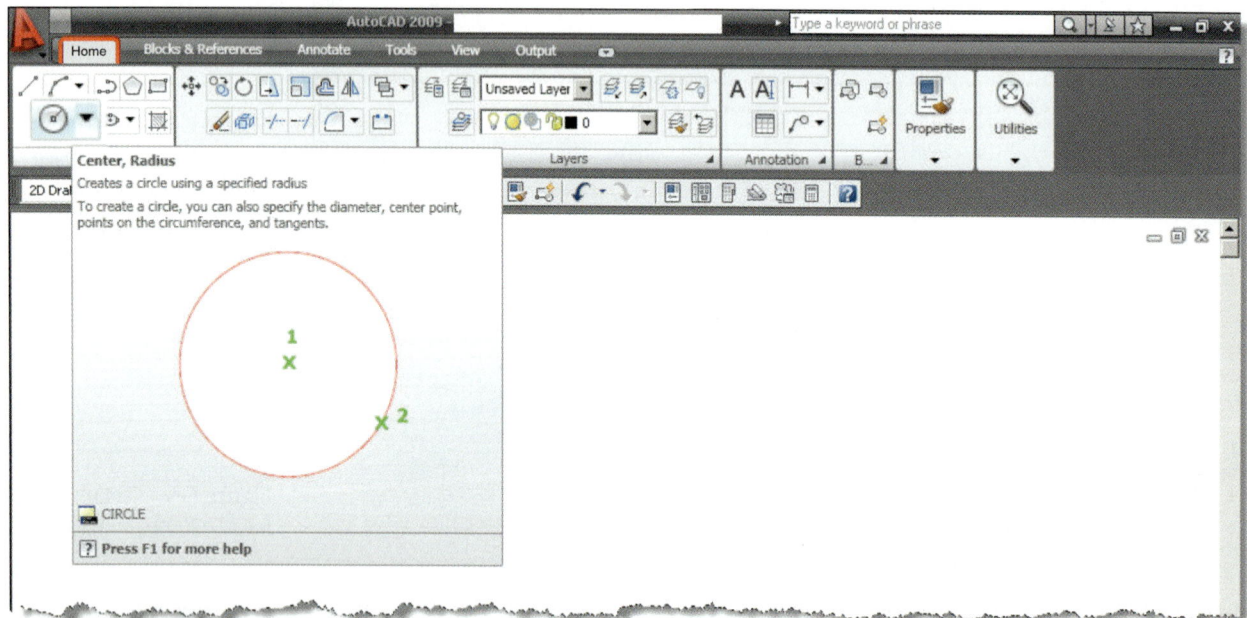

Figure E-8   Expanded tooltip for the CIRCLE command

icon for two or more seconds, an extended tooltip will appear as shown in Figure E-8. This tooltip offers additional information about the **CIRCLE** command.

Holding the cursor over the **CIRCLE** icon and pressing the **F1** function key on the keyboard opens the **Quick Reference** help screen which offers more information about the **CIRCLE** command as shown in Figure E-9.

By clicking on the down arrow located to the right of the **CIRCLE** icon, an expanded palette of **CIRCLE** creation options is displayed as shown in Figure E-10.

Another way to view all of the icons located on a toolbar panel, is to select the down arrow located in the lower right corner of the panel. Figure E-11 shows the expanded **Modify** toolbar panel after the down arrow has been selected.

Note:
The technique of holding the cursor over an icon and pressing **F1** can be used to access expanded information about the commands on other toolbar panels as well.

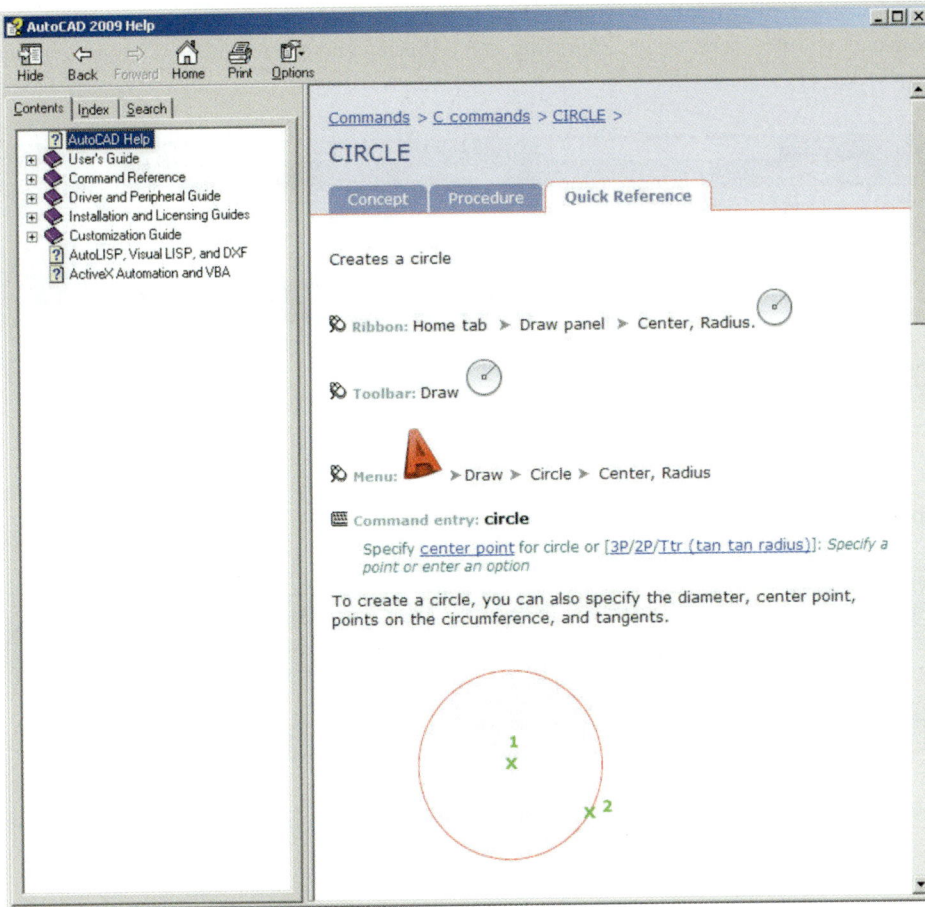

Figure E-9   Quick Reference help screen for the CIRCLE command

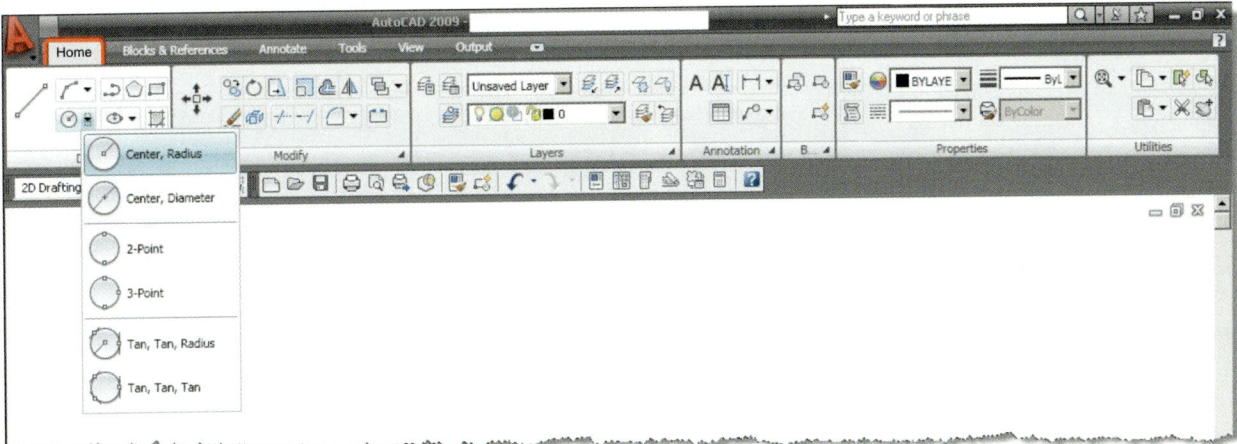

Figure E-10   Circle creation options

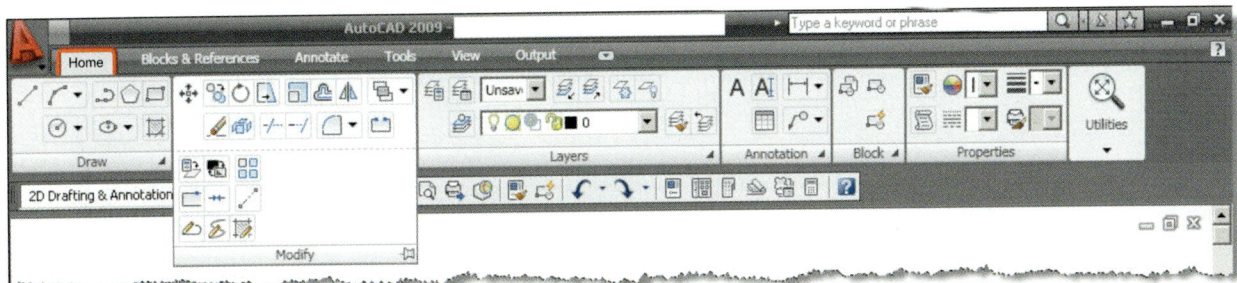

Figure E-11   The Home tab showing the expanded Modify toolbar

Selecting the **Blocks & References** tab displays a group of toolbar panels related to blocks, attributes, references, importing, data, linking and extraction, and content as shown in Figure E-12.

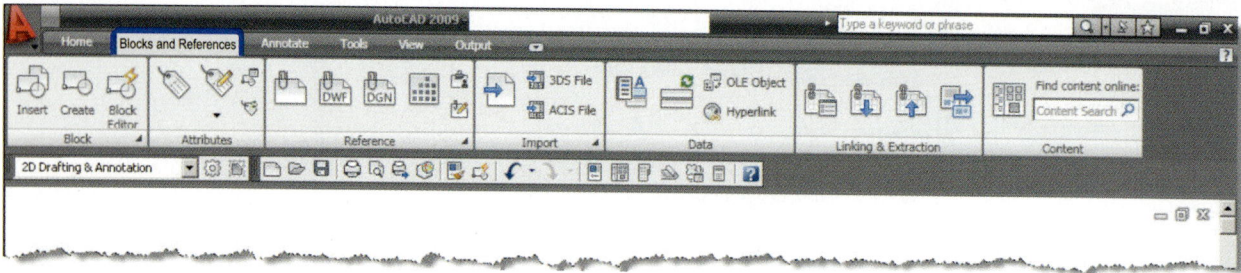

Figure E-12   Blocks & References tab

Selecting the **Annotate** tab displays a group of toolbar panels related to placing annotations on drawings as shown in Figure E-13.

Figure E-13   The Annotate tab

Selecting the **Tools** tab displays a group of toolbar panels related to action recording, inquiries, animations, drawing utilities, customization, applications, and layer standards as shown in Figure E-14.

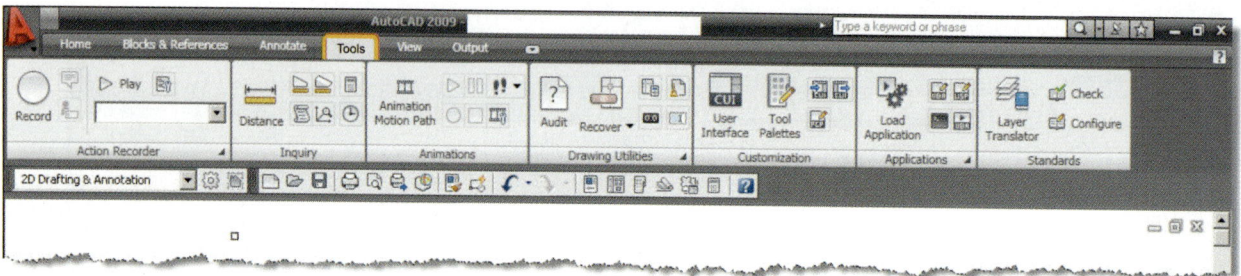

Figure E-14   The Tools tab

Selecting the **View** tab displays a group of toolbars related to the User Coordinate System, viewports, displaying tool palettes, changing an object's properties, opening a drawing, arranging multiple drawings, and setting windows elements as shown in Figure E-15.

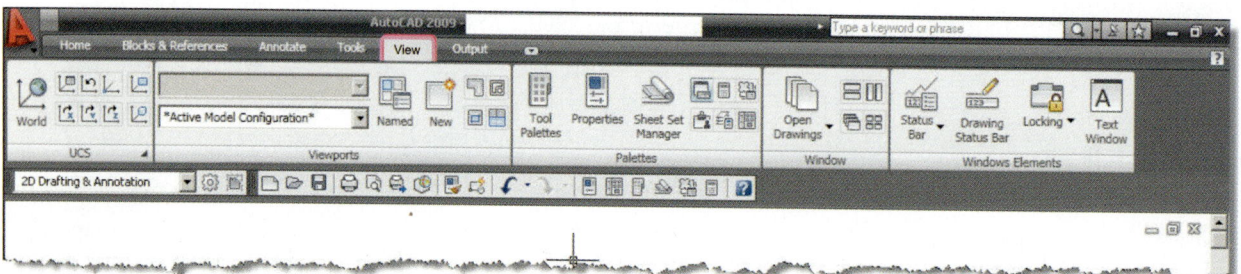

Figure E-15   The View tab

Selecting the **Output** tab displays a group of toolbars related to plotting, publishing, and exporting drawings as shown in Figure E-16.

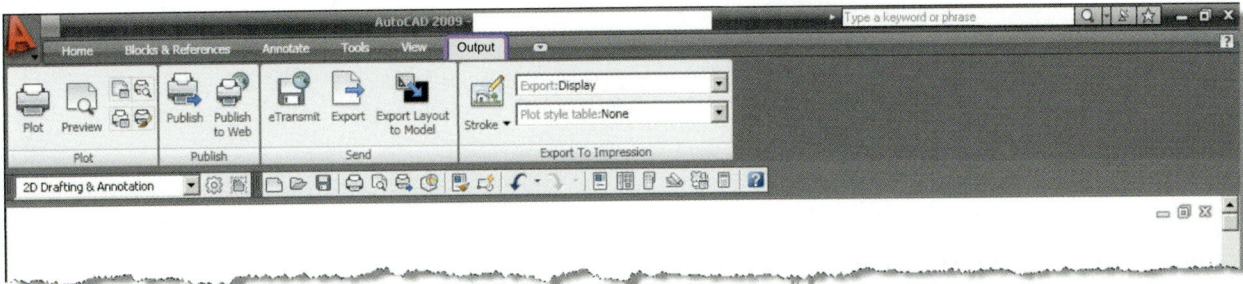

Figure E-16   The Output tab

Selecting the **Minimize to Panel Titles** button shown in Figure E-17 minimizes the ribbon to show only the tabs and toolbar panel names which results in a larger drawing area.

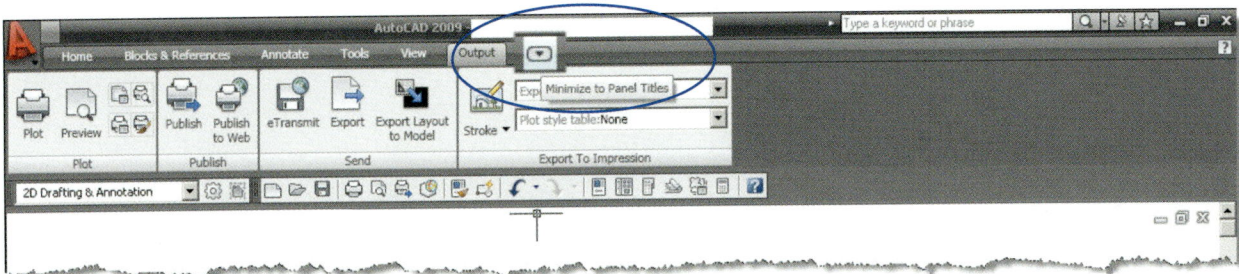

Figure E-17   The Minimize to Panel Titles button

Selecting the **Minimize to Tabs** button again minimizes both the ribbon and toolbar names as shown in Figure E-18.

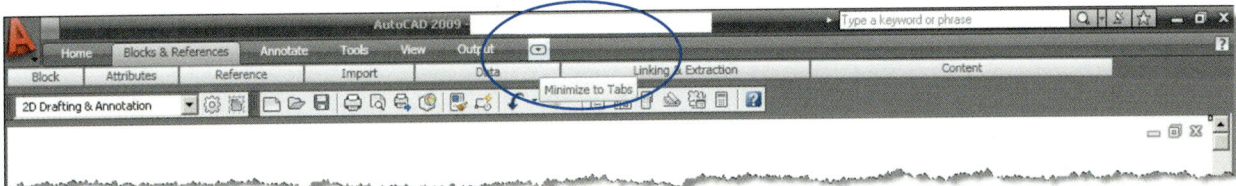

Figure E-18   The Minimize to Tabs button

Selecting the **Show Full Ribbon** button shown in Figure E-19 will reset the ribbon to its full size. Select the **Home** tab to reset the ribbon to show the original toolbar panels.

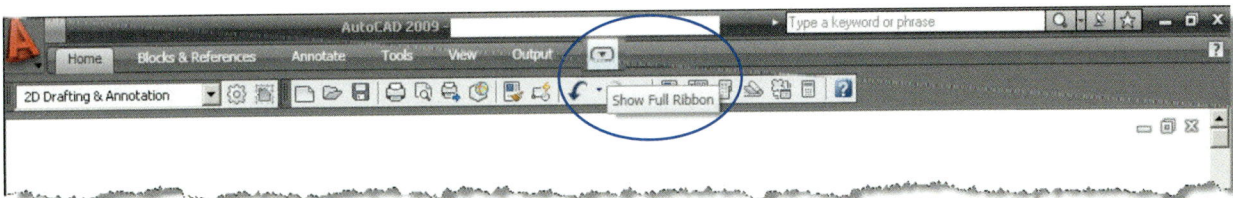

Figure E-19   The Show Full Ribbon button

Glossary

absolute coordinates: Points that are located along the X-, Y-, and Z-axes that are relative to a point defined as 0,0,0. In an AutoCAD drawing, 0,0,0 is usually located in the lower left corner of the graphics window.

actual size: The measured size of a finished part. This size determines whether the part passes a quality control inspection. See *quality control inspection*.

aeronautical or aerospace drafters: Drafters who prepare engineering drawings detailing plans and specifications used in the manufacture of aircraft and related equipment.

aligned text: Text placed on a technical drawing that faces the bottom and the right side of the sheet. This technique is common on architectural drawings but is not allowed on drawings employing the ASME dimensioning standard.

allowance: The minimum clearance, or maximum interference, between parts.

American Institute of Architects (AIA): The accrediting body for architects.

American National Standards Institute (ANSI): The national organization for the development of standards in the United States. ANSI represents the United States as a member of the International Organization for Standardization.

American Society of Mechanical Engineers (ASME): Publisher of standards for the creation of technical drawing in the United States. *ASME Y14.5M-1994 Dimensioning and Tolerancing* and *ASME Y14.2M Line Conventions and Lettering* are two standards important to drafters.

architect: A person who is licensed by the American Institute of Architects to practice architecture.

architectural drafters: Drafters who prepare the drawings used in construction industries.

assembly drawing: A drawing that illustrates how the separate parts of an assembly are related to each other, for example, how mating parts fit together.

auxiliary view: a view that shows the true shape of features that are not parallel to any of the principal projection planes (front, top, side, etc.).

baseline dimensions: A group of linear dimensions that are referenced from the same datum or baseline.

basic size: The theoretical perfect size or location of a feature. Basic dimensions are enclosed in a rectangle when shown on a drawing.

block: AutoCAD term that refers to a predefined object, or symbol, that is stored in the AutoCAD drawing file which can be inserted into the drawing whenever it is needed.

block library: A block library is a group of block definitions stored in a drawing file. For example, an architectural firm might create a block library of door and window symbols that are frequently used on floor plans. See *block*.

broken-out section: A type view that shows only a small section detail of an object versus a full, or half, section.

Building Information Model (BIM): A system of managing construction documents as a linked database that users of the system can tap into to assist them in performing their jobs.

CAD (Computer Aided Design): A term often used to describe the creation of technical drawings. Also a term for the software used to create technical drawings. Some popular CAD programs include AutoCAD, SolidWorks, Revit, Pro/ENGINEER, and Inventor.

Cartesian Coordinate System: A system of locating points along X- and Y-axes relative to a starting point representing "Zero X" and "Zero Y." Named for its originator, René Descartes.

checker: An experienced designer/drafter with expertise in manufacturing, drafting techniques, and dimensioning conventions who is responsible for reviewing and approving the drawings prepared by other drafters.

civil drafters and design technicians: Drafters who prepare construction drawings and topographical maps used in civil engineering projects.

Computer Aided Manufacturing (CAM):   A manufacturing system in which equipment and processes are controlled by computer commands.

construction documents (CDs):   Drawings used in the construction of residential and commercial buildings. These drawings may include floor plans, elevations, foundations, wall sections, and roof framing plans.

continuous dimensions: A dimensioning technique in which a linear dimension is placed using the second extension line origin of a selected dimension as its first extension line origin. This technique is also called *chain dimensioning*.

cutting plane: An imaginary plane that slices through an object to reveal its interior features in a section view.

datum: A theoretically perfect feature (plane, axis, or point) from which dimensions are referenced.

designer: Individual who assists engineers or architects with the design process. Designers are often former drafters who have proven their ability to take on more responsibility and decision-making duties.

detail drawings: Drawings that provide the information required to manufacture or purchase each part in an assembly including the necessary views, dimensions, notations, and specifications.

Dimension commands: The commands used to dimension an AutoCAD drawing. These commands are located on the **Dimensioning** toolbar and include **LINEAR, BASELINE, CONTINUE, ANGULAR, DIAMETER,** and **RADIUS.**

dimension standards: Dimensioning rules that have been created to standardize dimensioning styles and techniques. Dimensioning standards for mechanical drawings have been defined by the American Society of Mechanical Engineers (ASME) and the International Organization for Standardization (ISO). Dimensioning standards for construction documents are defined by the United States National CAD Standard.

dimensions: Annotations that are added to a technical drawing specifying the size and location of the features of an object. There are two types of dimensions: *size* and *location*. For example, the diameter of a hole is a size dimension, whereas the dimensions that indicate the placement of the center of the hole are location dimensions. Dimensional information may include notes concerning the material the object is manufactured from, special processes performed on the object (heat treating, polishing, etc.), and any other information needed during the manufacture of a part or the construction of a building.

drafter: An individual with specialized training in the creation of technical drawings.

drafting: A term often used to describe the creation of technical drawings.

Draw commands: The commands used to place geometry in an AutoCAD drawing. These commands are located on the **Draw** toolbar and include **LINE, CIRCLE, ARC,** and **MULTILINE TEXT.**

drawing limits: The limits of an AutoCAD drawing define its drawing area; this is comparable to selecting the sheet size for the drawing. When setting limits, you are prompted to specify the lower left and upper right corners of the drawing area. In most cases, the lower left corner will default to **0,0,** and the upper right corners will be defined by typing in the coordinates of the corresponding sheet size. For example, if using decimal units, an A size sheet limits would be **0,0** and **12,9;** a B size sheet limits would be **0,0** and **17,11;** a C size sheet limits would be **0,0** and **22,17;** and a D sheet limits would be **0,0** and **34,22.**

drawing units: The units of measurement to be used in the creation of an AutoCAD drawing. For example, in an architectural drawing, one unit may equal one foot, whereas in a mechanical drawing, one unit might equal one inch or one millimeter. The drawing units should be set before beginning a drawing or defining the drawing limits.

electrical drafters: Drafters who prepare diagrams used in the installation and repair of electrical equipment and building wiring.

electrical plans: Drawings that provide electrical contractors information about the type, location, and installation of electrical components (switches, lamps, ceiling fans, electrical outlets, cable TV jacks, etc.) used in the project. All of the information needed by the electrical contractor to wire the building should be provided by this plan.

electro/mechanical drafters: Drafters who split their duties between mechanical drafting and electrical/electronic drafting.

electronics drafters: Drafters who prepare schematic diagrams, printed circuit board artwork, integrated circuit layouts, and other graphics used in the design and maintenance of electronic (semiconductor) devices.

elevation drawings: Drawings that provide information about the exterior details of a building. This information may include roof pitch, exterior materials and finishes, overall heights of features, and window and door styles. All of the dimensions and notations required by workers on the jobsite should be included on this sheet.

engineering drawing: A term often used to describe the creation of technical drawings.

engineering graphics: A term often used to describe the creation of technical drawings.

Engineer's, Architect's, and Metric scales: Specialized rulers used to make precise measurements during the creation, or interpretation, of technical drawings.
Each scale is marked in increments related to the system of measurement appropriate to its field; for example, the Architect's scale is divided into increments representing feet and inches.

features: Geometric elements that are added to a base part. Features include holes, slots, arcs, fillets, rounds, angled planes, counterbored holes, and countersunk holes. Features are located on the object with location dimensions and are described with size dimensions.

fillet: A rounded inside corner of a part.

Finite Element Analysis (FEA): Software that is used in conjunction with CAD modeling software to perform advanced design analysis on 3D models. FEA allows engineers and designers to calculate such properties as an object's mass, center of gravity, strength, distribution of stresses, and bending moments.

First Angle Projection: A technique for arranging multiview drawings in which the left side view is drawn to the right of the front view and the top view is placed below the front view, and so on. Commonly used on drawings prepared outside North America.

floor plans: Drawings that provide home builders and contractors with the necessary information to lay out the building, including the locations of features such as walls, doors, electrical components (switches, lamps, etc.) and plumbing fixtures (tubs, commodes, sinks, etc.). Floor plans usually include all of the dimensions and notations required by the workers on the jobsite. Doors and windows are dimensioned to their centers and continuous (also known as *chain*) dimensioning is typically employed on floor plans.

foreshortening: A term used to describe the phenomenon that occurs when a feature on an inclined plane is not shown true size, or true shape, in a multiview drawing.

full section: A view of an object that shows its interior detail as if it has been cut in half.

Geometric Dimensioning and Tolerancing (GD&T): A dimensioning technique that is used to control the *form* (flatness, straightness, circularity, and cylindricity), *orientation* (perpendicularity, angularity, and parallelism), or *position* of a part's features.

graphic primitives: Geometric shapes such as boxes, cylinders, cones, spheres, wedges, and prisms that can be combined (unioned) or removed from one another (subtracted) to create more complicated shapes.

half section: A view of an object that shows its interior detail as if one-fourth of it has been removed.

inclined plane: A plane located on an object in a multiview drawing that is sloping and is not perpendicular to the line of sight of the viewer.

inquiry commands: Commands used to display information about AutoCAD entities such as distance between two points, area of a closed figure like a rectangle or circle, or the volume of a 3D object. These commands are located on the **Inquiry** toolbar.

International Organization for Standardization (ISO):   The international organization for the development of standards including technical drawing and dimensioning standards. ANSI represents the United States as a member of ISO. The ISO dimensioning standard is almost identical to the ASME dimensioning standard.

isometric drawing: A type of pictorial drawing in which receding lines are drawn at 30° relative to the horizon. Commonly used in the mechanical engineering field. See *pictorial drawing*.

layers: In AutoCAD drawings, lines and other entities are drawn on layers. Think of layers as sheets of clear glass layered one on top of the other. A layer can have its own properties such as color, linetype, or lineweight.

lead hardness scale: The scale that defines the hardness of leads used in technical pencils.

Least Material Condition (LMC):   The condition of a part when it contains the least amount of material. The LMC of an external feature, such as a shaft, is the lower limit of size defined by the tolerance. The LMC of an internal feature, such as a hole, is the upper limit of size defined by the tolerance.

limits: The maximum and minimum sizes of a feature as defined by its tolerances. For example, a feature with a nominal dimension of .50, with a tolerance of ±.02, has an upper limit of .52 and a lower limit of .48.

line types: Include *visible lines* which show the visible edges and features of an object, *hidden lines* which represent features that would not be visible, and *center lines* which locate the centers of features such as holes and arcs. Standard line types have been established by the American Society of Mechanical Engineers (ASME) in *ASME Y14.2M*.

lineweight: Refers to the width of the lines in a technical drawing. Standard lineweights have been established by the American Society of Mechanical Engineers (ASME) in *ASME Y14.2M*. In this standard, visible lines are drawn .6mm wide, while center and hidden lines are drawn .3mm wide.

Maximum Material Condition (MMC):   The condition of a part when it contains the greatest amount of material. The MMC of an external feature, such as a shaft, is the upper limit. The MMC of an internal feature, such as a hole, is the lower limit.

mechanical drafters: Drafters who prepare detail and assembly drawings of machinery and mechanical devices.

mechanical working drawings: Drawings used in the fabrication and assembly of machine parts.

miter line: In drawings created with orthographic projection techniques, a construction line drawn at 45° which enables information to be projected from the top view to the side view, and from the side view to the top view. See *orthographic projection*.

modeling commands: Used to create 3D models in an AutoCAD drawing. These commands are located on the **Modeling** toolbar and include **UNION, SUBTRACT, 3D ROTATE,** and **EXTRUDE**.

Modify commands: The commands used to modify the geometry of an AutoCAD drawing. These commands are located on the **Modify** toolbar and include **ERASE, MOVE, COPY, OFFSET, ROTATE,** and **SCALE**.

multiview drawing: A technique used by drafters and designers to depict a three-dimensional object (an object having height, width, and depth) as a group of related two-dimensional (having only width and height, or width and depth, or height and depth) views.

National Society of Professional Engineers: The accrediting agency for engineers.

nominal size: A dimension that describes the general size of a feature. Tolerances are applied to this dimension.

Object Snap settings: A technique used in the creation and editing of AutoCAD drawings that allows the user to snap to exact points on an object. Common object snap settings include snap to endpoint, snap to midpoint, snap to intersection, snap to center, and snap to quadrant.

offset section: A type of section in which the cutting plane line is *offset* at 90° angles to pass through features that would not lie on the path of a straight cutting plane line.

ordinate dimensioning: A dimensioning technique in which, a 0,0 (zero X, zero Y) datum point is defined on the object and the location the object's features are located along the X and Y axes as referenced from the 0,0 datum point. Ordinate dimensioning is useful for dimensioning parts that are to be manufactured by *Computer Aided Manufacturing* (CAM) machinery, such as a CAM drill press.

orthographic projection: The technique employed in the creation of multiview drawings to project geometric information (points, lines, planes, or other features) from one view to another.

parametric modeling: A method of creating 3D CAD models in which the geometry of the model is driven by the dimensions associated with the geometry. This allows designers to modify the features of a model by simply editing its dimensions. When the parametric dimension is changed, the 3D model updates to reflect the new dimension value.

partial auxiliary view: A type of auxiliary view in which an object's planes that are not perpendicular to the plane of projection, and would be foreshortened and not true size or shape, are not drawn.

Parts List: A table placed on a technical drawing that itemizes all of the parts in an assembly (sometimes referred to as a *Bill of Material or BOM*). The Parts List may include columns for part number, part name, description, quantity, and material.

perspective drawing: A type of pictorial drawing in which receding lines appear to converge at a vanishing point. Commonly used in the architectural field. See *pictorial drawing*.

pictorial drawing: A type of drawing in which an object appears to be three dimensional, that is, it appears to have width, height, and depth. But unlike an actual 3D model, a pictorial drawing is constructed using only X and Y coordinates. See *isometric drawing* and *perspective drawing*.

pipeline drafters and process piping drafters: Drafters who prepare drawings used in the construction and maintenance of oil refineries, oil production and exploration industries, chemical plants, and process piping systems such as those used in the manufacture of semiconductor devices.

polar coordinates: Coordinates defined by a length and an angle that are relative to the last point defined.

primary auxiliary view: A view that is adjacent to, and aligned with, a principal view of the object showing the true shape of features that are not parallel to any of the principal projection planes (front, top, side, etc.).

professional engineer (P.E.):   An engineer who is licensed by the National Society of Professional Engineers.

projection planes: A clear, two-dimensional plane (like a sheet of glass) that is placed between the object and the viewer. The image of the object is drawn as if it had been projected onto this plane.

properties: In an AutoCAD drawing, the properties of an object include its color, lineweight, layer, linetype, linetype scale, etc.

quality control inspection: A step in the manufacturing cycle performed by a quality control (QC) inspector using precise measuring equipment to determine the actual size of the part. The

QC inspector compares the actual size of the part to the dimensions noted on the technical drawing. Parts that measure within the allowable size limits will pass the QC inspection, whereas parts that measure outside the limits will be rejected.

reference dimension: A dimension that is included on a technical drawing for information only and is not necessary to manufacture, or inspect, the part. No tolerances are applied to reference dimensions. Reference dimensions are enclosed in parentheses.

regular views: In a multiview drawing, this term refers to an object's front, top, bottom, right, left, and back (or rear) views.

relative coordinates: Points that are located along the *X*-, *Y*-, and *Z*-axes that are relative to the last point defined.

removed section: A type of section in which the section view is not drawn in its normal "projected" position, but rather somewhere else on the sheet. Drafters must ensure that the cutting plane line and the resulting section are labeled alike in order to avoid confusion when referencing the removed section view(s).

revolved section: A type of section in which the cross-sectional view of the object is drawn on the object itself.

roof pitch: The angle of a roof expressed as a ratio of the vertical *rise* of the roof (measured in inches) to the horizontal *run* of the roof (measured in inches). Using this notation, a roof with a "four-twelve" pitch (labeled as **4/12** on the drawing) would rise 4″ for every 12″ of horizontal run. Roof pitch is noted on elevation drawings in a set of construction documents.

round: A rounded outside corner of a part.

scales: Instruments used to make precise measurements on a technical drawing. Scales are available in architectural, engineering, and metric units.

secondary auxiliary view: A view that is adjacent to, and aligned with, a primary auxiliary view.

section: A drawing technique in which an object is drawn as if part of its exterior has been removed to reveal its interior features and details.

section lines: Diagonal lines drawn that are placed on a section view to indicate the areas of the object that came in contact with the cutting plane line. On AutoCAD drawings, section lines are placed with the **HATCH** command.

sheet sizes: Technical drawings are created on standardized sheet sizes. Sheet size varies with the type of drawing and/or the unit of measurement used to create the drawing. The ASME standards governing sheet sizes are *Y141.1-2005* (decimal inch sizes) and *Y141.1M-2005* (metric sheet sizes).

technical drawing: Term used to describe the process of creating the drawings used in the field of engineering and architecture.

technical lettering: Freehand lettering that is added to a technical drawing or sketch. Technical lettering should be legible and consistent with regard to style.

text style: The characteristics of text used in a drawing such as font name, height, width factor, and oblique angle. These values are determined by the values set in AutoCAD's **Text Style** dialog box.

Third Angle Projection: A technique for arranging multiview drawings with the top view above the front view and the right side drawn to the right of the front view. Commonly used on drawings prepared in North America.

three-dimensional (3D) object: An object having height, width and depth.

tick marks: Short diagonal lines that are used (instead of arrowheads) to show the termination of dimensions on architectural drawings.

tolerances: The total permissible variation in the size and/or shape of the object's features as defined by applying tolerances to the nominal size dimension. The difference between the upper and lower size limits of the feature.

traditional drafting tools: The tools that were used to create technical drawings before CAD techniques became the standard. These tools include parallel straight-edges, drafting machines, drafting boards, drafting triangles, protractors, circle and ellipse templates, and technical pens and pencils.

two-dimensional (2D) object: An object having height and width, width and depth, or height and depth.

unidirectional text: Text placed on a technical drawing that faces only the bottom of the sheet. This technique is required when the ASME text standard is applied to a drawing.

United States National CAD Standard (NCS):   A standard developed for the preparation of technical drawings for the building design and construction industry.

User Coordinate System (UCS):   The point in an AutoCAD drawing where the X-, Y-, and Z-axes intersect (0,0,0). This point is noted in the graphics window with the UCS icon.

Index

Multiview Sketching-Sheet Number

Name: _____   Date: _____

3.	6.	9.
2.	**5.**	**8.**
1.	**4.**	**7.**

Multiview Sketching-Sheet Number _____   Name: _____   Date: _____

1.

2.

3.

4.

5.

6.

7.

8.

9.

Multiview Sketching–Sheet Number _____ Name: _____

Date: _____

1.

2.

3.

4.

5.

6.

7.

8.

9.

Multiview Sketching-Sheet Number _____ Name:_____

Date:_____

1.

2.

3.

4.

5.

6.

7.

8.

9.

Multiview Sketching-Sheet Number _____

Name: _____

Date: _____

3.	6.	9.
2.	5.	8.
1.	4.	7.

Multiview Sketching-Sheet Number _____ Name:_____ Date:_____

1.	2.	3.
4.	5.	6.
7.	8.	9.

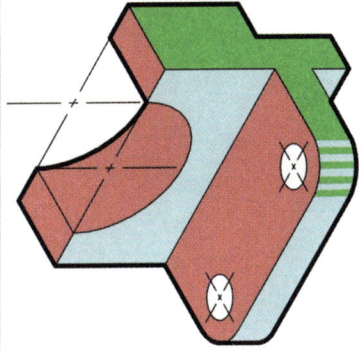